Capillary Gel Electrophoresis

Capillary Gel Electrophoresis

ANDRÁS GUTTMAN
Horváth Csaba Memorial Laboratory of Bioseparation Sciences, Research Center for Molecular Medicine, Faculty of Medicine, University of Debrecen, Debrecen, Hungary

LÁSZLÓ HAJBA
Translational Glycomics Research Group, Research Institute of Biomolecular and Chemical Engineering, University of Pannonia, Veszprém, Hungary

ELSEVIER

Elsevier
Radarweg 29, PO Box 211, 1000 AE Amsterdam, Netherlands
The Boulevard, Langford Lane, Kidlington, Oxford OX5 1GB, United Kingdom
50 Hampshire Street, 5th Floor, Cambridge, MA 02139, United States

Notices
Knowledge and best practice in this field are constantly changing. As new research and experience broaden our understanding, changes in research methods, professional practices, or medical treatment may become necessary.

Practitioners and researchers must always rely on their own experience and knowledge in evaluating and using any information, methods, compounds, or experiments described herein. In using such information or methods they should be mindful of their own safety and the safety of others, including parties for whom they have a professional responsibility.

To the fullest extent of the law, neither the Publisher nor the authors, contributors, or editors, assume any liability for any injury and/or damage to persons or property as a matter of products liability, negligence or otherwise, or from any use or operation of any methods, products, instructions, or ideas contained in the material herein.

British Library Cataloguing-in-Publication Data
A catalogue record for this book is available from the British Library

Library of Congress Cataloging-in-Publication Data
A catalog record for this book is available from the Library of Congress

ISBN: 978-0-444-52234-4

For Information on all Elsevier publications
visit our website at https://www.elsevier.com/books-and-journals

Publisher: Susan Dennis
Acquisitions Editor: Kathryn Eryilmaz
Editorial Project Manager: Ivy Dawn Torre
Production Project Manager: R. Vijay Bharath
Cover Designer: Mark Rogers

Typeset by MPS Limited, Chennai, India

Contents

Preface

At the beginning of the new millennium, entering from the age of genomics to the age of functional genomics, and other "omics" fields such as proteomics, glycomics, and metabolomics, we expect to see high-resolution separation techniques used in an integrated and automated fashion to solve formidable separation problems and provide the support for challenging analytical applications. Capillary gel electrophoresis greatly enhances the productivity of the analysis of biologically important molecules including nucleic acids, proteins/peptides, complex carbohydrates, and even small molecules by automating current manual procedures of slab gel electrophoresis and reducing both the analysis time and human intervention from sample loading to data analysis. As the advent of capillary gel electrophoresis has already made it possible to sequence the human genome [1,2], we anticipate this technique to reveal global changes in the "omics" field (genome, proteome, glycome, and metabolome level), bringing about the revolutionary transition in our views of living systems on the molecular basis.

This book covers all theoretical and practical aspects of capillary gel electrophoresis and provides an overview of the key application areas of nucleic acid, protein and complex carbohydrate analysis, affinity-based methodologies, as well as related microseparation methods such as ultrathin-layer gel electrophoresis and electric field–mediated separations on microchips including a cross section of references from all areas of the field published from the 1980s. This work also aims to give the reader a better understanding of how to utilize this technology and determine which actual method will provide appropriate technical solutions to problems that may have been perceived as more fundamental. It is interesting to note that Web of Science reports close to 4000 research articles on capillary gel electrophoresis and related topics and almost 500 review papers published.

References

[1] J.C. Venter, et al., The sequence of the human genome, Science 291 (2001) 1304–1351.

[2] L. Karger Barry, A. Guttman, DNA sequencing by CE, Electrophoresis 30 (Suppl 1) (2009) S196–S202.

CHAPTER ONE

Introduction

Gel electrophoresis is an electric field–mediated differential migration technique that separates charged molecules or particles. The basic setup of a gel electrophoresis apparatus comprises a gel-filled separation compartment, such as slab gel holding glass plates, or narrow bore tubings (e.g., capillary), connected to separation buffer-filled containers for the positive (anode) and negative (cathode) electrodes. These electrodes are connected to a high-voltage electric power supply that provides the necessary electric field to drive the separation. The schematic representations of vertical (Fig. 1.1, left panel) and horizontal (Fig. 1.1, right panel) slab gel electrophoresis systems, as well as a capillary electrophoresis system (Fig. 1.2), show the relevant gel compartments, buffer reservoirs, and electrodes.

After introducing the sample to the separation gel compartment (into the sample wells in slab gel format or into the gel-filled narrow bore tubing in tubular format), the electric field is applied and causes movement of the charged particles/molecules toward the corresponding electrodes [1]. Positively charged cations migrate toward the negative electrode, the cathode (−). Negatively charged anions migrate toward the positive electrode, the anode (+).

In gel electrophoresis, in both slab and tubular formats, the migration rates of the sample components are dependent on their charge to hydrodynamic volume ratio. This means that smaller molecules with the same charge move faster toward their corresponding electrodes than that of larger ones. In other words, molecules of the same hydrodynamic volumes but different charge states migrate toward their corresponding electrodes at different speeds, meaning, higher surface charge density higher speed, lower surface charge density —lower speed. The gel in the separation compartment provides the mesh for size-based migration and also acts as an anticonvective medium, controlling the development of such electrokinetic phenomena as electroosmotic flow (EOF) [2]. This also means that the pH of the gel electrophoresis buffer system should be carefully chosen to ensure that all solute molecules have the same charge type (positive or negative) in order to have them migrated toward the same direction as it is important for their detection, in both instances of slab or tubular gel

Capillary Gel Electrophoresis.
DOI: https://doi.org/10.1016/B978-0-444-52234-4.00008-8

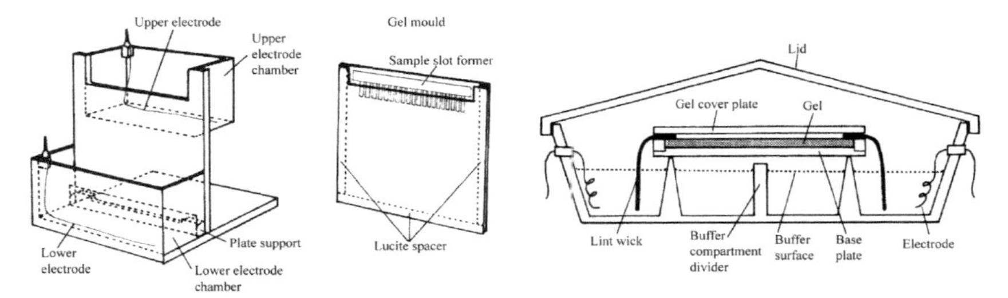

Figure 1.1 Vertical (left) and horizontal (right) slab gel electrophoresis apparatuses. *With permission from A.T. Andrews, Electrophoresis, Theory, Techniques and Biochemical and Clinical Applications, second ed., Claredon Press, Oxford, 1986 [1].*

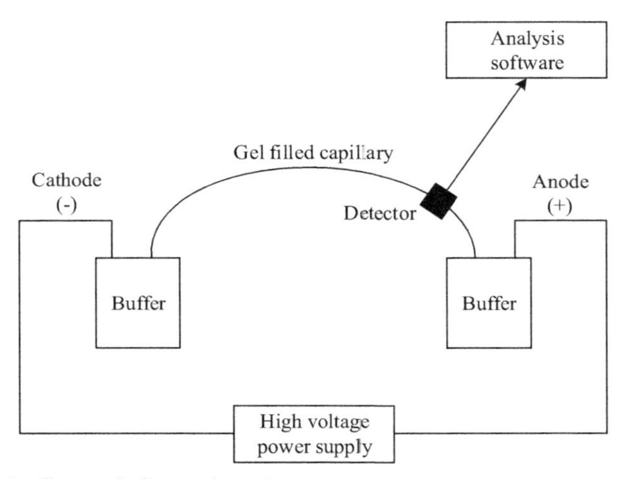

Figure 1.2 Capillary gel electrophoresis system.

electrophoresis. Oppositely charged sample components and/or analytes with no charge in the mixture do not even migrate into the separation compartment, consequently will not be detected.

Separation of neutral species in gel electrophoresis systems is possible only with covalent derivatization or noncovalent complexation with charged tags to ensure their electrophoretic migration within the separation chamber. Fig. 1.3 shows examples of electrophoretic separations in a slab gel (left panel) featuring spatial banding patterns and in a narrow bore capillary tubing (right panel), this latter with time-based peak patterns.

Electrophoresis separation attempts in glass tubes were reported as early as in the 19th century, but the first real breakthrough happened in the first half of the 20th century, when the Swedish chemist Arne Tiselius applied

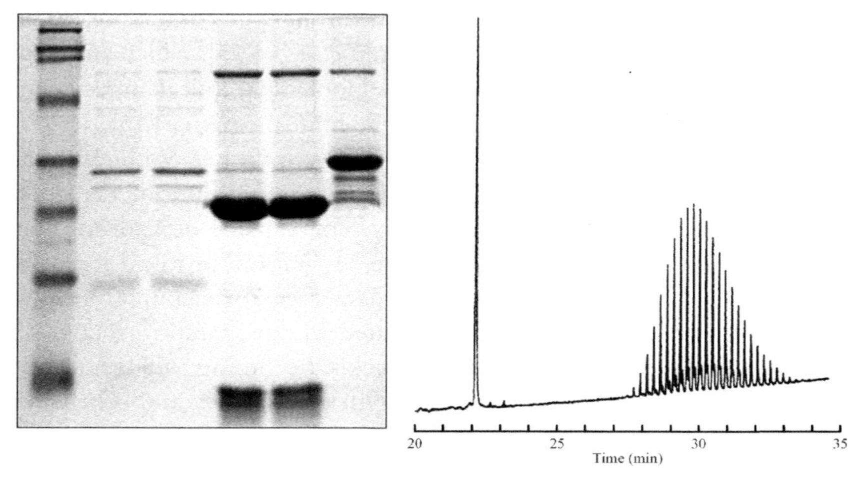

Figure 1.3 Visualization of separation in slab (left panel) and detection in capillary (right panel) gel electrophoresis.

free solution electrophoresis—that is, moving boundary—to serum protein analysis, for which he received the Nobel Prize in 1937 [3]. In less than two decades, just after the striking scientific discovery of the double helical structure of DNA by Watson and Crick in 1953 [4] and the following unveiling of the genetic code, gel electrophoresis became a standard and indispensable tool in the field of modern biochemistry and molecular biology. Since the 1960s, gel electrophoresis methods employing anticonvective sieving media, such as starch, agarose, and polyacrylamide gels, have become the norm for nucleic acid and protein analysis.

In the last decade of the 20th century, a novel, automated, and high-performance electric field–mediated differential migration technique, capillary gel electrophoresis (CGE), was introduced in almost every aspect of basic and applied biochemistry and molecular biology research. CGE replaced the labor-intensive and time-consuming slab gel electrophoresis methods with the major focus on the analysis of nucleic acids, proteins, and complex carbohydrates [5–10]. Using narrow bore fused silica capillaries, filled with cross-linked gels or noncross-linked polymer solutions, unprecedented high resolving power was achieved in separation of biologically important macromolecules. As an instrumental approach to electrophoresis, capillary electrophoresis offers online detection and full automation as depicted in Fig. 1.2. The method is ideally suited for handling microliter amounts of samples and the throughput is superior over conventional approaches [11–13]. High-throughput multicapillary systems can drastically reduce analysis time and cost [14,15], which are highly required by the medical diagnostic and biopharmaceutical industry and actually was the workhorse to sequence the human genome [16].

CGE separates complex mixtures in just minutes with excellent reproducibility, generating in this way a large amount of data. The availability of high-capacity computer systems capable of rigorous qualitative and quantitative analysis of the separation profiles enable to establish, store, and operate with large databases, for example, in DNA sequencing. This is why CGE and related microseparation techniques (e.g., microchip electrophoresis) are quickly becoming important separation and characterization tools in analytical chemistry and biochemistry, as well as in analytical biotechnology, molecular biology, and clinical research.

1.1 Short history of capillary gel electrophoresis

Narrow bore columns were reportedly used for gel electrophoresis of proteins as early as in 1965 by Grossbach [17] who employed 0.20–0.45 mm internal diameter (i.d.) cross-linked polyacrylamide gel-filled capillaries to separate the monomer and dimer forms of human serum albumin. The detection sensitivity of his method was in the range of several nanograms. In 1967, Hjerten [18] demonstrated the applicability and usefulness of high electric field strength in 3 mm i.d. capillaries.

A couple of years later, Neuhoff and coworkers reported on a micro disc electrophoresis method for quantitative assay of cellular level glucose-6-phosphate dehydrogenase variants in 5 μL volume capillaries [19] achieving 1 pg detection sensitivity. The method was apparently suitable to detect glucose-6-phosphate dehydrogenase from a single ovum cell of a mouse as shown in Fig. 1.4. In 1974, Virtanen described the benefits of smaller diameter capillary columns [20] that was utilized in the early 1980s by Hjerten [21] reportedly separating bovine serum proteins in a narrow bore (0.1–0.2 mm i.d.) gel-filled capillary column as shown in Fig. 1.5.

A real breakthrough in capillary electrophoresis as a general method came from the report of Jorgenson and Lukacs [22], who demonstrated the extremely high resolving power of capillary electrophoresis using less

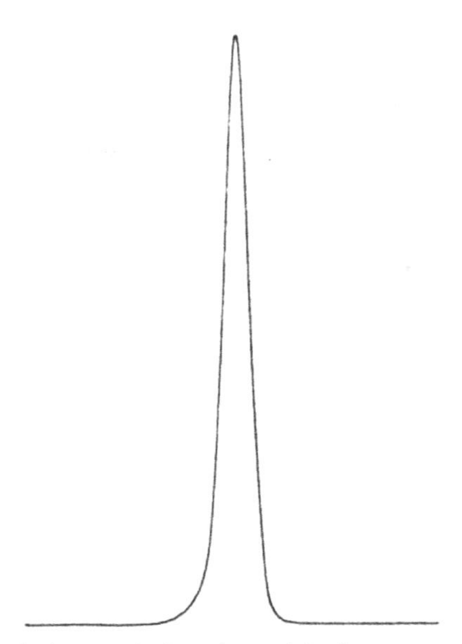

Figure 1.4 Detection of glucose-6-phosphate dehydrogenase in a single egg cell of a mouse, measured by micro disc electrophoresis. *With permission from T. Cremer, W. Dames, V. Neuhoff, Micro disc electrophoresis and quantitative assay of glucose-6-phosphate dehydrogenase at the cellular level, Hoppe-Seyler's Zeitschrift fuer Physiologische Chemie 353 (8) (1972) 1317–1329 [19].*

than 100 μm internal diameter columns. Another major improvement of the technique was the introduction of micellar electrokinetic capillary chromatography by Terabe et al. in 1984 [23] utilizing partitioning of the solute molecules between the background electrolyte and a pseudo-stationary phase comprised of micelles. An interesting combination of micellar electrokinetic chromatography and capillary polyacrylamide gel electrophoresis was used to separate large DNA molecules (restriction fragments, whole phage, viral, and plasmid DNA) by Brownlee and coworkers in the late 1980s [24]. On the theoretical side, Verheggen's group [25,26] derived a treatment on separation efficiency in electrophoresis methods considering the effect of concentration distributions finding asymmetrical zones when a concentration gradient was induced by differential electromigration representing inhomogeneity in the electric field. Later, Thormann [27] demonstrated asymmetrical broadening under overloading conditions due to electric field–mediated changes in solute mobilities, suggesting the use of significantly lower sample concentrations than that of the background electrolyte.

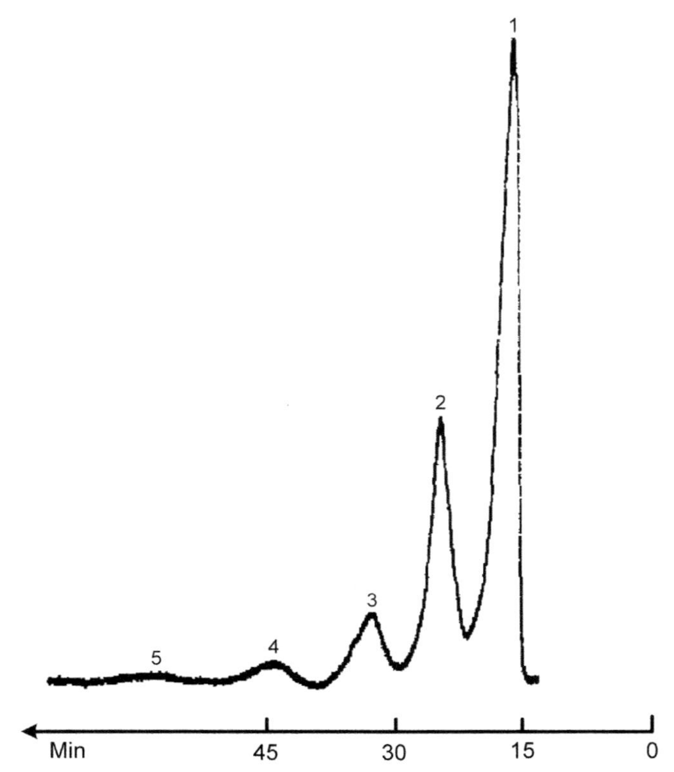

Figure 1.5 High-performance capillary electrophoresis separation of bovine serum albumin oligomers in polyacrylamide gel (peaks 1–5 monomer to pentamer). *With permission from S. Hjerten, High-performance electrophoresis: the electrophoretic counterpart of high-performance liquid chromatography. Journal of Chromatography 270 (1983) 1–6 [21].*

1.2 The capillary gel electrophoresis system

1.2.1 Basic setup

Here only a short overview of the basic setup of CGE systems will be given, as a more detailed discussion of the individual parts including sample introduction methods, detection systems, capillary columns, and temperature control and automation will be provided in later chapters of this book. A very basic schematic representation of a simple CGE system is depicted in Fig. 1.2, showing the separation capillary, the buffer reservoirs, the high-voltage power supply, and the detector. This latter one is connected to a data acquisition and handling system, such as a personal computer (PC). During CGE, the separation capillary is filled with a viscous separation matrix (cross-linked or noncross-linked polymer) and the buffer reservoirs containing the background electrolyte (with or without the separation matrix), all of them are usually in aqueous systems. The inlet side of the separation capillary is first immersed into the sample vial to introduce the sample by pressure (positive from the inlet end or negative from the outlet end) or by electric field (referred to as electrokinetic injection). Once the sample is introduced into the capillary tube, the separation voltage is applied with the appropriate polarity. CGE usually works in reversed polarity mode, meaning the cathode is at the injection and the anode is at the detection side, since most biopolymers are negatively charged (e.g., DNA, RNA, SDS–protein complexes, fluorophore-labeled carbohydrates), thus electromigrating toward the positive electrode (anode). With the electric field applied, the solute molecules start migrating in the capillary, separate according to their hydrodynamic volume to charge ratio, and pass the detection system that sends the corresponding signal to a data acquisition and processing module. In capillary electrophoresis at large, the resulting data output is referred to as electropherogram showing the separated peaks as a function of their migration time (Fig. 1.3, right panel). Each peak corresponds to one or more components, this latter in the case of comigration that should be always considered as a possibility. Time zero in the electropherogram is the end of the injection time and the start of the application of the separation voltage. However, with electrokinetic injection, the actual electromigration process starts already when the injection voltage is applied, even at a different strength. In spite of their similar look, electropherograms are different from chromatograms. Considering equal concentration and equal detector response for all sample components injected, in the case of liquid chromatography the longer the retention times the broader and smaller the peaks are, due to the chromatographic elution phenomena. In this instance the peak area represents the actual injected amount. In capillary electrophoresis, on the other hand, peak heights of the equal concentration/equal detector response sample components remain the same during the separation, but for later migrating species the time required to pass the detector window gets longer resulting in apparently broader peaks. Since this phenomenon is caused by the slower migration of the later migrating species, in this instance the area of these later migrating peaks does not represent their real injected amount. Commercial capillary electrophoresis system software automatically correct for this discrepancy. Finally, once the analyte molecules passed the detection window, similar to liquid chromatography, the separated components can be collected for further processing at the end of the separation capillary. This can be particularly important in the genomics field, as with the advent of polymerase chain reaction, even a small copy number of oligo/poly-nucleotides can be amplified for downstream processes like sequencing and cloning.

1.2.2 The separation capillary

In CGE the most frequently used column type is fused silica, usually coated on the outside by a few micron thick polyimide or polyacrylamide layer, this latter enabling window-free optical detection due to its transparency. This ensures durability and prevents breakage of the narrow bore glass tubing. Typical column lengths are between 30 and 60 cm, but in some instances even longer, up to 1 m, for example, in certain CE-MS settings. The internal diameters of the capillaries are usually ranging between 20 and 75 μm (outer diameter 365–375 μm), but for micropreparative applications it can go up to 100–200 μm. For UV-visible or fluorescent detection systems, the polyimide outer coating should be removed in a small section of the capillary (2–4 mm), providing in this way a window for online detection. This window can be easily made by burning or scraping off the polyimide layer along the required length, resulting in a transparent section for the light path of the detection system. CGE also supports off-column detection methods such as amperometric, electrochemical, and mass spectrometry (see later in Chapter 4, Instrumentation, under detection techniques). Choosing the right inner tube diameter of the separation capillary is an important step when designing CGE analysis. Smaller diameter tubes generate less Joule heat due to the lower current (higher resistance) flowing through the column; however, this might represent difficulties when filling the capillary with high-viscosity gels. Larger tube diameters, on the other hand, permit easier filling of thicker separation matrices, also allowing injection of larger sample amounts. However, due to the higher current (lower resistance) flowing through these columns, more Joule heat is generated that should be appropriately dissipated to obtain high separation time reproducibility and sharp peaks.

1.2.3 Power supply

The high-voltage power supply in the CGE setup (Fig. 1.2) provides the necessary electric field strength for the CGE separation in the narrow bore tubing. Power supplies can operate in constant voltage (isoelectrostatic), constant current (isorheic), or constant power (isoergic) modes [28] and the gradients of those with normal (anode at the injection side) or reversed (cathode at the injection side) polarity. The usual voltage range of such high-voltage power supplies is 0–30 kV with current output at the microampere range. In some instances, if necessary, it can go up as high as 60 kV or higher with the provision of moisture-free compartmentalization to prevent electric shortages. Depending on the conductivity of the background electrolyte, application of higher electric field strengths may result in as high as several watts output power. Therefore an efficient temperature control system is necessary to dissipate the generated heat (>1W). Among the separation modes in CGE, the constant voltage mode is the most frequently used, with stable power supply output to prevent voltage fluctuation–related migration time irreproducibility.

1.2.4 Injection

Sample injection in CGE is accomplished by the introduction of the sample components into the inlet end of the separation capillary. For gel-filled tubings, this can be done electrokinetically or by the application of pressure, as shown in the upper and lower panels in Fig. 1.6, respectively. An important difference between open tubular capillary electrophoresis and CGE is that injection of the solute molecules into gel-filled capillaries is dependent on the viscosity of the separation matrix. In the case of low-viscosity polymer-based sieving solutions, both electrokinetic and pressure (positive at the injection end or negative at the detection end) can be applied. Commercially available capillary electrophoresis instruments are

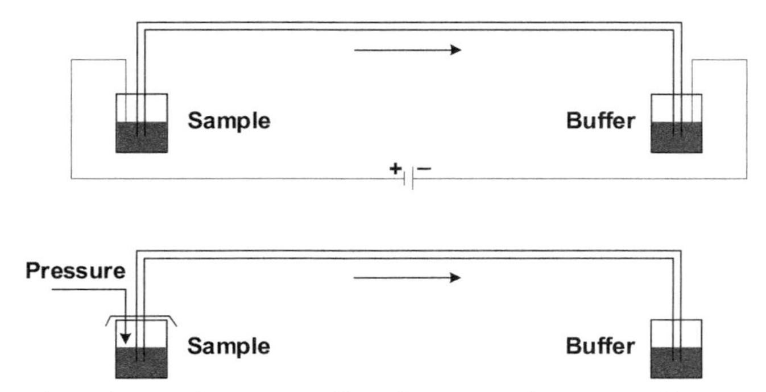

Figure 1.6 Electrokinetic (upper panel) and pressure (lower panel) injection modes in capillary gel electrophoresis.

equipped with pumps or utilize high-pressure nitrogen tanks to provide the necessary injection pressure (up to a few bars) that can replace a small portion of the separation matrix at the injection end of the capillary for the sample. In case of high-viscosity gels, electrokinetic injection may be the only option.

In all instances several microliter samples are needed in the sampling vial in order to be able to physically immerse the injection end of the capillary and the electrode, albeit only several nanoliters of the sample are actually introduced into the column. Manual CE systems mostly use electrokinetic injection mode; however, with really low-viscosity separation matrices hydrodynamic injection is also viable, that is, elevating the injection end of the column immersed into the sample vial and let gravity to drive the injection process. Pressure-based injection methods introduce the sample into the capillary in such a way that the injection plug represents the real concentration distribution of all sample components. The sample in this case is simply replacing a small section of the separation matrix in the inlet end of the column. Electrokinetic injection, on the other hand, results in biased injection, since higher mobility sample components are migrating faster into the separation capillary as their lower mobility counterparts, changing in this way the concentration representation of the sample components for the analysis. Sample injection in microchip CGE can be achieved via different injection designs such as T-type, double T-type, and cross [29]. In addition, various injection modes are available such as pinched, gated, optically gated, pressure/pneumatic, and double T, just to list the most important ones [30].

1.2.5 Detection

Detection systems used in capillary electrophoresis include UV-visible light absorbance, fluorescence (both laser and LED-induced), amperometric/electrochemical, conductivity, mass spectrometry, and radioactivity. modes as listed in Table 1.1. In CGE, laser or LED-induced fluorescence detection techniques are usually used for most single- and double-stranded DNA and carbohydrate samples after fluorophore labeling. UV detection is frequently used for SDS-protein analysis. However, if high detection sensitivity and/or specificity are required, fluorescence detection should be the method of choice by either utilizing the native fluorescence of proteins/peptides or after fluorophore labeling. Table 1.1 lists the detection limits of the various CGE detection methods. In general, detectors should be selected based on the optical or other properties of the

Table 1.1 Detection modes in capillary gel electrophoresis [31].[a]

Detector	LOD (mol)	LOD (mol/L)
Absorption UV—direct	10^{-12}–10^{-15}	10^{-6}
Absorption UV—indirect	10^{-12}–10^{-13}	10^{-5}–10^{-6}
Fluorescence[b]	10^{-15}–10^{-20}	10^{-10}–10^{-11}
Amperometric	10^{-18}	10^{-7}–10^{-8}
Conductometric	10^{-16}	10^{-7}–10^{-8}
Electroluminescence	10^{-16}	10^{-6}
Mass spectrometric	10^{-17}	10^{-6}–10^{-9}

[a]Considering nanoliter (nL) injection volumes, depending on the system.
[b]LOD depends on the derivatization method.

samples being analyzed and the sensitivity required for the analysis [32]. Imaging detectors are also very effective detection systems [33–35].

1.2.6 Data acquisition

As mentioned earlier, electropherograms are plots of the detector signal as a function of migration time. Data acquisition and handling is almost exclusively done by PC systems, irrelevant if they are commercial or home-built systems. Modern capillary electrophoresis software packages enable full data analysis of the electropherograms in regard to migration time determination, peak area calculation with the necessary migration time correction feature, peak efficiency (theoretical plate numbers), resolution, and peak asymmetry analysis, just to mention a few important ones. CGE analysis involves both qualitative and quantitative aspects. The qualitative aspect represents the determination of migration times of the separated peaks in a sample mixture. If pure standards are available, they can be used in spiking measurements to possibly identify the corresponding peaks. However, this approach requires the knowledge of the identity of these sample components and comigration with other components cannot be excluded. In special cases, such as carbohydrate analysis, publicly available databases can be utilized for structural elucidation based on the glucose unit values of the separated components [36]. On the quantitation side, electropherograms should be handled with care as in the case of electrokinetic injection the peak heights do not represent their real concentration distribution in the sample due to the injection bias toward faster migrating components. The best approach by all means is the use of internal quantification standards coinjected with the samples, as that method provides excellent quantification results [37].

The first capillary electrophoresis instrument commercialized in the late 1980s (Microphore 1000) featured a multichannel UV/fluorescence diode array detection system and automatic sample handling using 96-well microplates. (photo courtesy of Robert Weinberger)

1.3 Separation modes in gel-filled capillary columns

Similar to liquid chromatography, one can use various separation modes in CGE. The most frequently used are regular CGE, that is, zone electrophoresis in a sieving matrix. However, other modes such as capillary gel isoelectric focusing (CGIEF), capillary gel isotachophoresis (CGITP), and capillary affinity gel electrophoresis (CAGE), which are counterpart techniques of size exclusion chromatography, chromatofocusing, displacement chromatography, and affinity chromatography, respectively, are also utilized. CGE is a zone type separation method that includes molecular sieving. CGIEF is an isoelectric point (pI)-based focusing technique, CGITP is a moving boundary type, and CAGE is a complexation equilibrium–based method. All of these different separation modes of CGE complement each other and also can be used in parallel to solve special separation problems by taking advantage of their different selectivity provision. Indeed, if low-viscosity separation matrices are used, the same capillary column can be easily switched between these different separation modes, gaining in this way orthogonal separation results that are certainly easier than that of changing columns in chromatography settings. When high-viscosity gels are used, a similar change of columns is necessary as in LC methods.

1.3.1 CGE for molecular sieving

In conventional CGE mode the narrow bore tubing is filled with a gel or polymer network solution (random coil structure in Fig. 1.7, panel CGE). The buffer reservoirs are filled with the same sieving matrix or just with the background electrolyte that is the buffer system of the separation

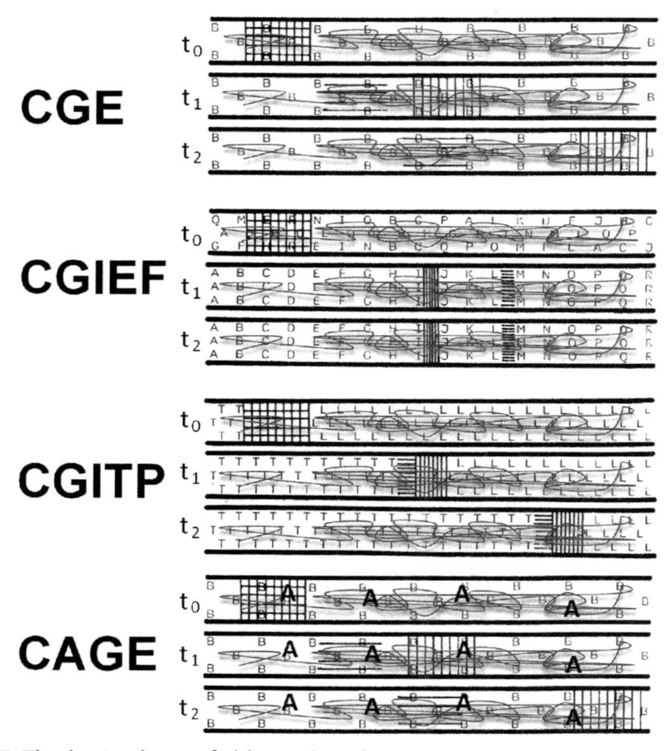

Figure 1.7 The basic electric field–mediated separation modes in gel-filled capillaries. *CGE,* Conventional capillary gel electrophoresis; *CGIEF,* capillary gel isoelectric focusing; *CGITP,* capillary gel isotachophoresis; *CAGE,* capillary affinity gel electrophoresis. In all schematics, the random coil structure represents the gel or polymer network. In CGE mode the sample components separate according to their size (B: background electrolyte). In CGIEF the sample components are focused according to their isoelectric point along the gel-filled capillary column (A–R are ampholite components providing the pH gradient). In CGITP mode the sample components are displaced in the gel-filled capillary according to their conductivity between the leading (L) and terminating (T) electrolytes. In CAGE mode the affinity agent (A) can be covalently bound, physically entrapped, or simply dissolved in the gel matrix. *Adapted from P. Bocek, M. Deml, P. Gebauer, V. Dolnik, Analytical Isotachophoresis, VCH, Weinheim, 1988 [38].*

matrix. In CGE the separation starts in zone electrophoresis mode and becomes size separation downstream due to the sieving effect of the gel or polymer network in the capillary. The sample is introduced as a narrow zone into the inlet portion of the column, usually in a different buffer to support discontinuous buffer system–mediated sample stacking during the injection process [39]. In other words, the sample is usually dissolved in a different buffer than that of the background electrolyte and does not contain the sieving matrix, providing discontinuity in both ways. In addition, due to the larger friction caused by the gel, electromigration of the sample components slow down at the interface of the injection plug and the separation matrix, forming an even sharper zone utilizing this stacking option. By the application of the separation voltage, the sample components and the co-ions of the background electrolyte start migrating toward the detection zone. As was mentioned earlier, in most instances the polarity of the electric field is reversed, that is, the cathode is at the injection side as this setting works properly for negatively charged polyionic biopolymer separations, such as nucleic acids, SDS-protein complexes, or fluorophore-labeled oligosaccharides. Because of the high buffer capacity background electrolyte of the gel-buffer system, the effect of the sample pH on the separation system pH is negligible. As the sample zone moves along the capillary, the different length biopolymers, with practically identical charge to hydrodynamic volume ratios, are getting separated because of the sieving effect of the separation matrix and migrate through the detection zone as discrete bands according to their size. Besides its sieving capability, the gel in the capillary also acts as an anticonvective medium influencing the EOF.

1.3.2 Capillary gel isoelectric focusing

Isoelectric focusing is a method of choice to focus amphoteric molecules into narrow zones, which otherwise would not separate based on their mobility differences. In this separation mode the solute molecules migrate in a preestablished pH gradient until they reach the pH position corresponding to their isoelectric point (pI). It is a real focusing technique as at that point the analyte molecules have zero net charge, so cannot be subject to further electrophoretic migration. At lower pH the species are positively charged and migrate toward the cathode. Consequently, at higher pH the species are negatively charged and migrate toward the anode. The pH gradient across the capillary is generated by a mixture of carrier ampholytes with pIs ranging from acidic to basic pHs. These ampholytes also have a good buffering capacity. The shape of the resulting pH gradient in CGIEF is usually linear, but in special cases can be logarithmic, exponential, etc. In practice as shown in Fig. 1.7, panel CGIEF, the ampholyte mixture is mixed with the separation polymer matrix and filled into the capillary column. The anode and cathode reservoirs are filled by acidic and basic solutions, respectively. By the application of the electric field across the capillary tubing the ampholyte molecules migrate according to their pI toward the corresponding electrodes and form the pH gradient required for the analysis. Each position along the axis of the capillary will have a different pH value, accommodating the isoelectric focusing step. In most instances the sample is blended with the ampholyte–polymer matrix mixture before filling in the capillary; therefore forming the pH gradient and starting the focusing steps are simultaneous. Detection of the focused bands is possible by means of an appropriate mobilization technique, such as pushing the gel with the focused species toward the detection window by pressurizing the inlet side of the capillary or by chemical mobilization [40]. Recent advances in the technology introduced imaging detection of the focused zones, requiring no mobilization [33,35].

1.3.3 Capillary gel isotachophoresis

Isotachophoresis is a technique that utilizes a discontinuous buffer system with a leading (L) and terminating (T) electrolyte [41,42]. The co-ions in the leading electrolyte migrate faster than that of any of the sample components; on the other hand, the co-ions of the terminating electrolyte migrate slower than that of any of the sample components (Fig. 1.7, panel CGITP). Important to note that the buffer capacity of the leading electrolyte should be appropriate for the electrophoretic separation. In CGITP the capillary tube is filled with the leading electrolyte containing the sieving polymer matrix. The inlet and outlet reservoirs are filled by the terminating and leading electrolytes, respectively, either containing the separation matrix or not. After introducing the sample at the inlet end of the gel-filled capillary, the separation voltage is applied. The higher mobility sample component will migrate faster followed by the lower mobility ones as shown in Fig. 1.7, panel CGITP. The buffer systems in CGITP are carefully chosen, that is, the leading electrolyte co-ion (L) has the highest mobility preceding any of the sample components, while in a similar fashion, the mobility of the trailing electrolyte co-ion is chosen as such that it never passes any of the sample components during the analysis. Upon application of the separation voltage, the different zones start a train-like movement lead by the co-ion of the leading electrolyte, followed by the sample components according to their mobilities and finally the co-ion of the terminating electrolyte. When the steady state is reached, all zones migrate connected to each other with the same velocity toward the detection point with their length defined by the conductivity of the actual zone. This also means that if by some reason any of the sample components migrate slower or faster than that of its zone, the higher or lower field drop of the neighboring zone will correct for it.

1.3.4 Capillary affinity gel electrophoresis

The versatility of gel electrophoresis can be extended via the incorporation of special selectivities into the migration process in affinity gel electrophoresis mode [43]. In CAGE, chemo- or bio-specific agents can be utilized to establish interaction with the analyte, resulting in changes in the electrophoretic migration of the complex formed. Both soluble and immobilized affinity-type ligand molecules can be used in the gel-buffer system, labeled with A in Fig. 1.7, panel CAGE. In the case of soluble ligands a broad range of affinity type interactions can be formed with the solute molecules. If the ligand is very small compared to the analyte and has no electric charge, the change in mobility of the resulting complex is negligible. A substantial effect on electrophoretic mobility occurs when the ligand and the analyte are comparable in size or if the ligand is highly or oppositely charged as of the solute molecule. In these latter cases the affinity interaction can result in a great change in the mobility of the analyte–ligand complex. Affinity ligands can also be immobilized in the separation gel system by physical entrapment or chemical bond [44]. The entrapping method can be used, for example, for chiral separation [45].

1.4 Comparison with slab gel electrophoresis

When CGE is compared to conventional slab gel electrophoresis techniques, either vertical or horizontal, one can observe significant differences [1,2]. First, the interpretation of the results is spatial in slab gel electrophoresis and time based in CGE as shown in Fig. 1.3. CGE also offers a lot of variability in the separation parameters as depicted in Table 1.2. While slab gel electrophoresis is still a labor-intensive manual approach requiring long preparation and separation times (up to hours), CGE is automated and offers rapid separation of the analyte mixtures (as low as seconds). Albeit the overall amount of sample required by both techniques is comparable (5–20 μL), while in slab gel mode the entire sample volume is injected for a single run, in CGE hundreds of runs can be made from the same volume as in that case the injection volume is in the nanoliter range. The capillaries with internal diameters of 25–100 μm usually applied in CGE allow the use of significantly higher voltages compared to slab gel electrophoresis owing to better Joule heat dissipation [46]. In addition, there is an interesting difference in peak/band capacity in these methods. While CGE offers very high peak capacity, it is still a one-dimensional separation tool and cannot accommodate the two-dimensional option of the slab gel technique. On the other hand, the multilane feature of the slab gel systems that increases sample throughput can be countered by the use of multicapillary systems, such as manifested in DNA sequencing devices [16]. Regarding equipment cost, commercially available automated CE systems are significantly more expensive than simple slab gel electrophoresis units, even considering the documentation part involved for this latter. Also, important to note that capillary electrophoresis can be directly interfaced with ESI mass spectrometry that is not an option for the slab gel platforms [47].

Table 1.2 Separation variables in capillary gel electrophoresis.

<table>
<tr>
<td>• Electrode polarity
• Applied voltage
• Capillary temperature
• Capillary dimensions
  • Internal diameter
  • Length</td>
<td>• Buffer system
  • pH
  • Ionic species
  • Ionic strength
  • Organic additives
  • Affinity modifiers</td>
<td>• Gel type
  • Chemically cross-linked
  • Entangled (physically cross-linked) polymer network below or above the critical micell concentration
  • Reversibly cross-linked</td>
</tr>
</table>

References

[1] A.T. Andrews, Electrophoresis, Theory, Techniques and Biochemical and Clinical Applications., second ed., Claredon Press, Oxford, 1986.
[2] S.F.Y. Li, Capillary Electrophoresis, Elsevier, Amsterdam, 1993.
[3] A. Tiselius, A new apparatus for electrophoretic; analysis of colloidal mixtures, Transactions of the Faraday Society 33 (1937) 524–531.
[4] J.D. Watson, F.H.C. Crick, Molecular structure of nucleic acids. A structure for deoxyribose nucleic acid, Nature (London, United Kingdom) 171 (1953) 737–738.
[5] B.L. Karger, A.S. Cohen, A. Guttman, High-performance capillary electrophoresis in the biological sciences, Journal of Chromatography 492 (1989) 585–614.
[6] N.A. Guzman, Capillary Electrophoresis Technology, Marcel Dekker, New York, 1993.
[7] J.P. Landers, Handbook of Capillary electrophoresis, CRC Press, Boca Raton, FL, 1994.
[8] K.D. Altria, Capillary Electrophoresis Guidebook, Principles, Operation and Aplications, Humana Press, Totowa, NJ, 1996.
[9] P. Camilleri, Capillary Electrophoresis, Theory and Practice, CRC Press, Boca Raton, FL, 1998.
[10] M.G. Khaledi, High Performance Capillary Electrophoresis, Theory, Techniques and Applications, John Wiley & Sons, New York, 1998.
[11] E. Szantai, Z. Ronai, M. Sasvari-Szekely, G. Bonn, A. Guttman, Multicapillary electrophoresis analysis of single-nucleotide sequence variations in the deoxycytidine kinase gene, Clinical Chemistry 52 (9) (2006) 1756–1762.
[12] C. Gay-Bellile, D. Bengoufa, P. Houze, D. Le Carrer, M. Benlakehal, B. Bousquet, et al., Bricon, Automated multicapillary electrophoresis for analysis of human serum proteins, Clinical Chemistry 49 (11) (2003) 1909–1915.
[13] A. Szekrenyes, U. Roth, M. Kerekgyarto, A. Szekely, I. Kurucz, K. Kowalewski, et al., High-throughput analysis of therapeutic and diagnostic monoclonal antibodies by multicapillary SDS gel electrophoresis in conjunction with covalent fluorescent labeling, Analytical and Bioanalytical Chemistry 404 (5) (2012) 1485–1494.
[14] M.S. Liu, V.D. Amirkhanian, DNA fragment analysis by an affordable multiple-channel capillary electrophoresis system, Electrophoresis 24 (1–2) (2003) 93–95.
[15] V. Amirkhanian, M. Lui, A. Guttman, E. Szantai, Cost benefit analysis of a multicapillary electrophoresis system, American Laboratory 38 (13) (2006) 26–28.
[16] L. Karger Barry, A. Guttman, DNA sequencing by CE, Electrophoresis 30 (Suppl. 1) (2009) S196–S202.
[17] U. Grossbach, Acrylamide gel electrophoresis in capillary columns, Biochimica et Biophysica Acta 107 (1) (1965) 180–182.
[18] S. Hjerten, Free zone electrophoresis, Chromatographic Reviews 9 (2) (1967) 122–219.
[19] T. Cremer, W. Dames, V. Neuhoff, Micro disc electrophoresis and quantitative assay of glucose-6-phosphate dehydrogenase at the cellular level, Hoppe-Seyler's Zeitschrift fuer Physiologische Chemie 353 (8) (1972) 1317–1329.
[20] Virtanen, R., Zone electrophoresis in a narrow-bore tube employing potentiometric detection. Theoretical and experimental study. Acta Polytechnica Scandinavica 123 (1974) 1–67
[21] S. Hjerten, High-performance electrophoresis: the electrophoretic counterpart of high-performance liquid chromatography, Journal of Chromatography 270 (1983) 1–6.
[22] J.W. Jorgenson, K.D. Lukacs, Capillary zone electrophoresis, Science (Washington, D. C.) 222 (1983) 266–272.

[23] S. Terabe, K. Otsuka, K. Ichikawa, A. Tsuchiya, T. Ando, Electrokinetic separations with micellar solutions and open-tubular capillaries, Analytical Chemistry 56 (1) (1984) 111–113.
[24] T.J. Kasper, M. Melera, P. Gozel, R.G. Brownlee, Separation and detection of DNA by capillary electrophoresis, Journal of Chromatography 458 (1988) 303–312.
[25] F.E.P. Mikkers, F.M. Everaerts, T.P.E.M. Verheggen, Concentration distributions in free zone electrophoresis, Journal of Chromatography 169 (1979) 1–10.
[26] F.E.P. Mikkers, F.M. Everaerts, T.P.E.M. Verheggen, High-performance zone electrophoresis, Journal of Chromatography 169 (1979) 11–20.
[27] W. Thormann, Description and detection of moving sample zones in zone electrophoresis: zone spreading due to the sample as a necessary discontinuous element, Electrophoresis (Weinheim, Fed. Repub. Ger.) 4 (6) (1983) 383–390.
[28] A. Guttman, N. Cooke, Effect of temperature on the separation of DNA restriction fragments in capillary gel electrophoresis, Journal of Chromatography 559 (1–2) (1991) 285–294.
[29] C.X. Zhang, A. Manz, Narrow sample channel injectors for capillary electrophoresis on microchips, Analytical Chemistry 73 (11) (2001) 2656–2662.
[30] B.W. Wenclawiak, R. Puschl, Sample injection for capillary electrophoresis on a micro fabricated device/on chip CE injection, Analytical Letters 39 (1) (2006) 3–16.
[31] P. Tůma, F. Opekar, Detectors in capillary electrophoresis, in: J.L. Anderson, et al. (Eds.), Analytical Separation Science, Wiley-VCH Verlag GmbH & Co. KGaA, 2015, pp. 607–628.
[32] K. Swinney, D.J. Bornhop, Detection in capillary electrophoresis, Electrophoresis 21 (7) (2000) 1239–1250.
[33] A. Goyon, M. Excoffier, M.-C. Janin-Bussat, B. Bobaly, S. Fekete, D. Guillarme, et al., Determination of isoelectric points and relative charge variants of 23 therapeutic monoclonal antibodies, Journal of Chromatography B 1065–1066 (2017) 119–128.
[34] M. Szarka, M. Szigeti, A. Guttman, Imaging laser-induced fluorescence detection at the Taylor cone of electrospray ionization mass spectrometry, Analytical Chemistry 91 (12) (2019) 7738–7743.
[35] S. Mack, D. Arnold, G. Bogdan, L. Bousse, L. Danan, V. Dolnik, et al., A novel microchip-based imaged CIEF-MS system for comprehensive characterization and identification of biopharmaceutical charge variants, Electrophoresis 40 (23–24) (2019) 3084–3091.
[36] S. Mittermayr, J. Bones, A. Guttman, Unraveling the glyco-puzzle: glycan structure identification by capillary electrophoresis, Analytical Chemistry 85 (9) (2013) 4228–4238.
[37] K.D. Altria, Improved performance in capillary electrophoresis using internal standards, LC GC Europe 15 (2002) 588–594.
[38] P. Bocek, M. Deml, P. Gebauer, V. Dolnik, Analytical Isotachophoresis, VCH, Weinheim, 1988.
[39] A. Chrambach, Analytical polyacrylamide gel electrophoresis in multiphasic buffer systems (\"disc electrophoresis\"), Electrokinetic Separation Methods, 1979, pp. 275–292.
[40] T. Wehr, M. Zhu, R. Rodriguez-Diaz, Capillary isoelectric focusing, Methods in Enzymology 270 (1996) 358–374.
[41] P. Bocek, M. Deml, P. Gebauer, V. Dolnik, Analytical Isotachophoresis, VCH Publishers, New York, 1988, p. 237.
[42] E. Kenndler, Capillary isotachophoresis, third ed., Encyclopedia of Chromatography, 1, CRC Press, 2010, pp. 298–299.

[43] V. Horejsi, M. Ticha, Qualitative and quantitative applications of affinity electrophoresis for the study of protein-ligand interactions. A review, Journal of Chromatography 376 (1986) 49–67.
[44] A. Guttman, N. Cooke, Capillary gel affinity electrophoresis of DNA fragments, Analytical Chemistry 63 (18) (1991) 2038–2042.
[45] A. Guttman, A. Paulus, A.S. Cohen, N. Grinberg, B.L. Karger, Use of complexing agents for selective separation in high-performance capillary electrophoresis. Chiral resolution via cyclodextrins incorporated within polyacrylamide gel columns, Journal of Chromatography 448 (1) (1988) 41–53.
[46] T. Nakazumi, Y. Hara, Separation of small DNAs by gel electrophoresis in a fused silica capillary coated with a negatively charged copolymer, Separations 4 (3) (2017) 28.
[47] R.D. Smith, J.A. Olivares, N.T. Nguyen, H.R. Udseth, Capillary zone electrophoresis-mass spectrometry using an electrospray ionization interface, Analytical Chemistry 60 (5) (1988) 436–441.

CHAPTER TWO

Basic principles of capillary gel electrophoresis

The fundamental migration equations of capillary gel electrophoresis (CGE) were delineated by Karger and coworkers in their 1989 seminal review paper [1], emphasizing the benefit of the applicability of high electric field strengths in order to obtain rapid analysis and high separation performance in narrow bore tubings. The importance of dissipating the so-called "Joule heat" generated by the high applied voltage and the various factors influencing band broadening and complexation equilibrium-based separations are also discussed in this chapter.

2.1 The electrophoretic migration

By the application of a uniform electric field (E) to an ion with a net charge of Q, the electrical force (F_e) is defined as

$$F_e = Q \cdot E \tag{2.1}$$

In gels or polymer networks, the applied electric field drives the electrophoretic migration of charged analytes, but a frictional force (F_f) acts in the opposite direction:

$$F_f = f \cdot \frac{dx}{dt} \tag{2.2}$$

where f is the translational friction coefficient and dx and dt are the distance and time increments, respectively.

Differences in size, shape, and overall charge of the solute molecules cause variances in their electrophoretic mobilities, providing the basis of their differential electromigration. When F_e and F_f are counterbalanced, the solute molecules migrate with a steady-state velocity of v,

$$v = \frac{dx}{dt} = E \cdot \frac{Q}{f} \tag{2.3}$$

Capillary Gel Electrophoresis.
DOI: https://doi.org/10.1016/B978-0-444-52234-4.00003-9

The electrophoretic mobility (μ) is defined as the velocity per unit field strength:

$$\mu = \frac{v}{E} \tag{2.4}$$

The translational friction coefficient is proportional to the viscosity (η) of the gel and can be expressed as

$$f = c \cdot \eta \tag{2.5}$$

The proportional constant c in Eq. (2.5) is influenced by the molecular configurations and, for example, defined as $6\pi r$ for small spherical molecules, where r is the radius of the molecule [2], or for cylindrical shape molecules such as sodium dodecyl sulfate (SDS)–protein complexes, it is the function of the sixth root of the molecular weight ($\mathrm{Mw}^{1/6}$) [3].

Considering Eqs. (2.3), (2.4) and (2.5), the electrophoretic mobility is expressed as

$$\mu = \frac{Q}{6\pi \mathrm{r} \cdot \eta} \tag{2.6}$$

The viscosity is governed by an activation energy barrier apparently related to the energy required by the solute molecule to pass through the obstacles of the gel or polymer matrix. The viscosity can be expressed by the Arrhenius equation [4] as

$$\eta = A \cdot e^{E_a/RT} \tag{2.7}$$

where A is the preexponential factor, E_a is the activation energy of the viscous flow, and R is the universal gas constant. The mobility of the solute molecules can then be expressed by combining Eqs. (2.6) and (2.7),

$$\mu = \frac{Q}{const} \cdot e^{-E_a/RT} \tag{2.8}$$

where *const* is a collection of constants also including c and A of Eqs. (2.5) and (2.7), respectively. In practice, the activation energy values are usually calculated from the slopes of the Arrhenius plots of the logarithmic electrophoretic mobility as a function of the reciprocal absolute temperature, as shown in [3,5–8] for different analytes. The Arrhenius plots can also be nonlinear, especially for nonspherical analytes as was reported earlier [9,10].

Size-based retardation of the solute molecules in gel or polymer network filled capillaries is a function of the sieving matrix concentration (P) and its physical interaction with the analytes is characterized by the retardation coefficient (K_R):

$$\mu = \mu_0 \cdot e^{-K_R \cdot P} \tag{2.9}$$

where μ is the apparent electrophoretic mobility and μ_0 is the free solution mobility of the solute (i.e., with no gel or polymer matrix in the capillary) [11]. When the average pore size of the matrix is in the same size range as the hydrodynamic radius of the migrating analyte molecule, classical sieving exists (referred to as the Ogston regime, Fig. 2.1, upper section [13,14]). In this case, at constant polymer concentration, the retardation coefficient (K_R) is an exponential function of the size of the analyte molecule [15].

$$\mu \approx e^{-Mw} \tag{2.10}$$

The mobility of the solute molecule, on the other hand, is an apparent exponential function of its size (usually defined by the molecular weight, Mw of the solute) [Eq. (2.10)]. In this case, the logarithmic mobility versus gel concentration plots are linear [16,17] crossing each other at zero gel concentration (see later in Section 4.3.1 under the Ferguson plot).

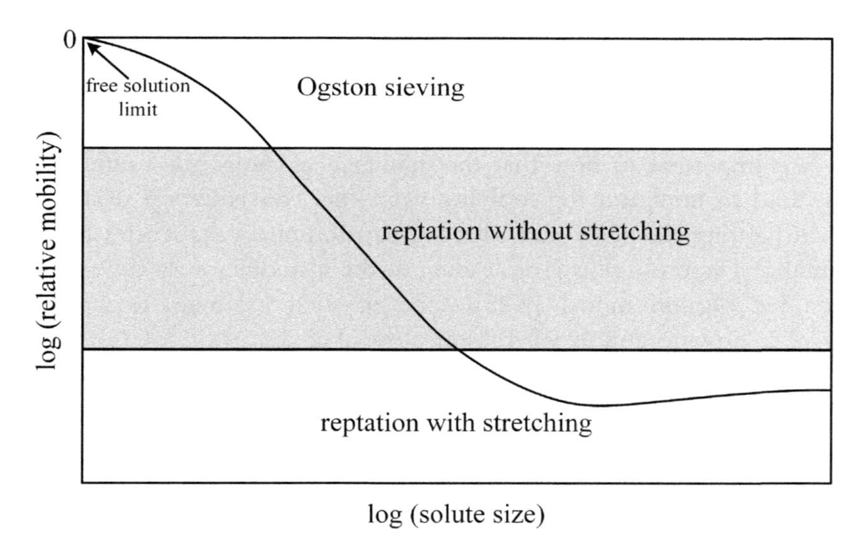

Figure 2.1 Schematic representation of the relationship between the logarithmic relative electrophoretic mobility and the logarithmic solute size.

The Ogston theory assumes that the migrating solute molecule behaves as an unperturbed spherical object with comparable size to the gel pores. However, large biopolymers such as DNA and protein molecules can migrate through polymer networks with pore sizes significantly smaller than that of their size [17]. This phenomenon is referred to as reptation (reptation regime, Fig. 2.1, middle section), which describes the electric field–mediated migration of large biopolymers as "snakelike" motion through the much smaller gel pores [12,18–21]. The reptation model suggests an inverse relationship between the size [represented by the molecular weight in Eq. (2.11)] and the mobility of the analyte molecules with a slope value being close or equal to unity, in a log–log interpretation:

$$\mu \sim 1/\mathrm{Mw} \tag{2.11}$$

At extremely high electric field strengths, reptation turns into biased reptation mode with apparently no separation between the migrating species (biased reptation regime, Fig. 2.1, lower section) and the resulting mobility of the analyte is described by

$$\mu \sim \left(1/\mathrm{Mw} + b \cdot E^{a}\right) \tag{2.12}$$

where b is a function of the mesh size of the sieving matrix as well as the charge and segment length of the migrating solute ions, and $1 < a < 2$. Again, Fig. 2.1 depicts the double logarithmic plot of solute mobility as the function of molecular size, to show the Ogston, reptation, and biased reptation regimes.

It is important to note that the influence of molecular conformation may lead to nonlinear log mobility versus gel concentration relationship when plotting Eq. (2.9) [22], that is, that nonspherical particles produce nonlinear Ferguson plots [16], leading to the introduction of the so-called extended Ogston model [9,23]. One of such examples is shown in Fig. 2.2 showing concave Ferguson plots obtained in SDS-CGE separation of a mixture of protein sizing standards ranging from 20 to 225 kDa in 2% and 4% cross-linker (borate) concentration dextran (2 M Mw) gels.

For different shape solute molecules, the $6\pi r$ term of Eq. (2.6) can be replaced by the molecular weight (Mw) on various exponents such as $\mathrm{Mw}^{1/3}$ as dimensional [13], $\mathrm{Mw}^{2/3}$ as surface area [25], and $\mathrm{Mw}^{1/2}$ as the radius of gyration equivalent [26], as well as by $\mathrm{Mw}^{1/6}$ for cylindrical shape molecules [3,27]. Considering the above, one can express the

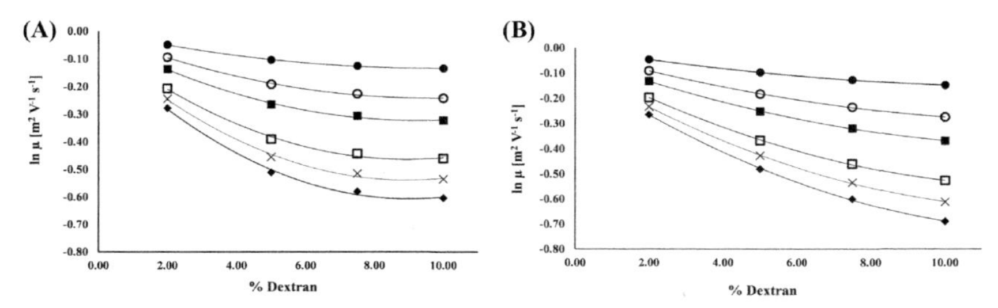

Figure 2.2 Ferguson plots of the EOF corrected logarithmic relative electrophoretic mobilities of the molecular weight sizing proteins as the function of monomer (dextran, 2 M) and cross-linker (boric acid) concentrations. (A) 2.0% and (B) 4.0% boric acid. Symbols: ● 20 kDa, ○ 35 kDa, ■ 50 kDa, □ 100 kDa, x 150 kDa, and ♦ 225 kDa protein standard (*EOF*, Electroosmotic flow). *With permission from A. Guttman, C. Filep, B.L. Karger, Fundamentals of capillary electrophoretic migration and separation of SDS proteins in borate cross-linked dextran gels, Analytical Chemistry 93 (26) (2021) 9267–9276 [24].*

electrophoretic mobility of the charged poly-ion by Eq. (2.13), where k represents the molecular characteristics:

$$\mu = \frac{Q}{Mw^k \eta} \tag{2.13}$$

When concave Ferguson plots are obtained as in the example of Fig. 2.2, the extended Ogston model should be used to describe the electromigration properties of the SDS–protein complexes in the sieving matrices used. Combining Eqs. (2.9) and (2.13), one can express the retardation coefficient by using parameters such as viscosity, molecular weight, and net charge [28],

$$K_R = \frac{1}{P} ln \frac{\eta \cdot \mu_0}{Q} + \frac{k}{P} \ln \text{Mw} \tag{2.14}$$

Based on Eq. (2.14), by delineating the retardation coefficient (K_R) as a function of the logarithmic molecular weight, the slope of the resulting plots would represent the molecular characteristics (k), which was earlier reported as $k = 1/6$ for such cylindrical, no posttranslational modification possessing objects as SDS–protein sizing standards [3]. The intersection of the Y axis holds information about the effect of matrix viscosity and the net molecular charge. However, k may be shape dependent, for example, in cases of heavily glycosylated or disulfide cross-linked intact proteins. Please note that the K_R value is also separation parameter dependent, thus factors such as temperature, buffer composition (including pH and ionic strength), and monomer to cross-linker ratio should be kept constant during measurements.

The K_R plot provides better size (Mw) and molecular characteristics (k) estimations for both linear and nonlinear Ferguson plots. Based on Eq. (2.14), the K_R values were plotted as the function of the logarithmic molecular weights in Fig. 2.3 at several cross-linker concentrations (2%–4% boric acid) using 2 M dextran gels. The molecular shape characteristics (k) values were derived from the slopes of the plots. With the use of the industry standard 4% borate cross-linker concentration, k was defined as 0.164, that is, confirming the earlier proposed cylindrical molecular shape [3]. However, with decreasing cross-linker concentrations, the k values decreased, for example, at 2% borate concentration $k = 0.113$, suggesting some shape changes in the SDS proteins during their electromigration through lower level cross-linked sieving matrices.

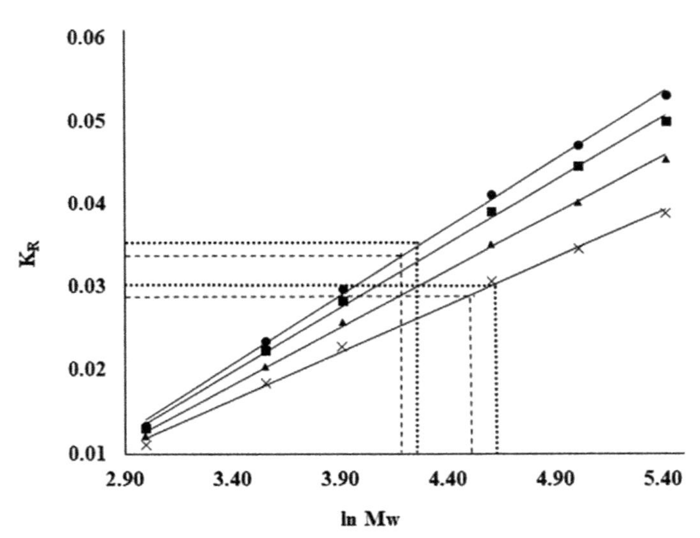

Figure 2.3 K_R plot of the retardation coefficient as the function of the logarithmic molecular weight of the protein Mw standards. The dotted and dashed lines show the Mw determination of the regular and de-*N*-glycosylated etanercept subunits for the 4% and 2% gels, respectively. Symbols: ● 4.0%, ■ 3.3%, ▲ 2.7%, and X 2.0% cross-linker containing dextran gels. *With permission from A. Guttman, C. Filep, Capillary sodium dodecyl sulfate gel electrophoresis of proteins: introducing the three dimensional Ferguson method, Analytica Chimica Acta 1183 (2021) 338958 [28].*

2.2 Efficiency and resolution

In CGE in the absence of electroosmotic flow (EOF) ($\mu_{EOF} = 0$), the electrophoretic mobility of a given solute (μ) is defined by Eqs. (2.3) and (2.4) and by the ratio between the charge of the analyte (Q), and the translational friction coefficient (f), as follows:

$$\mu = \frac{Q}{f} \tag{2.15}$$

The diffusion coefficient (D) of a given solute molecule is a direct function of the absolute temperature (T) and an inverse function of the translational friction coefficient [29]:

$$D = \frac{k \cdot T}{f} \tag{2.16}$$

where k is the Boltzmann constant.

The number of theoretical plates as expressed by Giddings [30] is

$$N = \mu \frac{E \cdot \ell}{2D} \tag{2.17}$$

where E is the applied electric field strength and ℓ is the effective length of the capillary column (from the injection point to the detection point). As one can see, an increase in electrophoretic mobility, applied electric field strength, and separation distance and/or a decrease in diffusion coefficient would all improve column efficiency. However, increasing the applied electric field strength, while speeds up analysis time, leads to extra heat generation (Joule heat) and consequently increased diffusion coefficient. The use of longer separation capillary, on the other hand, leads to increased separation time. Considering the effects of the different parameters in Eq. (2.17), careful optimization is always necessary to obtain the required separation efficiency with reasonable migration times. The effect of separation parameters on peak efficiency was addressed by Schwer and Kenndler in the early 1990s [31,32]. Along with their work, a peak dispersion model was proposed based on longitudinal diffusion and the initial width of the sample plug in the capillary, suggesting that anomalous theoretical plate values can be caused by Joule heating [33].

By combining Eqs. (2.15–2.17) results in

$$N = \frac{E \cdot \ell \cdot Q}{2k \cdot T} \tag{2.18}$$

According to Eq. (2.18), the theoretical plate number (peak efficiency) in CGE is a function of the applied electric field strength, effective capillary length, solute charge, and temperature. Note that the frictional coefficient, f, does not appear in Eq. (2.18). Thus as first shown by Giddings [30] and later discussed by Kenndler and Schwer [34], the number of theoretical plates is not dependent on the friction coefficient because its role in influencing the velocity of a charged species is counterbalanced by its role in altering diffusion. Consequently, a significant increase or decrease cannot be expected in peak efficiency because of the presence of the gel matrix. However, a number of nonidealities might be considerably affected by the presence of the sieving matrix, such as narrow initial sample zones could be introduced into a gel-filled capillary and band broadening due to nonuniformities in EOF is practically eliminated [35]. It was also shown that the efficiency in capillary electrophoresis defined by the theoretical plate height is dependent on their charge state in systems with no EOF [34]

The pore size and/or viscosity of the sieving matrix are also influenced by temperature, and this effect is manifested by changes in separation selectivity (α) [3,36,37]. In analogy to the selectivity definition from the chromatography field, considering no dead volume conditions in CGE, α is defined as follows:

$$\alpha = \frac{t_2}{t_1} \tag{2.19}$$

where t_1 and t_2 are the electrophoretic migration times of the migrating components to be separated.

As a first approximation, it is considered that temperature-mediated pore size and viscosity changes do not affect peak efficiency, yet separation performance is measured by the resolution [38]. Resolution (R_s) between two peaks can be calculated from the differences in their electrophoretic mobilities ($\Delta\mu$) [1] as

$$R_s = 0.18 \cdot \Delta\mu \sqrt{\frac{E \cdot \ell}{D \cdot \overline{\mu}}} \tag{2.20}$$

where, $\overline{\mu}$ is the mean mobility of the sample components of interest. Eqs. (2.18) and (2.20) suggest that higher applied electric field strength and lower solute diffusion coefficient would result in higher separation efficiency (N) and concomitantly higher resolution (R_s).

A practical way to calculate resolution in CGE is based on the migration times of the peaks of interest (t_1 and t_2) and their baseline peak widths (w_1 and w_2) as shown in

$$R_s = 2\frac{t_2 - t_1}{w_1 + w_2} \tag{2.21}$$

Please note that in both Eqs. (2.19) and (2.21), $t_2 > t_1$.

2.3 Band broadening in capillary gel electrophoresis

In CGE, it is important to understand the factors that determine resolution in order to optimize performance. As shown by Eq. (2.21), resolution is defined by the width of the peaks and the spacing between them. One of the major contributors to band broadening, besides the injection and detection extra column effects [39], is the longitudinal diffusion of the solute molecules in the capillary tube that increases peak widths [40].

Under nonstacking buffer system conditions, the resolution in CGE of nucleic acid molecules was shown to be determined by four factors: injection, diffusion, thermal gradients, and detection volume [41]. The relative contribution of each effect was detected as a function of DNA fragment length and this information was used in conjunction with empirically defined mobilities to predict resolution as a function of capillary length at fixed electric field strengths. The effect of the extremely high applied electric fields was also investigated in capillary electrophoresis with respect to band broadening, emphasizing the importance of efficient heat dissipation, which permits higher voltages to be employed without deleterious thermal effects [42]. Application of high electric fields also leads to a reduction in spacing between the peaks, highlighting that the tradeoff between speed and resolution is a very important practical aspect of high-voltage separations. This issue was addressed by investigating the band broadening and resolution of DNA fragments as they were separated through a fixed distance of gel at field strengths ranging from 50 to 400 V/cm. The study concluded that the bandwidths of DNA fragments decreased with the higher field strengths due to a reduction in diffusion-based broadening of the peaks. However, at higher electric field strengths, the bands started to broaden again due to the thermal gradient across the gel, suggesting the necessity of electric field optimization to obtain high resolution and the appropriate temperature controlling system.

In the pioneering work of Schomburg and coworkers, diffusion of oligonucleotides was measured in polyacrylamide gel-filled capillaries in the absence of the electric field [43] and the results were compared with data from a system with the presence of an electric field. The authors revealed that diffusion coefficients significantly changed by the application of the electric field and suggested that intermolecular interactions with the gel matrix might influence their electrophoretic behavior. Slater et al. calculated the longitudinal and transverse diffusion coefficients for nucleic acids undergoing gel electrophoresis in reptation mode [44] and suggested that both increased with the electric field strength. In other words, the normally equal transverse and longitudinal diffusion coefficients at regular field intensities change at the higher electric fields used in CGE. The former dominates at high field strengths suggesting important implications in method optimization. The authors also showed that field-dependent diffusion coefficients should be considered during electric field optimization, even if the effect of Joule heat is neglected. Peak dispersion for oligonucleotides as a function of field strength was measured by multipoint detection [45] and utilized to evaluate dispersion originated from sample introduction.

One option to increase peak resolution in CGE is the use of longer capillaries. However, this approach usually requires coiling the column to accommodate the space in the holding cartridges of commercial units. The effect of column coiling on peak efficiency for gel-filled capillaries was investigated by Wicar and coworkers [46]. While earlier experiments revealed that open tubular configurations were not affected by coiling, reduced peak efficiency was observed in gel-filled capillaries as depicted in Fig. 2.4. The authors argued that in rigid gels the effect was mostly caused by the prevention of diffusional relaxation and in softer gels, the anisotropy induced by mechanical stress was considered as the main contributor to the inferior efficiency.

Conductivity differences between the analyte and surrounding buffer zones cause nonuniform electric field strengths within the capillary. This band broadening effect is further enhanced by the longitudinal diffusion of the solute molecules. This so-called electrodispersion phenomenon was investigated not only at the theoretical, but also at the practical level [47]. A theoretical description for zone evolution within a capillary electrophoresis column as the functions of electrodispersion and solute-wall

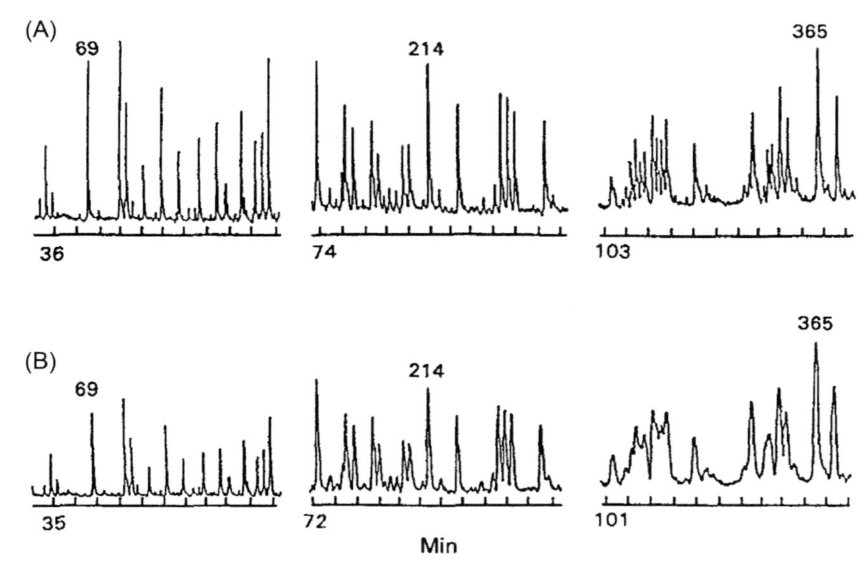

Figure 2.4 Separation of a Fam-labeled M13mp18 a2T DNA sequencing mixture in 9% linear polyacrylamide with 7 M urea. (A) Straight column and (B) column coiled to one loop of 32-mm diameter. *With permission from S. Wicar, M. Vilenchik, A. Belenkii, A.S. Cohen, B.L. Karger, Influence of coiling on performance in capillary electrophoresis using open tubular and polymer network columns, Journal of Microcolumn Separations 4 (4) (1993) 339–348 [46].*

interactions was also given [48] along with experimental approaches to minimize electrodispersion by buffer co-ion manipulation [49–51]. These efforts led to an approach of close mobility matching based on a model to determine the necessary experimental parameter adjustments to modify the mobility of the buffer co-ion to match with the mobility of the solute molecules [50] as shown in Fig. 2.5.

Vigh's group extensively investigated the mobility matching issue by synthesizing and characterizing a number of poly(ethylene glycol) monomethyl ether sulfates [52] and *N*-[poly(ethyleneglycol)monomethylether]-*N*-methylmorpholinium hydroxides [53] as respective anionic and cationic mobility matching co-ions. Application of these co-ions minimized electrodispersion mediated band broadening without any adverse effects on selectivity. Multiple electrolyte buffers, on the other hand, may affect peak symmetry as Bullock et al. [49] demonstrated by their efforts of simultaneous peak symmetry optimization for a mixture of different analytes with widely varying mobilities. Williams et al. [54] applied numerical simulation to predict peak shape distortions that can occur when the

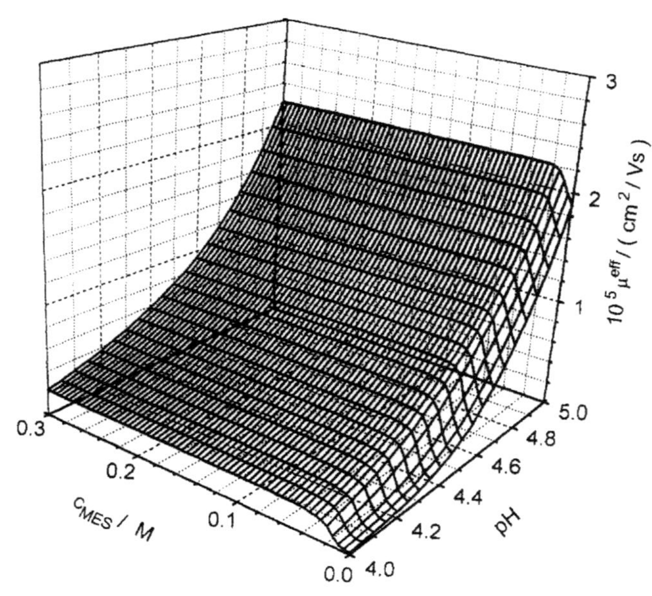

Figure 2.5 Calculated effective mobility of 2-(*N*-morpholino)ethanesulfonic acid (MES) as a function of pH and the analytical concentration of MES. *With permission from Y.Y. Rawjee, R.L. Williams, G. Vigh, Efficiency optimization in capillary electrophoretic chiral separations using dynamic mobility matching, Analytical Chemistry 66 (21) (1994) 3777–3781 [50].*

background electrolyte contains two strong electrolyte co-ions. They concluded that abnormal peak shape and peak disappearance can occur when the analyte peak and the non-co-migrating system peaks overlap.

Unless required for separation modulation, analyte adsorption to the inner capillary wall should be prevented because it may cause asymmetric band dispersion in milder cases, but in severe cases permanent adsorption of the solute molecules to the capillary wall and concomitant sample loss as well as column failing. Experimental and computer modeling of such adsorption phenomena was described in several publications [55–58]. One of the interesting models developed by Ermakov and Righetti [59] utilized typical experimental parameters to predict peak distortion (fronting and tailing) caused by conductivity differences between the sample and the surrounding background electrolyte. Others generated a computer program that simulated electropherograms depicting the impact of certain experimental parameters on electrodispersion [60,61]. Smith and coworkers developed a model to predict resolution under different electrophoretic conditions and suggested separation condition optimization to attain the best possible separation [41]. The study first systematically investigated the effect of electric field strength on band broadening of DNA fragments in CGE revealing that, in terms of bandwidth, the optimum electric field strength was relatively low (i.e., 100–250 V/cm) and also a function of the length of the DNA chain under investigation. At higher field strengths, thermal effects prevailed and caused broader zone widths.

2.4 Temperature effects and power dissipation

During CGE, when the electric current is passing through the gel-filled column, part of the electrical energy is converted to heat, a phenomenon that is usually referred to as Joule heating. This heat increases the temperature within the capillary and if not dissipated properly the separation efficiency can be degraded.

In CGE, the generated Joule heat (Q_j) is directly proportional with the applied power ($P = U \ x \ I$) and inversely proportional with the square of the radius of the capillary (r), the current (I), and the total column length (L, electrode to electrode) [62]:

$$Q_j = \frac{P}{r^2 \cdot I \cdot L} \tag{2.22}$$

Due to the temperature dependence of the electrophoretic mobility (and complex formation constant if applicable, e.g., in affinity gel electrophoresis), efficient temperature control during the electrophoretic separation is important in order to attain reliable results with good reproducibility. Modern, automated capillary electrophoresis instruments are equipped with effective liquid- or air-cooling systems to address temperature change-related issues.

As the theory predicts [Eqs. (2.18) and (2.20)], the application of higher field strengths in CGE should improve separation efficiency (N) and resolution (R_s) but the associated Joule heat always represent a limiting factor. Consequently, temperature control is an important topic on system performance. As early as in 1988, Knox has published a comprehensive treatment on thermal effects and band spreading in capillary electrophoresis, including gel-based separations [63]. Evenhuis et al. also studied the temperature distribution inside and outside of narrow bore fused-silica capillaries [64] as depicted in Fig. 2.6, considering homogenous heat generation in the central

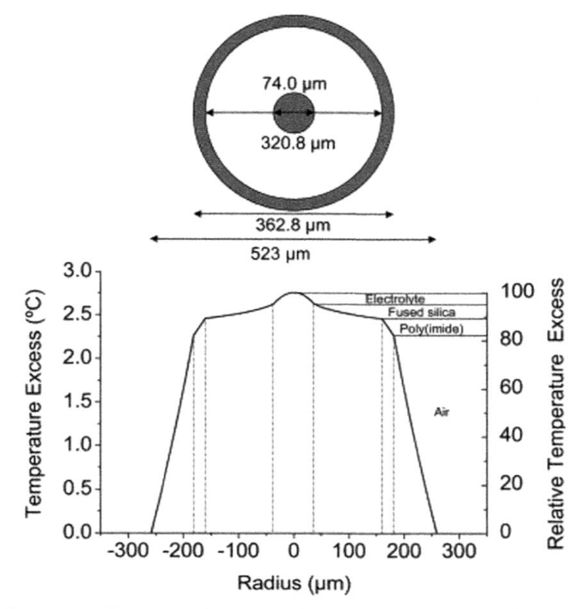

Figure 2.6 Schematic diagram showing the temperature profile for a fused-silica capillary during electrophoresis. *With permission from C.J. Evenhuis, R.M. Guijt, M. Macka, P.J. Marriott, P.R. Haddad, Temperature profiles and heat dissipation in capillary electrophoresis, Analytical Chemistry 78 (8) (2006) 2684–2693 [64].*

region of the column with a parabolic profile. The generated heat is then conducted through the fused-silica capillary wall and coating and dissipates in the surrounding medium.

It was concluded that the temperature change within the electrolyte contributes to efficiency variation but the temperature difference between the capillary and the surrounding media did not. The parabolic temperature profile inside the capillary apparently affects the viscosity, the partition ratios (important only with systems having complexation equilibrium, such as capillary affinity gel electrophoresis), and some kinetic process rates. This latter is important in CGE where diffusion is influenced by the polymer matrix, an effect analogous to viscosity with the appropriate coefficient. A proportional variation in mobility per temperature due to viscosity variation (migration rate) and kinetic processes was estimated as 0.04/Kelvin. A good overview of thermal effects from both experimental and theoretical point of views is given in [65,66].

At the early days it was quickly realized that Joule heating represents a major problem in capillary electrophoresis separations utilizing buffer systems with high ionic strengths, thus various cooling systems were developed. Two of the most popular ones were air- and liquid-cooled devices, both are currently used in most commercially available systems [67]. Other approaches, such as the one Nelson et al. developed showed improved separation efficiency with the use of a thermoelectric (Peltier) device (Fig. 2.7) to control the separation temperature within the capillary [62].

To really understand the effect of the applied electric field strength on heat generation, temperature should be measured inside of the capillary. An innovative method was published by Musheev et al. for such measurement inside narrow bore tubings during electrophoretic analysis [68]. In their approach, a probe peak was moved past the detector area and its initial width was measured. After a given time period, the peak was moved back across the detector area and the diffusion constant was quantified, which was apparently in close correlation with the inside temperature of the column. The same group experimentally determined the temperature differences across the cooled and noncooled parts of a capillary [69], revealing the up to 15°C higher temperature in the noncooled region of the column even when low conductivity buffer systems were used. It was anticipated that with high conductivity buffers, this temperature increase would be even greater. The results explained the possible occurrence of hot spots within the capillary that is usually detrimental for gel-filled

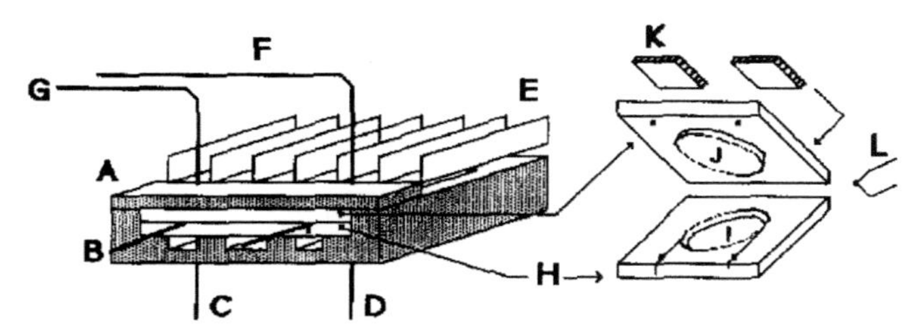

Figure 2.7 Expanded view of a thermoelectric cooling system for capillary electrophoresis. A = black plexiglass holder; B = fused-silica capillary ends; C = 100 μm source fiber optic; D = position of 100 μm reference fiber optic; E = heat sink for thermoelectric devices; F = reference cell 600 μm fiber optic; G = sample cell 600 μm fiber optic; H = alumina plates; I = lower plate with well for column loops; J = upper plate with extension to fit well; K = thermoelectric devices sandwiched between heat sink and alumina plate; L = 10 kΩ thermistor. *With permission from R.J. Nelson, A. Paulus, A.S. Cohen, A. Guttman, B.L. Karger, Use of Peltier thermoelectric devices to control column temperature in high-performance capillary electrophoresis, Journal of Chromatography 480 (1989) 111–127 [62].*

columns by causing bubble formation and concomitant electric current discontinuity.

Theoretical models for temperature effects, such as band broadening, as a function of electric field have also been established and experimentally evaluated. The model of Smith and Khaledi considered the effects of diffusion, adsorption, and thermal band broadening in capillaries with 10−100 μm internal diameters [70]. A numerical algorithm was developed to compute the radial temperature and ion mobility profiles within the capillary [71]. By calculating the theoretical plate height, significant zone broadening was suggested only when the temperature difference between the center and wall of the capillary exceeded 5°C. Burgi et al. compared the calculated and measured column temperatures and found a linear relationship when the power dissipation was less than 3 W [72]. Interestingly, the heat dissipation efficiency was found to be more dependent on the surrounding of the outside wall than that of inside of the capillary.

Kirkland and coworkers [73] investigated the effect of the ionic strengths of the background electrolyte on heat generation and band broadening in narrow bore capillaries by comparing 0.1 and 0.01 M concentration buffers. The significant efficiency loss at the higher ionic strength was explained as a result of the temperature gradient across the capillary. To alleviate the effect of high ionic strength buffer systems, Righetti and coworkers suggested CGE analysis to be performed in isoelectric buffers, such as histidine (His), that allowed the application of as high as 800 V/cm without any resolution loss due to excess Joule heat generation [74].

2.5 Complexation equilibrium

CGE requires the solute molecules to be charged or become charged via tagging with a charged molecule (covalent or noncovalent) to support their electromigration within the column. An example of the former one is DNA or RNA analysis, as each nucleotide holds a charge. Proteins, on the other hand, represent an example of the latter one, as to assure their size-based separation, complexation with a charged surfactant (e.g., SDS) is necessary.

If there is any special soluble additive in the separation matrix, such as a complexing ligand (L), and the solute is a negatively charged poly-ion (P^{n-}) this latter distributes between the complex ($PL_m^{(n \pm m)-}$) and the electrolyte. Depending on the charge of the ligand, it may increase (Eqs. (2.23a) and (2.23b), e.g., SDS–protein complex + negatively charged staining dye [75]) or decrease (Eqs. (2.24a) and (2.24b), e.g., DNA + positively charged intercalator dye [76,77]) the charge of the resulting complex:

$$P^{n-} + mL^{-} \Leftrightarrow PL_m{}^{(n+m)-} \tag{2.23a}$$

$$K = \frac{[PL_m{}^{(n+m)-}]}{[P^{n-}]\cdot[L^{-}]^m} \tag{2.23b}$$

$$P^{n-} + mL^{+} \Leftrightarrow PL_m{}^{(n-m)-} \tag{2.24a}$$

$$K = \frac{[PL_m{}^{(n-m)-}]}{[P^{n-}]\cdot[L^{+}]^m} \tag{2.24b}$$

where K is the complex formation constant, m is the number of ligand molecules in the complex, and n is the total number of charges on the solute ion. As a first approximation, n can be considered the same as the number of bases in the case of nucleic acids or the number of attached charged surfactant molecules in the case of SDS–protein complexes. The total number of charges of this latter is mainly defined by the number of surfactant molecules on the protein (one SDS per two amino acids) [78].

The electrophoretic velocity (v) of the complex is described as

$$v = \ell / t_M = \mu_P \cdot E \cdot R_P \tag{2.25}$$

where ℓ is the effective separation length (from the injection to the detection point), t_M is the migration time of the solute–ligand complex, μ_P is the electrophoretic mobility of the solute ion, and E is the applied electric field strength. The molar ratio of the complexed ligand, R_P, is given by Eq. (2.26) in the case of similarly charged ligand and analyte molecules:

$$R_P = \frac{[PL_m{}^{(n+m)-}]}{c_P} = \frac{K\cdot[P^{n-}]\cdot[L^{-}]^m}{c_P} \tag{2.26}$$

and by Eq. (2.27) for oppositely charged solute and ligand molecules:

$$R_P = \frac{[P^{n-}]}{c_P} = \frac{1}{1 + K[L^{+}]^m} \tag{2.27}$$

where c_P is the total solute concentration. Since the usually limited amount of ligand binds only a fraction of the solute ions, combining

Eq. (2.25) with Eqs. (2.26) and (2.27), respectively, results in Eq. (2.28) in the case of similarly charged ligand and analyte molecules:

$$v = \mu_P \cdot E \cdot K \cdot [L^-]^m \tag{2.28}$$

and Eq. (2.29) for oppositely charged solute ion and ligand molecules:

$$v = \mu_P \cdot E \cdot \frac{1}{1 + K[L^+]^m} \tag{2.29}$$

Eqs. (2.28) and (2.29) suggest the extent of increase (e.g., SDS–protein complex + negatively charged staining dye [75]) or decrease (e.g., DNA + positively charged intercalator dyes [76,77]) in electrophoretic velocity when the negatively or positively charged ligand binds the analyte, respectively. Please note that one should also take into consideration the mass change of the solute ion–ligand complex, that is, the additional mass of the ligand molecules that increases the resulting mass of the complex in both instances.

In case the ligand is immobilized to the sieving polymer the situation will be as follows. One of these examples is SDS-CGE in borate cross-linked dextran gels (Scheme 2.1) [79]. The apparent separation selectivity (α) in SDS-CGE using these kinds of sieving matrices is considered to be the product of the molecular weight sieving selectivity (α_{MW}) that is based on gel

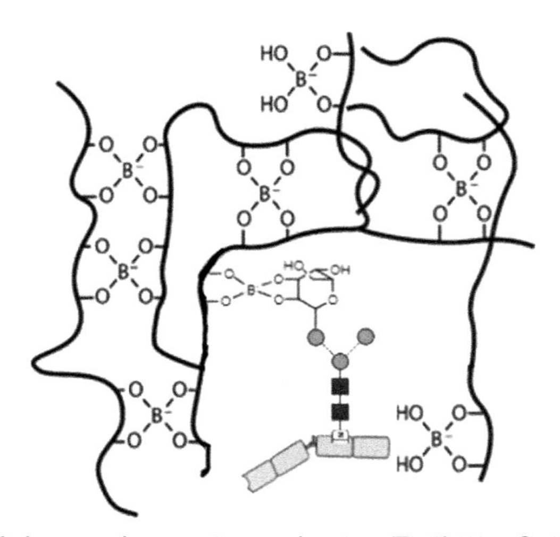

Scheme 2.1 Diol–borate–glycoprotein complexation. ■: GlcNAc; ●: Mannose; ▒-▒: IgG heavy chain. *With permission from C. Filep, A. Guttman, Effect of the monomer cross-linker ratio on the separation selectivity of monoclonal antibody subunits in sodium dodecyl sulfate capillary gel electrophoresis, Analytical Chemistry 93 (7) (2021) 3535–3541 [79].*

pore size and the complexation equilibrium-based selectivity (α_C) that is based on the dextran–borate–glycan adduct formation, if applicable,

$$\alpha = \alpha_{MW} \cdot \alpha_C \tag{2.30}$$

As discussed in the Introduction, glycans connected to the polypeptide chain of glycoproteins do not bind SDS, therefore available to form borate–glycoprotein complexes with appropriately oriented diol groups. On one hand, this can be with the free borate in the background electrolyte, which actually would increase the surface charge density of the complex, thus decreasing the mobility differences (selectivity) between the nonglycosylated and glycosylated forms. On the other hand, the intrachain borate adducts (1:1 form in Scheme 2.1) in the sieving matrix can also complex with the carbohydrate moieties of the analyte molecules. In the omalizumab test mAb used in this study, these can be alpha-oriented vicinal hydroxyl groups on the core fucose, terminal mannose, sialic acid, and terminal galactose residues. One of such examples is shown in Scheme 2.1, where the mannose-5 glycan possessing heavy chain (HC) complexes with the borate–dextran 1:1 adduct of the sieving matrix. This complexation influences the differential electromigration, as during the complexation equilibrium-based adduct formation, the HC will be transiently immobilized to the cross-linked dextran–borate gel. The separation window will be increased accordingly between the HC and the nonglycosylated heavy chain (ngHC) fragments of the monoclonal antibody sample.

When intrachain dextran–borate adducts (D_B) are present in the separation matrix in a 1:1 ratio, dextran–borate–glycoprotein complexation can readily occur. Therefore a fraction of the glycoprotein (P) will be in noncomplexed free (f) state (fraction R), while the rest of the molecules will be in the complex (c) form (fraction $1 - R$). Considering a given monomer–cross-linker ratio sieving matrix, the apparent mobility of the glycoprotein (μ) will be the sum of the free and complexed forms, as described in Eq. (2.31).

$$\mu = R\mu^f + (1 - R)\mu^c \tag{2.31}$$

The fraction R can be calculated from the complex formation constant (K). The number of glycoproteins complexed with the dextran–borate chain is n.

$$D_B + nP \rightleftarrows D_B \cdot P_n \tag{2.32}$$

$$K = \frac{[D_B \cdot P_n]}{[D_B][P]^n} \tag{2.33}$$

thus

$$R = \frac{1}{1 + K[D_B]} \tag{2.34a}$$

$$1 - R = \frac{K[D_B]}{1 + K[D_B]} \tag{2.34b}$$

Combination of Eqs. (2.31), (2.34a), and (2.34b) results in the general mobility equation for the glycosylated heavy chain (μ_{HC}) of the monoclonal antibody:

$$\mu_{HC} = \frac{\mu^f_{HC}}{1 + K[D_B]} + \frac{\mu^c_{HC} K[D_B]}{1 + K[D_B]} \tag{2.35}$$

The complexation-based separation selectivity (α_C), that is, without taking into account the MW sieving selectivity (α_{MW}), between the glycosylated (HC) and nonglycosylated (ngHC) heavy chain fragments can then be expressed as

$$\alpha_C = \frac{\mu_{ngHC}}{\mu_{HC}} = \frac{\mu_{ngHC}(1 + K[D_B])}{\mu^f_{HC} + \mu^c_{HC} K[D_B]} = \frac{\mu_{ngHC}(1 + K[D_B])}{\mu^f_{HC}\left(1 + K[D_B]\mu^c_{HC}/\mu^f_{HC}\right)} \tag{2.36}$$

The mobility of the free glycosylated HC is assumed to be significantly higher than that of its dextran–borate bound counterpart ($\mu^f_{HC} \gg \mu^c_{HC}$). Thus the resulting selectivity can be expressed as

$$\alpha_C = \frac{\mu_{ngHC}}{\mu^f_{HC}}(1 + K[D_B]) \tag{2.37}$$

if $K[D_B] >> 1$, then the selectivity equation is simplified:

$$\alpha_C = \frac{\mu_{ngHC}}{\mu^f_{HC}} \quad K[D_B] \tag{2.38}$$

On the other hand, if $K[D_B] << 1$

$$\alpha_C = \frac{\mu_{ngHC}}{\mu^f_{HC}} \tag{2.39}$$

Best separation is expected when $\mu^f_{HC} \gg \mu^c_{HC}$.

The resolution (Rs) between the ngHC and the HC fragments is expressed by Eq. (2.40), where N is the theoretical plate number, E is the applied electric field strength, and ℓ is the effective length of the separation capillary [80].

$$Rs = \frac{1}{4}\left(\frac{\alpha - 1}{\alpha}\right)\sqrt{N}\frac{\mu E}{\ell} \tag{2.40}$$

Therefore similar to chromatography, substantial influence on resolution is expected when α is close to unity.

2.6 Electric field-mediated migration of biopolymers in sieving matrices

One of the important issues to understand CGE of biopolymers is the mechanisms by which these large polyionic molecules migrate under the influence of the electric field and separate. The use of gel matrix or polymer network in the capillary column in addition to significantly decreasing or even extinguishing the EOF also alters the mobilities of the charged solute molecules. For biopolymers with an identical or very similar charge to mass ratio, for example, DNA or SDS–protein complexes, CGE is the method of choice for their effective separation, based on differences in molecular sizes, that is, hydrodynamic volume. It is important to note that both the transport properties and the separation selectivity are influenced by the sieving medium; therefore it is also essential to study the effects of the sieving matrix on the overall separation efficiency. Apparently, the reduced mobility caused by the presence of the polymer network corresponds to the increase in the friction coefficient. On the other hand, the increase in friction is associated with a reduced diffusion coefficient.

Righetti and coworkers investigated DNA fragment migration in the size range of 51–23,130 base pairs in CGE using linear polyacrylamide (LPA) solutions of 4%–10%. The log(mobility) versus log(size) plots clearly showed the three expected migration regimes of Ogston sieving up to 200 base pairs, reptation without stretching up to 3000–4000 base pairs, and reptation with partial stretching for the larger fragments as shown in Fig. 2.1. The authors suggested that during the polymerization of the sieving matrix, various chain length strands were synthesized that accommodated simultaneous separation of short and long DNA fragments based on the idea that shorter sieving polymer chains more effectively sieve shorter DNA fragments and the longer chains sieve the longer DNA fragments better [81].

A novel biased reptation theory was introduced by Slater and Noolandi [82] that allowed qualitative and semiquantitative studies of gel electrophoresis of biologically important polymers such as DNA molecules. They proved the stretching phenomenon of the end-to-end vector of very long charged reptating chains under electric field in a short period of time during a typical gel electrophoresis experiment. Their assumption considered that field-dependent mobility only weakly depended on the chain size (Fig. 2.8).

Carlsson et al. [83] reported on the oscillatory dynamics of biopolymer (DNA) movement during capillary electrophoresis in a diluted polymer matrix directly observed by fluorescence microscopy. As a result of entanglements with the polymer chains of the medium, the solute molecules changed rhythmically between extended and contracted configurations as moving through the gel. Even more interestingly, at lower polymer

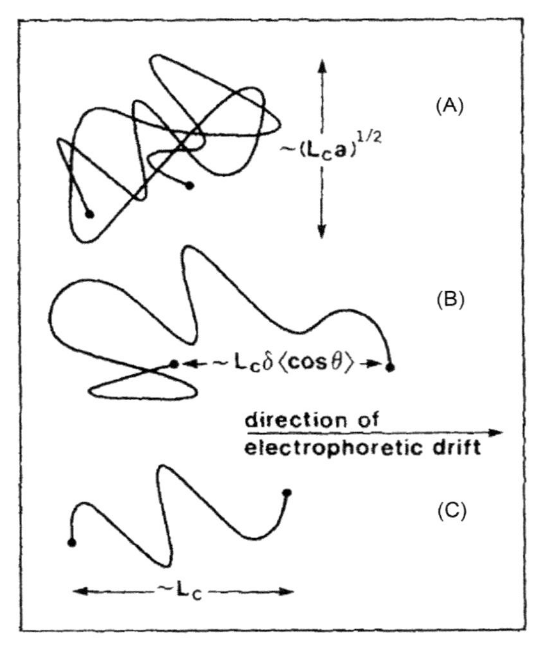

Figure 2.8 Some typical chain conformations of biopolymers in gel electrophoresis. (A) Without an electric field; (B) for small fields; and (C) for very large electric forces. *With permission from G.W. Slater, J. Noolandi, On the reptation theory of gel electrophoresis, Biopolymers 25 (3) (1986) 431–454 [82].*

concentrations, the DNA exhibited a rotatory motion through the solution as a result of the interplay between Brownian dynamics and interactions with the polymer chains under the electric field. Fig. 2.9 shows a semiquantitative picture of the most probable scenarios and typical biopolymer configurations from video recordings in sieving systems containing different polymer concentrations.

Video microscopy was used by Morris and coworkers to visualize biopolymer movement (nucleic acids) during CGE in hydroxyethyl cellulose (HEC) sieving medium demonstrating previously unobserved shape-changing interactions between the solute and the sieving matrix [84]. Typical polyionic biopolymers such as nucleic acids appear to become entangled with HEC at a single region only, in both diluted and fully entangled HEC solutions as shown in Fig. 2.10. This work provided additional support to the reports on rapid, high-resolution direct current and pulsed-field capillary electrophoretic separations of such biopolymers in ultradilute hydrophilic polymer-based sieving media.

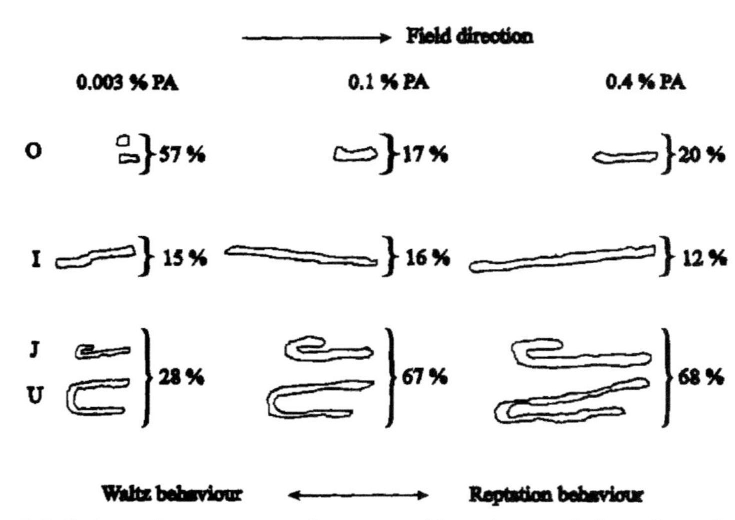

Figure 2.9 Estimated occurrence of compact (0) and extended configurations (*I*, *J*, and *U*) of a linear polyionic biopolymer (T2 DNA) during electrophoresis (6 V/cm) at three concentrations of dilute solutions of linear polyacrylamide (MW 1,800,000). Differences in shapes are indicated schematically. Solute configurations were recorded in every 3 s and the statistics was based on 250 observations for each polymer concentration. *With permission from C. Carlsson, A. Larsson, M. Jonsson, B. Norden, Dancing DNA in capillary solution electrophoresis, Journal of the American Chemical Society 117 (13) (1995) 3871–3872 [83].*

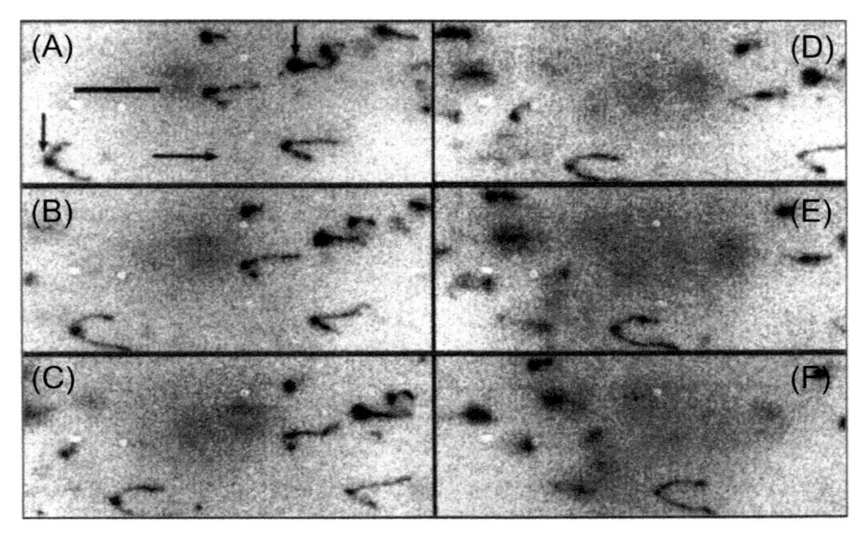

Figure 2.10 Time sequence (A–F) of yeast chromosomal DNA electrophoresis in 0.16% HEC/55% sucrose at 25 V/cm field. Frames at 4.0 s intervals. The downward arrows identify the apex or starting positions of two molecules. The movement of the apex of the U-shaped molecules is clearly visible. *With permission from X. Shi, R. W. Hammond, M.D. Morris, DNA conformational dynamics in polymer solutions above and below the entanglement limit, Analytical Chemistry 67 (6) (1995) 1132–1138 [84].*

A detailed mechanistic study on protein migration in semidilute polyethylene glycol (PEG) (MW 0.2–8 MDa) tested the predictions of the scaling theory with respect to the relation of retardation to the size and concentration of the sieving polymer and the radius of the solute molecules [85]. In the size range of 60–500 kDa of mostly spherical biopolymers (proteins), the retardation was a function of the sieving polymer concentration in close agreement with the scaling theory. All measured retardation values were independent of the electric field strength in the range of 37–370 V/cm.

A very interesting fundamental study was conducted by Archer and coworkers on the electrophoretic behavior of large linear (162 kbp) and three-arm-branched ($N_{Arm} = 48.5$ kbp) polyionic biopolymers (DNA) in LPA solution above the overlap concentration (c^*), but below the entanglement concentration (c_e), using fluorescence visualization to evaluate conformation and mobility [86]. It was observed that in unentangled semidilute solution ($c^* < c < c_e$), the conformational dynamics of linear molecules were U- or I-shaped while the branched ones migrated in a squid-like profile with the arms pointing the direction opposite to the electric field (Fig. 2.11). However, in spite of these apparent differences, the mobility of the linear and branched molecules of comparable size was nearly identical in semidilute, unentangled linear polyacrylamide sieving matrix. When the sieving polymer concentration was elevated above the entanglement threshold ($c > c_e$), while the conformation of the migrating linear and branched molecules exhibited only little changes compared to the unentangled matrix, the mobility of the branched structure significantly decreased with increasing sieving polymer concentration. Interestingly, the mobility of the linear biopolymer structure was almost independent of the concentration and the MW of the sieving polymer. Soane's group developed a transient entanglement coupling theory for DNA separations in very dilute polymer solutions [87] and concluded that the Ogston and reptation models were not adequate to characterize DNA migration in highly diluted polymer network solutions.

Later Viovy and Duke investigated the migration mode of double-stranded (ds) DNA molecules during CGE [20] and predicted that dsDNA of up to 1 kilobases (kb) or more may separate in nonentangled

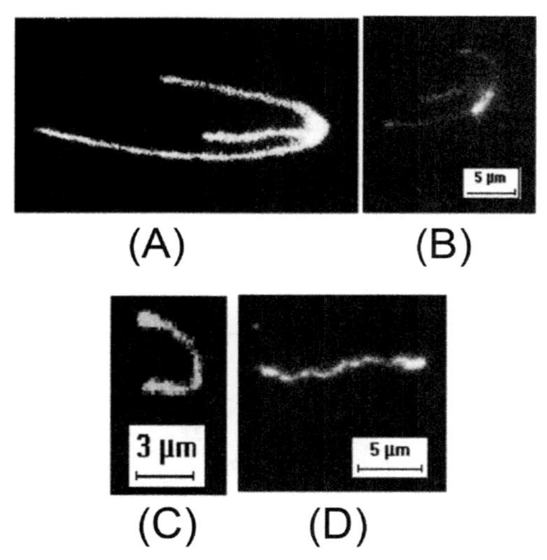

Figure 2.11 Photograph of (A) three-arm star DNA migrating in 0.5% linear polyacrylamide (LPA) (5–6 MDa), (B) the conformation of three-arm-branched DNA dragging the star core behind it in 1.5% LPA (2.05 MDa), (C) U-shaped conformation of T2 DNA in 0.75% LPA sieving matrix (2.05 MDa), and (D) I-shaped conformation of T2 DNA in 1.5% LPA (2.05 MDa). *With permission from S. Saha, D.M. Heuer, L.A. Archer, Electrophoretic mobility of linear and star-branched DNA in semidilute polymer solutions, Electrophoresis 27 (16) (2006) 3181–3194 [86].*

(dilute) solutions of high molecular weight polymers. At higher concentrations, that is in entangled solutions, and for DNA molecules larger than that of the pore size of the sieving matrix, they introduced a fluctuation-reptation model to predict separation size ranges as a function of electric field (E) and pore size (ζ_b). They also suggested a novel, not size-dependent migration mechanism, dubbed as constraint release (Fig. 2.12) in entangled polymer solutions.

Mobility in electric field-mediated separations is a function of the charge to mass ratio of the analyte molecules. In the case of uniformly charged polyelectrolytes either some type of sieving medium is required for their separation or the charge to mass ratio should be altered using a labeling technique, referred to as drag-tag. By the proposed separation mechanism, the highly charged DNA molecule is dragging polymer molecules through the separation medium [88]. Theoretical calculations and predictions were verified by independent reports, that is, that free solution separation can be obtained by attaching an electrically neutral, friction-generating molecule to the DNA prior to the electrophoretic separation process. In a similar manner, Sudor and Novotny [89] end-labeled

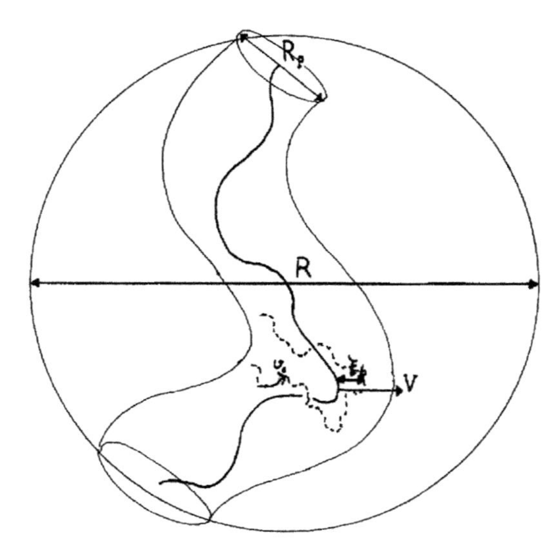

Figure 2.12 Representation of the constraint release process. The DNA is represented as a full, thick line. Only one polymer chain is represented (dotted line) for the sake of clarity, but actually the whole space is filled with polymer blobs. *With permission from J.L. Viovy, T. Duke, DNA electrophoresis in polymer solutions: Ogston sieving, repatation and constraint release, Electrophoresis 14 (1993) 322–329 [20].*

oligosaccharides with fluorescent reagents, which accommodated both size separation in free solution and fluorescence detection with high sensitivity. As this chapter suggests, optimization of CGE performance can be accomplished by manipulating several experimental parameters [41,90,91]. The most important ones include both physical parameters, such as capillary dimensions, separation voltages, and temperature, and separation chemistry like various concentration and/or size gel or polymer matrix, background electrolyte composition, and special additives to utilize complexation equilibrium in the resulting separation process [79].

2.7 Micropreparative capillary gel electrophoresis

In an early micropreparative effort by capillary gel electrophoresis, 800 ng of a 20-mer oligonucleotide was isolated in less than 20 min, as shown in Fig. 2.13. The crude 20-mer was injected using a Teflon sleeve of defined volume [92]. (step i) A gas-tight syringe with a fused-silica needle filled with a known concentration of sample was used to fully displace the buffer in the Teflon sleeve, thus providing a fixed and known volume of sample. Peaks *a*, *b*, and *c* correspond to those in the inset. (step ii) Collection and associated traces are shown at the bottom. As the positive end of the gel-filled capillary was transferred from one collecting tube to another, the field was broken and the current dropped to zero. Also, because of the movement of buffer ions into the collection tubes, the current decreased, causing a concomitant drop in the migration rate of all ionic species. This decrease occurred linearly, making it possible to calculate the degree of slowing for each species by a simple linear approximation, that is, peak *a* was determined to migrate in fractions 1 and 2, peak *b* in fraction 3, and peak *c* in fractions 4–6. (step iii) To assess sample removal from the column, the polarity of the electrodes was reversed for 10 min to drive any ionic species remaining on the column between the detector and the collection tube back past the detector again. The absence of a peak indicated that the collection was complete. (step iv) Reinjection of collected fraction 4. Fifty nanograms of collected fraction 4 (step ii) was injected and electrophoresed as described in step i. This fraction migrated at the same time as peak c in step i. The purified oligonucleotide was collected in water and used downstream as a probe in a regular dot-blot assay.

In another attempt, very precise fraction collection of DNA samples was possible with high resolution by applying electric field

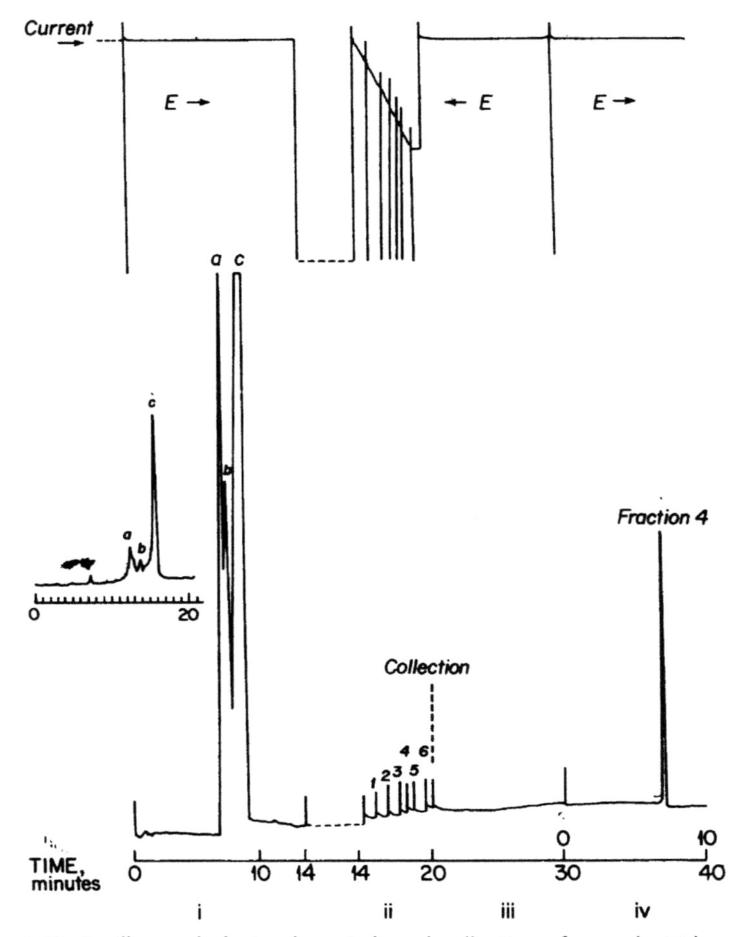

Figure 2.13 Capillary gel electrophoresis based collection of a crude 20-base oligodeoxynucleotide (5' GCCACGTCCAGATTTATCAG 3'). Capillary size: 300 mm × 0.075 μm i.d., running buffer: 0.1 M Tris/0.25 M borate/7 M urea (pH 8.3), gel: 8% T/3.3% C. Upper panel: electrical field during the steps; lower panel: preparative electropherogram. *With permission from A.S. Cohen, D.R. Najarian, A. Paulus, A. Guttman, J.A. Smith, B.L. Karger, Rapid separation and purification of oligonucleotides by high-performance capillary gel electrophoresis, Proceedings of the National Academy of Sciences of the United States of America 85 (24) (1988) 9660–9663 [92].*

programming [93]. In the approach, the sample was separated at high electric field where speed and resolution were maintained. In the collection step the electric field was reduced where the peak would broaden a little with no significant loss in resolution. Fig. 2.14 depicts the principle of this approach featuring the fraction collection of the

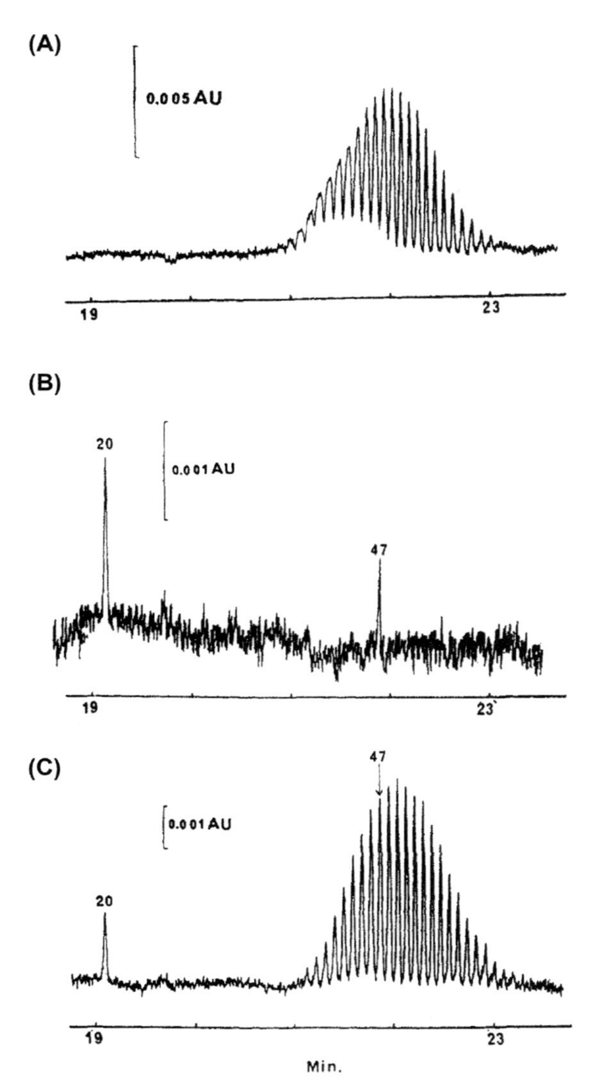

Figure 2.14 (A) Micropreparative capillary electrophoretic separation from a polydeoxy-adenylic acid test mixture, p(dA)$_{40-60}$, (B) analytical run of the isolated p (dA)$_{47}$ spiked with p(dA)$_{20}$, and (C) analytical run of p(dA)$_{40-60}$ spiked with p(dA)$_{20}$. The micropreparative and the analytical injections were made electrophoretically at 9 kV for 10 s (A) and 6 kV for 1 s (B and C), respectively. (A) The applied field was 300 V/cm for the run, then decreased to 30 V/cm for the collection, over 60 s. (B and C) The applied field was constant at 300 V/cm. *With permission from A. Guttman, A.S. Cohen, D.N. Heiger, B.L. Karger, Analytical and micropreparative ultrahigh resolution of oligonucleotides by polyacrylamide gel high-performance capillary electrophoresis, Analytical Chemistry 62 (2) (1990) 137–141 [93].*

47mer [$p(dA)_{47}$] from the polydeoxy-adenylic acid test mixture of $p(dA)_{40-60}$.

In the mid-1990s, the Karger group introduced a high-precision fraction collection setup comprising a fiber optics–based detector near the end of the capillary and a sheath flow assembly at the end of the capillary for continuous automatic fraction collection up to 60 fractions (1 μL or less) [94]. The same group built an instrument for high-throughput fraction collection with capillary array electrophoresis [95] for mutation detection. The 12-capillary array system had two-point detection using a side illumination as shown in Fig. 2.15. This instrument was used to separate and purify yeast genomic DNA fragments [96]. The DNA fractions, exiting the separation capillaries, were continuously collected in a 1536-well collection plate made of agarose gel that allowed recovery for cloning and sequencing. At the same time, Shi et al. developed an automated CGE DNA fraction collection method to create expressed sequence tag (EST) library based on a novel DNA fragment-pooling strategy [97]. Successful fraction collection of sequencing grade DNA

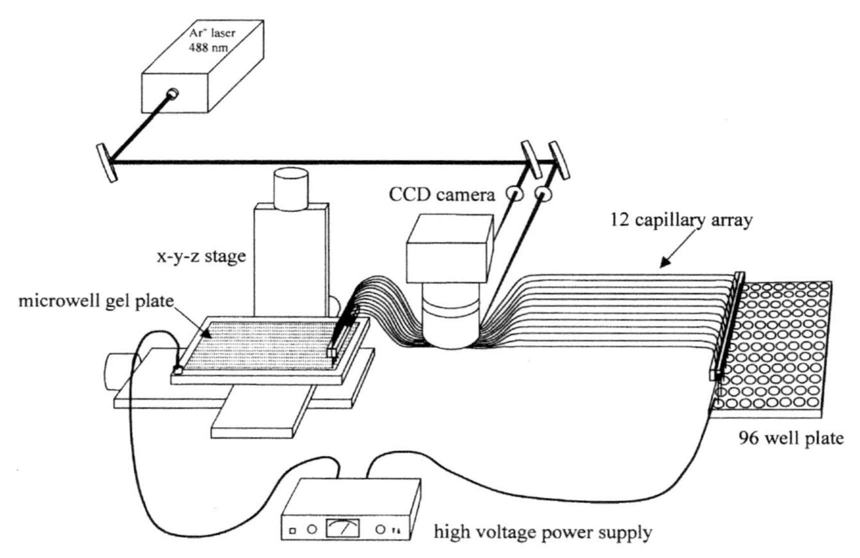

Figure 2.15 Capillary array instrument with two-point detection and side-entry illumination for high sensitivity and high-resolution multicapillary fraction collection applications. *With permission from M. Minarik, K. Kleparnik, M. Gilar, F. Foret, A.W. Miller, Z. Sosic, et al., Design of a fraction collector for capillary array electrophoresis, Electrophoresis 23 (1) (2002) 35–42 [95].*

fragments of complementary DNA transcripts was achieved. The collected DNA fragments were cloned and sequenced to create the EST library.

Others developed an automated multicapillary instrument with fraction collection capability for DNA mutation discovery by constant denaturant capillary electrophoresis [98], comprising a high sensitivity fluorescence detector, precise and stable temperature control ($\pm 0.01^\circ C$), and full automation (sample delivery, injection, matrix replacement, and fraction collection). The array of capillaries was divided into 6×4 narrow bore tubes, each group with independent temperature control from room temperature upto 90°C.

On the protein analysis side, Hjerten and Zhu used CGE to separate the ingredients of Actinomycetes culture filtrate butanol extract in a vertical capillary tube containing polyacrylamide gel and employed off-tube spectrophotometric detection [99] in analytical and micropreparative scale. An interesting economic performance analysis in preparative CGE was conducted by Guttman and Mazsaroff [100]. Nanomolar quantities of nucleic acids or proteins can be collected for subsequent microsequencing with the goal to maximize the production of a product with a given purity in the shortest time, that is, to achieve the highest throughput. The authors suggested a simple economic model for analyzing the yield of preparative capillary electrophoresis. A fraction collector was also built for capillary electrophoresis by Jorgensen's group enabling to collect proteins and peptides after separation [101].

References

[1] B.L. Karger, A.S. Cohen, A. Guttman, High-performance capillary electrophoresis in the biological sciences, Journal of Chromatography 492 (1989) 585–614.

[2] G.G. Stokes, On the theories of the internal friction in motion, and of the equilibrium and motion of elastic solids, Transactions of the Cambridge Philosophical Society 8 (1845) 287–305.

[3] C. Filep, A. Guttman, The effect of temperature in sodium dodecyl sulfate capillary gel electrophoresis of protein therapeutics, Analytical Chemistry 92 (5) (2020) 4023–4028.

[4] S.A. Arrhenius, Über die Dissociationswärme und den Einflusß der Temperatur auf den Dissociationsgrad der Elektrolyte, Zeitschrift für Physikalische Chemie 4 (1889) 96–116.

[5] M. Szoke, M. Sasvari-Szekely, A. Guttman, Ultra-thin-layer agarose gel electrophoresis. I. Effect of the gel concentration and temperature on the separation of DNA fragments, Journal of Chromatography A 830 (2) (1999) 465–471.

[6] H. Lu, E. Arriaga, D.Y. Chen, D. Figeys, N.J. Dovichi, Activation energy of single-stranded DNA moving through cross-linked polyacrylamide gels at 300 V/cm. Effect of temperature on sequencing rate in high-electric-field capillary gel electrophoresis, Journal of Chromatography A 680 (2) (1994) 503–510.

[7] A. Guttman, M. Kerekgyarto, G. Jarvas, Effect of separation temperature on structure specific glycan migration in capillary electrophoresis, Analytical Chemistry 87 (23) (2015) 11630–11634.
[8] M. Kerekgyarto, G. Jarvas, L. Novak, A. Guttman, Activation energy associated with the electromigration of oligosaccharides through viscosity modifier and polymeric additive containing background electrolytes, Electrophoresis 37 (4) (2016) 573–578.
[9] D. Tietz, Gel electrophoresis of intact subcellular particles, Journal of Chromatography 418 (1987) 305–344.
[10] D. Tietz, A. Chrambach, Concave Ferguson plots of DNA fragments and convex Ferguson plots of bacteriophages: evaluation of molecular and fiber properties, using desktop computers, Electrophoresis 13 (5) (1992) 286–294.
[11] A.T. Andrews, Electrophoresis, Theory, Techniques and Biochemical and Clinical Applications, second ed., Claredon Press, Oxford, 1986.
[12] J. Noolandi, Polymer dynamics in electrophoresis of DNA, Annual Review of Physical Chemistry 43 (1992) 237–256.
[13] A.G. Ogston, The spaces in a uniform random suspension of fibres, Transactions of the Faraday Society 54 (1958) 1754–1757.
[14] A. Guttman, On the separation mechanism of capillary sodium dodecyl sulfate-gel electrophoresis of proteins, Electrophoresis 16 (4) (1995) 611–616.
[15] P.D. Grossman, S. Menchen, D. Hershey, Quantitative analysis of DNA-sequencing electrophoresis, Genetic Analysis: Techniques and Applications 9 (1) (1992) 9–16.
[16] K.A. Ferguson, Starch-gel electrophoresis—application to the classification of pituitary proteins and polypeptides, Metabolism 13 (Suppl) (1964) 985–1002.
[17] O.J. Lumpkin, P. Dejardin, B.H. Zimm, Theory of gel electrophoresis of DNA, Biopolymers 24 (8) (1985) 1573–1593.
[18] P.G. De Gennes, Scaling Concept in Polymer Physics, Cornell University Press, Ithaca, NY, 1979.
[19] G.W. Slater, J. Noolandi, The biased reptation model of DNA gel electrophoresis: mobility vs molecular size and gel concentration, Biopolymers 28 (10) (1989) 1781–1791.
[20] J.L. Viovy, T. Duke, DNA electrophoresis in polymer solutions: Ogston sieving, repatation and constraint release, Electrophoresis 14 (1993) 322–329.
[21] G.B. Salieb-Beugelaar, K.D. Dorfman, A. van den Berg, J.C. Eijkel, Electrophoretic separation of DNA in gels and nanostructures, Lab on a Chip 9 (17) (2009) 2508–2523.
[22] J.K. Bryan, Molecular weights of protein multimers from polyacrylamide gel electrophoresis, Analytical Biochemistry 78 (2) (1977) 513–519.
[23] D. Rodbard, A. Chrambach, Unified theory for gel electrophoresis and gel filtration, Proceedings of the National Academy of Sciences of the United States of America 65 (4) (1970) 970–977.
[24] A. Guttman, C. Filep, B.L. Karger, Fundamentals of capillary electrophoretic migration and separation of SDS proteins in borate cross-linked dextran gels, Analytical Chemistry 93 (26) (2021) 9267–9276.
[25] R.E. Offord, Electrophoretic mobilities of peptides on paper and their use in the determination of amide groups, Nature 211 (5049) (1966) 591–593.
[26] E.C. Rickard, M.M. Strohl, R.G. Nielsen, Correlation of electrophoretic mobilities from capillary electrophoresis with physicochemical properties of proteins and peptides, Analytical Biochemistry 197 (1) (1991) 197–207.
[27] L. He, B. Niemeyer, A novel correlation for protein diffusion coefficients based on molecular weight and radius of gyration, Biotechnology Progress 19 (2) (2003) 544–548.
[28] C. Filep, A. Guttman, Capillary sodium dodecyl sulfate gel electrophoresis of proteins: introducing the three dimensional Ferguson method, Analytica Chimica Acta 1183 (2021) 338958.

[29] A. Einstein, Motion of suspended particles in stationary liquids required from the molecular kinetic theory of heat, Annalen der Physik (Weinheim, Germany) 17 (1905) 549–560.
[30] J.C. Giddings, Harnessing electrical forces for separation. Capillary zone electrophoresis, isoelectric focusing, field-flow fractionation, split-flow thin-cell continuous-separation and other techniques, Journal of Chromatography 480 (1989) 21–33.
[31] C. Schwer, E. Kenndler, Peak broadening in capillary zone electrophoresis with electro-osmotic flow: dependence of plate number and resolution on charge number, Chromatographia 33 (7–8) (1992) 331–335.
[32] E. Kenndler, C. Schwer, Peak dispersion and separation efficiency in high-performance zone electrophoresis with gel-filled capillaries, Journal of Chromatography, 595 (1992) 313–318.
[33] S.L. Delinger, J.M. Davis, Influence of analyte plug width on plate number in capillary electrophoresis, Analytical Chemistry 64 (17) (1992) 1947–1959.
[34] E. Kenndler, C. Schwer, Nondependence of diffusion-controlled peak dispersion on diffusion coefficient and ionic mobility in capillary zone electrophoresis without electroosmotic flow, Analytical Chemistry 63 (21) (1991) 2499–2502.
[35] P.G. Righetti, Macroporous gels: facts and misfacts, Journal of Chromatography A 698 (1–2) (1995) 3–17.
[36] A. Guttman, N. Cooke, Effect of temperature on the separation of DNA restriction fragments in capillary gel electrophoresis, Journal of Chromatography 559 (1–2) (1991) 285–294.
[37] A. Guttman, J. Horvath, N. Cooke, Influence of temperature on the sieving effect of different polymer matrixes in capillary SDS gel electrophoresis of proteins, Analytical Chemistry 65 (3) (1993) 199–203.
[38] A. Guttman, N. Cooke, Practical aspects of chiral separation of pharmaceuticals by capillary electrophoresis: I. Separation optimization, Journal of Chromatography 680 (1994) 157–162.
[39] R. Kuhn, S. Hofstetter-Kuhn, Capillary Electrophoresis: Principles and Practice, Springer Laboratory, Berlin, 1993.
[40] S. Terabe, K. Otsuka, T. Ando, Band broadening in electrokinetic chromatography with micellar solutions and open-tubular capillaries, Analytical Chemistry 61 (3) (1989) 251–260.
[41] J.A. Luckey, T.B. Norris, L.M. Smith, Analysis of resolution in DNA sequencing by capillary gel electrophoresis, Journal of Physical Chemistry 97 (12) (1993) 3067–3075.
[42] J.A. Luckey, L.M. Smith, Optimization of electric field strength for DNA sequencing in capillary gel electrophoresis, Analytical Chemistry 65 (20) (1993) 2841–2850.
[43] H.F. Yin, M.H. Kleemiss, J.A. Lux, G. Schomburg, Diffusion coefficients of oligonucleotides in capillary gel electrophoresis, Journal of Microcolumn Separations 3 (4) (1991) 331–335.
[44] G.W. Slater, P. Mayer, P.D. Grossman, Diffusion, Joule heating, and band broadening in capillary gel electrophoresis of DNA, Electrophoresis 16 (1) (1995) 75–83.
[45] P. Sun, R.A. Hartwick, The effect of electric fields on the dispersion of oligonucleotides using a multi-point detection method in capillary gel electrophoresis, Journal of Liquid Chromatography 17 (9) (1994) 1861–1875.
[46] S. Wicar, M. Vilenchik, A. Belenkii, A.S. Cohen, B.L. Karger, Influence of coiling on performance in capillary electrophoresis using open tubular and polymer network columns, Journal of Microcolumn Separations 4 (4) (1993) 339–348.
[47] S.V. Ermakov, M.S. Bello, P.G. Righetti, Numerical algorithms for capillary electrophoresis, Journal of Chromatography A 661 (1–2) (1994) 265–278.

[48] M.S. Bello, M.Y. Zhukov, P.G. Righetti, Combined effects of nonlinear electrophoresis and nonlinear chromatography on concentration profiles in capillary electrophoresis, Journal of Chromatography A 693 (1) (1995) 113–130.
[49] J. Bullock, J. Strasters, J. Snider, Effect of multiple electrolyte buffers on peak symmetry, resolution, and sensitivity in capillary electrophoresis, Analytical Chemistry 67 (18) (1995) 3246–3252.
[50] Y.Y. Rawjee, R.L. Williams, G. Vigh, Efficiency optimization in capillary electrophoretic chiral separations using dynamic mobility matching, Analytical Chemistry 66 (21) (1994) 3777–3781.
[51] L.G. Song, Q.Y. Ou, W.L. Yu, G.F. Xu, Effect of high-concentrations of salts in samples on capillary electrophoresis of anions, Journal of Chromatography A 696 (2) (1995) 307–319.
[52] R.L. Williams, G. Vigh, Polyethylene glycol monomethyl ether sulfate-based background electrolytes in capillary electrophoresis, Journal of Chromatography A 744 (1–2) (1996) 75–80.
[53] R.L. Williams, G. Vigh, N-(polyethyleneglycol monomethyl ether)-N-methylmorpholinium-based background electrolytes in capillary electrophoresis, Journal of Chromatography A 763 (1–2) (1997) 253–259.
[54] R.L. Williams, B. Childs, E.V. Dose, G. Guiochon, G. Vigh, Peak shape distortions in the capillary electrophoretic separations of strong electrolytes when the background electrolyte contains two strong electrolyte co-ions, Analytical Chemistry 69 (7) (1997) 1347–1354.
[55] S.V. Ermakov, M.Y. Zhukov, L. Capelli, P.G. Righetti, Wall adsorption in capillary electrophoresis—experimental study and computer-simulation, Journal of Chromatography A 699 (1–2) (1995) 297–313.
[56] B. Gas, M. Stedry, A. Rizzi, E. Kenndler, Dynamics of peak dispersion in capillary zone electrophoresis including wall adsorption. 1. Theoretical-model and results of simulation, Electrophoresis 16 (6) (1995) 958–967.
[57] M.R. Schure, A.M. Lenhoff, Consequences of wall adsorption in capillary electrophoresis: theory and simulation, Analytical Chemistry 65 (21) (1993) 3024–3037.
[58] I. Pagonabarraga, J. Bafaluy, J.M. Rubi, Adsorption of colloidal particles in the presence of external fields, Physical Review Letters 75 (3) (1995) 461–464.
[59] S.V. Ermakov, P.G. Righetti, Computer-simulation for capillary zone electrophoresis—a quantitative approach, Journal of Chromatography A 667 (1–2) (1994) 257–270.
[60] J.L. Beckers, Calculation of the composition of sample zones in capillary zone electrophoresis. 2. Simulated electropherograms, Journal of Chromatography A 696 (2) (1995) 285–294.
[61] J.L. Beckers, Calculation of the composition of sample zones in capillary zone electrophoresis. 1. Mathematical-Model, Journal of Chromatography A 693 (2) (1995) 347–357.
[62] R.J. Nelson, A. Paulus, A.S. Cohen, A. Guttman, B.L. Karger, Use of Peltier thermoelectric devices to control column temperature in high-performance capillary electrophoresis, Journal of Chromatography 480 (1989) 111–127.
[63] J.H. Knox, Thermal effects and band spreading in capillary electroseparation, Chromatographia 26 (1988) 329–337.
[64] C.J. Evenhuis, R.M. Guijt, M. Macka, P.J. Marriott, P.R. Haddad, Temperature profiles and heat dissipation in capillary electrophoresis, Analytical Chemistry 78 (8) (2006) 2684–2693.
[65] J.H. Knox, K.A. Mccormack, Temperature effects in capillary electrophoresis. 1. Internal capillary temperature and effect upon performance, Chromatographia 38 (3–4) (1994) 207–214.

[66] J.H. Knox, K.A. Mccormack, Temperature effects in capillary electrophoresis. 2. Some theoretical calculations and predictions, Chromatographia 38 (3–4) (1994) 215–221.
[67] S.F.Y. Li, Capillary Electrophoresis, Elsevier, Amsterdam, 1993.
[68] M.U. Musheev, S. Javaherian, V. Okhonin, S.N. Krylov, Diffusion as a tool of measuring temperature inside a capillary, Analytical Chemistry 80 (17) (2008) 6752–6757.
[69] M.U. Musheev, Y. Filiptsev, S.N. Krylov, Temperature difference between the cooled and the noncooled parts of an electrolyte in capillary electrophoresis, Analytical Chemistry 82 (20) (2010) 8692–8695.
[70] S.C. Smith, M.G. Khaledi, Prediction of the migration behavior of organic acids in micellar electrokinetic chromatography, Journal of Chromatography 632 (1–2) (1993) 177–184.
[71] J.M. Davis, Influence of thermal variation of diffusion-coefficient on nonequilibrium plate height in capillary zone electrophoresis, Journal of Chromatography 517 (1990) 521–547.
[72] D.S. Burgi, K. Salomon, R.L. Chien, Methods for calculating the internal temperature of capillary columns during capillary electrophoresis, Journal of Liquid Chromatography 14 (5) (1991) 847–867.
[73] E. Grushka, R.M. McCormick, J.J. Kirkland, Effect of temperature gradients on the efficiency of capillary zone electrophoresis separations, Analytical Chemistry 61 (3) (1989) 241–246.
[74] C. Gelfi, M. Perego, P.G. Righetti, Capillary electrophoresis of oligonucleotides in sieving liquid polymers in isoelectric buffers, Electrophoresis 17 (9) (1996) 1470–1475.
[75] Z. Csapo, A. Gerstner, M. Sasvari-Szekely, A. Guttman, Automated ultra-thin-layer SDS gel electrophoresis of proteins using noncovalent fluorescent labeling, Analytical Chemistry 72 (11) (2000) 2519–2525.
[76] A. Guttman, N. Cooke, Capillary gel affinity electrophoresis of DNA fragments, Analytical Chemistry 63 (18) (1991) 2038–2042.
[77] S. Nathakarnkitkool, P.J. Oefner, G. Bartsch, M.A. Chin, G.K. Bonn, High-resolution capillary electrophoretic analysis of DNA in free solution, Electrophoresis 13 (1–2) (1992) 18–31.
[78] K. Shirahama, K. Tsujii, T. Takagi, Free-boundary electrophoresis of sodium dodecyl sulfate-protein polypeptide complexes with special reference to SDS-polyacrylamide gel electrophoresis, Journal of Biochemistry 75 (2) (1974) 309–319.
[79] C. Filep, A. Guttman, Effect of the monomer cross-linker ratio on the separation selectivity of monoclonal antibody subunits in sodium dodecyl sulfate capillary gel electrophoresis, Analytical Chemistry 93 (7) (2021) 3535–3541.
[80] A. Guttman, A. Paulus, A.S. Cohen, N. Grinberg, B.L. Karger, Use of complexing agents for selective separation in high-performance capillary electrophoresis. Chiral resolution via cyclodextrins incorporated within polyacrylamide gel columns, Journal of Chromatography 448 (1) (1988) 41–53.
[81] M. Chiari, M. Nesi, P.G. Righetti, Movement of DNA fragments during capillary zone electrophoresis in liquid polyacrylamide, Journal of Chromatography A 652 (1) (1993) 31–39.
[82] G.W. Slater, J. Noolandi, On the reptation theory of gel electrophoresis, Biopolymers 25 (3) (1986) 431–454.
[83] C. Carlsson, A. Larsson, M. Jonsson, B. Norden, Dancing DNA in capillary solution electrophoresis, Journal of the American Chemical Society 117 (13) (1995) 3871–3872.
[84] X. Shi, R.W. Hammond, M.D. Morris, DNA conformational dynamics in polymer solutions above and below the entanglement limit, Analytical Chemistry 67 (6) (1995) 1132–1138.

[85] S.P. Radko, A. Chrambach, Electrophoresis of proteins in semidilute polyethylene glycol solutions: mechanism of retardation, Biopolymers 42 (2) (1997) 183–189.
[86] S. Saha, D.M. Heuer, L.A. Archer, Electrophoretic mobility of linear and star-branched DNA in semidilute polymer solutions, Electrophoresis 27 (16) (2006) 3181–3194.
[87] A.E. Barron, H.W. Blanch, D.S. Soane, A transient entanglement coupling mechanism for DNA separation by capillary electrophoresis in ultradilute polymer solutions, Electrophoresis 15 (5) (1994) 597–615.
[88] S.J. Hubert, G.W. Slater, Theory of capillary electrophoretic separations of DNA-polymer complexes, Electrophoresis 16 (11) (1995) 2137–2142.
[89] J. Sudor, M.V. Novotny, End-label, free-solution capillary electrophoresis of highly charged oligosaccharides, Analytical Chemistry 67 (22) (1995) 4205–4209.
[90] D.A. Mcgregor, E.S. Yeung, Optimization of capillary electrophoretic separation of DNA fragments based on polymer filled capillaries, Journal of Chromatography A 652 (1) (1993) 67–73.
[91] T. Manabe, N. Chen, S. Terabe, M. Yohda, I. Endo, Effects of linear polyacrylamide concentrations and applied voltages on the separation of oligonucleotides and DNA sequencing fragments by capillary electrophoresis, Analytical Chemistry 66 (23) (1994) 4243–4252.
[92] A.S. Cohen, D.R. Najarian, A. Paulus, A. Guttman, J.A. Smith, B.L. Karger, Rapid separation and purification of oligonucleotides by high-performance capillary gel electrophoresis, Proceedings of the National Academy of Sciences of the United States of America 85 (24) (1988) 9660–9663.
[93] A. Guttman, A.S. Cohen, D.N. Heiger, B.L. Karger, Analytical and micropreparative ultrahigh resolution of oligonucleotides by polyacrylamide gel high-performance capillary electrophoresis, Analytical Chemistry 62 (2) (1990) 137–141.
[94] O. Muller, F. Foret, B.L. Karger, Design of a high-precision fraction collector for capillary electrophoresis, Analytical Chemistry 67 (17) (1995) 2974–2980.
[95] M. Minarik, K. Kleparnik, M. Gilar, F. Foret, A.W. Miller, Z. Sosic, et al., Design of a fraction collector for capillary array electrophoresis, Electrophoresis 23 (1) (2002) 35–42.
[96] J. Berka, M.C. Ruiz-Martinez, R. Hammond, M. Minarik, F. Foret, Z. Sosic, et al., Application of high-resolution capillary array electrophoresis with automated fraction collection for GeneCalling trade mark analysis of the yeast genomic DNA, Electrophoresis 24 (4) (2003) 639–647.
[97] L. Shi, J. Khandurina, Z. Ronai, B.-Y. Li, W.K. Kwan, X. Wang, et al., Micropreparative capillary gel electrophoresis of DNA: rapid expressed sequence tag library construction, Electrophoresis 24 (1–2) (2003) 86–92.
[98] Q. Li, C. Deka, B.J. Glassner, K. Arnold, X.C. Li-Sucholeiki, A. Tomita-Mitchell, et al., Design of an automated multicapillary instrument with fraction collection for DNA mutation discovery by constant denaturant capillary electrophoresis (CDCE), Journal of Separation Science 28 (12) (2005) 1375–1389.
[99] S. Hjerten, M.D. Zhu, Analytical and micropreparative high-performance electrophoresis, Protides of the Biological Fluids 33 (1985) 537–540.
[100] A. Guttman, I. Mazsaroff, Economical performance analysis in preparative capillary gel electrophoresis, in: New Approaches in Chromatography '93 [Paper Budapest Chromatography Conference] (1993) 63–75.
[101] D.J. Rose, J.W. Jorgenson, Fraction collector for capillary zone electrophoresis, Journal of Chromatography 438 (1) (1988) 23–34.

CHAPTER THREE

Separation matrix and column technology

3.1 Gels and polymer networks

Gels may vary from viscous fluids to solids [1]. Physical gels can be strong or weak gels. Under large deformation forces, strong gels maintain their solid state, while weak gels, on the other hand, behave as liquids. Entangled polymer networks are a third class of gels formed by topological interaction of polymer chains in cases when the product of polymer concentration and molecular mass is high enough. This results in a kind of quasigel form as its mechanical properties are different and in most instances behave as high-viscosity liquids [2].

The usual sieving matrices in size separation—based capillary electrophoresis are gels or in a wider category, polymer networks. Gels are coherent colloidal suspensions, consisting of materials with solid-state mechanical properties and constituents in which the solute and solvent are spreading continuously throughout the system [3]. In other words, the gel (or polymer network) can be contemplated as a container holding a large amount of solvent, thus have the characteristics of both liquids and solids. Gels are considered open materials in nonequilibrium state and can be homogeneous or heterogeneous (Fig. 3.1). With respect to size, there are microgels and macrogels (Fig. 3.1, right panel). In a very swollen gel, the diffusion coefficients of small molecules are very high.

One of the most important phenomena of gels and polymer networks is phase transition that can be caused by changes in solvent composition, temperature, pH, ion composition, and electric fields, just to mention the most important ones [4]. Indeed, to obtain gels, one should start with a solution of monomers and process it via polymerization in such a way that gelation occurs. The discussion is still open about the borderline of sols and gels that actually is defined by the gel point. Gelation can occur either by chemical linking (chemical gelation, Fig. 3.2, left panel) or by physical adhesion (physical gelation, Fig. 3.2, middle panel). Chemical

Capillary Gel Electrophoresis.
DOI: https://doi.org/10.1016/B978-0-444-52234-4.00002-7

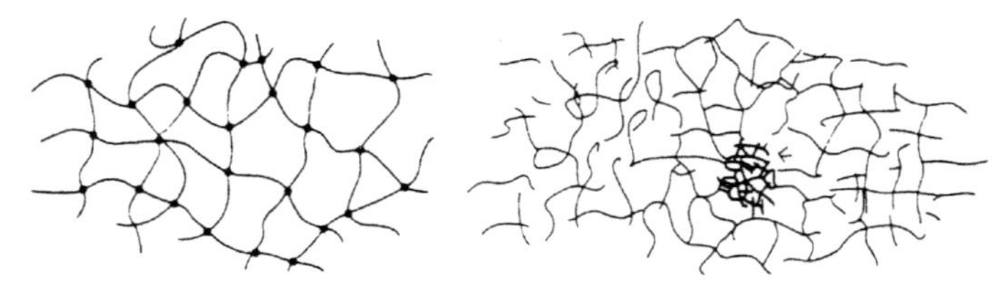

Figure 3.1 Homogeneous (left) and heterogeneous (right) gels. The heterogeneous gel shows a microgel formation within a macrogel.

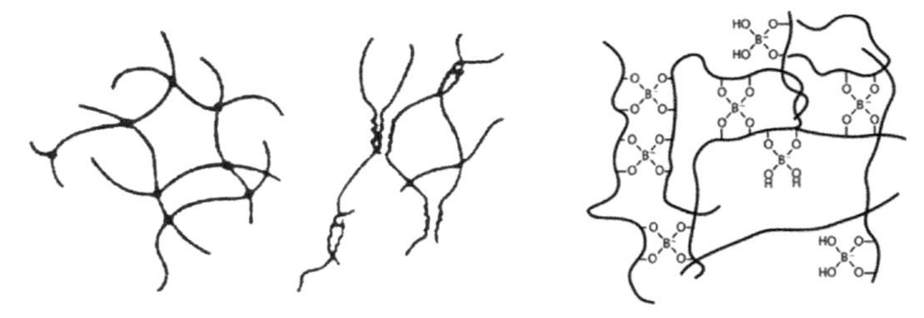

Figure 3.2 Schematic representation of chemical (left panel), physical (middle panel), and transitionally cross-linked chemical (right panel) gels.

gels can be classified based on the covalent bond formation into the categories of condensation, addition, and cross-linking [5]. Gels formed by permanent chemical bonds cannot be dissolved again; therefore chemical gels are considered irreversible gels. Physical gels can be classified into strong physical gels, for example, microcrystals, double/triple helices, as well as weak gels where the three-dimensional network is held together by hydrogen bonds, ionic and hydrophobic association, and agglomeration. Physical gels have reversible links between the polymer chains, therefore adjustable by changing the gelation parameters as listed above. Transitionally cross-linked chemical gels such as borate–dextran [6,7] represent a third category (Fig. 3.2, right panel), extensively used in size separation of sodium dodecyl sulfate (SDS) protein complexes [8] by capillary electrophoresis.

Chemically cross-linked gels with covalent bonds can be formed by heat, catalysts, light, radiation, or plasma treatment. Cross-linking can be accomplished during the initial polymerization process or by treating the already existing polymer chains. Thermal polymerization can be divided into two main approaches of polyaddition- or polycondensation-based reactions. When monomers of more than three functional groups are polymerized alone or with other monomers, branching occurs and a three-dimensional network is formed [9].

The most frequently used method for covalent matrix preparation in electrophoresis is free radical polymerization. Usually vinyl monomers are used, such as acrylamide, cross-linked by bisacrylamide or other multifunctional crosslinkers [10]. Please note that cross-linking is a random process, dependent on factors such as monomer/crosslinker ratio; hence homogeneous or heterogeneous networks can be formed (Fig. 3.1). Free radical polymerization requires initiators and sometimes catalysts, which, in the case of polyacrylamide gel preparation, are *N,N,N',N'*-tetramethylethylenediamine (TEMED) and ammonium persulfate (APS), respectively, in most instances. In free radical polymerization, it is also important to control the oxygen content of the reagent solutions as it acts as a scavenger. Therefore it should be removed from the reaction mixture before the polymerization process (e.g., by vacuum, He or N_2 purge, and sonication). Other polymerization techniques such as photopolymerization, radiation polymerization, and plasma polymerization can also be applied. Regardless of the methods of polymerization, in chemical gels, the gel point is when the longest relaxation time becomes infinite [11]. It is proportional to the third power of the molecular mass; therefore according to the Flory—Stockmayer definition [9], the molecular mass of the polymer should be infinite too as Eq. (3.1) and Eq. (3.2) delineate.

$$p_A \cdot p_B = \frac{1}{r \cdot (f_A - 1)(f_B - 1)} \tag{3.1}$$

where p_A and p_B are the fractions of all possible A and B bonds, respectively, r (<1) is the ratio of reactive sites of A and B on each monomer, f_A and f_B are defined in Eq. (3.2).

$$f_A = \frac{\sum_i (f_{Ai}^2 \cdot N_{Ai})}{\sum_i (f_{Ai} \cdot N_{Ai})} \quad \text{and} \quad f_B = \frac{\sum_i (f_{Bi}^2 \cdot N_{Bi})}{\sum_i (f_{Bi} \cdot N_{Bi})} \tag{3.2}$$

where N_{Ai} and N_{Bi} are the number of moles of Ai and Bi, containing f_{Ai} and f_{Bi} functional groups for each type of A and B functional molecules. If the starting concentrations of A and B reactive sites are the same, then $p_A \cdot p_B$ can be condensed to p_{gel} and values for the fraction of all bonds at which an infinite network will form can be found.

Intermolecular interactions such as hydrogen-, coordination-, static-, hydrophobic, and van der Waals bonds can form physical gels. Polyvinyl alcohol (PVA) gels are good examples of hydrogen bond—based cross-linking. Cross-linking by hydrogen bonds can also be formed between different polymers, like polyacrylamide and PVA or polyethylene glycol (PEG) and PVA. In coordination bonding for example, the side chains of a synthetic polymer form coordination complexes with multivalent metallic ions [12]. Polyelectrolytes with opposite charges can have static interactions to form gels [13]. Finally, block copolymers of acrylic acid and acrylic acid stearyl represent examples of cross-linking by van der Waals forces [14]. It is important to note that there are often cases when the form of physical interaction, that is, the nature of cross-linking is not known.

The primary difference between classical polyacrylamide or agarose gel electrophoresis and capillary electrophoresis with gels or polymer networks is the use of narrow-bore fused-silica capillary columns containing the sieving medium. In the early days, capillaries were filled with cross-linked gels similar to that used in traditional polyacrylamide slab or rod gel electrophoresis (PAGE) and have proven to be particularly promising in size-based separation of DNA and protein molecules [15,16]. While gel-filled capillaries were originally prepared by the users, later more and more manufacturers offered prefilled and ready-to-use CGE columns. The three-dimensional structure of these gels acts as molecular sieves, supporting size separation of polyionic biopolymers when they migrate through the media. However, unlike in slab PAGE, where in most instances cross-linked gels are used, in capillary gel electrophoresis (CGE) both cross-linked and various viscosity (high or low) noncross-linked polymer networks can be applied. In traditional slab gel electrophoresis, the monomer/crosslinker mixture of the separation matrix is filled into the separation platform (in most instances between two glass or plastic plates) and the sieving matrix is polymerized in situ usually by the user prior to use. In CGE, narrow-bore capillaries [$<100\ \mu m$ inner diameter (i.d.)] are used and depending on the preferred separation medium, the matrix can be polymerized in situ inside the capillary or outside of the capillary and filled in/replenished by applying appropriate pressure on a polymer network-filled container, which is connected to the separation capillary. The separation is carried out under high applied voltages (tens of thousands of volts), resulting in rapid separations with good efficiencies.

The three types of gels usually used in capillary electrophoresis are the high-viscosity chemically cross-linked gels, transitionally cross-linked chemical gels, and low-viscosity polymer networks (noncross-linked, physical gels). It is important to note that transitionally cross-linked chemical and physical gels offer several advantages over their chemical cross-linked counterparts, such as higher stability, lower viscosity (i.e., easy replenishment in the capillary), and simple manufacturing. Transitionally cross-linked chemical and noncross-linked physical gels are less sensitive to changes in temperature and may accommodate pressure injection. These two latter gel types are extensively employed for the separation of various biopolymers [17–20]. In the case of size separation of proteins, mostly SDS-based background electrolytes are used [15], but attempts were made to employ positive surfactants such as cetyltrimethylammonium bromide [21] and higher hydrophobicity negatively charged ones like sodium hexadecyl sulfate as shown in Fig. 3.3 [22].

Peak dispersion and separation efficiency in gel-filled capillaries were investigated by Kenndler and Schwer [23]. They suggested that the effective charge number is the only analyte-specific parameter that influences the plate height or the plate number; therefore peak broadening is independent of the diffusion coefficient and mobility, because these parameters can be substituted by physical and instrumental constants. Consequently, the theoretical plate number is a linear function of the effective charge number of the analyte molecules.

3.1.1 Cross-linked (chemical) gels

Chemical gels are covalently cross-linked polymer networks, usually featuring high viscosities and well-defined pore structures. Polyacrylamide is

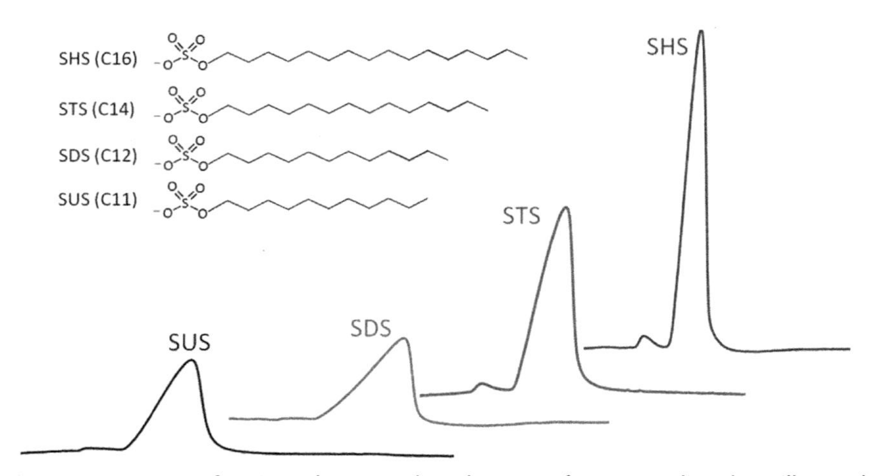

Figure 3.3 Impact of various detergent lengths on surfactant mediated capillary gel electrophoresis of proteins. Inset: Structures of the detergents added to the gel buffer solutions. *With permission from J. Beckman, Y. Song, Y. Gu, S. Voronov, N. Chennamsetty, S. Krystek, et al., Purity determination by capillary electrophoresis sodium hexadecyl sulfate (CE-SHS): a novel application for therapeutic protein characterization, Analytical Chemistry 90 (4) (2018) 2542–2547 [22].*

the most widely used chemical gel material in electric field–mediated differential migration methods, typically cross-linked with *N,N*-methylene-bisacrylamide (BIS) using APS as an initiator and TEMED as a catalyst in a free radical reaction. The pore size of the resulting cross-linked gel is determined by the relative concentration of the monomer and crosslinker used during polymerization (%T: total monomer concentration; and %C: crosslinker concentration as a percent of the total monomer plus crosslinker concentration [24]). Generally used cross-linked polyacrylamide gels of ~3.0%–10.0% T and ~3.0%–5.0% are very rigid and cannot be removed from the capillary tubing after polymerization. For stabilization, most chemical gels are covalently attached to the inner fused-silica surface of the capillary via a bifunctional reagent [25]. Samples can only be introduced into cross-linked gel-filled capillaries by electrokinetic injection, that is, by simply starting the electrophoresis process from the sample vial. While this approach usually results in sharp peaks with possible sample preconcentration (if the separation and sample buffers are chosen properly [26]), the method suffers from poor peak area reproducibility due to sample depletion if the same sample vial is used multiple times. Chemical gels are sensitive to changes in temperature, pH, and high voltage, which all may lead to the formation of small bubbles due to the so-called phase transition phenomena, for example, gel shrinking [1] during polymerization or sometimes during high-voltage separations. On the other hand, chemical gels offer high resolving power, especially for the separation of lower molecular weight biopolymers.

The first report on high-resolution (sequencing grade) single nucleotide separation of oligonucleotides (pdA_{40-60} DNA ladder, Fig. 3.4) by Guttman et al. [27] opened up new horizons in modern CGE-based DNA sequencing, and actually made possible the faster than anticipated completion of the human genome project [28]. In continuation of the successful introduction of gel-filled capillaries for high-resolution separation of DNA molecules, in 1990 Swerdlow and Gesteland reported on the development of the first CGE-based DNA sequencer [29]. This novel approach demonstrated the enormous separation power of CGE, capable of resolving DNA molecules differing by only one nucleotide, even with long-chain fragments, exactly what was required for DNA sequencing.

Chemical gels have been used in the early days of capillary electrophoresis as sieving media, mainly to separate shorter single-stranded oligonucleotide molecules and also to separate proteins by capillary SDS gel electrophoresis. The pioneering work of Karger and coworkers [15]

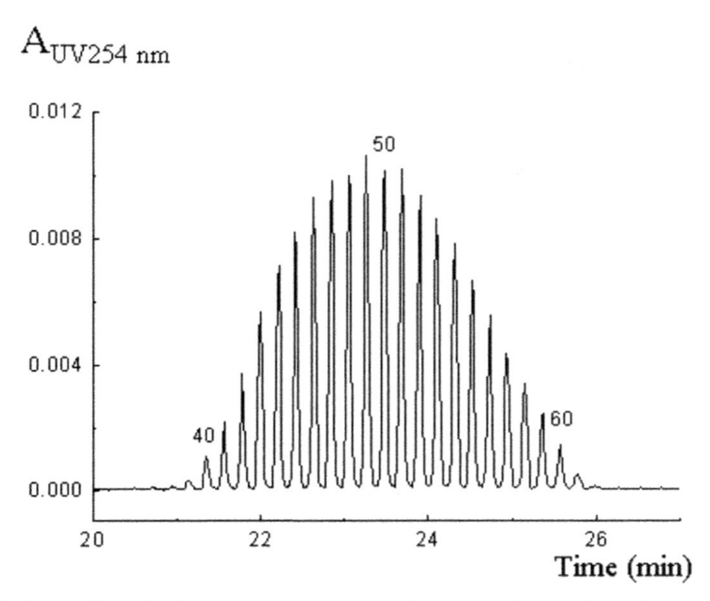

Figure 3.4 The first single base resolution capillary gel electrophoresis–based DNA separation (pdA$_{40-60}$). Conditions: Polyacrylamide gel (7.5% T/3.3% C) in 7 M urea, 100 mM Tris and 250 mM borate (pH 8.3), $E = 350$ V/cm, detection: 254 nm, capillary: $\ell = 30$ cm (effective), 75 μm i.d. *With permission from A. Guttman, A. Paulus, A.S. Cohen, B.L. Karger, H. Rodriguez, W.S. Hancock, High performance, capillary gel electrophoresis: high resolution and micropreparative applications, in: Electrophoresis'88, Copenhagen VCH, Weinheim, 1988 [27].*

demonstrated the possibility of polymerizing cross-linked SDS polyacrylamide gels into narrow-bore capillaries and shown the excellent separation power of this system by resolving the two chains of insulin in less than 8 minutes as depicted in Fig. 3.5, using a 10% T/3.3% C polyacrylamide gel, containing 8M urea and 0.1% SDS.

One interesting example of later developed gel formulations is the work of Vegvari and Hjerten who introduced a special polyacrylamide gel formulation, cross-linked with allyl-beta-cyclodextrin for the separation of DNA fragments [30]. The resolving power of their gel was almost independent of the concentration of the crosslinker, in contrast to regularly cross-linked polyacrylamide gels with *N,N'*-methylene-bisacrylamide. Paulus et al. [31] also used cross-linked polyacrylamide gel-filled capillaries to separate oligonucleotides. They obtained a linear plot of migration time versus base number with an identical slope for three oligonucleotide samples that enabled them to do molecular mass calibration.

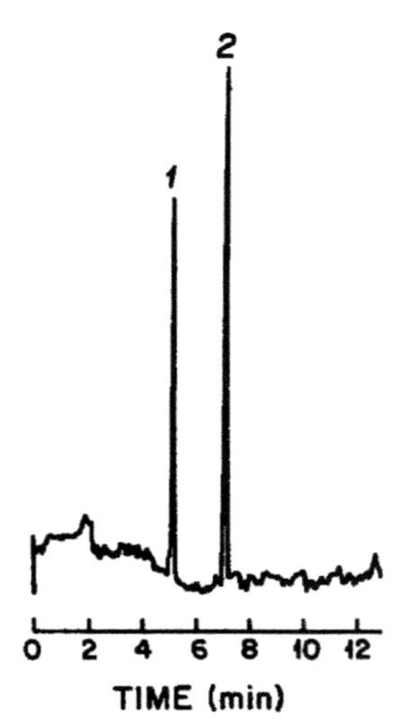

Figure 3.5 High-performance capillary SDS gel electrophoresis separation of the two chains of insulin. Conditions: Effective separation length: 10 cm, $E = 400$ V/cm, $T = 10\%$, $C = 3.3\%$, buffer: 90 mM Tris-phosphate (pH 8.6), 0.1% SDS. *With permission from A.S. Cohen, B.L. Karger, High-performance sodium dodecyl sulfate polyacrylamide gel capillary electrophoresis of peptides and proteins, Journal of Chromatography 397 (1987) 409–417 [15].*

Righetti's group conducted extensive research on structural modification of acrylamide to enhance the hydrophilicity, improve resistance to hydrolysis, and alter pore size [32,33] via elucidation of the polymerization mechanism and the chemical characteristics of the gels [34,35]. Among others, they studied the kinetics of riboflavin-catalyzed photopolymerization that apparently shortened the polymerization times by increasing polymerization efficiency [36,37]. Methylene blue was another additive that acted as a good photopolymerization agent for acrylamide [38]. Others studied the kinetics of acrylamide polymerization in gel electrophoresis capillaries by Raman microprobes and found that the formation of cross-linked 3.5% T/3.3% C polyacrylamide gel (Fig. 3.6) was >99% after 2 hours [39]. The rate of acrylamide polymerization was also investigated in gel-filled capillary columns as a function of APS concentration to prevent bubble formation during polymerization and under high electric field separation conditions. The use of optimal APS concentration resulted in a gel that was capable of single base resolution of short-chain length oligodexoyadenylic acids [40].

Chrambach and coworkers investigated the effect of DNA conformation on the electrophoretic mobility in gel-filled capillaries [41] and found

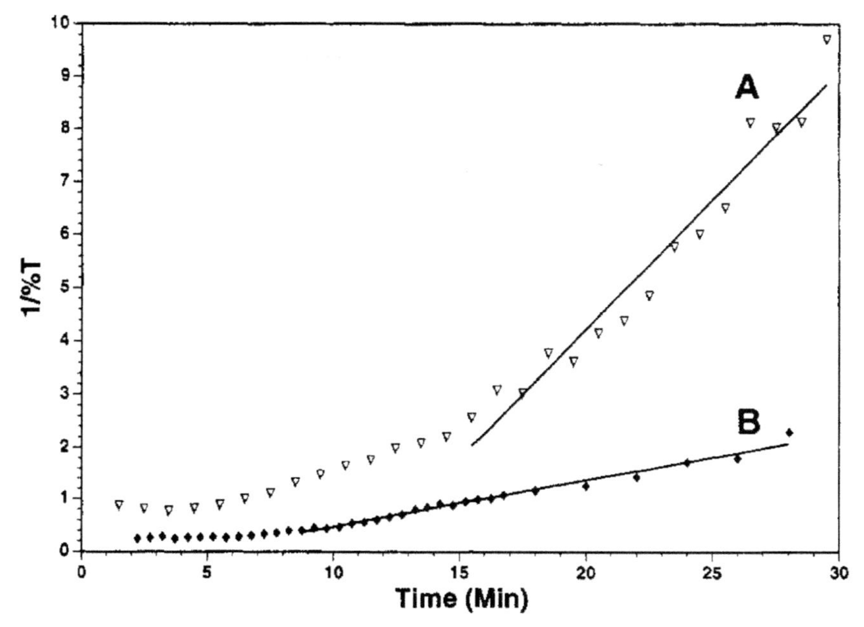

Figure 3.6 Reaction kinetic plots for $T = 3.5\%$, $C = 3.3\%$ gel polymerizations (A) in bulk and (B) in an electrophoresis capillary. Reciprocal acrylamide concentration is plotted in units of 1/%T. *With permission from T.L. Rapp, W.K. Kowalchyk, K.L. Davis, E.A. Todd, K.L. Liu, M.D. Morris, Acrylamide polymerization kinetics in gel electrophoresis capillaries. A Raman microprobe study, Analytical Chemistry 64 (20) (1992) 2434–2437 [39].*

that intramolecular base pairing at the 3′ end of single-stranded oligonucleotides and the type of the 3′ nucleotide may influence the resulting electrophoretic mobility. One of the major problems with cross-linked polyacrylamide gel–based singe-stranded DNA analysis at ambient temperature was peak compression. The addition of a chaotropic agent of 10% formamide or up to 8M urea to polyacrylamide gels minimized such compressions [42]. Konrad and Pentoney investigated this effect by changing some of the experimental parameters such as electric field strength and capillary temperature, to increase resolution when DNA conformation effects were observed [43]. Kobayashi and coworkers studied oligonucleotide migration in gel-filled capillary columns and reported that the migration time of any known sequence without a secondary structure can be estimated with good accuracy (<0.5-mer of cytidine). In addition, deviation between the measured and calculated migration times could be used to provide secondary structure information [44]. However, the authors were not sure whether there were hydrogen bonds or base stacking between the guanine (G) and cytosine (C) residues to maintain the secondary structure in the presence of 7M urea when a small number of the G and C base pairs were present. The decrease in migration time and the increment of the number of G and C base pairs exhibited direct correlation (Fig. 3.7). This strongly suggested that hydrogen bonds or base stacking played an important role in affecting migration time, probably via secondary structure formation.

Fabrication methods of polyacrylamide-filled capillary columns were examined by Gelfi et al. using various optimization approaches for DNA analysis, for example, PCR fragment and genetic disease studies in the range of 50–500 base pairs (bp) [45]. The group developed a special class of gels in the late 1990s favoring substituted derivatives of polyacrylamide. These *N*-substituted acrylamides showed great characteristics in their thermal and hydrolytic stability as well as separation performance compared to

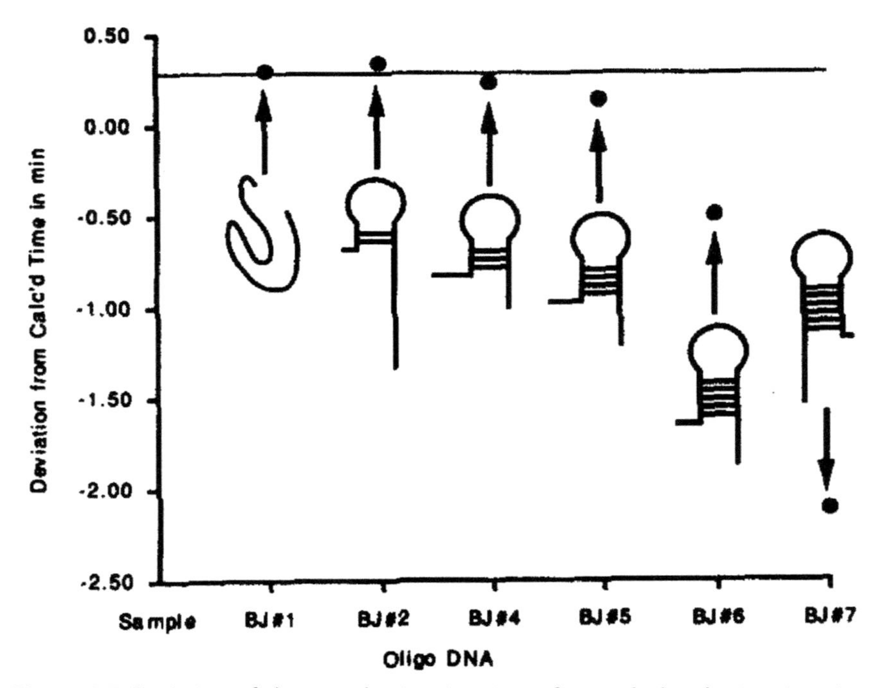

Figure 3.7 Deviation of the actual migration times from calculated migration time versus the number of possible hydrogen bonds. *With permission from T. Satow, T. Akiyama, A. Machida, Y. Utagawa, H. Kobayashi, Simultaneous determination of the migration coefficient of each base in heterogeneous oligo-DNA by gel filled capillary electrophoresis, Journal of Chromatography 652 (1) (1993) 23–30 [44].*

regular acrylamide- or alkylcellulose-based matrices [46–48]. Even gradient gels in step [49] and continuous [50] fashion were prepared using cross-linked polyacrylamide (Fig. 3.8). Preparation of highly condensed polyacrylamide gel-filled capillaries was also demonstrated [51] to produce

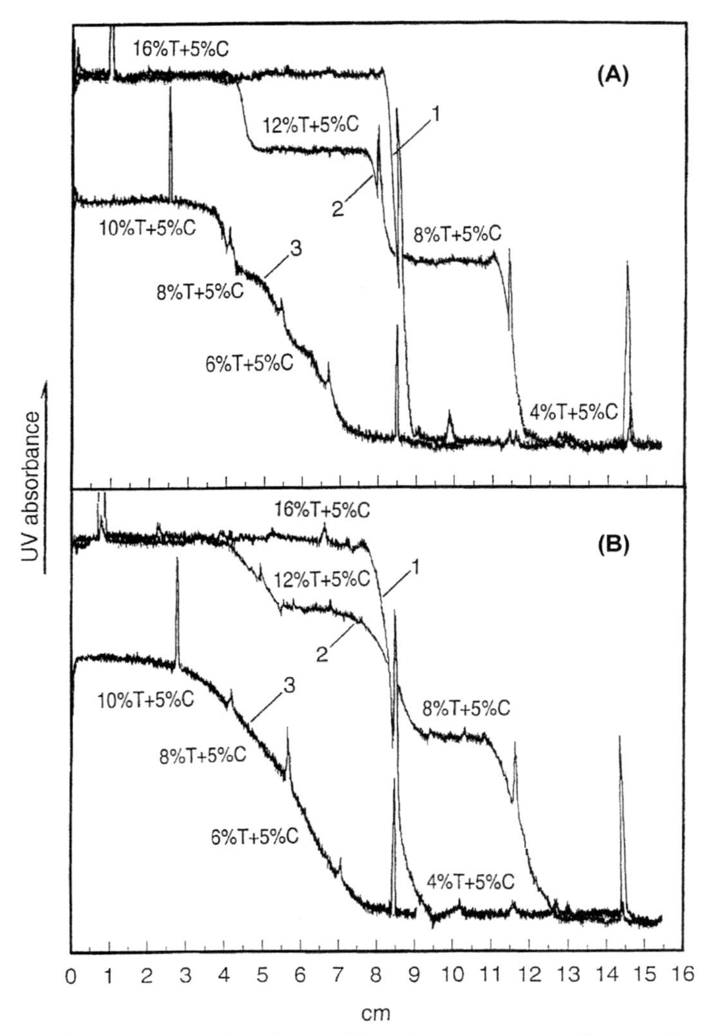

Figure 3.8 Influences of step length and filling directions on capillary gradient gel formation scanned from 75 μm i.d. capillaries. Peaks show the filling marks, some partially washed off. The filling direction was (A) $C_1 \rightarrow C_n$ or (B) $C_n \rightarrow C_1$. *With permission from Y. Chen, F.-L. Wang, U. Schwarz, Polyacrylamide gradient gel-filled capillaries with low detection background, Journal of Chromatography A 772 (1–2) (1997) 129–135 [50].*

void-free columns of small i.d. (25 μm) with monomer concentrations up to 30% T/5% C. The capillaries were ready to be used after 5 hours from the initial preparation, good for >20 injections with gels immobilized at the capillary tips after polymerization and for >70 injections with gels immobilized during polymerization. The device the authors developed was capable to prepare partially or step gradient-filled capillaries as Fig. 3.9 exhibits.

Swerdlow et al. studied the stability of capillary gels, primarily for automated DNA sequencing applications. Typical gel instability issues such as a decrease in current under a constant voltage were found to be caused by bubble formation. Ionic depletion was another factor that resulted in gel instability and elimination of the template from the DNA sequencing sample allowed sample loading without complications [52]. Liu and coworkers [53] introduced an alternative sieving matrix in the form of replaceable cross-linked polyacrylamide for SDS-CGE analysis of proteins. The polymer solution was easily replaceable in the capillary

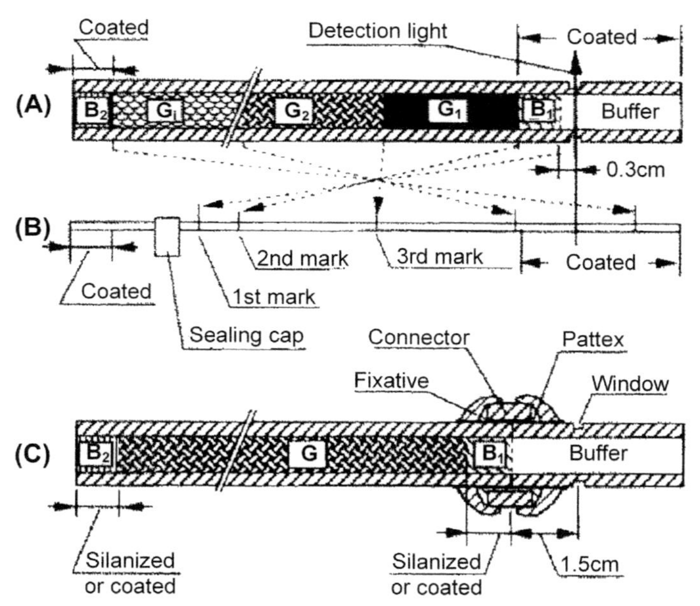

Figure 3.9 Schematic configuration of (A) a partially or step gradient gel-filled capillary and (C) a coupled capillary where the boundaries between the different gels were defined by the positions of the initial solutions. (B) Capillary ready to fill. *With permission from Y. Chen, J.-V. Hoeltje, U. Schwarz, Preparation of highly condensed polyacrylamide gel-filled capillaries with low detection background, Journal of Chromatography A 685 (1) (1994) 121–129 [51].*

columns at 80 psi. Unfortunately, the light absorption of this polymer was quite high for UV detection but with laser-induced fluorescence systems it did not represent a problem.

3.1.1.1 Transitionally cross-linked sieving matrices

The most frequently employed industry standard matrix for capillary electrophoresis—based size separation of proteins is the cross-linked gel composition of dextran monomer (2M molecular weight) with borate crosslinker, in the presence of SDS [54]. This formulation also allows easy replenishment of the separation matrix in the narrow-bore capillary [55].

Borate cross-linking of polyols has long been used in the fracking industry [6] before being implemented in capillary SDS gel electrophoresis separation of proteins [54,56]. At neutral pH, borate ions readily form complexes with vicinal and even isolated vicinal (one carbon atom between the OH holding carbons) cis-diol groups of polyhydroxy compounds, provided that the orientation of the hydroxyl groups is in the cis position [20,57]. In the case of glucose polymers such as dextran with α1—6 building blocks, six-member borate complexes can be formed in 1:1 (one glucose:one borate, left panel in Fig. 3.10) or 2:1 (two glucoses: one borate, middle panel in Fig. 3.10) ratios [58], thus forming intrachain and interchain adducts, respectively. The right panel in Fig. 3.10 shows these possible complexation forms of the entangled dextran chain—borate adducts. Due to the quite rigid polymer structure of dextran, formation of

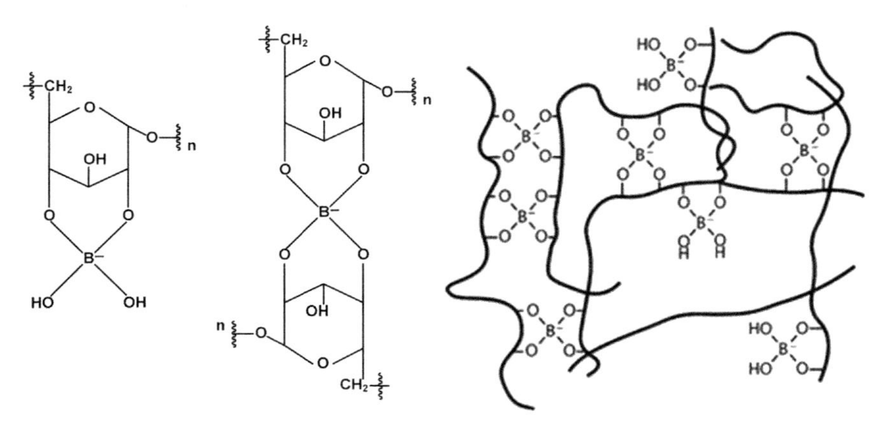

Figure 3.10 Six-member diol—borate complexation in 1:1 (left panel) and 2:1 (middle panel) ratios and their ability to form interchain and intrachain adducts with long-chain dextran polymers (right panel).

intrachain 2:1 cross-linked complexes (loops) is practically negligible [59], thus interchain linkages prevail. The lifetime of the borate–diol complex is in the millisecond range, that is, breaking and reforming of the complex constantly, depending on pH, ionic strength, and temperature [6]. This enables the formation of reversible gels, which are very similar to chemically cross-linked gels because of their high-frequency lifetime [7]. The continuous breaking reconstruction is important for maintaining the sieving capability of the matrix after possible distortion by the migrating large SDS–protein complexes [55], while allowing easy replenishment of the separation matrix in the capillary column. This phenomenon also alleviated the local EOF-mediated bubble formation effect, otherwise represented a serious problem in permanently cross-linked polyacrylamide gels in narrow-bore capillaries [60,61]. An increase in borate concentration for a fixed dextran concentration increases the number of interchain bonds (cross-linking level) and concomitantly decreases the pore size of the resulting gel. However, with excess borate in the system, intrachain borate complexation will also occur (Fig. 3.10, right panel), and the resulting negatively charged patches may cause the chains to repel and consequently open the local pore structure [62].

Using the similar complexation phenomena, hydroxyalkyl-methylcelluloses with borate buffers were used by Cheng and Mitchelson [63] to create physically entangled polymer solutions with transient borate cross-linking capable of separating nucleic acid polymers by CGE. The authors observed that the addition of glycerol to the entangled hydroxypropylmethyl cellulose (HPMC) solution markedly improved the separation of dsDNA fragments in the size range of 100 bp to about 1 kb. This range of DNA fragment sizes was well suited to many PCR product-based studies. The authors attributed the improved performance to the borate content of the buffer system as it can form dimeric 1:2 borate:didiol complexes with both glycerol and HPMC, acting as a central linkage to allow an entangled solution with different pore sizes to form, as shown in Fig. 3.11.

Other additives, such as mannitol, were attempted to enhance the sieving ability of HPMC for dsDNA fragment analysis [64,65], suggesting that chains are formed through hydrogen bonds among mannitol, HPMC, and borate, reshaping the network and decreasing the pore size.

3.1.2 Noncross-linked polymer solutions (physical gels)

Promising advances in capillary electrophoresis separation of biopolymers originated from the exploration of noncross-linked separation matrices.

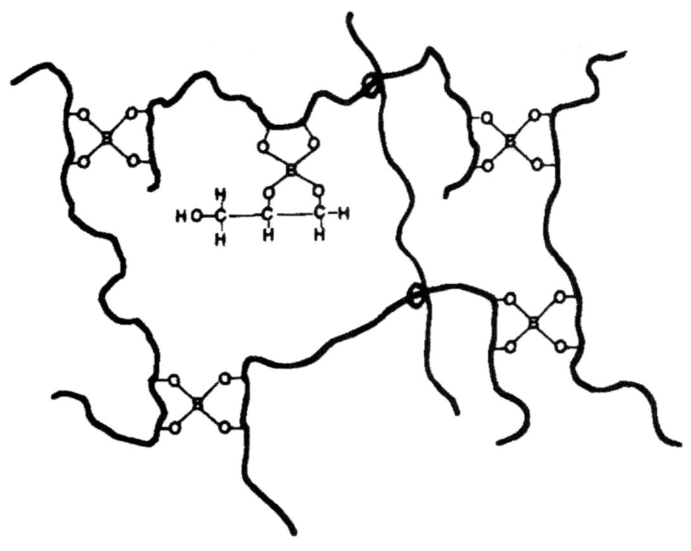

Figure 3.11 1:2 Borate:didiol complexes between glycerol:HPMC:borate, which may form in an entangled glycerol–HPMC–borate buffer system. *With permission from J. Cheng, K.R. Mitchelson, Glycerol-enhanced separation of DNA fragments in entangled solution capillary electrophoresis, Analytical Chemistry 66 (23) (1994) 4210–4214 [63].*

These so-called physical gels or polymer networks became very popular in the 1990s [26,66–68] and still widely used these days for the analysis of biologically important polymers. Examples are low concentration entangled linear polymers, such as polyacrylamide, polyethylene oxide (PEO), derivatized celluloses, pollulane, and PVA. These noncross-linked, usually linear polymers are not attached to the inside wall of the capillary and feature very flexible dynamic pore structures (Fig. 3.2, middle panel). Actually, the pore size of these matrices is defined by dynamic interactions between the polymer chains and can be varied at any time by changing the capillary temperature, separation voltage, salt concentration, or pH. These noncross-linked physical gels are not heat sensitive, and even if a thin layer is adsorbed to the capillary wall (see dynamic coating), the separation matrix can be easily replaced in the capillary by applying positive pressure at the inlet or negative pressure at the outlet side [68]. Thus fresh separation medium can be used for each consecutive analysis, preventing any cross-contamination from previous sample separations [69].

The theoretical treatment on the use of physical gels, otherwise also referred to as entangled polymer solutions, for DNA restriction fragment analysis was introduced by Grossman and Soane [70,71]. Their rheological studies confirmed that the entanglement threshold of hydroxyethyl cellulose (HEC) was 0.003 g/mL in agreement with their theoretical prediction, that is, its mash size was an order of magnitude smaller than that in agarose gels (Fig. 3.12). They showed the agreement between the experimental value and the predicted one using the De Gennes relationship [Eq. (3.3)] [72]:

$$\varphi^* \approx N^{-\frac{4}{5}} \tag{3.3}$$

where Φ^* is the entanglement threshold and N is the number of segments in the polymer chain.

To apply physical gels to a wide range of size separations, the appropriate dynamic pores size should be used. On the other hand, high concentration polymers should be used to obtain a small dynamic pore size that is also associated with increased viscosity. Working near the entanglement threshold fulfills the ideal case of low viscosity and smaller mesh. Eq. (3.4) delineates that smaller pores can be obtained by using shorter polymers (ξ average mesh size) [71].

$$\xi(\varphi^*) \approx aN^{0.6} \tag{3.4}$$

Physical gels support both electrokinetic and pressure injection methods; however, electrokinetic injection causes bias as shown in Fig. 3.13 [73]. During the pressure injection process, introduction of the sample occurs by applying constant pressure (positive at the inlet or negative at the outlet side) for a short period of time having the capillary connected to the sample vial.

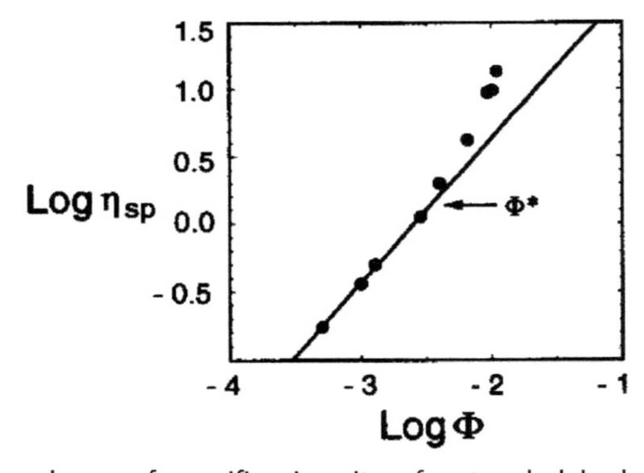

Figure 3.12 Dependence of specific viscosity of entangled hydroxyethyl cellulose (HEC) solution on HEC volume fraction. *With permission from P.D. Grossman, D.S. Soane, Capillary electrophoresis of DNA in entangled polymer solutions, Journal of Chromatography 559 (1–2) (1991) 257–266 [70].*

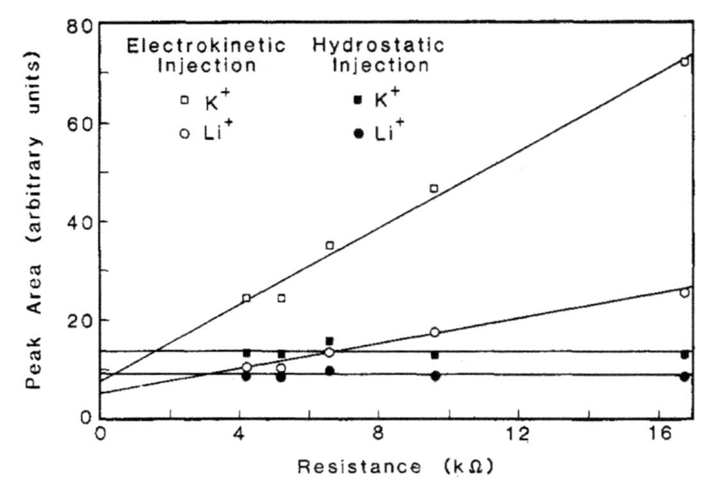

Figure 3.13 Plot of peak areas as a function of sample solution resistance for both electrokinetic and pressure injection. Electrokinetic injection causes a bias that is linear with sample solution resistance (which is inversely proportional to electrolyte concentration). *With permission from X. Huang, M.J. Gordon, R.N. Zare, Bias in quantitative capillary zone electrophoresis caused by electrokinetic sample injection, Analytical Chemistry 60 (4) (1988) 375–377 [73].*

In this case the injected sample replaces the separation matrix. This method also enables sample stacking by preinjecting appropriately designed multiphasic buffer systems and offers excellent run-to-run peak area reproducibility, supporting routine quantitative analysis. Electrokinetic injection, on the other hand, may not provide precise quantitative peak representation due to injection bias mediated by the different electrophoretic mobilities of the individual sample components.

The underlying phenomenon of a transient entanglement coupling mechanism in physical gels for DNA separation by capillary electrophoresis in ultra-dilute polymer solutions was investigated by Barron et al. [74]. Large DNA molecules of 2.0–23.1 kbp were rapidly separated by CE in ultra-dilute polymer solutions of $<0.002\%$ by applying high voltages. At this extremely low polymer concentration, the separation mechanism appeared to be significantly different from that postulated to occur in cross-linked gels. The Ogston and reptation models typically used to describe gel electrophoresis of polyionic biopolymers were not found to be appropriate to characterize DNA separations in such diluted noncross-linked polymer solutions. The authors proposed a transient entanglement coupling mechanism as the basis of the electrophoretic separation of DNA molecules in this system, based on physical polymer/nucleotide interactions [75,76]; however, borate complexation–based transient cross-linking mediated sieving by the polyhydroxy polymer was not considered at that time [77].

The transient entanglement coupling mechanism in diluted polymer solutions suggested no a priori upper size limit for DNA separation by CGE. Fig. 3.14 gives a schematic illustration of the motion of DNA molecules in ultra-dilute HEC solution with the approximate relative sizes of HEC and small and large DNA shown. Using the Kratky–Porod model for a stiff, worm-like coil, the radius of gyration was calculated. The authors concluded that when transient entanglement coupling occurred, the larger

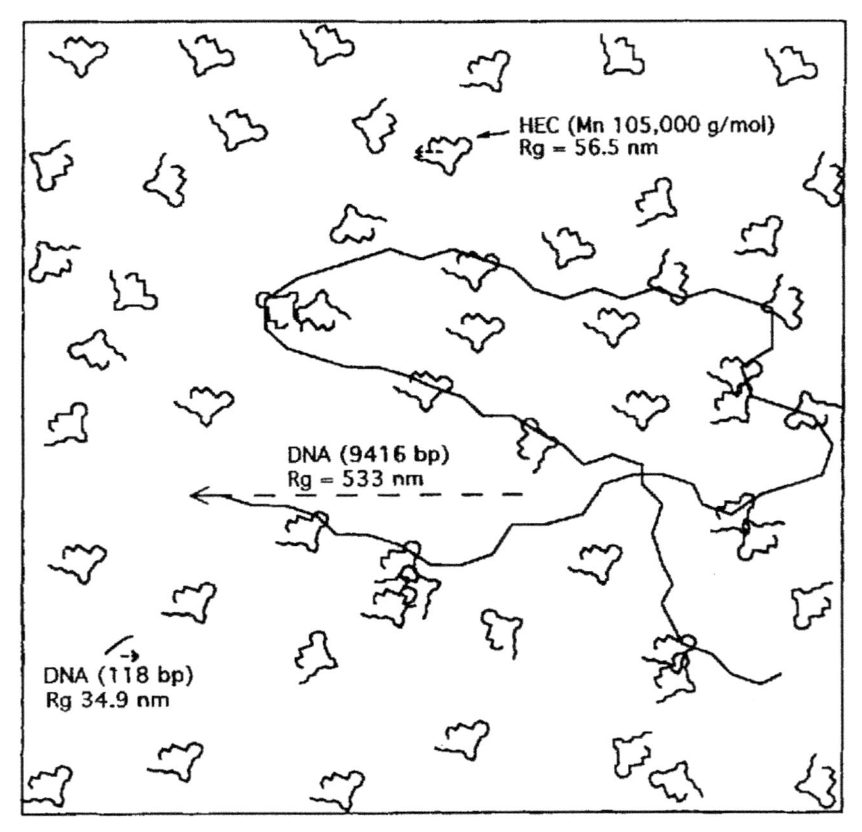

Figure 3.14 A schematic illustration of DNA motion in ultra-dilute HEC solution by transient entanglement coupling mechanism. *With permission from A.E. Barron, H.W. Blanch, D.S. Soane, A transient entanglement coupling mechanism for DNA separation by capillary electrophoresis in ultradilute polymer solutions, Electrophoresis 15 (5) (1994) 597–615 [74].*

DNA molecule dragged the uncharged HEC molecules along during the electrophoresis process, decreasing in this way its electrophoretic mobility in a size-dependent manner, resulting in size-based separation.

Another interesting approach for sieving effect modification utilized various ion-pairing reagents [78] to improve the resolution for a wide range of DNA fragment sizes. These reagents interacted with the linear polymer and changed their physical properties. Enhanced retention of DNA fragments was observed by increasing the ion-pairing contents of the sieving matrix and different ion-pairing reagents produced different degrees of retention.

Righetti's group reported CGE separation of dsDNA fragments in poly (*N*-acryloylaminoethoxyethanol) network [79] using 8%–18% polymer concentration. Double logarithmic plots of mobility versus size (in base pairs, bp) exhibited the three migration regimes referred to in Chapter 2, Basic Principles of Capillary Gel Electrophoresis, that is, the Ogston (i.e., the solute molecules are considered as spherical globules) up to 200 bp, reptation without stretching up to 3–4000 bp and reptation with partial stretching for larger fragments as shown in Fig. 3.15. 10%–12% polymer concentrations were

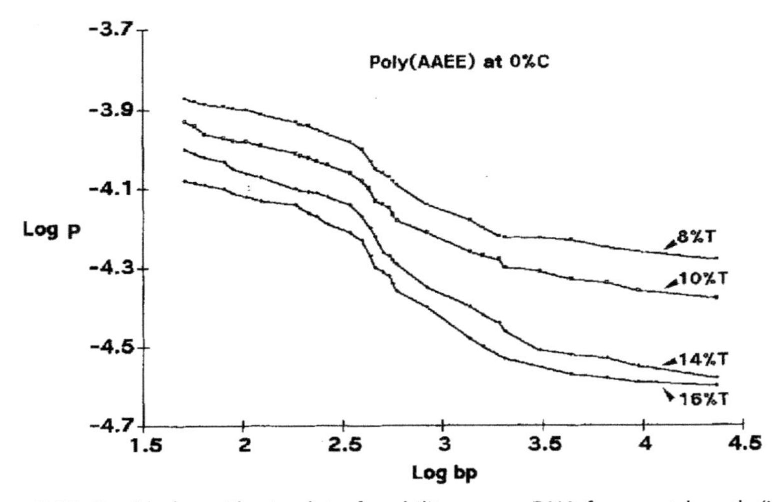

Figure 3.15 Double logarithmic plot of mobility versus DNA fragment length (in bp, ranging in size from 51 to 21,226). The plot has been calculated for different poly (AAEE) concentrations, from 8% to 18%T. Note the sharp transitions from the Ogston regime to reptation without stretching at approximately 200 bp and to reptation with (partial) stretching at 3–4000 bp. *With permission from M. Chiari, M. Nesi, P.G. Righetti, Capillary zone electrophoresis of DNA fragments in a novel polymer network: poly(N-acryloylaminoethoxyethanol), Electrophoresis 15 (5) (1994) 616–622 [79].*

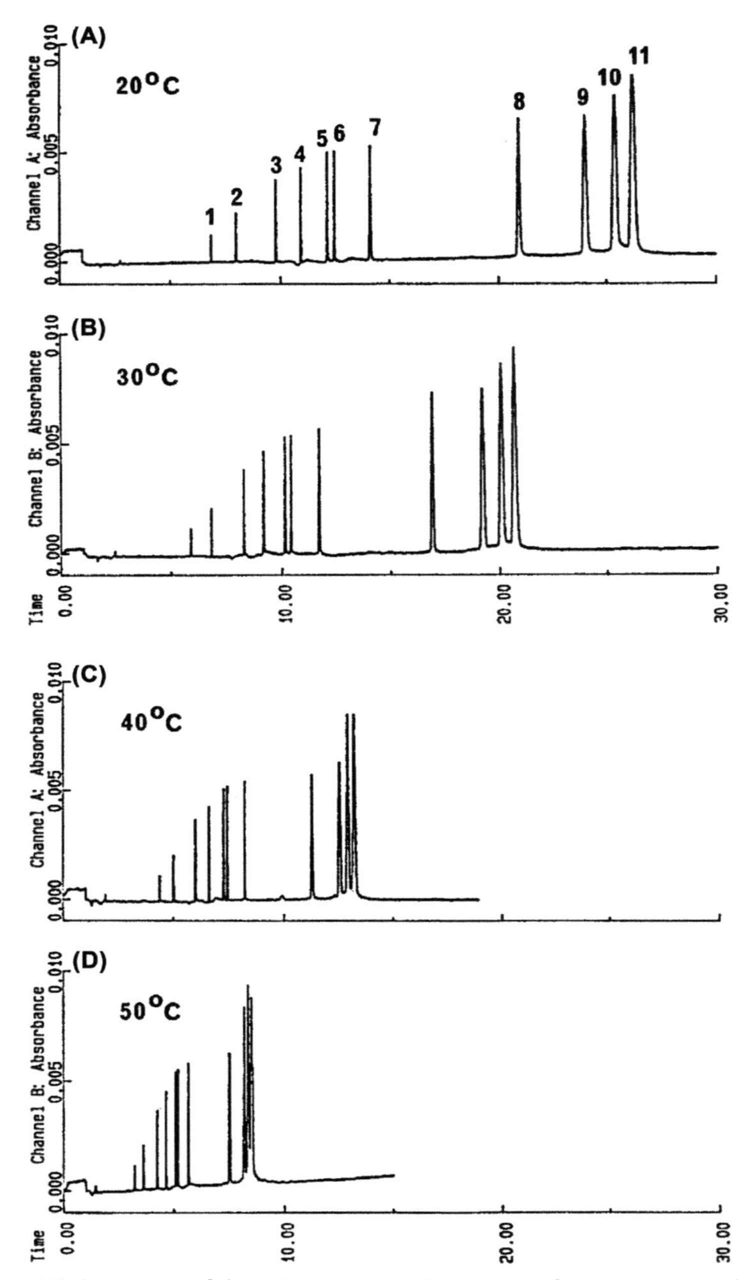

Figure 3.16 Separation of the φX174 DNA *Hae*III restriction fragment mixture by capillary gel electrophoresis at different temperatures in isoelectrostatic (constant (*Continued*)

found to be optimal to allow good resolution for most dsDNA fragments. The authors also hypothesized that in situ polymerization allowed formation of a large distribution of polymer sizes, thus facilitated simultaneous separation of short and long fragments based on the principle that shorter polymer chains sieve shorter DNA fragments better and vice versa. This polymer was also reported as 500 times more resistant to hydrolysis due to its higher hydrophilicity than that of acrylamide, that is, beneficial for long-time use.

3.1.2.1 Linear polyacrylamide

Heiger et al. utilized noncross-linked polyacrylamide with coated capillary columns and applied for the separation of double-stranded DNA fragments ranging up to several thousand base pairs [66]. The same group studied the sequence-dependent migration behavior of dsDNA fragments in capillary electrophoresis using linear polyacrylamide (LPA) matrices [80] and found significant conformation effects under high electric field strengths. Guttman and Cooke applied two different separation modes (isoelectrostatic and isorheic) in high-performance CGE to investigate the effect of temperature between 20°C and 50°C on the separation of φX 174 DNA *Hae*III restriction fragments in a LPA sieving matrix (Fig. 3.16). In isoelectrostatic (constant voltage) separation mode, the migration time and the resolution decreased with increasing temperature, while inisorheic (constant current) separation mode the increasing column temperature resulted in a maximum in migration time and resolution [81]. Later, Kenndler and coworkers have demonstrated the usefulness of LPA gels for the separation of SDS–protein complexes ranging 17.8–77 kDa in 60 minutes [82]. Regnier's group [83] and others [84–86] successfully employed LPA gels for the separation of not only standard proteins but also samples of biological origin and biotherapeutics.

Karger's group [87] carried out a detailed evaluation and optimization of the experimental parameters in CGE, including LPA concentration and molecular weight, column temperature, DNA sequencing chemistry, and base-calling software. As mentioned earlier, large-scale

◀ voltage) separation mode: (A) 20°C; (B) 30°C; (C) 40°C; (D) 50°C. Peaks: 1 = 72; 2 = 118; 3 = 194; 4 = 234; 5 = 271; 6 = 281; 7 = 310; 8 = 603; 9 = 872; 10 = 1,078; 11 = 1353 base pairs. *With permission from A. Guttman, N. Cooke, Effect of temperature on the separation of DNA restriction fragments in capillary gel electrophoresis, Journal of Chromatography 559 (1–2) (1991) 285–294 [81].*

DNA sequencing in capillary array electrophoresis was possible with the use of high-performance but low-viscosity separation matrices, which can be easily replaced after each run as was first reported by Guttman [68]. Such replaceable high-molecular-weight LPA was generated by means of inverse emulsion polymerization. The resulting approximately 9 MDa polymer allowed to obtain >1000 bases sequencing read length in 80 minutes (Fig. 3.17) with base-calling accuracy better than 97%. The viscosity of the polymer exponentially decreased under pressure, allowing easy separation matrix replacement between runs.

Figeys and Dovichi [88] evaluated the stability of noncross-linked polyacrylamide-filled capillaries containing 7M urea and found them stable for at least 4 months when stored at room temperature. Interestingly, capillaries used the day after polymerization exhibited a significant increase in migration time with successive separations (Fig. 3.18). Capillaries used for more than 1 week after the polymerization process, on the other hand, exhibited very reproducible migration times. This phenomenon was observed in both cases with in-column capillary polymerization and in instances when the polymer was polymerized outside of the capillary and filled in for use.

Polyacrylamide-based separation matrices have been used extensively for DNA analysis, including PCR products, sequencing fragments, and other oligonucleotides. Polyacrylamide, however, is not the ideal matrix for DNA analysis as it requires stable neutral capillary coating to eliminate electroosmotic flow (EOF) and more importantly to prevent DNA-wall interactions. Large MW LPA chains prone to shear and deform when pushed into the capillary. Once deformed, one has to wait for relaxation that slows down the analysis process. An attractive alternative is polydimethylacrylamide that is more hydrophobic than that of LPA, thus provides a dynamic self-coating for the inner capillary wall [89].

Schomburg and coworkers evaluated the diffusion of oligonucleotides in polyacrylamide gel–filled capillaries and revealed a relationship between the diffusion coefficients and the size of oligonucleotides. They suggested that the slower diffusion of larger oligonucleotides cannot explain the high separation efficiencies of CGE. It became obvious that besides the diffusion coefficients of oligonucleotides their intermolecular interaction with the polyacrylamide gel may be considerably changed at high electric field strengths [90].

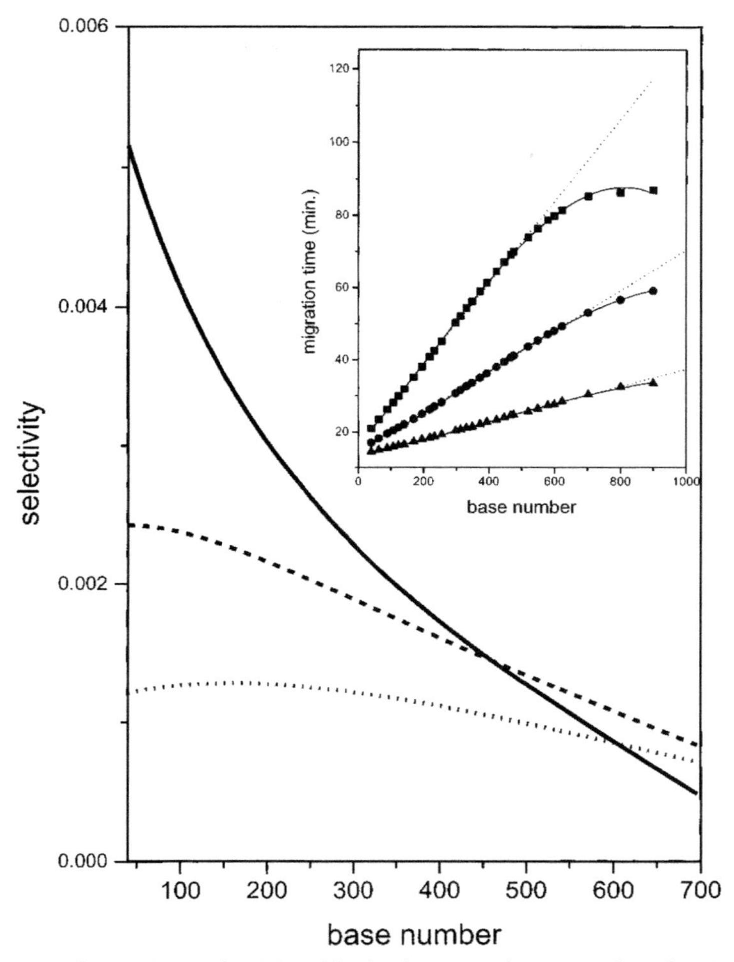

Figure 3.17 Separation selectivity ($\triangle\mu/\mu_{av}$) versus base number for in-house polymerized LPA solutions of 4 (-), 2 (- -), and 1%T (w/v) (. . .). The inset shows the corresponding plots of migration time versus base number for the 4 (■), 2 (●), and 1%T (w/v) (▲) matrices. Conditions: PVA-coated capillary (100 μm i.d.), running buffer 50 mM Tris/50 mM TAPS/3.5M urea: 30% (w/v) formamide, electric field 200 V/cm, electrokinetic injection from DMSO for 10 seconds at 200 V/cm of FAM-labeled and ddTTP-terminated sequencing reaction products. *With permission from E. Carrilho, M.C. Ruiz-Martinez, J. Berka, I. Smirnov, W. Goetzinger, A.W. Miller, et al., Rapid DNA sequencing of more than 1000 bases per run by capillary electrophoresis using replaceable linear polyacrylamide solutions, Analytical Chemistry 68 (19) (1996) 3305–3313 [87].*

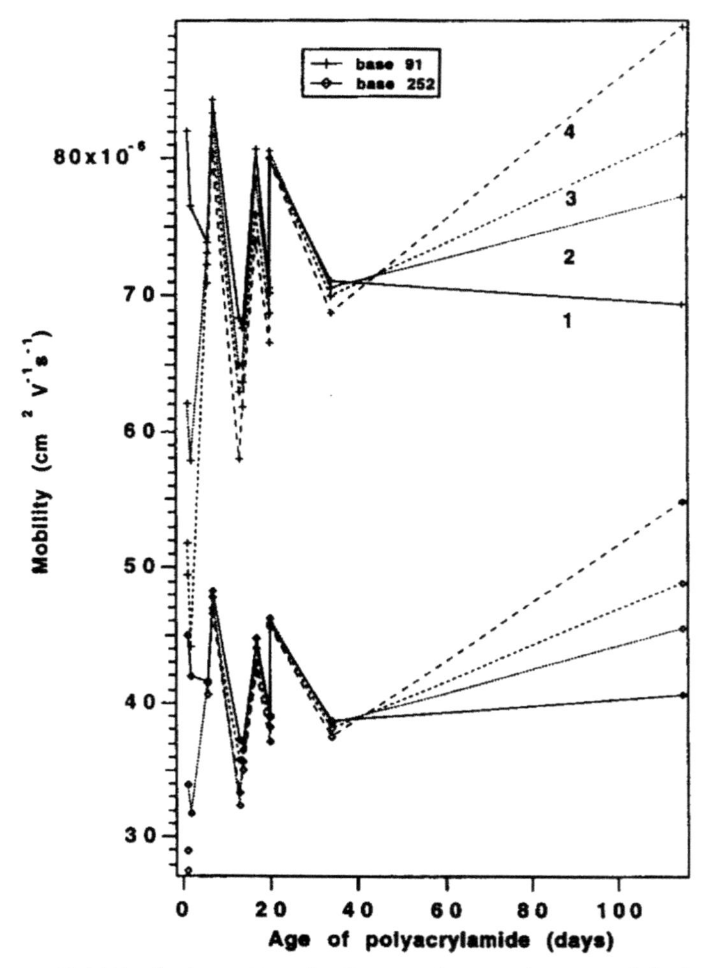

Figure 3.18 Mobilities for bases 91 and 252 versus the age of noncross-linked polyacrylamide for four subsequent sequencing runs. The run number is indicated beside each line. *With permission from D. Figeys, N.J. Dovichi, Effect of the age of non-cross-linked polyacrylamide on the separation of DNA sequencing samples, Journal of Chromatography A 717 (1–2) (1995) 105–111 [88].*

3.1.2.2 Derivatized celluloses

Derivatized celluloses were introduced in the early 1990s by the groups of Brownlee and coworkers [91] and Bonn and coworkers [92] as efficient sieving matrices to separate DNA fragments in capillary electrophoresis. Others separated standard SDS proteins ranging from 14 to 97 kDa with hydroxypropyl cellulose polymers in less than 30 minutes with no apparent influence of the various buffer compositions on sieving performance [93]. Morris and coworkers have investigated the migration of dsDNA molecules in semidilute HEC solution and demonstrated the segmental DNA motion allowing quantitative description of the changing shape of DNA during its physical interaction with the sieving polymer [94]. With a 3D view of the migrating DNA molecules, they observed U-shape conformations oriented at an angle to the microscope plane, as well as ambiguities and artifacts resulting from loss of information of DNA segments not in focus. A polymer coil shrinking theory was compared with the existing entangled polymer solution theory and showed promising potential of semidilute HPMC solutions for DNA fragment analysis [95]. Gibson and Sepaniak reported that axial diffusion in high and low molecular mass methylcellulose polymers, even when adjusted for kinetic conditions using the Einstein relationship [96], did not account for the total observed band variances [97]. Separation of DNA restriction fragments in HEC solutions was also reported by Grossman and Soane [70,71]. They calculated the effective mesh size as being one order of magnitude smaller in agarose gels than that in polymer entanglement and confirmed their theory by electrophoresis.

Barron et al. studied capillary electrophoresis of DNA fragments using diluted and semidiluted HEC as noncross-linked separation matrices. The effects of HEC molecular weight and concentration were investigated with respect to resolution with the attempt to relate these parameters to the entanglement threshold concentration of the polymer, which were measured by plotting the logarithmic specific viscosity of the solution as a function of HEC weight fractions [98]. Since their finding was not in agreement with classical scaling arguments, the authors presented a relationship to predict the observed entanglement threshold of HEC in solution as a function of the average molecular weight (Fig. 3.19). Surprisingly, excellent separation of DNA restriction fragments (ranging from 72 to 1353 base pairs) was obtained by capillary electrophoresis in HEC concentrations significantly below the entanglement threshold. Therefore the authors concluded that the presence of a fully entangled network was not a prerequisite for good DNA separation. However, since they used borate-based background electrolytes that can form complexes with the vicinal OH groups of the HEC as explained above [19], it may transitionally cross-linked the linear polymer strands, providing in this way some sieving capability.

Mitnik et al. [99] conducted a study on the separation of duplex DNA molecules in hydroxypropyl cellulose matrix with a molecular mass of 10^6,

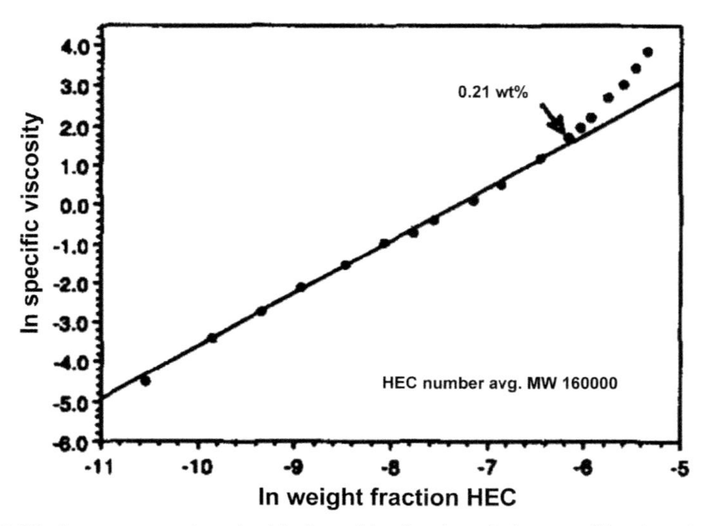

Figure 3.19 A representative double logarithmic plot of the specific viscosity versus HEC weight fraction to define the entanglement threshold of the sieving polymer. HEC average molecular weight: 160,000 g/mol. *With permission from A.E. Barron, D.S. Soane, H.W. Blanch, Capillary electrophoresis of DNA in uncrosslinked polymer solutions, Journal of Chromatography 652 (1) (1993) 3–16 [98].*

evaluating a variety of concentrations from 0.1% to 1% and different electric field strengths of 6–540 V/cm. Their data showed that at >0.4% polymer concentrations and low applied electric field strengths, the separation mechanism was similar to that occurring in gels with a pore size smaller than the persistence length of the DNA molecules. The separation mechanism was confirmed by direct observation using fluorescence video microscopy. Morris and coworkers also used video microscopy to investigate the shape-changing entanglement between DNA and HEC [100]. They visually demonstrated the entanglement between yeast chromosomal DNA (approximately several hundred thousand base pairs) and one or several discrete HEC molecules as postulated earlier in ultra-dilute polymer solutions (0.032%).

Minarik et al. investigated the migration regimes of polystyrene sulfonates (PSS) in HEC polymer networks and optimized the separation parameters [101]. One of the interesting findings was that separation below the entanglement threshold resulted in a relative mobility decrease with increasing number of N_{PPS} monomer units, as shown in Fig. 3.20. This indicated possible solute-chain interaction effects similar to that was observed earlier in double-stranded DNA separation.

3.1.2.3 Polyethylene oxide, polyvinyl alcohol, and polyvinylpyrrolidone

PEO and polyvinylpyrrolidone (PVP) (Scheme 3.1) solutions both proved to be good separation matrices in capillary electrophoresis–based DNA

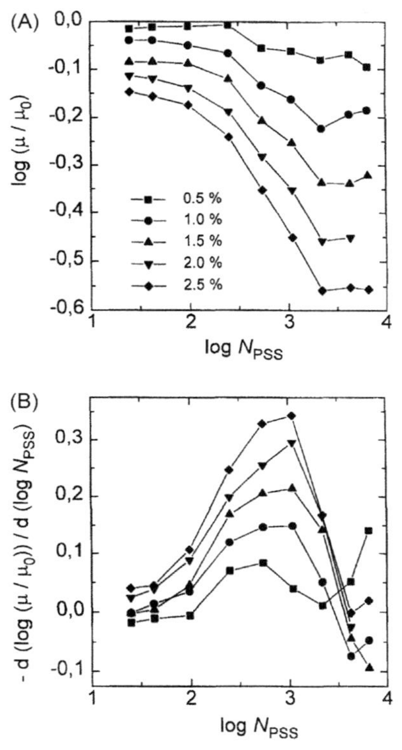

Figure 3.20 Migration regimes: relative mobility (μ/μ_0) versus number of monomer units of polystyrenesulfonates (N_{PSS}) in HEC solutions (A) with various concentrations (% w/v) and (B) its derivative expressing separation selectivity. Conditions: MW (HEC) = 35,900, electric field strength, 190 V/cm. *With permission from M. Minarik, B. Gas, E. Kenndler, Size-based separation of polyelectrolytes by capillary zone electrophoresis: migration regimes and selectivity of poly(styrenesulphonates) in solutions of derivatized cellulose, Electrophoresis 18(1) (1997) 98–103 [101].*

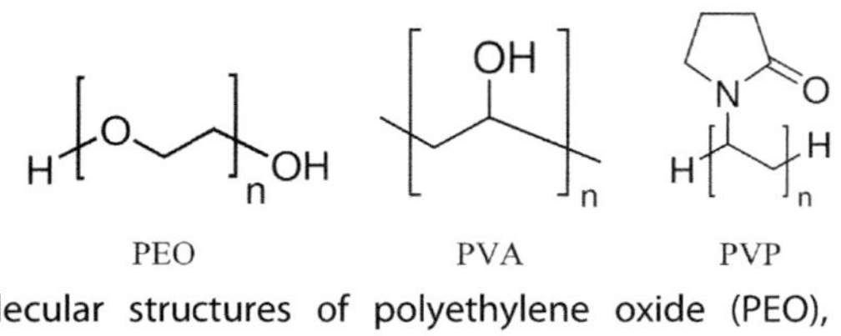

PEO PVA PVP

Scheme 3.1 Molecular structures of polyethylene oxide (PEO), polyvinyl alcohol (PVA), and polyvinylpyrrolidone (PVP).

separations, also featuring self-coating properties [102–106]. Dye-induced mobility shifts were, however, reported to affect DNA sequencing in PEO sieving matrices [107]. A versatile low-viscosity PEO-based sieving polymer was developed for nondenaturing DNA separations in capillary array electrophoresis [108]. Rapid molecular diagnostics option of 21-hydroxylase deficiency was investigated by Barta et al. [109] detecting the most common mutations in the 21-hydroxylase gene using primer extension technique and capillary electrophoresis with PVP sieving and wall coating matrix. The Cy5-labeled primer and the two possible primer extension products (mutant and wild type) were completely separated in 90 seconds in a 10 cm effective length capillary (Fig. 3.21).

Very high peak efficiencies were obtained by CE analysis of bacteria using PEO as polymeric buffer additive [110]. PVA (Scheme 3.1) with a molecular mass of 133,000 was found to be a useful polymer for capillary SDS gel electrophoresis of proteins [111]. The entanglement threshold of this particular polymer was 3% (w/v), thus solutions from 4% to 6% PVA provided excellent resolution in the 14,400–94,000 protein molecular mass interval. Among the advantages of PVA are excellent UV transparency down to 200 nm and very low viscosities. A unique wall effect was also reported by the authors, that is, by decreasing the i.d. of the capillary from 75 to 25 μm, the useful sieving range fall below 1%, well below the entanglement threshold.

UV-transparent replaceable polymer networks of PEG and dextran were evaluated for the separation of SDS–protein complexes according to their molecular mass by Ganzler et al. [112]. Due to the low to moderate viscosity of these polymers, routine replacement within the separation capillary was possible leading to the option of hundreds of injections with a single column as depicted in Fig. 3.22.

3.2 Alternative matrices (composite gels, pluronics, agarose, sol–gel systems, block polymers)

Besides the regularly used cross-linked and entangled linear polymers, some alternative matrices also proved to be useful for size separation of biopolymers in narrow-bore capillaries [113]. One of these groups comprised of thermoresponsive copolymers, using hydrophobic and hydrophilic blocks, such as pluronics [114–116] and polyisopropylacrylamide grafted with PEO

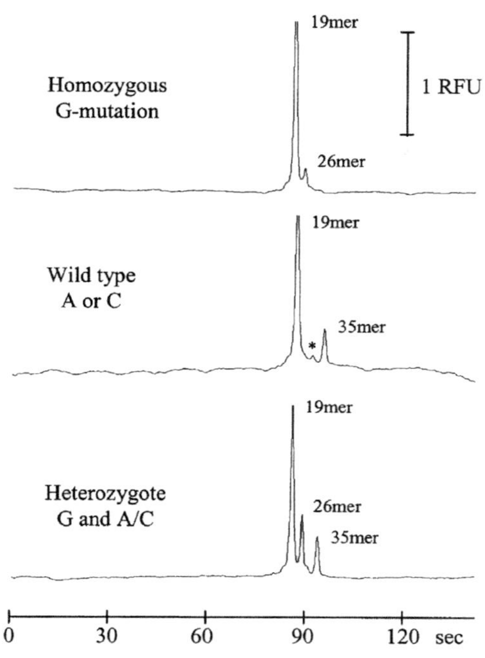

Figure 3.21 Ultra-fast SNP analysis by primer extension and capillary gel electrophoresis. Upper panel: electropherogram of the mutant homozygote (G-mutation) with the 19-mer primer peak and the 26-mer product; middle panel: electropherogram of the wild type with the primer and the 35-mer product; lower panel: electropherogram of the heterozygote with all three peaks (19-, 26-, and 35-mers). Conditions: capillary: $\ell = 10$ cm (effective), $L = 30$ cm, i.d. 75 μm, separation matrix and running buffer: 10% PVP (MW 1,300,000) in 1 × TBE, applied voltage: 20 kV, injection: 30 seconds/10 kV, temperature: 30°C. *With permission from C. Barta, Z. Ronai, M. Sasvari-Szekely, A. Guttman, Rapid single nucleotide polymorphism analysis by primer extension and capillary electrophoresis using polyvinyl pyrrolidone matrix, Electrophoresis 22 (4) 2001 779–782 [109].*

chains [117], which exhibited promising results. The grafted copolymer of poly(*N*-isopropylacrylamide)-γ-polyethyleneoxide (PNIPAM-g-PEO) [117] with self-coating ability and slightly adjustable viscosity properties was applied for rapid and high-resolution separation of dsDNA fragments. Fig. 3.23 shows the separation of the ΦX174/*Hae*III digest fragments applying three different concentrations of PNIPAM-g-PEO polymer solutions (4, 8, and 12% w/v) as separation matrix at room temperature. The best resolution with optimal separation speed was achieved with 8% polymer concentration. These matrices have pronounced temperature-dependent viscosity transition points, implying promising highly flexible implementation options. In

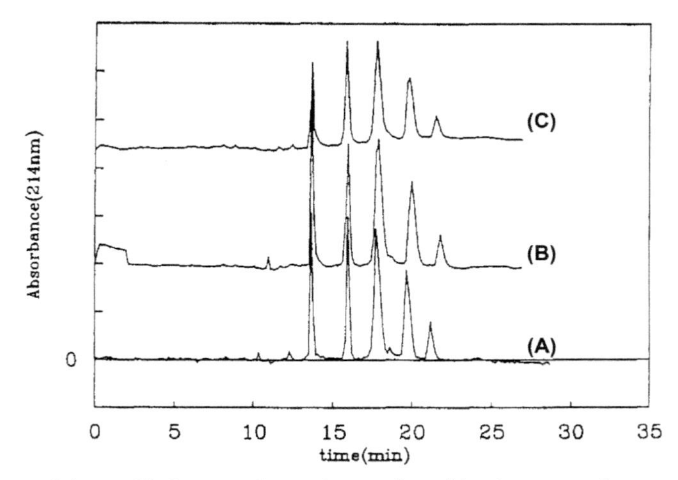

Figure 3.22 Column lifetime studies using replaceable dextran polymer matrix: (A) 100th injection, (B) 200th injection, (C) 300th injection. *With permission from K. Ganzler, K.S. Greve, A.S. Cohen, B.L. Karger, A. Guttman, N.C. Cooke, High-performance capillary electrophoresis of SDS-protein complexes using UV-transparent polymer networks, Analytical Chemistry 64 (22) (1992) 2665–2671 [112].*

particular, thermoresponsive polymers can offer some practical advantages in CGE, such as easier handling and loading of the viscous polymer solutions without the requirement of a high-pressure manifold.

Pluronic copolymer liquid crystals were successfully used as unique, replaceable media in CGE [116]. Separation of peptide/protein model mixtures was attempted in surface-modified capillaries filled with pluronic liquid crystals acting in secondary partition mechanism [118,119].

Barron's group [120] has constructed an interesting "viscosity switch" material responding to changes in temperature, pH, or ionic strength. These matrices were based on copolymers of acrylamide derivatives with variable hydrophobicity and featured a reversible temperature-controlled viscosity switching option from high-viscosity solutions at room temperature to low-viscosity colloid dispersions at elevated temperatures. High resolving power and good DNA sequencing performance (up to 463 bases in 78 minutes) were achieved with these sieving media. Sudor et al., introduced a novel block copolymer thermoassociating matrix for DNA sequencing [121]. These comb polymers were made of hydrophilic polyacrylamide backbone, grafted with poly-*N*-isopropylacrylamide side chains, and characterized by lower critical solution temperature. These matrices combined easy loading with high sieving performance due to

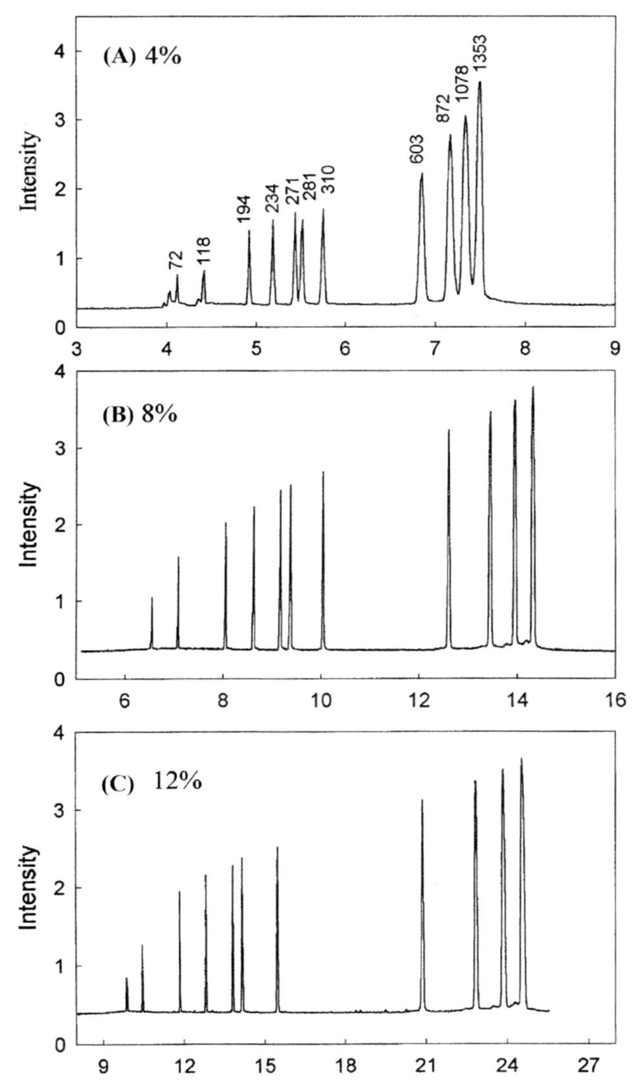

Figure 3.23 CGE separation of the ΦX174/*Hae*III digest in PNIPAM-g-PEO (A) 4% w/v; (B) 8% w/v; and (C) 12% w/v. Experimental conditions: 100 μm i.d. capillary; effective capillary length, 10 cm; 1X TBE; 3 mg/mL EtBr; electric field strength, 200 V/cm; electrokinetic injection, 300 V/cm for 10 seconds; room temperature. Peak indentification from left to right in bp: 72; 118; 194; 234; 271; 281; 310; 603; 872; 1078; and 1353. *With permission from D. Liang, L. Song, S. Zhou, V.S. Zaitsev, and B. Chu, Poly(N-isopropylacrylamide)-g-poly(ethyleneoxide) for high resolution and high speed separation of DNA by capillary electrophoresis, Electrophoresis 20 (1999) 2856–2863 [117].*

switching between a low- and a high-viscosity state by temperature change. Thermothickening properties of these polymers, due to formation of transient intermolecular cross-links at higher temperatures, offered clear advantages for DNA sequencing, for example, easy handling of low-viscosity solutions at low temperatures and excellent sieving characteristics of high-viscosity solutions at elevated temperatures. The rheological behavior and separation performance were in correlation with their microstructure. Sequencing read lengths as high as 800 bases were attained in 1 hour using these polymers.

The effect of temperature and viscosity of pollulan-based sieving medium on electrophoretic behavior of SDS proteins in capillary electrophoresis was studied by Nakatani et al. [122]. A low-viscosity polysaccharide matrix, TreviSol has been characterized in capillary electrophoresis and exhibited good separation ability for larger DNA fragments [123]. The same group evaluated composite agarose (SP-AG)/HEC matrices for the separation of DNA fragments by capillary electrophoresis [124]. High-resolution capillary electrophoresis separation of the 1 kbp DNA ladder standard was obtained by the increasing concentrations of HEC (MW: 125,000) from 0% to 0.4% w/w added to 0.4% w/w SP-AG agarose in Tris-Hydrochloric acid-EDTA (THE) buffer (Fig. 3.24). These composite matrices provided enhanced separation selectivity, especially for DNA fragments larger than 100 bp. In another approach, a viscosity adjustable PEO99/PEO69/PEO99 block copolymer [115] and a triblock polymer of PEO–polypropyleneoxide–PEO [125] were successfully applied to obtain high-speed separation of DNA fragments by capillary electrophoresis.

Chiari et al. synthesized sugar-bearing (gluconic acid) polyacrylamide copolymers that resulted in high sieving capacity and provided similar performance to HEC with comparable viscosity with the additional advantage of self-coating capability [126]. They also copolymerized poly(*N*,*N*-dimethylacrylamide) with hydrophilic monomers to improve separation performance [127]. A copolymer of acrylamide and β-D-glucopyranoside was used as a low-viscosity and high-capacity sieving matrix for the separation of dsDNA molecules. The growth of the polymer chains was controlled by the different reactivity of the two monomers. The chain length was inversely proportional to the number of glucose residues incorporated into the copolymer [128]. Purified galactomannans from guaran, tara gum, and locust bean gum were also attempted as sieving medium for DNA sequencing in capillary electrophoresis. The separation efficiency exceeded 1 million theoretical plates for DNA fragments less than 600 bases long [129].

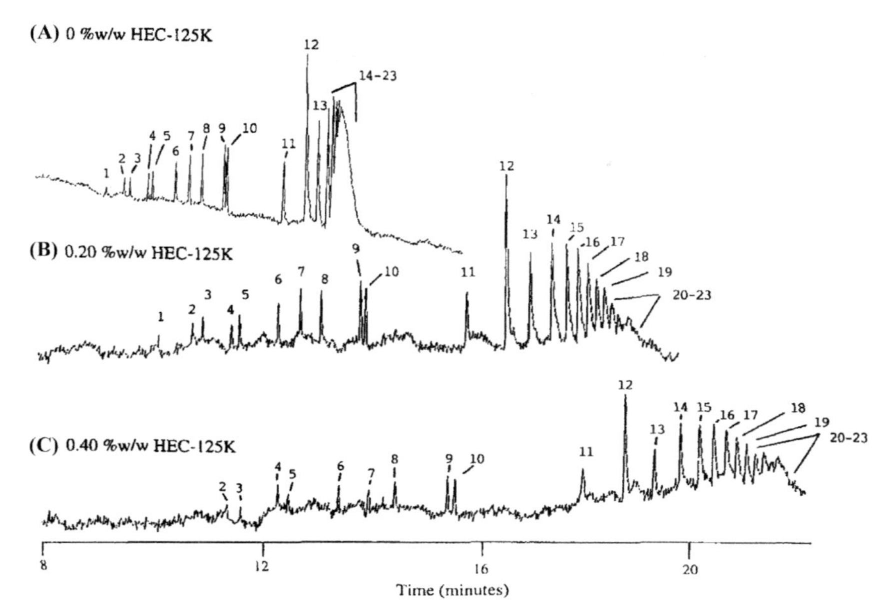

Figure 3.24 The electrophoretic separation of the 1 kbp DNA ladder standard at different HEC (MW: 125,000) concentrations (% w/w) added to 0.40% w/w SP-AG agarose in THE buffer, pH 7.3. Separations were performed at 30.0 ± 1°C, −248 V/cm applied electric field strength in a 50 μm × 58 cm (44 cm effective length) uncoated capillary with hydrodynamic injection at 50 mbar for 10 seconds. Peak assignments in bp: (1) 75; (2) 134; (3) 154; (4) 201; (5) 220; (6) 298; (7) 344; (8) 396; (9) 506; (10) 517; (11) 1018; (12) 1636; (13) 2036; (14) 3054; (15) 4072; (16) 5090; (17) 6108; (18) 7126; (19) 8144; (20) 9162; (21) 10,180; (22) 11,198; and (23) 12,216. *With permission from B.A. Siles, D.E. Anderson, N.S. Buchanan, M.F. Warder, The characterization of composite agarose/hydroxyethylcellulose matrixes for the separation of DNA fragments using capillary electrophoresis, Electrophoresis 18 (11) (1997) 1980–1989 [124].*

Solutions of monomeric nonionic surfactants, *n*-alkyl polyoxyethylene ethers behave as dynamic polymer structures and, therefore, can be used as sieving matrices for DNA fragment analysis in capillary columns [105]. Surfactant solutions offer several advantages over regularly used linear polymers. Some of the most important ones are the ease of matrix preparation, solution homogeneity, stable structure, low viscosity, and self-coating properties to reduce EOF. Good and rapid separation of both dsDNA fragments and single-stranded sequencing ladders were reported. The authors also developed a model to describe the migration behavior of DNA fragments in these solutions and concluded that the separation limit can be extended at low surfactant concentrations. In practice, this approach resulted in separation of DNA sequencing fragments up to 600 bases <1 hour. Separation of dsDNA fragments was also successfully attempted in a transient interpenetrating network of PVP with polyacrylamide [130] and poly(*N*,*N*-dimethylacrylamide) [131]. Viscosity measurements revealed that these interpenetrating networks had significantly higher viscosity than that of the simple mixture containing the same amount of components. The particularly good sieving ability of the system was attributed to the increase in the number of entanglements by the more extended polymer chains. The authors used laser light scattering to show that highly entangled interpenetrating networks were formed, which stabilized polymer chain entanglements.

3.2.1 Agarose-based gel compositions

Besides acrylamide, cellulose, and PEO-based polymer matrices, agarose-filled capillaries were also attempted to separate oligonucleotides, but the selectivity in this case was a complex function of the separation temperature [132]. The upper DNA size limit for separation in agarose was approximately 12 kb. Agarose gel—filled capillaries were extensively studied by Bocek and Chrambach [133] to separate dsDNA molecules. Electrophoretic separation of the components of a 1 kb ladder of DNA at 40°C using a 1.7% solution of agarose as a sieving medium is depicted in Fig. 3.25. A biphasic relationship was attained when the logarithmic base pair numbers were plotted against the mobility in the 1.7% agarose solution, showing higher resolution for DNA sizes of <1 kb. On the other hand, the resolving power increased for DNA >1 kb with increasing agarose concentration (up to 2.6%) with concave Ferguson plots [134]. Similar concave Ferguson plots were obtained by Tietz and Chrambach [135] shown in Fig. 3.26, which were consistent with the assumption that the effective equivalent radius of the DNA molecule diminished with increasing agarose concentration.

UV-transparent, replaceable agarose gels were used for molecular sieving in capillary electrophoresis of proteins and nucleic acids by Hjerten et al. [136]. Gels of methoxylated agarose and other low-melting agarose derivatives compared favorably with cross-linked polyacrylamide gels for CGE of proteins and DNA. These agarose gels can be pushed out of the capillary after each run and replaced by an agarose solution at 35°C—40°C, followed by gelation at lower temperature. The authors obtained similar results with this replaceable matrix in comparison with cross-linked polyacrylamide in

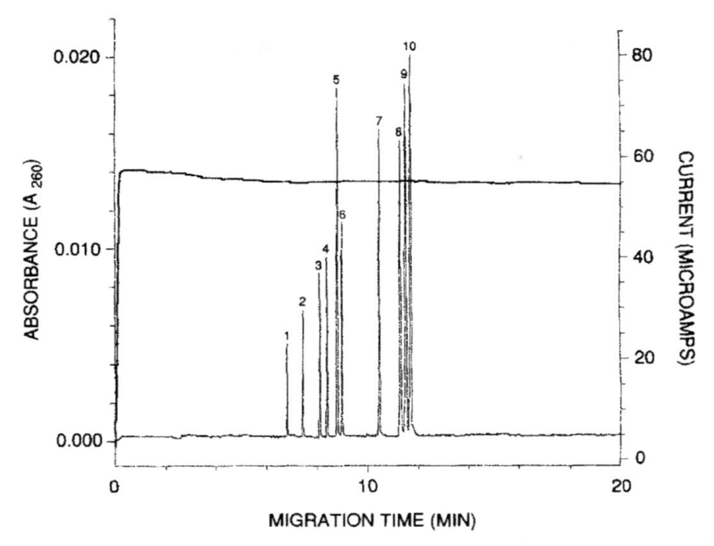

Figure 3.25 Separation of the ΦXI74 RFI *Hae*III restriction fragment mixture by capillary electrophoresis. Conditions 1.7% Seaplaque GTG agarose solution in 1 × TBE buffer; 40°C; 185 V/cm. Peaks (bp): (1) 72; (2) 118; (3) 194; (4) 234; (5) 271 and 281; (6) 310; (7) 603; (8) 872; (9) 1078; (10) 1353. *With permission from P. Bocek, A. Chrambach, Capillary electrophoresis of DNA in agarose solutions at 40 DegC, Electrophoresis 12 (12) (1991) 1059–1061 [133].*

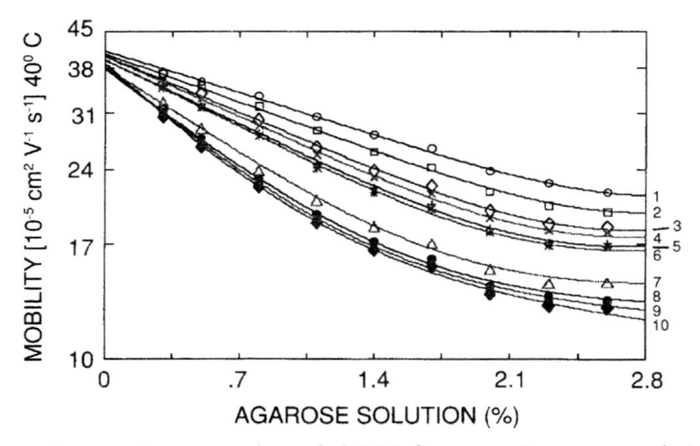

Figure 3.26 Concave Ferguson plots of dsDNA fragments in agarose solutions, 40°C, TBE buffer, using capillary electrophoresis. Agarose: Seaplaque GTG (FMC). Sample: ΦXI74 RF DNA/*Hae*III fragments component numbers are defined by DNA length (bp) and geometric mean radius (nm) as follows: (1) 72, 3.14; (2) 118, 3.7; (3) 194, 4.37; (4) 234, 4.65; (5) 271 and 281, 4.92; (6) 310, 5.11; (7) 603, 6.38; (8) 872, 7.21; (9) 1078, 7.74; (10) 1353, 8.35, respectively. *With permission from D. Tietz, A. Chrambach, Concave Ferguson plots of DNA fragments and convex Ferguson plots of bacteriophages: evaluation of molecular and fiber properties, using desktop computers, Electrophoresis 13 (5) (1992) 286–294 [135].*

separating proteins and DNA fragments. The same group applied field inversion CGE to separate DNA fragments up to 23 kbp in a low-melting, low-gelling agarose gel [137]. The authors investigated the influence of the voltage pulse amplitudes, pulse times, and gel concentrations on the separation performance and found it similar to that of polyacrylamide gels. Kozulic reported on excellent separation of small DNA fragments at low total polymer concentration (1%) cross-linked agarose [138]. In another interesting study, low-melting point agarose was used to analyze restriction enzyme-treated DNA in a borate-containing buffer [139]. Chemically modified agarose gels or composite agarose–noncross-linked polymer gels capable of resolving several base pair differences in DNA fragments of several hundred base pairs in length have also been developed [140]. The resolving power of capillary agarose gel electrophoresis was compared to conventional agarose slab gel electrophoresis with differences in temperature and field strength, but otherwise similar conditions based on the time required to reach a desired degree of resolution by the two methods [141]. Based on the experimental data, a resolution parameter was developed that was applicable to both methods. Agarose-filled capillaries were used by Borejdo and Burlacu [142] to determine the distribution of actin filament lengths and the orientation. The 12.5 cm long and 0.8 mm i.d. capillaries were filled with 1% agarose. The microscopic observation of the filaments extracted from different positions of the gel showed size-based mobility distribution, with the degree of orientation increasing with the applied electric field strength. Production of agarose gel–filled capillaries and their applications to separate DNA restriction fragments were described by Schomburg's group [143]. The same sieving matrix was also used to separate sulfated disaccharides with high efficiency.

3.2.2 Other alternative matrices

A range of sponge-like media, called "electrophoresis sponges," was reported by the group of Hood and coworkers [144] with differences from electrophoresis gels being mechanically stronger, providing a permanent structure of directly measurable pore size dimensions. The media were similar to capillary electrophoresis sieving matrices in respect to pore size range, but differed in that they provided a large number of channels, with a corresponding high loading capacity for simultaneous runs in multiple channels and they were compatible directly with multidimensional separations, such as high-resolution two-dimensional electrophoresis. Pore sizes of such material ranged from the subnanometer to 100 μm scale.

One- and two-dimensional electrophoresis of proteins have been achieved, for example, with high resolution of the charge variants of the haptoglobin beta chain, using sponge-based isoelectric focusing.

A two-component polymer mixture of PEO and polydextran was applied as separation media for capillary electrophoresis by Soini et al. [145]. The effects of the relative concentration of the two polymers were investigated on the analysis of small molecules. Fig. 3.27 depicts the separation enhancement in CGE for a mixture of small drugs (cimetidine, famotidine, diltiazem, and prazosin) by increasing the concentration of the PEO polymer in the dextran-based matrix.

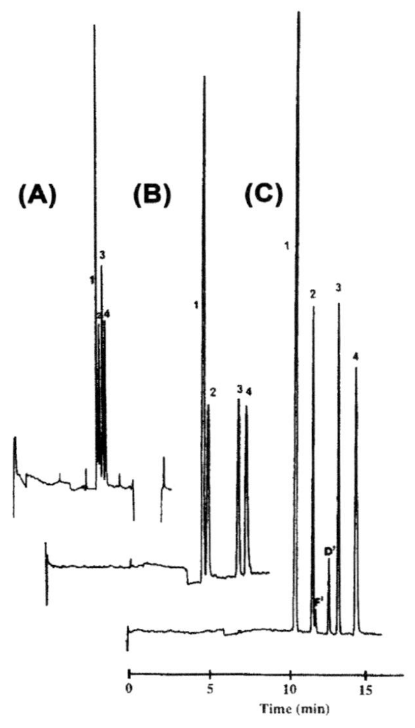

Figure 3.27 Effect of the relative concentration of polymeric additives on CGE resolution. (A) 5% (w/w) dextran, MW = 18,300; (B) 0.02% (w/w) PEO = MW = 300,000; (C) 5% (w/w) dextran (MW = 18,300) and 0.02% (w/w) PEO (MW = 300,000). Sample: 1 = cimetidine; 2 = famotidine; F′ = degradation product of famotidine; 3 = diltiazem; D′ = degradation product of diltiazem; 4 = prazosin. *With permission from H. Soini, M. L. Riekkola, M.V. Novotny, Mixed polymer networks in the direct analysis of pharmaceuticals in urine by capillary electrophoresis, Journal of Chromatography A 680 (2) (1994) 623–634 [145].*

Another alternative polymer reported for the separation of DNA fragments was *N*-acryloylaminopropanol (AAP) [48] with the proposal that the distal -OH groups in the AAP molecule can form transient H-bonds with the DNA double helix for enhanced DNA analysis. The optimal length for polyacrylamides was in the 250,000–400,000 Da range and for poly(AAP) it was 450,000 Da. The naturally occurring polysaccharide glucomannan seemed to be a good sieving additive to separate DNA restriction fragments in CGE [146]. The relationship between polymer concentration and the mesh size of the entangled network was measured by dynamic light scattering on dilute HEC solutions and the results were found to be in good agreement with predictions of the scaling concept of de Gennes [147].

Pullulan is a branched polysaccharide that can form a polymer network for denaturing CGE separation. The entanglement threshold of pullulan solution was around 0.5% w/v reported by Nakatani et al. [148], who used this matrix for the SDS-CGE analysis of proteins. The same authors studied the effect of temperature and viscosity on the electrophoretic behavior of SDS–protein complexes in the MW range of 14,400–116,000 Da and derived equations relating the mobility to capillary temperature and the viscosity of the pullulan sieving solution [122]. Fig. 3.28 depicts the relation between the reciprocal separation temperature and logarithmic electrophoretic mobility (Arrhenius plots), which were apparently parallel. Siles et al. [149,150] applied a commercially available dry formulation of polysaccharide matrix (Trevi-Gel) to CGE for the analysis of dsDNA fragments. This polymer solution provided excellent migration time reproducibility with high separation efficiency for dsDNA fragments ranging 500–7000 bp.

Copolymerization of polymers with different physical and chemical properties apparently provided an interesting means to achieve excellent sieving effect (e.g., long read length for DNA sequencing) with good wall coating capabilities. Examples of these include hydrophobic, end-capped polymer *n*-dodecane-poly(ethylene oxide)-*n*-dodecane [151], pluronic copolymer liquid crystals [116], mannitol [64], poly-*N*-isopropylacrylamide-g-poly(ethylene oxide) [117], PVP [103], polyacrylamide-poly(β-D-glucopyranoside) [128], poly(ethylene oxide)-poly(propylene oxide)-poly (ethylene oxide) [114], just to list a few interesting ones used with more or less success in CGE analysis.

Barron's group studied the impact of polymer hydrophobicity on the properties and performance of DNA sequencing gels for capillary electrophoresis [152]. Their study highlighted the significance of polymer hydrophobicity for high-performance DNA sequencing matrices to form robust

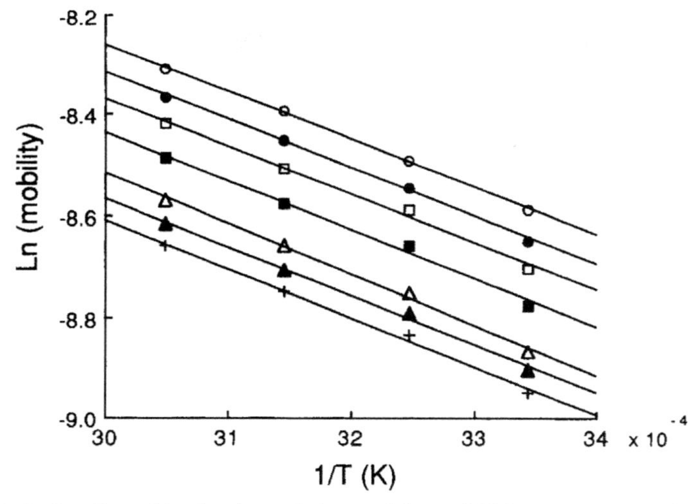

Figure 3.28 Semilogarithmic plots of the mobility of SDS proteins as a function of the reciprocal capillary temperature. ○, α-lactalbumin; ●, trypsin inhibitor; □, carbonic anhydrase; ■, ovalbumin; △, bovine serum albumin; ▲ phosphorylase b; +, β-galactosidase. *With permission from M. Nakatani, A. Shibukawa, T. Nakagawa, Effect of temperature and viscosity of sieving medium on electrophoretic behavior of sodium dodecyl sulfate-proteins on capillary electrophoresis in presence of pullulan, Electrophoresis 17 (7) (1996) 1210–1213 [122].*

highly entangled polymer networks and to minimize hydrophobic interactions between the polymers and the fluorophore-labeled sequencing fragments. Copolymerization was accomplished with free radical polymerization and the resulting copolymer was purified. Read lengths, on the other hand, decreased with increasing diethylacrylamide concentration. Acrylamide and dimethylacrylamide copolymers with molar ratios of 1:1, 2:1, and 3:1 [153] were capable of separating DNA sequencing fragments of up to almost 1000 bases in 80 minutes.

Several in-house synthesized poly(*N*,*N*-dimethylacrylamide) grafted polyacrylamide copolymers were studied as replaceable separation matrices for protein analysis by Taverna and coworkers [154]. Size-dependent retardation of native proteins in these copolymers was proved by Ferguson analysis showing that such copolymers combined the good coating property of poly (*N*,*N*-dimethylacrylamide) and the sieving property of the grafted polyacrylamide (PAA). A nanostructured copolymer matrix has successfully separated oligonucleotides with high resolution by CE using a very short separation channel. A triblock copolymer of E45B14E45B20-5000 with E, B, and denoting oxyethylene, oxybutylene, and segment length, respectively, exhibited a special temperature-dependent viscosity-adjustable property as well as a dynamic coating ability in CGE. The B-block was hydrophilic at low temperatures, for example, 4°C, and the polymer solution had very low viscosity even in a 32.5% w/v solution. At room temperatures, the B-block became hydrophobic due to the breakdown of hydrogen bonds between the blocks and water, and the polymer matrix formed a body-centered cubic structure at high concentrations. Oligonucleotide sizing markers ranging from 8 to 32 bases were successfully separated with one-base resolution in a 1.5 cm long separation channel by E45B14E45 in its gel-like state. The E block was responsible for the coating feature to eliminate EOF. In practice, the capillaries were first filled with the copolymer at 4°C (less viscous state). Then the temperature was increased to from the gel and accommodate the separation of oligonucleotides.

Sequence-dependent separation of polymorphic single-stranded DNA molecules was achieved in guanosine gels via capillary gel electrokinetic chromatography (CGEKC) [155]. The gel was formed by guanosine-5′-monophosphate (GMP) and allowed better resolution for ssDNA fragments of equal length than that of any other CE-based approach. Fig. 3.29 shows the achievement of some resolution among the four strands only with CGEKC. The resolving power of the gel was improved with increasing the GMP concentration. Separations of larger fragments (e.g., 76-mers) were also shown. In another publication, CGE was applied to the analysis of G-quartet-forming nucleotides [156]. G-quartets are structural motifs formed by guanine-rich sequences commonly occurring in the human genome. The G-quartet formation of in-house designed oligonucleotides was validated by CGE. Another interesting approach was the utilization of molecular imprinted polymers, which reportedly recognized specific DNA sequence motifs and applied for double-stranded DNA separation [157] in point mutation analysis.

Nanostructured gels were developed for dsDNA analysis by capillary electrophoresis using triblock copolymers as a sieving material [158] utilizing a sol–gel transition for easy introduction of the sieving matrix into the capillary at room temperature and the formation of a macrolattice structure at high temperature for the separation. The ϕX174 DNA-Hae III digest mixture was separated within 15 minutes in a 7-cm separation channel using a 2-D hexagonal triblock copolymer matrix. Multiwalled carbon nanotubes functionalized by poly(*N*,*N*-dimethylacrylamide) and combined with LPA (3.3 MDa)

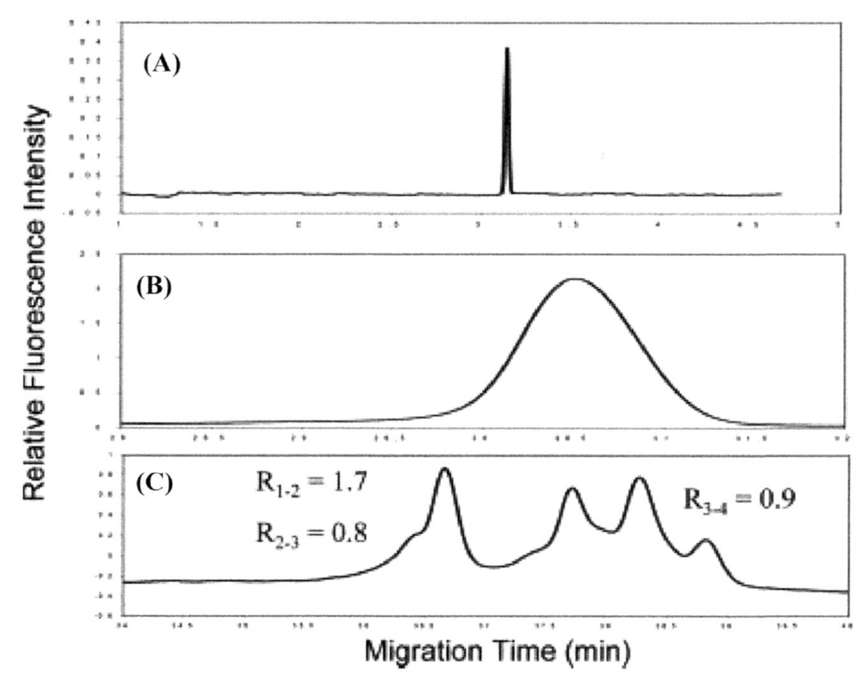

Figure 3.29 Comparison of the separation of four 76-mers. (A) Capillary zone electrophoresis (CZE) with 0.04 M KCl in 15 mM Tris–HCl (pH 7.0). 25 μm i.d. capillary, 580 V/cm, 55°C. (B) CGE with commercial sieving gel using run conditions recommended by the gel manufacturer. (C) CGEKC using G-gel phase containing 0.14 M GMP and 0.04 M KCl in 15 mM Tris–HCl (pH 7.0). 25 μm i.d. capillary, 580 V/cm, 55°C. *CGE*, capillary gel electrophoresis; *CGEKC*, capillary gel electrokinetic chromatography. *With permission from W.S. Case, K.D. Glinert, S. LaBarge, L.B. McGown, Guanosine gel for sequence-dependent separation of polymorphic ssDNA, Electrophoresis 28 (17) (2007) 3008–3016 [155].*

formed a polymer/nanotube double-network composite sieving matrix [159]. The authors used atom transfer radical polymerization to graft the polydimethylacrylamide on the multiwalled carbon nanotubes. The length of the PDMA polymer was optimized for DNA sequencing and the results showed that this matrix offered high resolution, rapid separation times, and excellent reproducibility.

Holland's group introduced nondenaturing nanogels as an effective separation matrix for protein size analysis in CGE [160]. Their results demonstrated the usefulness of 20%–30% nanogels to separate proteins with a molecular weight range of 20.5 – 79.5 kDa (radius sizes from 1.8 to 4.8 nm). The thermally responsive nanogels the authors introduced

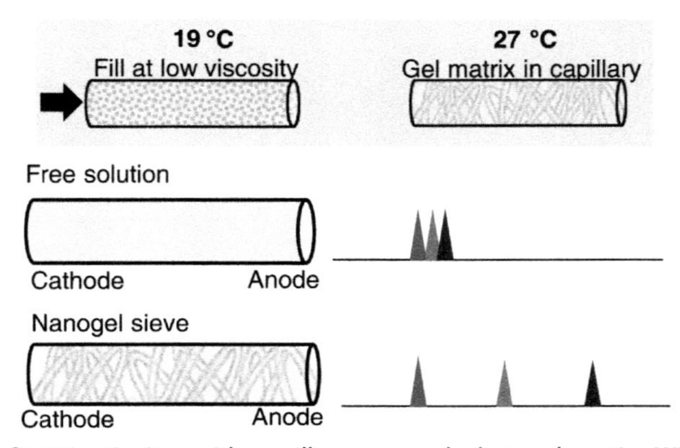

Figure 3.30 Protein sieving with capillary nanogel electrophoresis. *With permission from C.L. Crihfield, L.A. Holland, Protein sieving with capillary nanogel electrophoresis, Analytical Chemistry 93 (3) (2021) 1537–1543 [160].*

were easily replenished in the capillary tubing by a simple temperature switch between high and low viscosity as shown in Fig. 3.30.

Rustandi and coworkers [161] reported size separation of large RNA molecules by CGE under strongly denaturing, nonaqueous (formamide) conditions. They compared the separation of an RNA ladder up to 6000 nucleotides in aqueous and nonaqueous solutions of PEO, HEC, and polyacrylamide (PAA). The influence of polymer type, size, and concentration was all evaluated on separation performance. Fig. 3.31 compares the separation of the RNA ladder of 200–6000 nucleotides in aqueous and nonaqueous 0.06% w/v HEC (1.3 MDa) containing sieving matrix in a PVA-coated capillary. As one can observe, the resolution between peaks 4 and 5 significantly increased by using the nonaqueous background electrolyte-based sieving matrix.

An alternative to gel electrophoresis was reported with the use of an optically transparent film consisting of 300 nm silica colloids on glass as a transparent crystalline layer with an effective pore size of 45 nm [162]. The electromigration behavior of lambda-DNA molecules (48,502 bp) through this material was apparently analogous to gel electrophoresis, including phenomena such as chain extension, hooking of chains around the matrix, and hernia formation. The DNA chains electromigrated in a length-dependent fashion, proving that this approach offered an alternative to polymeric gels for high-speed electrophoresis. In another interesting approach, gold nanoparticles were added to an entangled network of LPA and polydimethylacrylamide to form a polymer/metal composite

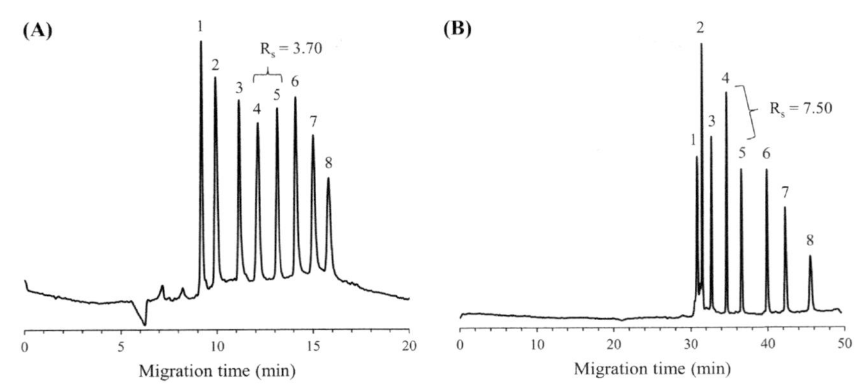

Figure 3.31 Comparison of aqueous and nonaqueous CGE separations of an RNA ladder in a PVA-coated capillary. (A) 0.06% w/v HEC (1.3 MDa) in an aqueous buffer, (B) 0.06% w/v HEC (1.3 MDa) in a nonaqueous (formamide) background electrolyte. Peaks: (1) 200, (2) 500, (3) 1000, (4) 1500, (5) 2000, (6) 3000, (7) 4000, and (8) 6000 nucleotides. *With permission from T. Lu, L.J. Klein, S. Ha, R.R. Rustandi, High-resolution capillary electrophoresis separation of large RNA under non-aqueous conditions, Journal of Chromatography A 1618 (2020) 460875 [161].*

matrix to improve the separation of single-stranded DNA molecules [163]. The interactions between the gold nanoparticles and the polymer chains along with the formation of the physical gel resulted in a separation matrix, which was less viscous than the gel without nanoparticles but with a similar resolving power. Other advantages of this type of gels were high efficiency, reproducibility, and readiness to facilitate automation.

3.3 Capillary coatings

Due to possible analyte-wall interactions and the separation influencing effect of the EOF, in most instances coated (permanent or dynamic) narrow-bore tubes are employed in CGE. At the usual background electrolyte pH of biopolymer separations, the interaction of solute molecules with the negatively charged and hydrophobic inner surface of the fused-silica capillary used in CE is one of the main reasons for poor separation profiles that are particularly prominent for highly basic analytes such as basic proteins. Uncoated columns possess strong electroosmotic character (EOF) at higher than neutral pHs (Fig. 3.32) and may also exhibit strong wall absorption of various sample types [164]. Apparently, gel or polymer solution-filled bare fused-silica capillaries, that is, without

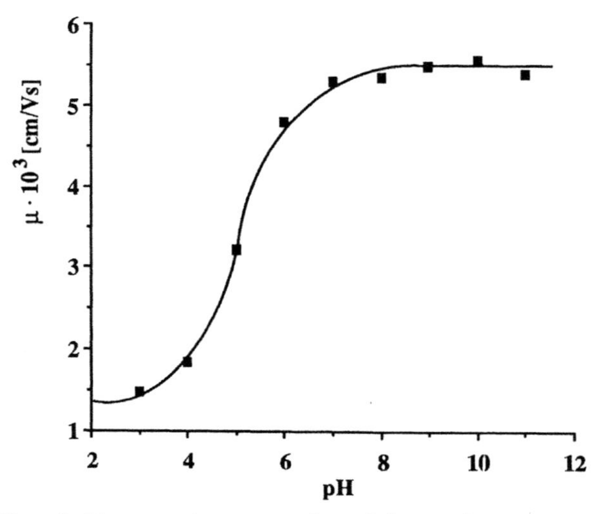

Figure 3.32 Plot of pH versus electroosmotic mobility. *With permission from R. Kuhn, S. Hofstetter-Kuhn, Capillary Electrophoresis: Principles and Practice, Springer Laboratory, Berlin, 1993 [164].*

any inner surface coating, only last for a few runs with real biological samples before significant deterioration in their performance with a few exceptions, such as SDS-CGE with borate cross-linked dextran gels [19], in which case the capillary is massively reconditioned before each run.

As a theoretical treatment, the Smoluchowski equation for the gel layer—liquid interface potential was modified to calculate the streaming potential [165] incorporating the influence of the ionic diffusion along the capillary axis and the liquid flow inside a porous gel layer at the capillary walls. The analysis showed that the diffusion influence was significant at Peclet numbers <50. The effect of liquid flow inside the porous gel layer was considered to be negligible when the gel layer porosity was sufficiently low. In CGE, the main purpose of such coatings was the elimination of the EOF.

Appropriate capillary coatings modify (increase, reduce, or eliminate) the EOF phenomenon and also prevent sample interactions with the inner silanol group covered surface of the column. With a proper coating in place, the capillary can be used at higher pHs for longer time periods and extensively washed to alleviate any solute molecules possibly bond to the coating. Coated capillaries feature good run-to-run migration time reproducibility, high peak efficiency, and consequently higher peak capacity. For this reason, most CGE separations are conducted in coated capillaries. Both physically and chemically coated capillaries proved useful in CE with gels or polymer solutions, as summarized in several important early reviews discussing the art of permanent and dynamic coatings [166–168]. While permanent coatings are, in most cases, covalently attached to the glass surface, dynamic coatings are usually applied by rinsing the capillary with the coating reagent. In this latter instance the coating solution is washed out after a short incubation time leaving a tiny adsorbed layer on the capillary wall. The stability and durability of such coatings can vary, but the easy recoating option of the dynamic type before each run alleviates this issue, thus increasing the reproducibility by using a fresh coating every time. One of the best ways of applying dynamic column coverage is to simply dissolving the coating materials in the background electrolyte [166]. The most commonly used dynamic coating material for capillaries is polyacrylamide, PEG, various cellulose derivatives, and PVP, as was recently reviewed by Hajba and Guttman [168].

Novotny and coworkers evaluated the separation efficiency in CE by using untreated and surface-treated open tubular and gel-filled capillaries [169]. Four CBQCA-amino acid derivatives were used as test compounds to assess the different types of capillary tubes. As Fig. 3.33 shows, in untreated open tubular capillaries the EOF was the driving force and the corresponding migration order in panel A was arginine, histidine, leucine, and glycine. With the use of LPA-coated capillary wall or gel-filled tubings, the EOF was eliminated, thus the polarity of the separation voltage had to be reversed and the charged solutes migrated in terms of their electrophoretic mobilities in reversed migration order (panels B and C; glycine, leucine, histidine, and arginine).

Others investigated the separation or SDS–protein complexes using noncross-linked LPA gels in both coated and uncoated fused-silica capillaries [83]. The viscosity of the polyacrylamide solution appeared to be one of the major factors affecting column stability. However, uncoated capillaries provided better resolution as well as stability and reproducibility than surface-coated capillaries when the concentration of the LPA sieving matrix was greater than 4%, as shown in Fig. 3.34.

A very simple approach to cover capillary surfaces was reported by Gordon et al. [170] using ethylene glycol in the background electrolyte system. Their protocol apparently worked for proteins with different molecular weights and pI values and the method was successfully applied to the analysis of human serum proteins [171]. Wehr and coworkers [172]

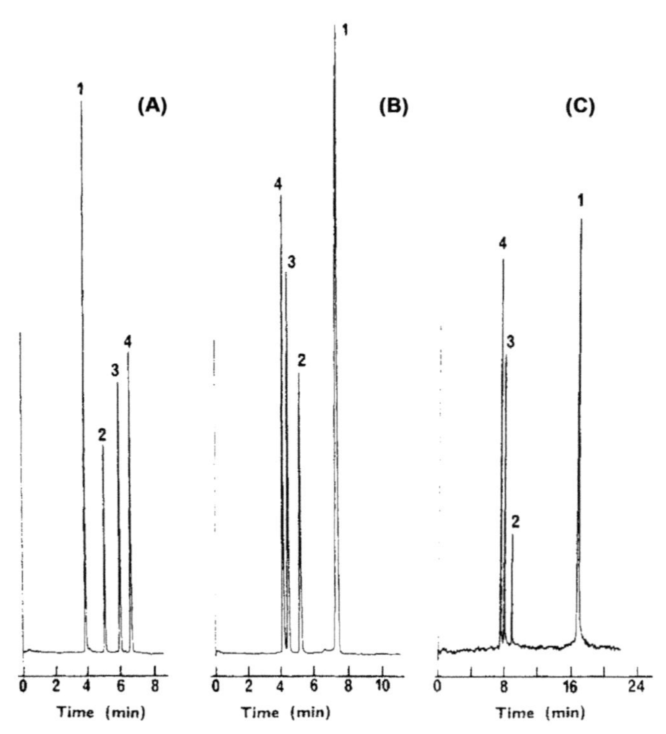

Figure 3.33 Electropherograms of a mixture of CBQCA-amino acids in (A) untreated capillary; (B) linear polyacrylamide-coated capillary, and (C) 4% linear polyacrylamide gel-filled capillary. Peaks: (1) CBQCA-Arg; (2) CBQCA-His; (3) CBQCA-Leu; and (4) CBQCA-Gly. *With permission from J. Liu, V. Dolnik, Y.Z. Hsieh, M. Novotny, Experimental evaluation of the separation efficiency in capillary electrophoresis using open tubular and gel-filled columns, Analytical Chemistry 64 (13) (1992) 1328–1336 [169].*

reported on three interesting surface coating methods primarily to prevent protein adsorption, but also amenable to support CGE analysis. The first one used a linear hydrophilic polymer coating to cover the inner surface of the capillary. The second one was a simple deactivation process of the silanol groups by acidic wash between runs and the third one utilized various additives to the background electrolyte. Bentrop et al. [173] applied thermal polymerization to coat the inside surface of capillaries with poly (methylglutamate) to be used for separation of proteins at intermediate pH values. Strege and Lagu [174] employed a methylcellulose sieving matrix and polyacrylamide coating for the separation of large dsDNA fragments (1 kbp DNA ladder and λ DNA-*Hind*III digest). Schomburg's group attempted the use of PEG coating to adequately decrease EOF and

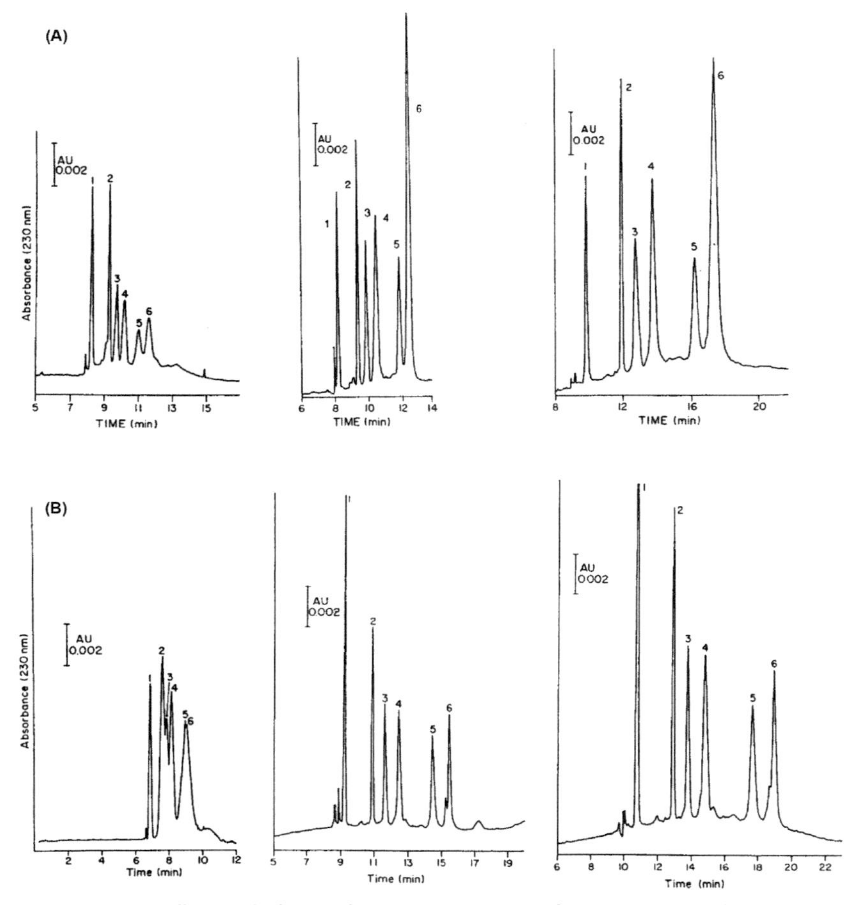

Figure 3.34 Capillary gel electrophoretic separation of protein molecular mass standards using coated (A) and uncoated capillaries (B) with 3% (left), 4% (middle), and 6% (right) linear polyacrylamide gel concentration. Peaks: 1 = α-lactalbumin (bovine milk) (MW 14,400); 2 = carbonic anhydrase (MW 29,000); 3 = glyceraldehyde-3-phosphatedehydrogenase (MW 36,000); 4 = albumin (chicken egg) (MW 45,000); 5 = albumin (bovine) (MW 66,000); 6 = conalbumin (MW 78,000). *With permission from D. Wu, F.E. Regnier, Sodium dodecyl sulfate-capillary gel electrophoresis of proteins using non-cross-linked polyacrylamide, Journal of Chromatography 608 (1–2) (1992) 349–356 [83].*

improved the resolution of polydeoxycytidines [175]. McCormick used phosphate treatment to modify the inside wall of fused-silica capillaries and found a significant impact on peak asymmetry and peak width of protein bands in capillary electrophoresis [176].

3.3.1 Covalent (chemical) coatings

Chemical modification of the inner capillary surface is one of the most frequently used ways to generate appropriate surface coating and eliminate EOF for CGE. In the mid-1980s, Hjerten introduced a special silanization technique using γ-methacryloxypropyltrimethoxysilane and subsequent cross-linking of the surface-bound methylacryl groups with LPA to eliminate EOF as well as to minimize wall adsorption of the analyte molecules [25]. Later, Cohen and Karger used similar polyacrylamide-coated tubes for SDS-CGE separation of various protein mixtures [15]. The coated capillary they used exhibited no apparent wall interaction with the SDS–protein complexes in the analyte mixture and resulted in good separation with sharp peaks. To decrease hydrolysis mediated deterioration of the coating at higher pHs, Novotny's group introduced a polyacrylamide coating similar to that described earlier, but bound to the silica wall via Si–C, rather than Si–O–Si bonds [171]. The Si–C bond assumed to be more hydrolytically stable, resulting in improved stability over the wide pH range of 2–10.5.

In the case of permanent coatings, the applied materials are irreversibly attached to the inner wall of the fused-silica capillary tube, either by physical adsorption or by covalent bonding [177]. Covalent modification of the silanol groups of the fused-silica surface usually made with a cross-linked or noncross-linked hydrophilic polymer. Preparation of covalent coatings requires expertise in chemistry and physics, and the process can be labor intensive and time-consuming. It was already emphasized in the 1980s that the surface tension in narrow-bore capillary columns tends to produce droplets of the derivatizing agents representing one of the physical barriers of the process [178]. The other problem with covalent coatings is degradation due to, for example, the pH of the background electrolytes and the efficiency of coating can change from batch-to-batch.

One of the major problems of capillary coatings is the occurrence of microspots, in which the silica surfaces are poorly coated, that is, exposed to the background electrolyte. This problem can be alleviated with cross-linking the coating polymer molecules connected to the capillary wall [179]. The authors reported on a simple device and associated coating protocol to produce reliable and reproducible coatings. In another report by Crippa and coworkers [180] described capillary coating by in-house synthesized trimethoxysilane-modified polydimethylacrylamide. After incubation with this solution at 60°C, the extending silyl groups formed condensation bonds with the silanols enabling subsequent formation of strong covalent linkage between the copolymer and the capillary wall.

Various existing liquid chromatography and gas chromatography type coatings were also attempted with gel or polymer network-filled capillaries as described in [168,181,182]. Ganzler et al. reported a dextran coating and filling the column with dextran polymer for protein analysis by SDS-CGE [112]. Their coated capillary was quite stable and featured good run-to-run migration time reproducibility up to hundreds of runs. Later, Shieh et al. developed a novel approach by combining the advantages of both dynamic and covalent coatings [183]. A thin layer of polyacrylamide was dynamically coated to the inner surface of the capillary followed by allylamine treatment for cross-linking [184]. The capillary coated in this way showed stable performance with biological samples for more than 200 injections but required short acid rinses between runs. Covalently coated capillaries also offer high-speed separations with little or no requirement for preseparation capillary equilibration, which can otherwise be a time-consuming process.

Following the same principle, noncross-linked PEG polymer was used to covalently coat capillary surfaces to suppress EOF by Poppe and coworkers [185]. They modified the inner wall of fused-silica capillaries with γ-glycidoxypropyltrimethoxysilane and PEG 600 to decrease wall adsorption and EOF in capillary zone electrophoresis. Nakatani et al. reported on SDS polyacrylamide CGE of proteins using stable LPA-coated capillary column [186]. The fused-silica capillary was coated through Si—C bonds (Fig. 3.35), which suppressed the EOF and reduced protein adsorption. Compared with conventional fused-silica capillary coating through 3-methacryloxypropyltrimethoxysilane resulting in a siloxane linkage of Si—O—Si—C, the Si—C bond was more stable even at alkaline conditions as was also mentioned above.

High-resolution DNA restriction digest analysis with CGE was accomplished by Paulus and Huesken using 6% noncross-linked polyacrylamide polymerized inside the capillary, which was coated with a methacryloxysilane to ensure binding of the polyacrylamide to the capillary wall [187]. For fused-silica capillaries, polyvinylmethylsiloxanediol linked to LPA proved to be an excellent coating material offering very high efficiencies in protein separations with excellent migration time reproducibility [188]. In the case of PEG, on the other hand, cross-linked with dicumyl peroxide in fused-silica capillaries, it was found that coating thickness was the

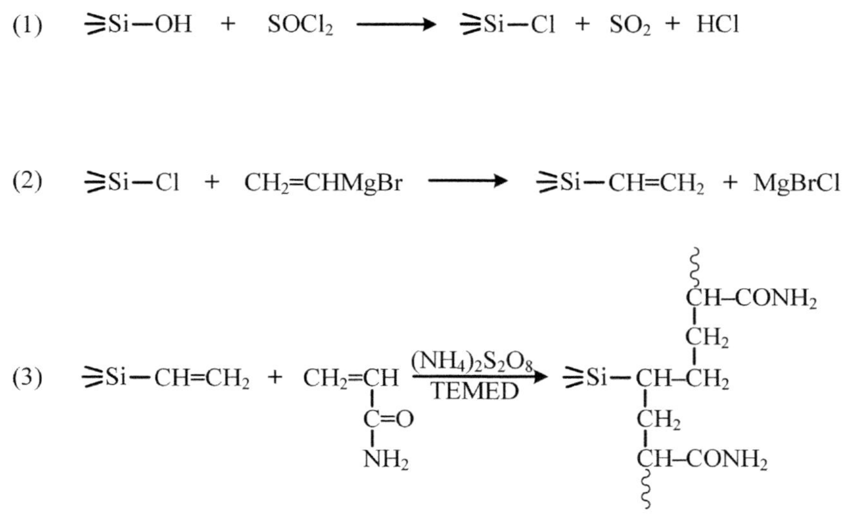

(1) $\geqslant Si{-}OH + SOCl_2 \longrightarrow \geqslant Si{-}Cl + SO_2 + HCl$

(2) $\geqslant Si{-}Cl + CH_2{=}CHMgBr \longrightarrow \geqslant Si{-}CH{=}CH_2 + MgBrCl$

(3) $\geqslant Si{-}CH{=}CH_2 + CH_2{=}CH(C{=}O)NH_2 \xrightarrow[TEMED]{(NH_4)_2S_2O_8} \geqslant Si{-}CH(CH_2{-}CH{-}CONH_2{\sim})(CH_2{-}CH_2{-}CH{-}CONH_2{\sim})$

Figure 3.35 Reaction scheme for capillary coating with linear polyacrylamide via Si–C linkage.

parameter, which mostly influenced the characteristics of the capillary, with a thickness of ~38 nm giving the best results [189]. Deposition of an epoxy [190] or siloxane [191] coat prior to polymer coating resulted in enhanced deactivation of the capillary surface and also provided covalent attachment sites for a hydrophilic polymer top coating layer. Ren and coworkers [190] bonded polyethylenepropylene glycol to an epoxy resin intermediate surface layer and obtained a pH stable coating for over 1000 runs in the range of pH 4–11. A surface-bound hydroxylated polyether coating was developed by Nashabeh and El Rassi [192], first depositing a glyceropropylpolysiloxane sublayer covalently attached to the inner silica surface. Then a top layer of polysiloxane polyether was attached to the wall at both ends with possible branching and interconnection. The long polyether chains provided a good shield to the unreacted surface silanols. Belder et al. developed a rapid method to generate permanent hydrophilic coatings for capillary electrophoresis [193] by treating the surface with a solution of glutaraldehyde as cross-linking agent followed by a PVA solution, which resulted in immobilization of the polymer at the capillary surface. Chemically bonded polydimethylacrylamide coating was also reported to provide neutral surface in a wide pH range [194] even at pH 10. This coating held up to 150 runs without apparent deterioration in separation performance.

LPA was used to pretreat and stabilize the inner surface of capillary columns to obtain excellent separation of polyuridines up to hundreds of bases [60]. The use of nonionic coatings is very important in CGE to eliminate any EOF that would push the gel out of the column and decrease separation performance. In addition, masking the charged sites of the inner surface of the capillary prevents nonspecific bindings to occur during the separation of biopolymers, especially proteins. Low-viscosity LPA solutions featured excellent fractionation capability for DNA sequencing fragments when the capillary walls were covalently coated with polyacrylamide to eliminate EOF in order to attain the highest separation efficiency [195].

Brownlee and coworkers separated DNA restriction fragments and polymerase chain reaction (PCR) products by capillary electrophoresis using polysiloxane-coated capillary with polymeric buffer additives [91]. The authors also reported on the generation Ferguson plots for the DNA fragments (Fig. 3.36) at various polymer (HPMC-4000) concentrations proving in this way the actual molecular sieving, namely, better sieving can be achieved at higher polymer concentrations. The addition of the intercalating agent ethidium bromide to the buffer system resulted in better resolution in the price of longer separation times.

A fully automated capillary coating procedure was introduced by Bodnar et al. [196] utilizing an automated capillary electrophoresis instrument to introduce the surface activating material and the monomers along with the catalyst and initiator into the narrow-bore capillary column. The automated covalent LPA coating and regeneration process supported long-term stability of fused-silica capillaries for protein analysis. The stability of the resulting capillary coatings was evaluated by a large number of separations.

3.3.2 Physical (noncovalent) surface coverage

To avoid the extra work and chemical reactions involved in producing covalently coated capillary columns, various kinds of noncovalent coatings have been developed, such as anionic, polymeric, and nonionic/zwitterionic types [166,168]. During the dynamic surface covering procedure the coating material can either be rinsed through the capillary prior to filling it up with the running buffer or simply added to the background electrolyte. In both instances a thin layer of coating material is adsorbed onto the inner capillary surface before or during the separation process as depicted in Fig. 3.37. Dynamic coatings are easily applied to any bare fused-silica

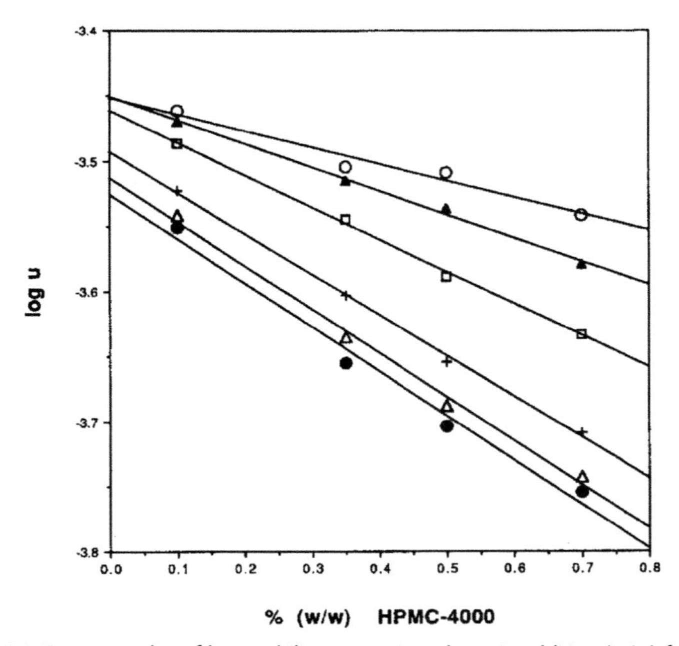

Figure 3.36 Ferguson plot of log mobility versus % polymeric additive (w/w) for a buffer containing HPMC-4000 as the sieving component in polysiloxane-coated capillary. Mobilities of selected ΦX 174 Hae III digest fragments were used to generate the plot. ○ = 118 bp; ▲ = 194 bp; □ = 310 bp; + = 603 bp; △ = 872 bp; ● = 1,353 bp. *With permission from H.E. Schwartz, K. Ulfelder, F.J. Sunzeri, M.P. Busch, R.G. Brownlee, Analysis of DNA restriction fragments and polymerase chain reaction products towards detection of the AIDS (HIV-1) virus in blood, Journal of Chromatography 559 (1–2) (1991) 267–283 [91].*

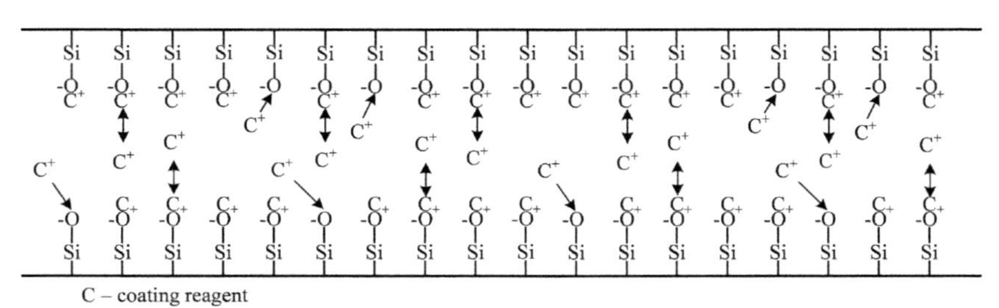

Figure 3.37 Schematics of the dynamic capillary coating procedure. *With permission from L. Hajba, A. Guttman, Recent advances in column coatings for capillary electrophoresis of proteins, TrAC Trends in Analytical Chemistry 90 (2017) 38–44 [168].*

capillary and readily regenerated whenever necessary, even before or after each run. In the early 1990s, rapid protein separations with good resolution have been demonstrated the advantages of the dynamically coated capillaries [68,183]. However, they required longer equilibration times prior to use and sometimes have not been so efficient and compatible with real biological samples.

High-resolution capillary electrophoresis separations of proteins and peptides were achieved by coating the inner wall of 75-micron ID fused-silica capillaries with 40–140 nm polystyrene particles, which have been derivatized with alpha-omega-diamines such as ethylenediamine or 1,10-diaminodecane [197]. Stable and irreversibly adsorbed coating was obtained upon deprotonation of the capillary surface with aqueous sodium hydroxide and subsequent flushing with a suspension of the positively charged particles. Landers and coworkers compared various surface derivatization agents in conjunction with coating polymers for DNA heteroduplex analysis of the breast cancer susceptibility gene (BRCA1) [198]. Their most effective coating combination was chlorodimethyloctylsilane with PVP polymer that was used to separate the relevant DNA fragments in HEC matrix within 10 minutes. Applications of short-chain polydimethylacrylamide as sieving and dynamic wall coating medium were also beneficial for the electrophoretic separation and mutation analysis of DNA fragments in bare fused-silica capillaries [199,200].

Procedures for capillary coatings using cellulose derivatives [201,202], cross-linked dextrans [203], and poly(ethylene oxide)-poly(propylene oxide)-poly(ethylene oxide) triblock copolymers [204] are extensively described in the literature. PEO networks are attractive matrices for CGE applications due to their low viscosity and good separation performance. A low-viscosity PEO sieving medium of approximately 5 cP at 25°C was used to separate DNA fragments up to 40 kbp without the need of pulsed or alternating electric field mode [108]. The matrix contained a mixture of MW 1×10^6 PEO combined with MW 8×10^6 PEO, which was compatible with a variety of intercalating dyes and capillary wall coating methods. A fully automated noncovalent coating technique was introduced by Chiari et al. [200] by adsorbing polydimethylacrylamide-co-allyl glycidyl ether onto the capillary surface in less than 30 minutes. The process did not require organic solvents, viscous solutions, or elevated temperature and the coating was stable even under harsh conditions such as alkaline pH, elevated temperature, and denaturant conditions that usually harm most adsorbed coatings. In addition, this coating could be rapidly regenerated by a simple wash with strong alkaline solution.

Advances in CGE of DNA focused on coating and sieving materials. Atomic force microscopy was applied to investigate the stability of polydimethylacrylamide and polydiethylacrylamide materials that in addition to their sieving capabilities also proved to have good coating properties

[205]. The authors used this combination for the point mutation analysis of the C677T loci in the human methylenetetrahydrofolate reductase gene.

3.4 Techniques for preparing gel-filled capillaries

During the early 1990s the main focus of CGE was to find new strategies for the polymerization and stabilization of gels in narrow-bore fused-silica capillaries in particular interest to avoid bubble formation during the polymerization reaction caused by phase transition [206]. The shrinking problem during acrylamide/bisacrylamide gel polymerization in narrow-bore capillary columns causes bubble formation as shown in Fig. 3.38, making the gel-filled capillary unusable. In addition, when the inner surface of the capillary is not coated properly, residual zeta potential is building up underneath the gel matrix resulting in extrusion of some part of the gel out of the capillary at the cathode end mediated by the developing EOF [60,207]. The production of bubble-free gel-filled capillary columns using a coating procedure involving treatment with 3-methacryloxypropyltrimethoxysilane only and followed by bonding a layer of noncross-linked polyacrylamide to the surface by regular APS and TEMED polymerization process was developed by the Schomburg group [60,207]. The authors suggested that stable gels can be obtained in capillaries with or without surface pretreatment; however, the use of a bifunctional reagent such as 3-methacryloxypropyltrimethoxysilane only

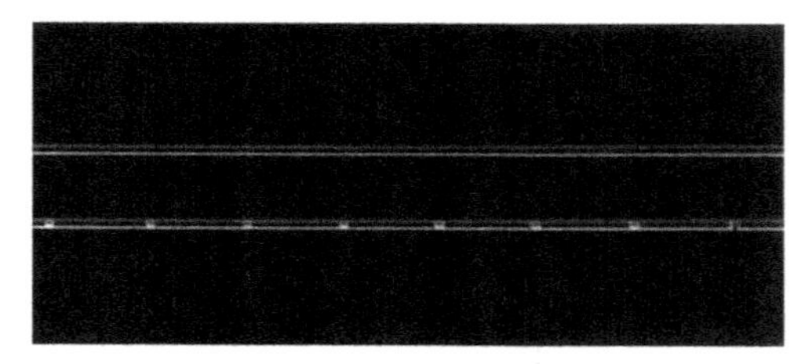

Figure 3.38 Photograph of polyacrylamide gel–filled capillaries. (A) Gels free of bubbles were obtained without surface pretreatment. (B) Equidistant bubbles were generated when the surface was pretreated with 3-methacryloxypropyltrimethoxysilane. *With permission from H.F. Yin, J.A. Lu, G. Schomburg, Production of polyacrylamide gel filled capillaries for capillary gel electrophoresis (CGE): influence of capillary surface pretreatment on performance and stability, Journal of High Resolution Chromatography 13 (9) (1990) 624–627 [60].*

does not stop the unavoidable shrinking process during gel polymerization and, therefore, does not prevail bubble formation. On the other hand, bubble formation can be avoided by applying a noncross-linked polyacrylamide surface coating.

In the early days, similar cross-linked polyacrylamide gels with %T and %C were used in capillary electrophoresis as in slab gel electrophoresis. Paulus and Ohms [208] used the mixture of $pd(T)_{10-50}$ to compare capillary polyacrylamide gel (7.5% T/3.3% C) electrophoresis with slab gel electrophoresis, shown in Fig. 3.39. They found that the slab gel pattern was similar to that obtained with the gel-filled capillary column, also suggesting quantification capability for the latter one. These authors used the standard free radical polymerization process with APS and TEMED.

Baba et al. [209] prepared a 5% T/5% C cross-linked polyacrylamide gel that was polymerized within the capillary for 2 hours. The resolving power of their gel was demonstrated by the separation of a mixture of 250 polyadenylic acids in 60 minutes with high efficiencies in comparison to HPLC data [210]. With an interesting approach, Schomburg and coworkers [175] used γ-radiation initiation for the polymerization reaction and separated a $pd(C)_{24-30}$ mixture on the resulting polyacrylamide gel (6% T/3% C)−filled capillary column. The Novotny group [211]

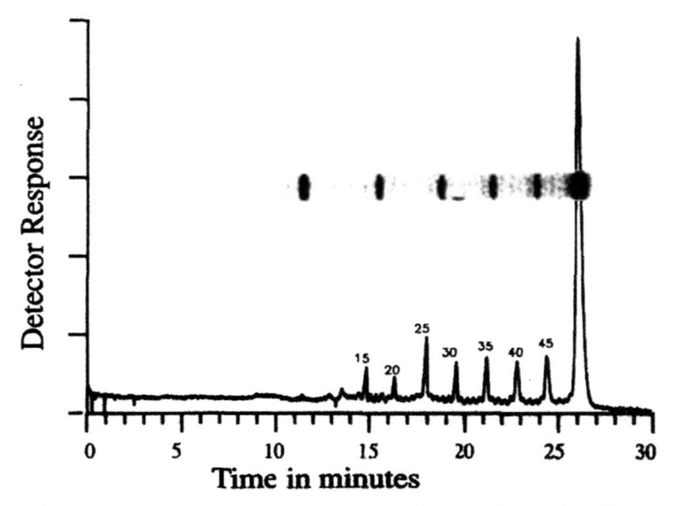

Figure 3.39 Comparative separation of deoxypolythymidylic acids $pd(T)_{10-50}$ on capillary gel and slab gel electrophoresis. *With permission from A. Paulus, J.I. Ohms, Analysis of oligonucleotides by capillary gel electrophoresis, Journal of Chromatography 507 (1990) 113−123 [208].*

described a novel procedure to fabricate polyacrylamide gel–filled capillaries using the principle of isotachophoretic polymerization. The polymerization initiator in this case was introduced into the capillary by electromigration resulting in gradual polymerization of the acrylamide monomer along the longitudinal axis of the tubing as depicted in Fig. 3.40. In this way the heat generation by the polymerization reaction was significantly slower alleviating the otherwise frequently occurring bubble formation problems. This simple and reproducible method was readily applicable to simultaneous capillary production. The authors used such a polymerized gel–filled capillary to analyze mouse urine with capillary SDS polyacrylamide gel electrophoresis using 6%T/3%C gel.

Alternatively, polymerization of the sieving matrix can be initiated by the addition of riboflavin to the acrylamide/bisacrylamide solution prior to filling the capillary [212]. In this case a UV-transparent coating was used that allowed the UV light through to start the polymerization reaction. The method was reportedly beneficial with regard to no bubble formation in the gel that was a problem with conventional persulfate and TEMED-based methods. The same group also measured the diffusion coefficients of four oligonucleotides in such riboflavin-initiated polyacrylamide gels [213] as shown in Fig. 3.41 and the results are delineated in Table 3.1.

The kinetics of riboflavin-mediated photopolymerization of acrylamide was re-examined by Gelfi et al. [37]. They optimized the reaction conditions to obtain >95% fabrication efficiency by conducting photopolymerization at 70°C for 1 hour utilizing a 105 W UV-A lamp. The same group checked the efficiency of methylene blue–based photocatalysis versus persulfate catalysis to fabricate polyacrylamide gels for capillary

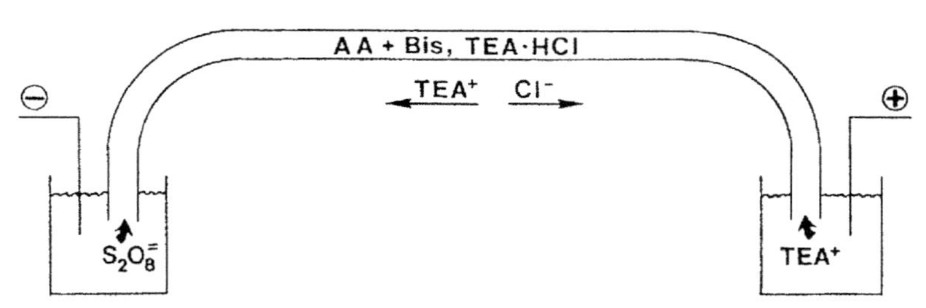

Figure 3.40 Schematic representation of isothachophoretic polymerization process of cross-linked polyacrylamide gels. AA, Acrylamide (monomer); Bis, bisacrylamide (crosslinker); S_2O_8, initiator; TEA, triethanolamine (catalyst). *With permission from V. Dolnik, K.A. Cobb, M. Novotny, Preparation of polyacrylamide gel-filled capillaries for capillary electrophoresis, Journal of Microcolumn Separations 3 (2) (1991) 155–159 [211].*

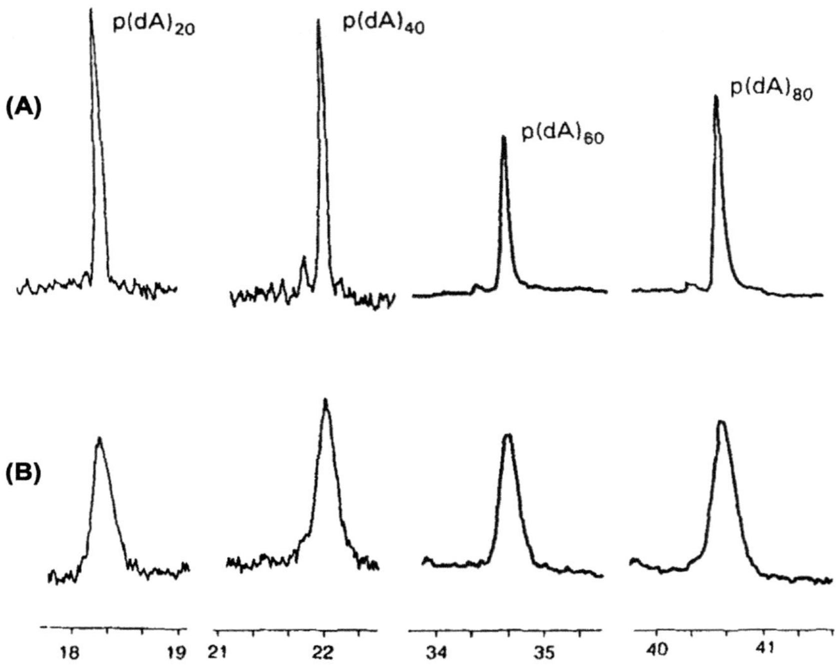

Figure 3.41 Determination of diffusion coefficients by CGE analysis of p(dA)20, 40, 60, and 80 DNA fragments. (A) Electropherograms without interruption and (B) electropherograms with interruption [5 hours, 17 minutes for p(dA)20 and p(dA)40; 6 hours, for p(dA)60 and P(dA)80]. *With permission from G.J.M. Bruin, T. Wang, X. Xu, J. C. Kraak, H. Poppe, Preparation of polyacrylamide gel-filled capillaries by photopolymerization for capillary electrophoresis, Journal of Microcolumn Separations 4 (5) (1992) 439–448 [213].*

Table 3.1 Electrophoretic mobilities, diffusion coefficients, and peak standard deviations for oligonucleotides with different base numbers observed in Fig. 3.41.

	$p(dA)_{20}$	$p(dA)_{40}$	$p(dA)_{60}$	$p(dA)_{80}$
Mobility $\times 10^5$ (cm^2/V s)	12.31	10.30	8.94	7.64
Injection length (mm)	0.61	0.51	0.67	0.29
$\sigma_{z,\text{inj}}$ (mm)	0.18	0.15	0.19	0.08
Diffusion coefficient $D_m \times 10^6$ (cm^2/s)	0.27	0.16	0.115	0.09
$\sigma_{z,\text{diff}}$ (mm)	0.24	0.20	0.21	0.21
$D_m/(\mu/z)$ (V)	0.046	0.064	0.078	0.094

With permission from G.J.M. Bruin, T. Wang, X. Xu, J.C. Kraak, H. Poppe, Preparation of polyacrylamide gel-filled capillaries by photopolymerization for capillary electrophoresis, Journal of Microcolumn Separations 4 (5) (1992) 439–448 [213].

electrophoresis. While any remaining oxygen in the reaction mixture caused quenching during persulfate-based catalysis, it essentially did not alter the photopolymerization-initiated process. The authors emphasized that the presence of high concentration of urea (e.g., 8M) significantly accelerated persulfate-driven polymerization. On the other hand, they found that the persulfate-catalyzed reaction was strongly quenched by the presence of organic solvents, while the photopolymerization process was essentially unaffected. Methylene blue–based polymerization systems were also unaffected by the pH of the reaction mixture in the interval of pH 4–10, so such a system was proposed in the field of polyacrylamide catalysis [38]. The group also studied the kinetics of monomer incorporation into polyacrylamide gels [38] in a photopolymerization system comprising methylene blue in the presence of a redox system, Na-toluenesulfinate (reducer), and diphenyliodinium chloride (Ph_2ICl, oxidizer). The efficiency of DNA separation in dilute polymer matrices was enhanced by using appropriate capillary coatings or at least surface deactivation as described in [214]. Surface deactivated capillary columns proved useful in the analysis mRNA transcripts amplified by PCR [92].

Toward the development of new formulation polyacrylamide matrices, Righetti's group introduced several novel aspects of polyacrylamide gel preparation including a series of mono- and disubstituted acrylamide monomers. These potential candidates of a novel class of matrices exhibited high hydrophilicity, high resistance to hydrolysis, and larger pore size than conventional gels. A series of crosslinkers and their contributions to the stability and hydrophilicity of the gel was assessed and reported in Ref. [33]. The approach was capable of producing extremely large-pore size gels as the polymerization was conducted in the presence hydrophilic polymers (typically high molecular mass PEGs). In the presence of these polymers in up to 2%–2.5%, the so-called "lateral-chain aggregation" occurred as shown in Fig. 3.42, producing extremely large expansion in pore size, for example, while a 5%T/4% C gel typically has an average pore diameter of 5–6 nm, the same gel, in the presence of a laterally aggregating agent exhibited an average pore size of 500 nm.

Cross-linked polyacrylamide-filled capillary columns for polynucleotide analysis were fabricated by in situ radical initiator polymerization paying special attention to avoid bubble formation using a special injection equipment [215]. The authors evaluated the performance of gel-filled capillaries in respect to stability, migration time reproducibility, feasibility of the method, and resolving power for polynucleotides.

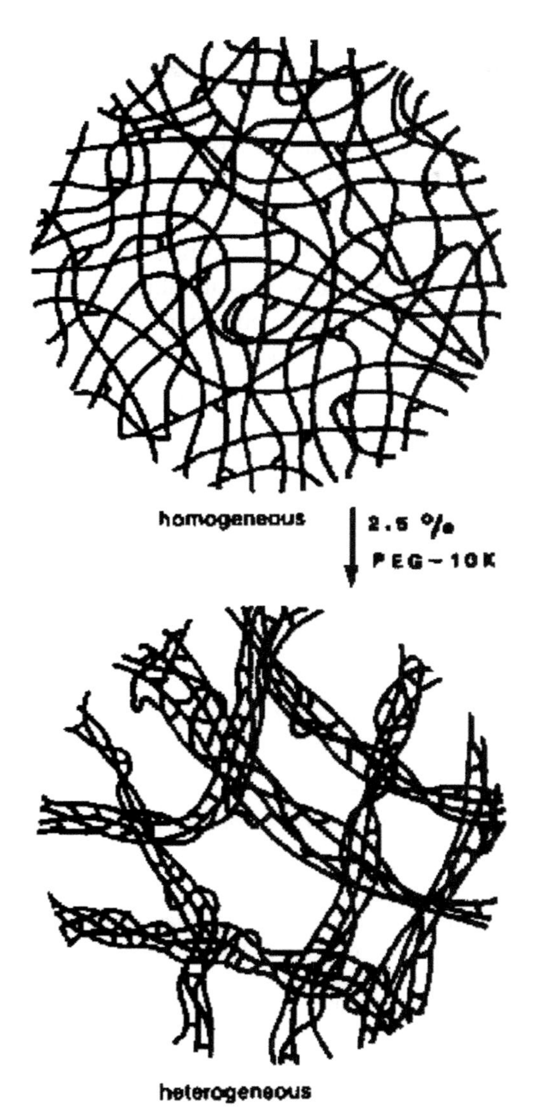

Figure 3.42 Hypothetical model for "laterally aggregated" gels. The upper panel represents the structure of a control gel (a random meshwork of fibers), a homogeneous gel in the sense that all fibers have random orientations in the three-dimensional space. The lower panel represents a gel polymerized in the presence of hydrophilic polymer additives, in which case owing to bundling of individual fibers (chain clustering), the average porosity was greatly increased. *With permission from P. G. Righetti, M. Chiari, M. Nesi, S. Caglio, Towards new formulations for polyacrylamide matrixes, as investigated by capillary zone electrophoresis, Journal of Chromatography 638 (2) (1993) 165–178 [33].*

Barron et al. showed that while both uncoated and coated capillaries can be used for the separation of large DNA fragments in the size range of 2.0–23.1 kbp in diluted cellulosic polymer solutions, uncoated capillaries provided better resolution, especially when ultra-diluted polymers were used [216]. The phenomenon was explained as the EOF in uncoated capillaries increases the residence time of the sample in the capillary (i.e., acts as using a longer capillary) without a noteworthy contribution to band broadening. The separation of large DNA fragments in ultra-diluted polymer solutions was primarily based on the entanglement interactions between the cellulosic polymers and DNA restriction fragments; however, borate complexation–mediated cross-linking in the separation matrix was not considered. PEO with a large chain length range offers easy preparation of homogeneous solutions to provide highly reproducible performance with adequate resolution [217]. The resolution can be improved with the use of mixed polymer matrices, that is, different chain length ingredients with relatively low viscosity.

References

[1] T. Tanaka, Gels, Sci. Am. 244 (1981) 124–138.
[2] T. Hirotsu, Theory of swelling, in: Y. Osada, K. Kajiwara (Eds.), Gels Handbook, Academic Press, London, 2001, pp. 65–97.
[3] P.H. Hermans, Gels, in: H.R. Kruyt (Ed.), Colloid Science, in: Vol. 2, Elsevier, Amsterdam, 1949, pp. 483–651.
[4] P.G. de Gennes, Scaling Concepts in Polymer Physics, Cornell University Press, Ithaca, NY:, 1979.
[5] F. Redaelli, M. Sorbona, F. Rossi, Synthesis and processing of hydrogels for medical applications, in: G. Perale, J. Hilborn (Eds.), Bioresorbable Polymers for Biomedical Applications, Woodhead Publishing, 2017, pp. 205–228.
[6] P.C. Harris, Chemistry and rheology of borate-crosslinked fluids at temperatures to 300°F, Journal of Petroleum Technology 45 (1993) 264–269.
[7] E. Pezron, A. Ricard, L. Leibler, Rheology of galactomannan-borax gels, Journal of Polymer Science: Part B: Polymer Physics 28 (1990) 2445–2461.
[8] C.E. Sanger-van de Griend, CE-SDS method development, validation, and best practice—An overview, Electrophoresis 40 (18–19) (2019) 2361–2374.
[9] P.J. Flory, Polymer Chemistry, Cornell University Press, Ithaca, NY, 1953.
[10] A.T. Andrews, Electrophoresis, Theory, Techniques and Biochemical and Clinical Applications, second ed., Claredon Press, Oxford, 1986.
[11] M. Doi, S.F. Edwards, The Theory of Polymer Dynamics, Oxford University Press, Oxford, 1986.
[12] H.M. James, E. Guth, Theory of the elastic properties of rubber, The Journal of Chemical Physics 11 (1943) 455–481.
[13] C. Haber, S.A. Ruiz, D. Wirtz, Shape anisotropy of a single random-walk polymer, Proceedings of the National Academy of Sciences of the United States of America 97 (20) (2000) 10792–10795.
[14] D. Stauffer, Introduction to Percolation Theory, Taylor and Francis, London, 1985.

[15] A.S. Cohen, B.L. Karger, High-performance sodium dodecyl sulfate polyacrylamide gel capillary electrophoresis of peptides and proteins, Journal of Chromatography 397 (1987) 409–417.
[16] B.L. Karger, A.S. Cohen, Capillary gel electrophoresis columns, in: European Patent Application, Northeastern University, USA, 1989, 18 pp.
[17] P.G. Righetti, C. Gelfi, Electrophoresis gel media: the state of the art, Journal of Chromatography, B: Biomedical Sciences and Applications 699 (1–2) (1997) 63–75.
[18] P.G. Righetti, C. Gelfi, Recent advances in capillary electrophoresis of DNA fragments and PCR products in poly(N-substituted acrylamides), Analytical Biochemistry 244 (2) (1997) 195–207.
[19] C. Filep, A. Guttman, The effect of temperature in sodium dodecyl sulfate capillary gel electrophoresis of protein therapeutics, Analytical Chemistry 92 (5) (2020) 4023–4028.
[20] K.R. Mitchelson, J. Cheng, Capillary electrophoresis with glycerol as an additive, Methods in Molecular Biology 162 (2001) 259–277.
[21] V. Dolnik, W.A. Gurske, Size separation of proteins by capillary zone electrophoresis with cationic hitchhiking, Electrophoresis 32 (20) (2011) 2884–2892.
[22] J. Beckman, Y. Song, Y. Gu, S. Voronov, N. Chennamsetty, S. Krystek, et al., Purity determination by capillary electrophoresis sodium hexadecyl sulfate (CE-SHS): a novel application for therapeutic protein characterization, Analytical Chemistry 90 (4) (2018) 2542–2547.
[23] E. Kenndler, C. Schwer, Peak dispersion and separation efficiency in high-performance zone electrophoresis with gel-filled capillaries, Journal of Chromatography 595 (1–2) (1992) 313–318.
[24] S. Hjerten, Molecular sieve chromatography on polyacrylamide gels, prepared according to a simplified method, Archives of Biochemistry and Biophysics 1 (Suppl)) (1962) 147–151.
[25] S. Hjerten, High-performance electrophoresis. Elimination of electroendosmosis and solute adsorption, Journal of Chromatography 347 (2) (1985) 191–198.
[26] B.L. Karger, Y.H. Chu, F. Foret, Capillary electrophoresis of proteins and nucleic acids, Annual Review of Biophysics and Biomolecular Structure 24 (1995) 579–610.
[27] A. Guttman, A. Paulus, A.S. Cohen, B.L. Karger, H. Rodriguez, W.S. Hancock, High performance, capillary gel electrophoresis: high resolution and micropreparative applications, Electrophoresis'88, Copenhagen VCH, Weinheim, 1988.
[28] J.C. Venter, et al., The sequence of the human genome, Science (New York, N.Y.) 291 (2001) 1304–1351.
[29] H. Swerdlow, R. Gesteland, Capillary gel electrophoresis for rapid, high resolution DNA sequencing, Nucleic Acids Research 18 (6) (1990) 1415–1419.
[30] A. Vegvari, S. Hjerten, Stable homogeneous gel for molecular-sieving of DNA fragments in capillary electrophoresis, Journal of Chromatography A 960 (1–2) (2002) 221–227.
[31] A. Paulus, E. Gassmann, M.J. Field, Calibration of polyacrylamide gel columns for the separation of oligonucleotides by capillary electrophoresis, Electrophoresis 11 (9) (1990) 702–708.
[32] C. Gelfi, A. Alloni, P. Debesi, P.G. Righetti, Investigation of the properties of acrylamide bifunctional monomers (cross-linkers) by capillary zone electrophoresis, Journal of Chromatography 608 (1–2) (1992) 343–348.
[33] P.G. Righetti, M. Chiari, M. Nesi, S. Caglio, Towards new formulations for polyacrylamide matrixes, as investigated by capillary zone electrophoresis, Journal of Chromatography 638 (2) (1993) 165–178.
[34] C. Gelfi, P. Debesi, A. Alloni, P.G. Righetti, Investigation of the properties of novel acrylamido monomers by capillary zone electrophoresis, Journal of Chromatography 608 (1–2) (1992) 333–341.

[35] P.G. Righetti, S. Caglio, On the kinetics of monomer incorporation into polyacrylamide gels, as investigated by capillary zone electrophoresis, Electrophoresis 14 (7) (1993) 573–582.
[36] M. Chiari, C. Micheletti, P.G. Righetti, G. Poli, Polyacrylamide gel polymerization under nonoxidizing conditions, as monitored by capillary zone electrophoresis, Journal of Chromatography 598 (2) (1992) 287–297.
[37] C. Gelfi, P. De Besi, A. Alloni, P.G. Righetti, T. Lyubimova, V.A. Briskman, Kinetics of acrylamide photopolymerization as investigated by capillary zone electrophoresis, Journal of Chromatography 598 (2) (1992) 277–285.
[38] S. Caglio, P.G. Righetti, On the efficiency of methylene blue vs persulfate catalysis of polyacrylamide gels, as investigated by capillary zone electrophoresis, Electrophoresis 14 (10) (1993) 997–1003.
[39] T.L. Rapp, W.K. Kowalchyk, K.L. Davis, E.A. Todd, K.L. Liu, M.D. Morris, Acrylamide polymerization kinetics in gel electrophoresis capillaries. A Raman microprobe study, Analytical Chemistry 64 (20) (1992) 2434–2437.
[40] T. Manabe, S. Terabe, Gel electrophoresis of oligonucleotides employing untreated fused silica capillaries, Seibutsu Butsuri Kagaku 37 (2) (1993) 123–128.
[41] D. Wheeler, D. Tietz, A. Chrambach, Information on DNA conformation derived from transverse pore gradient gel electrophoresis in conjunction with an advanced data analysis applied to capillary electrophoresis in polymer media, Electrophoresis 13 (9–10) (1992) 604–608.
[42] M.J. Rocheleau, R.J. Grey, D.Y. Chen, H.R. Harke, N.J. Dovichi, Formamide modified polyacrylamide gels for DNA sequencing by capillary gel electrophoresis, Electrophoresis 13 (8) (1992) 484–486.
[43] K.D. Konrad, S.L. Pentoney Jr., Contribution of secondary structure to DNA mobility in capillary gels, Electrophoresis 14 (5–6) (1993) 502–508.
[44] T. Satow, T. Akiyama, A. Machida, Y. Utagawa, H. Kobayashi, Simultaneous determination of the migration coefficient of each base in heterogeneous oligo-DNA by gel filled capillary electrophoresis, Journal of Chromatography 652 (1) (1993) 23–30.
[45] C. Gelfi, A. Orsi, F. Leoncini, P.G. Righetti, Fluidified polyacrylamides as molecular-sieves in capillary zone electrophoresis of DNA fragments, Journal of Chromatography A 689 (1) (1995) 97–105.
[46] E. Simo-Alfonso, C. Gelfi, M. Lucisano, P. Giorgio Righetti, Performance of a series of novel N-substituted acrylamides in capillary electrophoresis of DNA fragments, Journal of Chromatography A 756 (1–2) (1996) 255–261.
[47] C. Gelfi, M. Perego, F. Libbra, P.G. Righetti, Comparison of behavior of N-substituted acrylamides and celluloses on double-stranded DNA separations by capillary electrophoresis at 25 degrees and 60 degrees C, Electrophoresis 17 (8) (1996) 1342–1347.
[48] C. Gelfi, E. Simo-Alfonso, R. Sebastiano, A. Citterio, P.G. Righetti, Novel acrylamido monomers with higher hydrophilicity and improved hydrolytic stability: III. DNA separations by capillary electrophoresis in poly(N-acryloylaminopropanol), Electrophoresis 17 (4) (1996) 738–743.
[49] C.-C. Wang, S.C. Beale, Preparation of linear polyacrylamide gel step gradients for capillary electrophoresis, Journal of Chromatography A 756 (1–2) (1996) 245–253.
[50] Y. Chen, F.-L. Wang, U. Schwarz, Polyacrylamide gradient gel-filled capillaries with low detection background, Journal of Chromatography A 772 (1–2) (1997) 129–135.
[51] Y. Chen, J.-V. Hoeltje, U. Schwarz, Preparation of highly condensed polyacrylamide gel-filled capillaries with low detection background, Journal of Chromatography A 685 (1) (1994) 121–129.
[52] H. Swerdlow, K.E. Dew-Jager, K. Brady, R. Grey, N.J. Dovichi, R. Gesteland, Stability of capillary gels for automated sequencing of DNA, Electrophoresis 13 (8) (1992) 475–483.

[53] J.J. Lu, S. Liu, Q. Pu, Replaceable cross-linked polyacrylamide for high performance separation of proteins, Journal of Proteome Research 4 (3) (2005) 1012–1016.
[54] Y. Liu, P. Reddy, C.K. Ratnayjake, E.V. Koh, Methods and compositions for capillary electrophoresis (CE). United States Patent 7,831,317B2, 2003.
[55] C. Filep, A. Guttman, Effect of the monomer cross-linker ratio on the separation selectivity of monoclonal antibody subunits in sodium dodecyl sulfate capillary gel electrophoresis, Analytical Chemistry 93 (7) (2021) 3535–3541.
[56] S.R. Bean, G.L. Lookhart, Sodium dodecyl sulfate capillary electrophoresis of wheat proteins. 1. Uncoated capillaries, Journal of Agricultural and Food Chemistry 47 (10) (1999) 4246–4255.
[57] P. Sacco, F. Furlani, M. Cok, A. Travan, M. Borgogna, E. Marsich, et al., Boric acid induced transient cross-links in lactose-modified chitosan (chitlac), Biomacromolecules 18 (12) (2017) 4206–4213.
[58] H. Deuel, H. Neukom, Biological and synthetic polymer networks, Die Makromolekulare Chemie 3 (1949) 113–126.
[59] Y. Miyazaki, K. Yoshimura, Y. Miura, H. Sakashita, K. Ishimaru, 11B NMR investigation of the complexation behavior of borate with polysaccharides in aqueous solution, Polyhedron 22 (2003) 909–916.
[60] H.F. Yin, J.A. Lu, G. Schomburg, Production of polyacrylamide gel filled capillaries for capillary gel electrophoresis (CGE): influence of capillary surface pretreatment on performance and stability, Journal of High Resolution Chromatography 13 (9) (1990) 624–627.
[61] L. Karger Barry, A. Guttman, DNA sequencing by CE, Electrophoresis 30 (Suppl 1)) (2009) S196–S202.
[62] P. Sacco, F. Furlani, S. Paoletti, I. Donati, pH-assisted gelation of lactose-modified chitosan, Biomacromolecules 20 (8) (2019) 3070–3075.
[63] J. Cheng, K.R. Mitchelson, Glycerol-enhanced separation of DNA fragments in entangled solution capillary electrophoresis, Analytical Chemistry 66 (23) (1994) 4210–4214.
[64] Y. Shen, Q. Xu, F.T. Han, K. Ding, F. Song, Y. Fan, et al., Application of capillary nongel sieving electrophoresis for gene analysis, Electrophoresis 20 (9) (1999) 1822–1828.
[65] F. Han, B. Lin, Influence of mannitol additive on DNA separation by capillary non-gel sieving electrophoresis, Se Pu = Chinese Journal of Chromatography/Zhongguo hua xue hui 16 (1998) 489–491.
[66] D.N. Heiger, A.S. Cohen, B.L. Karger, Separation of DNA restriction fragments by high performance capillary electrophoresis with low and zero crosslinked polyacrylamide using continuous and pulsed electric fields, Journal of Chromatography 516 (1) (1990) 33–48.
[67] A. Guttman, Capillary sodium dodecyl sulfate-gel electrophoresis of proteins, Electrophoresis 17 (8) (1996) 1333–1341.
[68] A. Guttman, Capillary electrophoresis using replaceable gels, in: USPTO, USA, 1994.
[69] J. Sudor, F. Foret, P. Bocek, Pressure refilled polyacrylamide columns for the separation of oligonucleotides by capillary electrophoresis, Electrophoresis 12 (12) (1991) 1056–1058.
[70] P.D. Grossman, D.S. Soane, Capillary electrophoresis of DNA in entangled polymer solutions, Journal of Chromatography 559 (1–2) (1991) 257–266.
[71] P.D. Grossman, D.S. Soane, Experimental and theoretical studies of DNA separations by capillary electrophoresis in entangled polymer solutions, Biopolymers 31 (10) (1991) 1221–1228.
[72] P.G. De Gennes, Scaling Concept in Polymer Physics, Cornell University Press, Ithaca, NY, 1979.

[73] X. Huang, M.J. Gordon, R.N. Zare, Bias in quantitative capillary zone electrophoresis caused by electrokinetic sample injection, Analytical Chemistry 60 (4) (1988) 375–377.
[74] A.E. Barron, H.W. Blanch, D.S. Soane, A transient entanglement coupling mechanism for DNA separation by capillary electrophoresis in ultradilute polymer solutions, Electrophoresis 15 (5) (1994) 597–615.
[75] A.E. Barron, W.M. Sunada, H.W. Blanch, Capillary electrophoresis of DNA in uncrosslinked polymer solutions: evidence for a new mechanism of DNA separation, Biotechnology and Bioengineering 52 (2) (1996) 259–270.
[76] A.E. Barron, W.M. Sunada, H.W. Blanch, The effects of polymer properties on DNA separations by capillary electrophoresis in uncross-linked polymer solutions, Electrophoresis 17 (4) (1996) 744–757.
[77] N.C. Stellwagen, C. Gelfi, P.G. Righetti, DNA and buffers: the hidden danger of complex formation, Biopolymers 54 (2) (2000) 137–142.
[78] R.P. Singhal, J. Xian, Separation of DNA restriction fragments by polymer-solution capillary zone electrophoresis: influence of polymer concentration and ion-pairing reagents, Journal of Chromatography 652 (1) (1993) 47–56.
[79] M. Chiari, M. Nesi, P.G. Righetti, Capillary zone electrophoresis of DNA fragments in a novel polymer network: poly(N-acryloylaminoethoxyethanol), Electrophoresis 15 (5) (1994) 616–622.
[80] J. Berka, Y.F. Pariat, O. Muller, K. Hebenbrock, D.N. Heiger, F. Foret, et al., Sequence dependent migration behavior of double-stranded DNA in capillary electrophoresis, Electrophoresis 16 (3) (1995) 377–388.
[81] A. Guttman, N. Cooke, Effect of temperature on the separation of DNA restriction fragments in capillary gel electrophoresis, Journal of Chromatography 559 (1–2) (1991) 285–294.
[82] A. Widhalm, C. Schwer, D. Blaas, E. Kenndler, Capillary zone electrophoresis with a linear, noncrosslinked polyacrylamide gel: separation of proteins according to molecular mass, Journal of Chromatography 549 (1–2) (1991) 446–451.
[83] D. Wu, F.E. Regnier, Sodium dodecyl sulfate-capillary gel electrophoresis of proteins using non-cross-linked polyacrylamide, Journal of Chromatography 608 (1–2) (1992) 349–356.
[84] W.E. Werner, D.M. Demorest, J.E. Wiktorowicz, Automated Ferguson analysis of glycoproteins by capillary electrophoresis using a replaceable sieving matrix, Electrophoresis 14 (8) (1993) 759–763.
[85] K. Hebenbrock, K. Schuegerl, R. Freitag, Analysis of plasmid-DNA and cell protein of recombinant *Escherichia coli* using capillary gel electrophoresis, Electrophoresis 14 (8) (1993) 753–758.
[86] C.A.G. De Jong, J. Risley, A.K. Lee, S.S. Zhao, D.D.Y. Chen, Separation of recombinant therapeutic proteins using capillary gel electrophoresis and capillary isoelectric focusing, in: N.T. Tran, M. Taverna (Eds.), Capillary Electrophoresis of Proteins and Peptides: Methods and Protocols, Springer, New York, NY, 2016, pp. 137–149.
[87] E. Carrilho, M.C. Ruiz-Martinez, J. Berka, I. Smirnov, W. Goetzinger, A.W. Miller, et al., Rapid DNA sequencing of more than 1000 bases per run by capillary electrophoresis using replaceable linear polyacrylamide solutions, Analytical Chemistry 68 (19) (1996) 3305–3313.
[88] D. Figeys, N.J. Dovichi, Effect of the age of non-cross-linked polyacrylamide on the separation of DNA sequencing samples, Journal of Chromatography A 717 (1 – 2) (1995) 105–111.
[89] R.S. Madabhushi, Separation of 4-color DNA sequencing extension products in noncovalently coated capillaries using low viscosity polymer solutions, Electrophoresis 19 (2) (1998) 224–230.

[90] H.F. Yin, M.H. Kleemiss, J.A. Lux, G. Schomburg, Diffusion coefficients of oligonucleotides in capillary gel electrophoresis, Journal of Microcolumn Separations 3 (4) (1991) 331–335.
[91] H.E. Schwartz, K. Ulfelder, F.J. Sunzeri, M.P. Busch, R.G. Brownlee, Analysis of DNA restriction fragments and polymerase chain reaction products towards detection of the AIDS (HIV-1) virus in blood, Journal of Chromatography 559 (1–2) (1991) 267–283.
[92] S. Nathakarnkitkool, P.J. Oefner, G. Bartsch, M.A. Chin, G.K. Bonn, High-resolution capillary electrophoretic analysis of DNA in free solution, Electrophoresis 13 (1–2) (1992) 18–31.
[93] S. Hu, Z. Zhang, L.M. Cook, E.J. Carpenter, N.J. Dovichi, Separation of proteins by sodium dodecylsulfate capillary electrophoresis in hydroxypropylcellulose sieving matrix with laser-induced fluorescence detection, Journal of Chromatography A 894 (2000) 291–296.
[94] O. de Carmejane, Y. Yamaguchi, T.I. Todorov, M.D. Morris, Three-dimensional observation of electrophoretic migration of dsDNA in semidilute hydroxyethylcellulose solution, Electrophoresis 22 (2001) 2433–2441.
[95] Y. Jin, B. Lin, Y.S. Fung, Electrophoretic migration behavior of DNA fragments in polymer solution, Electrophoresis 22 (2001) 2150–2158.
[96] A. Einstein, Motion of suspended particles in stationary liquids required from the molecular kinetic theory of heat, Annalen der Physik 17 (1905) 549–560.
[97] T.J. Gibson, M.J. Sepaniak, Examination of band dispersion during size-selective capillary electrophoresis separations of DNA fragments, Journal of Chromatography B 695 (1997) 103–111.
[98] A.E. Barron, D.S. Soane, H.W. Blanch, Capillary electrophoresis of DNA in uncrosslinked polymer solutions, Journal of Chromatography 652 (1) (1993) 3–16.
[99] L. Mitnik, L. Salome, J.L. Viovy, C. Heller, Systematic study of field and concentration effects in capillary electrophoresis of DNA in polymer solutions, Journal of Chromatography A 710 (2) (1995) 309–321.
[100] X. Shi, R.W. Hammond, M.D. Morris, DNA conformational dynamics in polymer solutions above and below the entanglement limit, Analytical Chemistry 67 (6) (1995) 1132–1138.
[101] M. Minarik, B. Gas, E. Kenndler, Size-based separation of polyelectrolytes by capillary zone electrophoresis: migration regimes and selectivity of poly(styrenesulphonates) in solutions of derivatized cellulose, Electrophoresis 18 (1) (1997) 98–103.
[102] Y. Kim, E.S. Yeung, Separation of DNA sequencing fragments up to 1000 bases by using poly(ethylene oxide)-filled capillary electrophoresis, Journal of Chromatography A 781 (1 – 2) (1997) 315–325.
[103] Q. Gao, E.S. Yeung, A matrix for DNA separation: genotyping and sequencing using poly(vinylpyrrolidone) solution in uncoated capillaries, Analytical Chemistry 70 (7) (1998) 1382–1388.
[104] W. Wei, E.S. Yeung, Improvements in DNA sequencing by capillary electrophoresis at elevated temperature using poly(ethylene oxide) as a sieving matrix, Journal of Chromatography B 745 (2000) 221–230.
[105] W. Wei, E.S. Yeung, DNA capillary electrophoresis in entangled dynamic polymers of surfactant molecules, Analytical Chemistry 73 (8) (2001) 1776–1783.
[106] J.M. Song, E.S. Yeung, Optimization of DNA electrophoretic behavior in poly (vinyl pyrrolidone) sieving matrix for DNA sequencing, Electrophoresis 22 (2001) 748–754.
[107] H. Tan, E.S. Yeung, Characterization of dye-induced mobility shifts affecting DNA sequencing in poly(ethylene oxide) sieving matrix, Electrophoresis 18 (1997) 2893–2900.

[108] R.S. Madabhushi, M. Vainer, V. Dolnik, S. Enad, D.L. Barker, D.W. Harris, et al., Versatile low-viscosity sieving matrixes for nondenaturing DNA separations using capillary array electrophoresis, Electrophoresis 18 (1) (1997) 104–111.
[109] C. Barta, Z. Ronai, M. Sasvari-Szekely, A. Guttman, Rapid single nucleotide polymorphism analysis by primer extension and capillary electrophoresis using polyvinyl pyrrolidone matrix, Electrophoresis 22 (4) (2001) 779–782.
[110] M. Girod, D.W. Armstrong, Monitoring the migration behavior of living microorganisms in capillary electrophoresis using laser-induced fluorescence detection with a charge-coupled device imaging system, Electrophoresis 23 (13) (2002) 2048–2056.
[111] E. Simo-Alfonso, M. Conti, C. Gelfi, P.G. Righetti, Sodium dodecyl sulfate capillary electrophoresis of proteins in entangled solutions of poly(vinyl alcohol), Journal of Chromatography A 689 (1) (1995) 85–96.
[112] K. Ganzler, K.S. Greve, A.S. Cohen, B.L. Karger, A. Guttman, N.C. Cooke, High-performance capillary electrophoresis of SDS-protein complexes using UV-transparent polymer networks, Analytical Chemistry 64 (22) (1992) 2665–2671.
[113] M.N. Albarghouthi, A.E. Barron, Polymeric matrices for DNA sequencing by capillary electrophoresis, Electrophoresis 21 (18) (2000) 4096–4111.
[114] D. Liang, B. Chu, High-speed separation of DNA fragments by capillary electrophoresis in poly(ethylene oxide)-poly(propylene oxide)-poly(ethylene oxide) triblock polymer, Electrophoresis 19 (14) (1998) 2447–2453.
[115] C. Wu, T. Liu, B. Chu, Viscosity-adjustable block copolymer for DNA separation by capillary electrophoresis, Electrophoresis 19 (1998) 231–241.
[116] R.L. Rill, Y. Liu, D.H. Van Winkle, B.R. Locke, Pluronic copolymer liquid crystals: unique, replaceable media for capillary gel electrophoresis, Journal of Chromatography A 817 (1 + 2) (1998) 287–295.
[117] D. Liang, L. Song, S. Zhou, V.S. Zaitsev, B. Chu, Poly(N-isopropylacrylamide)-g-poly(ethyleneoxide) for high resolution and high speed separation of DNA by capillary electrophoresis, Electrophoresis 20 (1999) 2856–2863.
[118] I. Miksik, Z. Deyl, Application of pluronic copolymer liquid crystals for the capillary electrophoretic separation of collagen type I cyanogen bromide fragments, Journal of Chromatography B 739 (2000) 109–116.
[119] I. Miksik, Z. Deyl, V. Kasicka, Capillary electrophoretic separation of proteins and peptides using pluronic liquid crystals and surface-modified capillaries, Journal of Chromatography B 741 (2000) 37–42.
[120] B.A. Buchholz, E.A.S. Doherty, M.N. Albarghouthi, F.M. Bogdan, J.M. Zahn, A. E. Barron, Microchannel DNA sequencing matrices with a thermally controlled "Viscosity Switch", Analytical Chemistry 73 (2) (2001) 157–164.
[121] J. Sudor, V. Barbier, S. Thirot, D. Godfrin, D. Hourdet, M. Millequant, et al., New block-copolymer thermo-associating matrices for DNA sequencing: effect of molecular structure on rheology and resolution, Electrophoresis 22 (4) (2001) 720–728.
[122] M. Nakatani, A. Shibukawa, T. Nakagawa, Effect of temperature and viscosity of sieving medium on electrophoretic behavior of sodium dodecyl sulfate-proteins on capillary electrophoresis in presence of pullulan, Electrophoresis 17 (7) (1996) 1210–1213.
[123] B.A. Siles, G.B. Collier, The characterization of a new size-sieving polymeric matrix for the separation of DNA fragments using capillary electrophoresis, Journal of Capillary Electrophoresis 3 (1996) 313–321.
[124] B.A. Siles, D.E. Anderson, N.S. Buchanan, M.F. Warder, The characterization of composite agarose/hydroxyethylcellulose matrixes for the separation of DNA fragments using capillary electrophoresis, Electrophoresis 18 (11) (1997) 1980–1989.

[125] D. Liang, B. Chu, High speed separation of DNA fragments by capillary electrophoresis in poly(ethylene oxide)-poly(propylene oxide)-poly(ethylene oxide) triblock polymer, Electrophoresis 19 (14) (1998) 2447–2453.
[126] M. Chiari, M. Cretich, S. Riva, M. Casali, Performances of new sugar-bearing poly (acrylamide) copolymers as DNA sieving matrices and capillary coatings for electrophoresis, Electrophoresis 22 (4) (2001) 699–706.
[127] M. Chiari, M. Cretich, R. Consonni, Separation of DNA fragments in hydroxylated poly(dimethylacrylamide) copolymers, Electrophoresis 23 (4) (2002) 536–541.
[128] M. Chiari, F. Damin, A. Melis, R. Consonni, Separation of oligonucleotides and DNA fragments by capillary electrophoresis in dynamically and permanently coated capillaries, using a copolymer of acrylamide and b-D-glucopyranoside as a new low viscosity matrix with high sieving capacity, Electrophoresis 19 (18) (1998) 3154–3159.
[129] V. Dolnik, W.A. Gurske, A. Padua, Galactomannans as a sieving matrix in capillary electrophoresis, Electrophoresis 22 (2001) 707–719.
[130] L. Song, T. Liu, D. Liang, D. Fang, B. Chu, Separation of double-stranded DNA fragments by capillary electrophoresis in interpenetrating networks of polyacrylamide and polyvinylpyrrolidone, Electrophoresis 22 (17) (2001) 3688–3698.
[131] Y. Wang, D. Liang, J. Hao, D. Fang, B. Chu, Separation of double-stranded DNA fragments by capillary electrophoresis using polyvinylpyrrolidone and poly(N, N-dimethylacrylamide) transient interpenetrating network, Electrophoresis 23 (10) (2002) 1460–1466.
[132] K. Kleparnik, S. Fanali, P. Bocek, Selectivity of the separation of DNA fragments by capillary zone electrophoresis in low-melting-point agarose sol, Journal of Chromatography 638 (2) (1993) 283–292.
[133] P. Bocek, A. Chrambach, Capillary electrophoresis of DNA in agarose solutions at 40 DegC, Electrophoresis 12 (12) (1991) 1059–1061.
[134] P. Bocek, A. Chrambach, Capillary electrophoresis in agarose solutions: extension of size separations to DNA of 12 kb in length, Electrophoresis 13 (1–2) (1992) 31–34.
[135] D. Tietz, A. Chrambach, Concave Ferguson plots of DNA fragments and convex Ferguson plots of bacteriophages: evaluation of molecular and fiber properties, using desktop computers, Electrophoresis 13 (5) (1992) 286–294.
[136] S. Hjerten, T. Srichaiyo, A. Palm, UV-transparent, replaceable agarose gels for molecular-sieve (capillary) electrophoresis of proteins and nucleic acids, Biomedical Chromatography 8 (2) (1994) 73–76.
[137] N. Chen, L. Wu, A. Palm, T. Srichaiyo, S. Hjerten, High-performance field inversion capillary electrophoresis of 0.1–23 kbp DNA fragments with low-gelling, replaceable agarose gels, Electrophoresis 17 (9) (1996) 1443–1450.
[138] B. Kozulic, On the “"door-corridor” model of gel electrophoresis. II. Developments related to new gels, capillary gel electrophoresis and gel chromatography, Applied and Theoretical Electrophoresis 4 (3) (1994) 137–148.
[139] K. Yamagata, Y. Shirasaki, Gradient Buffer in Capillary Electrophoresis of Nucleic Acids and Other Substances, Shimadzu Corp., Japan, 1991. Application: JP, 5 pp.
[140] D. Soto, S. Sukumar, Improved detection of mutations in the p53 gene in human tumors as single-stranded conformation polymorphs and double-stranded heteroduplex DNA, PCR Methods and Applications 2 (1) (1992) 96–98.
[141] G.H. Weiss, M. Garner, E. Yarmola, P. Bocek, A. Chrambach, A comparison of resolution of DNA fragments between agarose gel and capillary zone electrophoresis in agarose solutions, Electrophoresis 16 (8) (1995) 1345–1353.
[142] J. Borejdo, S. Burlacu, Distribution of actin filament lengths and their orientation measured by gel electrophoresis in capillaries, Journal of Muscle Research and Cell Motility 12 (4) (1991) 394–407.

[143] S.R. Motsch, M.H. Kleemiss, G. Schomburg, Production and application of capillaries filled with agarose gel for electrophoresis, Journal of High Resolution Chromatography 14 (9) (1991) 629–632.

[144] M.G. Harrington, K.H. Lee, J.E. Bailey, L.E. Hood, Sponge-like electrophoresis media: mechanically strong materials compatible with organic solvents, polymer solutions and two-dimensional electrophoresis, Electrophoresis 15 (2) (1994) 187–194.

[145] H. Soini, M.L. Riekkola, M.V. Novotny, Mixed polymer networks in the direct analysis of pharmaceuticals in urine by capillary electrophoresis, Journal of Chromatography A 680 (2) (1994) 623–634.

[146] T. Izumi, M. Yamaguchi, K. Yoneda, T. Isobe, T. Okuyama, T. Shinoda, Use of glucomannan for the separation of DNA fragments by capillary electrophoresis, Journal of Chromatography 652 (1) (1993) 41–46.

[147] P.D. Grossman, T. Hino, D.S. Soane, Dynamic light-scattering-studies of hydroxyethyl cellulose solutions used as sieving media for electrophoretic separations, Journal of Chromatography 608 (1–2) (1992) 79–83.

[148] M. Nakatani, A. Shibukawa, T. Nakagawa, Separation mechanism of pullulan solution-filled capillary electrophoresis of sodium dodecyl sulfate-proteins, Electrophoresis 17 (10) (1996) 1584–1586.

[149] B.A. Siles, Z.E. Nackerdien, G. Bruce, Collier, Analysis of DNA fragmentation using a dynamic size-sieving polymer solution in capillary electrophoresis, Journal of Chromatography A 771 (1 + 2) (1997) 319–329.

[150] B.A. Siles, G.B. Collier, D.J. Reeder, W.E. May, The use of a new gel matrix for the separation of DNA fragments: a comparison study between slab gel electrophoresis and capillary electrophoresis, Applied and Theoretical Electrophoresis 6 (1) (1996) 15–22.

[151] S. Magnusdottir, J.-L. Viovy, J. Francois, High resolution capillary electrophoretic separation of oligonucleotides in low-viscosity, hydrophobically end-capped polyethylene oxide with cubic order, Electrophoresis 19 (10) (1998) 1699–1703.

[152] M.N. Albarghouthi, B.A. Buchholz, E.A.S. Doherty, F.M. Bogdan, H. Zhou, A.E. Barron, Impact of polymer hydrophobicity on the properties and performance of DNA sequencing matrices for capillary electrophoresis, Electrophoresis 22 (4) (2001) 737–747.

[153] L. Song, D. Liang, J. Kielescawa, J. Liang, E. Tjoe, D. Fang, et al., DNA sequencing by capillary electrophoresis using copolymers of acrylamide and N, N-dimethylacrylamide, Electrophoresis 22 (4) (2001) 729–736.

[154] J. Zhang, N.T. Tran, J. Weber, C. Slim, J.L. Viovy, M. Taverna, Poly(N, N-dimethylacrylamide)-grafted polyacrylamide: a self-coating copolymer for sieving separation of native proteins by CE, Electrophoresis 27 (15) (2006) 3086–3092.

[155] W.S. Case, K.D. Glinert, S. LaBarge, L.B. McGown, Guanosine gel for sequence-dependent separation of polymorphic ssDNA, Electrophoresis 28 (17) (2007) 3008–3016.

[156] A. Szilagyi, G.K. Bonn, A. Guttman, Capillary gel electrophoresis analysis of G-quartet forming oligonucleotides used in DNA-protein interaction studies, Journal of Chromatography A 1161 (1–2) (2007) 15–21.

[157] M. Ogiso, N. Minoura, T. Shinbo, T. Shimizu, DNA detection system using molecularly imprinted polymer as the gel matrix in electrophoresis, Biosensors & Bioelectronics 22 (9–10) (2007) 1974–1981.

[158] F. Wan, J. Zhang, A. Lau, S. Tan, C. Burger, B. Chu, Nanostructured copolymer gels for dsDNA separation by CE, Electrophoresis 29 (23) (2008) 4704–4713.

[159] D. Zhou, L. Yang, R. Yang, W. Song, S. Peng, Y. Wang, Novel quasi-interpenetrating network/functionalized multi-walled carbon nanotubes double-network composite matrices for DNA sequencing by CE, Electrophoresis 29 (23) (2008) 4637–4645.

[160] C.L. Crihfield, L.A. Holland, Protein sieving with capillary nanogel electrophoresis, Analytical Chemistry 93 (3) (2021) 1537–1543.
[161] T. Lu, L.J. Klein, S. Ha, R.R. Rustandi, High-resolution capillary electrophoresis separation of large RNA under non-aqueous conditions, Journal of Chromatography A 1618 (2020) 460875.
[162] H. Zhang, M.J. Wirth, Electromigration of single molecules of DNA in a crystalline array of 300-nm silica colloids, Analytical Chemistry 77 (5) (2005) 1237–1242.
[163] D. Zhou, Y. Wang, R. Yang, W. Zhang, R. Shi, Effects of novel quasi-interpenetrating network/gold nanoparticles composite matrices on DNA sequencing performances by CE, Electrophoresis 28 (17) (2007) 2998–3007.
[164] R. Kuhn, S. Hofstetter-Kuhn, Capillary Electrophoresis: Principles and Practice, Berlin, Springer Laboratory, 1993.
[165] V.M. Starov, Y.E. Solomentsev, Influence of gel layers on electrokinetic phenomena. 1. Streaming potential, Journal of Colloid and Interface Science 158 (1) (1993) 159–165.
[166] P.G. Righetti, C. Gelfi, B. Verzola, L. Castelletti, The state of the art of dynamic coatings, Electrophoresis 22 (4) (2001) 603–611.
[167] J. Horvath, V. Dolnik, Polymer wall coatings for capillary electrophoresis, Electrophoresis 22 (4) (2001) 644–655.
[168] L. Hajba, A. Guttman, Recent advances in column coatings for capillary electrophoresis of proteins, TrAC Trends in Analytical Chemistry 90 (2017) 38–44.
[169] J. Liu, V. Dolnik, Y.Z. Hsieh, M. Novotny, Experimental evaluation of the separation efficiency in capillary electrophoresis using open tubular and gel-filled columns, Analytical Chemistry 64 (13) (1992) 1328–1336.
[170] M.J. Gordon, K.J. Lee, A.A. Arias, R.N. Zare, Protocol for resolving protein mixtures in capillary zone electrophoresis, Analytical Chemistry 63 (1) (1991) 69–72.
[171] K.A. Cobb, V. Dolnik, M. Novotny, Electrophoretic separations of proteins in capillaries with hydrolytically stable surface-structures, Analytical Chemistry 62 (22) (1990) 2478–2483.
[172] M.D. Zhu, R. Rodriguez, D. Hansen, T. Wehr, Capillary electrophoresis of proteins under alkaline conditions, Journal of Chromatography 516 (1) (1990) 123–131.
[173] D. Bentrop, J. Kohr, H. Engelhardt, Poly(methylglutamate)-coated surfaces in HPLC and CE, Chromatographia 32 (3–4) (1991) 171–178.
[174] M. Strege, A. Lagu, Separation of DNA restriction fragments by capillary electrophoresis using coated fused silica capillaries, Analytical Chemistry 63 (13) (1991) 1233–1236.
[175] J.A. Lux, H.F. Yin, G. Schomburg, A simple method for the production of gel-filled capillaries for capillary gel electrophoresis, Journal of High Resolution Chromatography 13 (6) (1990) 436–437.
[176] R.M. McCormick, Capillary zone electrophoretic separation of peptides and proteins using low pH buffers in modified silica capillaries, Analytical Chemistry 60 (21) (1988) 2322–2328.
[177] C.Y. Liu, Stationary phases for capillary electrophoresis and capillary electrochromatography, Electrophoresis 22 (4) (2001) 612–628.
[178] M.L. Lee, B.W. Wright, Preparation of glass-capillary columns for gas-chromatography, Journal of Chromatography 184 (3) (1980) 235–312.
[179] L. Gao, S. Liu, Cross-linked polyacrylamide coating for capillary isoelectric focusing, Analytical Chemistry 76 (24) (2004) 7179–7186.
[180] M. Cretich, M. Chiari, G. Pirri, A. Crippa, Electroosmotic flow suppression in capillary electrophoresis: chemisorption of trimethoxy silane-modified polydimethylacrylamide, Electrophoresis 26 (10) (2005) 1913–1919.

[181] S.F.Y. Li, Capillary Electrophoresis, Elsevier, Amsterdam, 1993.
[182] C. Barta, M. Sasvari-Szekely, A. Guttman, Simultaneous analysis of various mutations on the 21-hydroxylase gene by multi-allele specific amplification and capillary gel electrophoresis, Journal of Chromatography A 817 (1–2) (1998) 281–286.
[183] P.C.H. Shieh, D. Hoang, A. Guttman, N. Cooke, Capillary sodium dodecyl sulfate gel electrophoresis of proteins. I. Reproducibility and stability, Journal of Chromatography A 676 (1) (1994) 219–226.
[184] P. Shieh, Coated capillary columns and electrophoretic separation methods for their use, USA, 1995.
[185] G.J.M. Bruin, J.P. Chang, R.H. Kuhlman, K. Zegers, J.C. Kraak, H. Poppe, Capillary zone electrophoretic separations of proteins in polyethylene glycol-modified capillaries, Journal of Chromatography 471 (1989) 429–436.
[186] M. Nakatani, A. Shibukawa, T. Nakagawa, Sodium dodecyl sulfate-polyacrylamide solution-filled capillary electrophoresis of proteins using stable linear polyacrylamide-coated capillary, Biological & Pharmaceutical Bulletin 16 (12) (1993) 1185–1188.
[187] A. Paulus, D. Huesken, DNA digest analysis with capillary electrophoresis, Electrophoresis 14 (1–2) (1993) 27–35.
[188] D. Schmalzing, C.A. Piggee, F. Foret, E. Carrilho, B.L. Karger, Characterization and performance of a neutral hydrophilic coating for the capillary electrophoretic separation of biopolymers, Journal of Chromatography 652 (1) (1993) 149–159.
[189] M. Mizuno, K. Tochigi, M. Taki, Polyethylene-glycol coated capillary for capillary electrophoresis, Bunseki Kagaku 41 (10) (1992) 485–489.
[190] X.L. Ren, Y.F. Shen, M.L. Lee, Poly(ethylene-propylene glycol)-modified fused-silica columns for capillary electrophoresis using epoxy resin as intermediate coating, Journal of Chromatography A 741 (1) (1996) 115–122.
[191] A. Fridstroem, N. Lundell, L. Nyholm, K.E. Markides, Polymethacryloxypropylhydrosiloxane deactivation as pretreatment of polymer-coated fused silica columns for capillary electrophoresis, Journal of Microcolumn Separations 9 (2) (1997) 73–80.
[192] W. Nashabeh, Z. El Rassi, Capillary zone electrophoresis of proteins with hydrophilic fused-silica capillaries, Journal of Chromatography 559 (1–2) (1991) 367–383.
[193] D. Belder, A. Deege, H. Husmann, F. Kohler, M. Ludwig, Cross-linked poly(vinyl alcohol) as permanent hydrophilic column coating for capillary electrophoresis, Electrophoresis 22 (17) (2001) 3813–3818.
[194] H. Wan, M. Ohman, L.G. Blomberg, Bonded dimethylacrylamide as a permanent coating for capillary electrophoresis, Journal of Chromatography. A 924 (1–2) (2001) 59–70.
[195] M. Chiari, M. Nesi, M. Fazio, P.G. Righetti, Capillary electrophoresis of macromolecules in 'syrupy' solutions: facts and misfacts, Electrophoresis 13 (9–10) (1992) 690–697.
[196] J. Bodnar, L. Hajba, A. Guttman, A fully automated linear polyacrylamide coating and regeneration method for capillary electrophoresis of proteins, Electrophoresis 37 (23–24) (2016) 3154–3159.
[197] G. Kleindienst, C.G. Huber, D. Gjerde, T.L. Yengoyan, G.K. Bonn, Capillary electrophoresis of peptides and proteins in fused-silica capillaries coated with derivatized polystyrene nanoparticles, Electrophoresis 19 (1998) 262–269.
[198] H. Tian, L.C. Brody, D. Mao, J.P. Landers, Effective capillary electrophoresis-based heteroduplex analysis through optimization of surface coating and polymer networks, Analytical Chemistry 72 (21) (2000) 5483–5492.

[199] J. Ren, A. Ulvik, H. Refsum, P.M. Ueland, Applications of short-chain polydimethylacrylamide as sieving medium for the electrophoretic separation of DNA fragments and mutation analysis in uncoated capillaries, Analytical Biochemistry 276 (1999) 188–194.
[200] M. Chiari, M. Cretich, J. Horvath, A new absorbed coating for DNA fragment analysis by capillary electrophoresis, Electrophoresis 21 (2000) 1521–1526.
[201] M.X. Huang, J. Plocek, M.V. Novotny, Hydrolytically stable cellulose-derivative coatings for capillary electrophoresis of peptides, proteins and glycoconjugates, Electrophoresis 16 (3) (1995) 396–401.
[202] J.L. Liao, J. Abramson, S. Hjerten, A highly stable methyl cellulose coating for capillary electrophoresis, Journal of Capillary Electrophoresis 2 (4) (1995) 191–196.
[203] Y. Mechref, Z. Elrassi, Fused-silica capillaries with surface-bound dextran layer cross-linked with diepoxypolyethylene glycol for capillary electrophoresis of biological substances at reduced electroosmotic flow, Electrophoresis 16 (4) (1995) 617–624.
[204] C.L. Ng, H.K. Lee, S.F.Y. Li, Prevention of protein adsorption on surfaces by polyethylene oxide-polypropylene oxide-polyethylene oxide triblock copolymers in capillary electrophoresis, Journal of Chromatography A 659 (2) (1994) 427–434.
[205] P. Zhang, J. Ren, Study of polydimethylacrylamide- and polydiethylacrylamide-adsorbed coatings on fused silica capillaries and their application in genetic analysis, Analytica Chimica Acta 507 (2) (2004) 179–184.
[206] T. Tanaka, S.T. Sun, Y. Hirokawa, S. Katayama, J. Kucera, Y. Hirose, et al., Mechanical instability of gels at the phase transition, Nature (London, UK) 325 (6107) (1987) 796–798.
[207] G. Schomburg, Problems and achievements in the instrumentation and column technology for chromatography and capillary electrophoresis, Chromatographia 30 (9–10) (1990) 500–508.
[208] A. Paulus, J.I. Ohms, Analysis of oligonucleotides by capillary gel electrophoresis, Journal of Chromatography 507 (1990) 113–123.
[209] Y. Baba, T. Matsuura, K. Wakamoto, M. Tsuhako, A simple method for the preparation of polyacrylamide gel filled capillaries for high performance separation of polynucleotides by using capillary gel electrophoresis, Chemistry Letters 3 (1991) 371–374.
[210] Y. Baba, T. Matsuura, K. Wakamoto, M. Tsuhako, Comparison of high-performance liquid chromatography with capillary gel electrophoresis in single-base resolution of polynucleotides, Journal of Chromatography 558 (1) (1991) 273–284.
[211] V. Dolnik, K.A. Cobb, M. Novotny, Preparation of polyacrylamide gel-filled capillaries for capillary electrophoresis, Journal of Microcolumn Separations 3 (2) (1991) 155–159.
[212] T. Wang, G.J. Bruin, J.C. Kraak, H. Poppe, Preparation of polyacrylamide gel-filled fused-silica capillaries by photopolymerization with riboflavin as the initiator, Analytical Chemistry 63 (19) (1991) 2207–2208.
[213] G.J.M. Bruin, T. Wang, X. Xu, J.C. Kraak, H. Poppe, Preparation of polyacrylamide gel-filled capillaries by photopolymerization for capillary electrophoresis, Journal of Microcolumn Separations 4 (5) (1992) 439–448.
[214] W.A. MacCrehan, H.T. Rasmussen, D.M. Northrop, Size-selective capillary electrophoresis (SSCE) separation of DNA fragments, Journal of Liquid Chromatography 15 (6–7) (1992) 1063–1080.
[215] Y. Baba, T. Matsuura, K. Wakamoto, Y. Morita, Y. Nishitsu, M. Tsuhako, Preparation of polyacrylamide gel filled capillaries for ultrahigh resolution of polynucleotides by capillary gel electrophoresis, Analytical Chemistry 64 (11) (1992) 1221–1225.

[216] A.E. Barron, W.M. Sunada, H.W. Blanch, The use of coated and uncoated capillaries for the electrophoretic separation of DNA in dilute polymer solutions, Electrophoresis 16 (1) (1995) 64–74.
[217] H.-T. Chang, E.S. Yeung, Poly(ethyleneoxide) for high-resolution and high-speed separation of DNA by capillary electrophoresis, Journal of Chromatography B: Biomedical Applications 669 (1) (1995) 113–123.

CHAPTER FOUR

Instrumentation

As was briefly discussed in Chapter 1, Introduction, the very basic components of a simple capillary gel electrophoresis (CGE) system (Fig. 1.2) comprise the separation capillary, the inlet and outlet buffer reservoirs, the sample vials, the high-voltage power supply, and the detector, which is connected to a data acquisition and handling system, such as a personal computer (PC). Most commercial systems also provide a thermostating option to adjust the capillary temperature for separation optimization and dissipate excess Joule heat if necessary. In most instances, the separation capillary is pretreated before filling in the separation matrix. Even with the use of covalently coated capillaries (Section 3.3.1), at least a wash step is added to the analysis protocol. After filling the separation capillary with the sieving matrix (gel), the sample is injected from the inlet side by either hydrodynamic or electrokinetic means. Then the inlet and outlet ends of the gel-filled separation capillary are immersed into the buffer reservoirs, which in most instances contain the same gel-buffer system as the capillary. By the application of the electric field, the sample components are migrating according to their hydrodynamic volume to charge ratio toward the corresponding electrodes. If fraction collection is involved in the protocol, the outlet reservoir is replaced by a sampling vial for the duration of the collection process. The sampling vial either contains the separation buffer or just water. Since CGE is mostly used for the separation of negatively charged biopolymers such as nucleic acids, SDS—protein complexes, or tagged carbohydrates, the polarity of the applied electric field is reversed, that is, the cathode is at the inlet side and the anode is at the outlet side. Size separation of the analyte molecules is provided by the sieving capability of the separation gel. The resulting electropherogram is recorded and analyzed by a data acquisition system.

4.1 Sample introduction methods

In liquid-phase separation systems, the sample is usually injected onto the separation column, such as in high-performance liquid chromatography

Capillary Gel Electrophoresis.
DOI: https://doi.org/10.1016/B978-0-444-52234-4.00005-2

(HPLC) using a valve injector system with a precise volume. In CGE there is no flow due to the anticonvective characteristics of the sieving matrix, thus the sample components are either introduced by positive or negative pressure, that is, replacing a small portion of the gel in the column or by starting the electrophoresis process from the sample vial. Hydrodynamic injection is only a viable option if the viscosity of the sieving matrix allows. In case of higher viscosity matrices, electrophoretic injection is the only option. Please note that in that case the electrokinetic term does not cover electroosmosis as one of the electrokinetic phenomena, only electrophoresis. In both instances of hydrodynamic or electrophoretic injection, only a small amount of sample is introduced into the narrow bore fused silica column; however, the sample vial still should contain several microliter samples in order to accommodate the insertion of the separation capillary and the platinum electrode for the process. Again, in the case of hydrodynamic injection, a small plug of the sample solution is pushed into the inlet side of the separation capillary. The usual volume of this sample plug is in the low nanoliter range. Please note that this is several orders of magnitude less than that of in HPLC. Upon immersing the inlet end of the capillary to the running gel-buffer reservoir after the injection process, the electrophoresis is started by the application of the electric field. During electrokinetic injection, the electric field–mediated separation process is already starting from the sample vial and the duration of this setup defines the amount of sample introduced into the separation gel in the narrow bore column. The applied electric field strength of electrokinetic injection is usually different from that of during the separation step, in most instances lower. Similar to hydrodynamic injection, once the sample is introduced into the separation capillary, the inlet end of the column is immersed into the inlet gel-buffer reservoir and the electrophoresis separation process is started by applying the separation voltage. In electrokinetic sample introduction mode, the term "injection volume" is not appropriate, since it is mediated by starting the electrophoretic process. Therefore the introduced sample amount is dependent on the injection voltage and the conductivities of the sample and gel-buffer system.

In both injection methods, the length of the sample plug has a great influence on the resulting sample zone during electromigration. Minimizing this sample zone is beneficial for obtaining high separation efficiencies and concomitantly high resolution with a given selectivity. Spreading of the sample zone defines the width of the electromigrating solute components, that is, the variance (σ^2). The injection variance part of this (σ^2_{inj}) is calculated from the length of the injection plug (ℓ) as suggested by Terabe et al. [1],

$$\sigma^2_{inj} = \ell^2_{inj}/12 \tag{4.1}$$

As a rule of thumb, the length of the injection sample plug should be less than 1% of the capillary length to obtain the highest possible efficiency. In addition, the sample concentration should be at least 100 × less than that of the gel-buffer system to avoid mobility matching mediated fronting or tailing effects [2]. To obtain the total variance, the injection variance is added to the other separation associated variances such as diffusion (σ^2_{diff}) and detection (σ^2_{det}) and other dispersion effects, for example, gel matrix inconsistencies (σ^2_{other}),

$$\sigma^2 = \sigma^2_{inj} + \sigma^2_{det} + \sigma^2_{diff} + \sigma^2_{other} \tag{4.2}$$

Alternative injection methods for CGE have been developed in the early 1990s including a special electrode/receptacle that reduced the volume of the required sample to only a few microliters [3]. This was further developed in the so-called "nanovial" format [4,5].

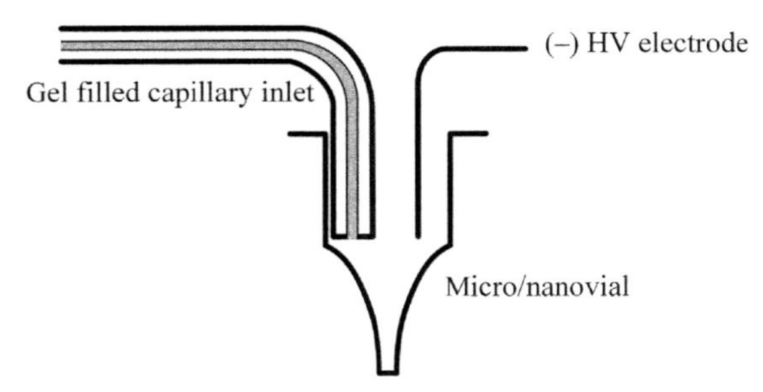

Scheme 4.1 Schematic drawing of a micro/nanovial design. Adapted from S.F.Y. Li, Capillary Electrophoresis: Principles, Practice and Application, Journal of Chromatographic Library 52 1992. pp. 48 [5].

Special, less frequently applied injection systems have also been reported. In one example, the utilization of a standard HPLC syringe was attempted with an adjustable split-vent to inject small sample volumes into the separation column [6]. In another example, a unique four-step injection process was proposed by Schomburg and coworkers [7], utilizing a sample introduction piston in the grounding unit that was lowered after switching off the membrane pump for a programmable time interval. The hydrostatic flow introduced the defined sample volume into the separation capillary. Performing injection from a sample vial by pressure or using the electrokinetic approach, sampling via microdialysis and directly from solid tissues have also been suggested as options [8]. Uchiyama and coworkers applied an inkjet setting for quantitative sample introduction for capillary electrophoresis (CE) coupled with stacking and sweeping online concentration techniques [9]. The injection amount was precisely controlled by manipulating the number of ejected sample droplets. Compared to conventional injection methods, the inkjet injection method resulted in the same electropherogram as for hydrodynamic injection mode. In addition, no sampling bias was observed that otherwise represent and issue with electrokinetic injection modes.

Other interesting high-throughput injection approaches include the work of Williams et al. [10] who developed a sequential injection protocol to increase sample throughput for oligonucleotides by approximately 4-fold over that of traditional CGE. The method was further developed by Britz-McKibbin and coworkers introducing multisegment injection [11] and later toward large-scale analysis of therapeutic antibody N-glycans by separation window-dependent multiple injection [12].

To accommodate CE injection with high-throughput approaches, an ultra-slim laminated capillary array unit was developed capable of simultaneously analyzing samples from 16 wells [13]. The authors placed two cylindrical capillaries between two polyimide sheets and laminated a 16-lane capillary array that was capable of direct injection from a sample plate. Using a replaceable sieving matrix, high-speed ssDNA fragment analysis was demonstrated in the size range of 50–500 bases.

4.1.1 Electrokinetic injection into high-viscosity gels

In case of using high-viscosity gel formulations in the separation capillary, hydrodynamic injection is usually not applicable, thus electrokinetic injection should be used. As emphasized earlier, among the electrokinetic phenomena, in this case only electrophoresis is applicable as in most gel formulations the electroosmosis is suppressed by the anticonvective properties of the separation gels. During this injection type, the inlet end of the capillary tubing is immersed into the sample vial and by applying the appropriate voltage the separation is starting in a way that the sample components migrate into the gel-filled capillary (Fig. 4.1). After introducing the required amount of sample into the narrow bore column, the

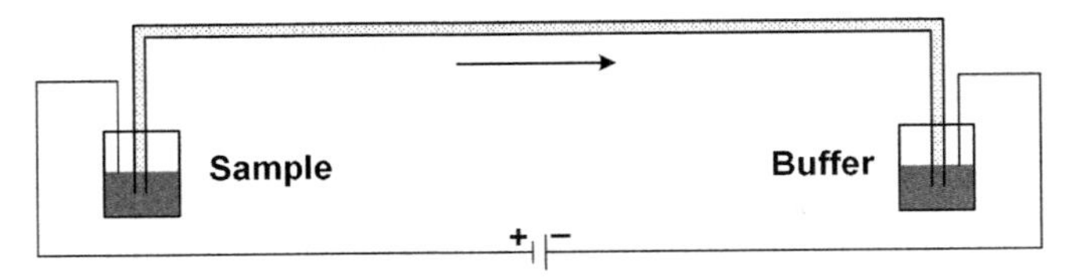

Figure 4.1 Electrokinetic injection in capillary gel electrophoresis.

electrophoresis process is continued by changing the inlet reservoir for the gel-buffer system or the actual separation buffer, whichever is preferred. Please note that CGE is mostly used for size separation of negatively charged polyionic molecules (e.g., DNA, RNA, SDS–protein, and labeled glycans); therefore the applied separation field is in reversed polarity mode, that is, the components electromigrate from the negative towards the positive electrode. The injection amount (QI) in electrokinetic injection into gel-filled capillary columns can be calculated by Eq. (4.3),

$$QI = Ectr^2\pi\mu \tag{4.3}$$

where E is the applied electric field strength, c is the sample concentration, t is the injection time, and μ is the electrophoretic mobility of the solute molecule. Xu et al. investigated the effect of the loading conditions (electric field strength, time) during electrokinetic injections on the injected amount and zone width in CGE of DNA [14]. They found that the zone width in CGE mainly depended on injection zone and other effects such as diffusion adsorption–desorption were not significant.

Please note that in CGE utilizing reversed polarity injection/separation mode (from negative to the positive electrode) with no or very little electroosmotic flow, neutral and positively charged components do not enter the capillary during electrokinetic injection. However, as Eq. (4.3) suggests, each component will enter the gel-filled capillary according to its actual electrophoretic mobility. In other words, higher surface charge density components will be overrepresented, while lower surface charge density solute molecules will be underrepresented. Zare and coworkers [15] thoroughly investigated this injection bias, caused by electrokinetic sample introduction, and concluded that it was proportional to the effective mobilities of the sample ions. As this bias was found to be quite systematic, the authors suggested a correction factor for differential mobilities by migration time normalization. Experimental and theoretical treatment of the injection bias during DNA fragment analysis by CGE in liquefied agarose gel was described by Kleparnik et al. [16] evaluating both electrokinetic and hydrodynamic injection methods. The observed size-based injection bias was explained by the different fragment mobilities in the sieving medium due to some possible intrinsic EOF in the agarose gel used. Even if the electrophoretic mobilities of the different size DNA fragments were the same in free solution, size-based bias was brought about by the different mobilities in a sieving medium as shown in Fig. 4.2.

Carrilho and coworkers studied the influence of sample composition with varying electrokinetic injection conditions with respect to reproducibility and separation performance [17]. By using a simplex optimization approach, best conditions were found after nine experiments. Bartlett and coworkers investigated the issues related to electrokinetic injection for oligonucleotides [18] observing the following points: (1) The relationship between the injection amount and the sample solution resistance is not always as linear for oligonucleotides as for small molecules. (2) Injecting a water plug prior to the oligonucleotide sample dramatically improved reproducibility. (3) The injected amount of oligonucleotides significantly increased by optimizing the composition and concentration of the sieving gel.

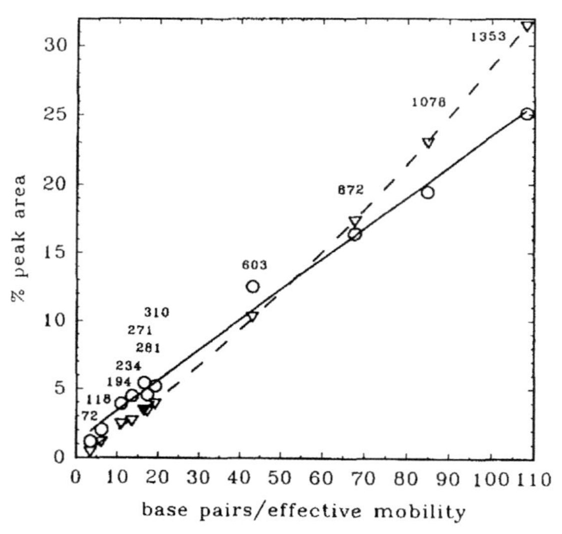

Figure 4.2 Dependence of the peak area percentage on the ratio of number of base pairs (bp) to the effective electrophoretic mobility. ∇—after pressure and o—after electromigration-based injections. *With permission from K. Kleparnik, M. Garner, P. Bocek, Injection bias of DNA fragments in capillary electrophoresis with sieving, Journal of Chromatography A 698 (1–2) (1995) 375–383 [16].*

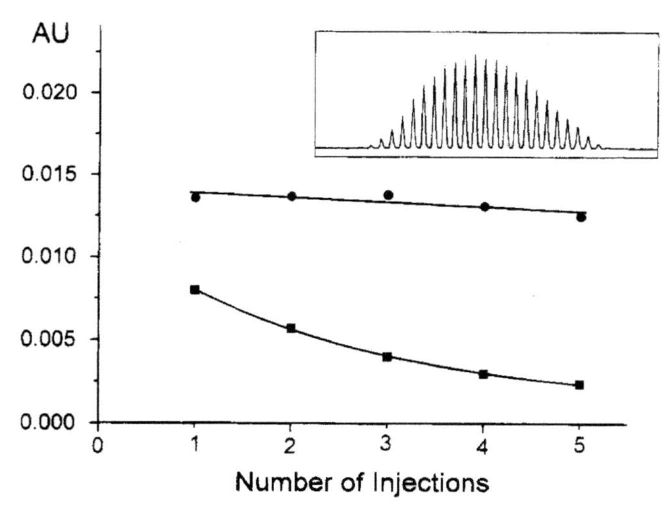

Figure 4.3 Peak height of the $p(dA)_{50}$ oligonucleotide versus the number of consecutive injections. Inset: Capillary gel electrophoresis separation of the $p(dA)_{40-60}$ test mixture. ■—regular electrokinetic injection and ●—water preinjection. *With permission from A. Guttman, H.E. Schwartz, Artifacts related to sample introduction in capillary gel electrophoresis affecting separation performance and quantitation, Analytical Chemistry 67 (13) (1995) 2279–2283 [19].*

One of the major issues of electrokinetic injection is sample depletion in case of consecutive injection from the same sample vial. Progressively smaller peak heights were reportedly obtained with each injection of an oligonucleotide mixture into a gel-filled capillary [19]. As Fig. 4.3 shows, an exponential decrease was observed in UV absorbance of the largest peak, $p(dA)_{50}$, of the polydeoxyadenylic acid sample mixture over five consecutive runs. Interestingly, a water preinjection step greatly alleviated this phenomenon.

The two-step injection procedure including the water preinjection resulted not only in dramatically increased precision (important for quantitative studies) but also in increased sample loading as will be discussed in detail in the following section.

4.1.2 Hydrodynamic injection into low-viscosity polymer solutions

Low-viscosity sieving matrices allow the application of hydrodynamic (pressure or vacuum) injection in addition to electrokinetic injection. During hydrodynamic injection, the pressure is applied to the sample containing vial and the sample is pushed into the capillary replacing the corresponding

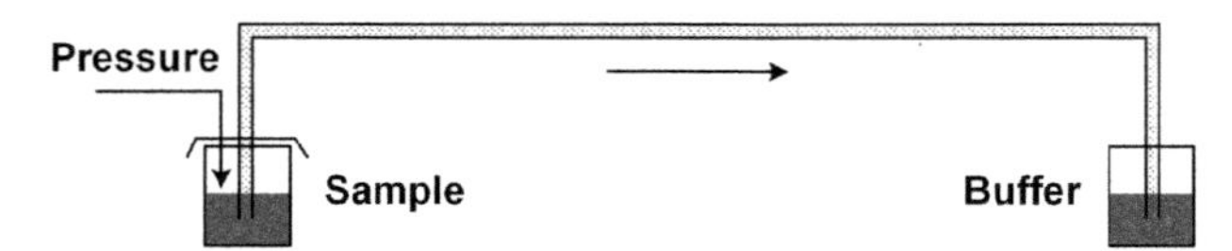

Figure 4.4 Pressure injection in capillary gel electrophoresis.

portion of the sieving matrix (Fig. 4.4). The injection volume (V) in this case can be readily calculated by the Hagen–Poiseuille equation [20]:

$$V = \frac{\Delta P r^2 \pi t}{8\eta L} \tag{4.4}$$

where ΔP is the pressure applied for t time from the injection side, r is the inner radius of the capillary column, η is the viscosity of the separation matrix, and L is the total length of the separation column. The length of the injection plug can be derived by dividing the injection volume (V) by the surface area of the capillary column ($r^2\pi$). Pressure injection into narrow bore capillary columns results in a parabolic profile, thus increases injection-related dispersion.

In case of vacuum injection, negative pressure is applied at the opposite end of the capillary to draw the sample into the column. Since in CGE, the separation capillary is filled by viscous gels, siphoning effects are negligible. Hydrodynamic injection is not biased and reproducible introduces the exact sample composition into the capillary [21]. However, leakage in the sampling vial seal may alter the applied pressure, leading to reproducibility issues. Indeed, one of the problems with hydrodynamic injection is the difficulty of precisely controlling the positive or negative pressure and the precision of the system may be degraded with the age of the instrument [22]. Bernard and Loge used poly(2-ethyl-2-oxazoline) (MW 500,000 g/mole) polymer solutions as a low-viscosity sieving matrix with concentrations between 6% and 12% w/v for SDS-CGE of proteins with hydrodynamic injection [23]. This sieving matrix had excellent separation efficiency and because of its low viscosity it readily supported hydrodynamic injection of the samples.

4.1.3 Sample preconcentration

In the instances of very dilute samples, it is important to apply some kind of a sample preconcentration technique before CE analysis to obtain the reasonable detection limit for the analyte molecules. One of the traditional methods utilizes electric field enhancement in low conductivity zones, in which case sample ions are "stacked" against the interface between the low and high conductivity buffer zones, therefore enter the separation section of the gel-filled capillary as a very sharp band (Fig. 4.5). In other words, electric field–amplified injection is based on the principle that higher electric fields result in higher migration velocity. At lower conductivity zones the potential drop is greater, causing a larger driving force for the electromigration. At higher conductivity zones, on the other hand, the velocity of the solute molecules is slower due to the shallower potential drop. Therefore if the sample is in a lower conductivity zone than that of the separation buffer, the solute molecules will rapidly migrate until hitting the boundary between the two zones, then their migration slows down because of the lower potential drop in the higher conductivity zone. At that point, their velocities decrease allowing significant preconcentration of up to several 1000-fold as demonstrated by Terabe and coworkers [24].

Simultaneous injection and preconcentration of charged sample ions were demonstrated by careful selection of separation conditions [25] where the actual enrichment factor (EF) was proportional to the concentration ratio between the gel and sample buffers, thus could be calculated accordingly. An alternative field enhancement method was also suggested by introducing a short water plug before sample injection, which also resulted in several-100-fold increase in the injected amount [26]. Isotachophoretic (ITP) effects, developing in the

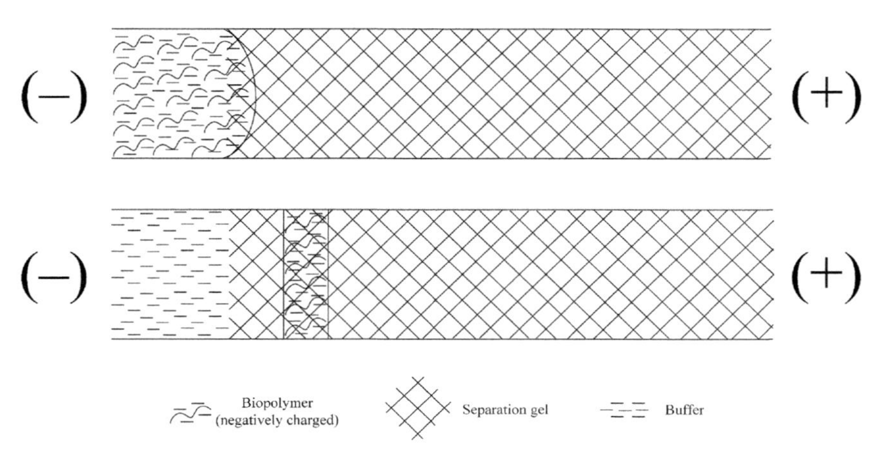

Figure 4.5 Sample stacking process in capillary gel electrophoresis. The upper panel shows the status after the pressure injection process and the lower panel the stacking-based preconcentration.

buffer zone prior and after the water plug, however, may alter their pH values [27]. The buffer ingredients, concentration, and the applied electric field strength can all influence the stacking effect and should be considered carefully. In addition, too long of a water plug leads to an extremely large voltage drop in that zone, concomitantly increasing the temperature, possibly even up to the boiling point [28]. This effect may cause thermal decomposition of sensitive sample components and bubble formation in the gel. Zhang and Meagher applied sodium dodecyl sulfate capillary gel electrophoresis (SDS-CGE) with head-column field-amplified sample stacking, an online sample preconcentration technique for the analysis of only 25 ng adeno-associated virus capsid proteins [29]. The separation sensitivity was enhanced by three orders of magnitude with the applied method, compared to the conventional SDS-CGE approach.

ITP preconcentration is another frequently used injection technique applied for dilute samples. ITP enables injection of large sample volumes, even as large as a significant portion of the separation capillary column itself. This sample injection method requires a careful selection of appropriate leading electrolytes with co-ions having greater and trailing electrolytes with co-ions having lower electrophoretic velocities than that of the sample ions, respectively. With appropriate leading and trailing electrolyte compositions, the analyte zone width can be narrowed when the electric field is applied. Another background electrolyte parameter to optimize for injection is the pH. In discontinuous buffer systems, where the pHs of the sample and the background electrolyte are different, a special pH junction develops causing the solute molecules to focus into narrow zones, based on their isoelectric points. It is important to mention that these latter methods cannot always be used with CGE mode and sometimes they require special settings, that is, partial gel filling of the capillary. A selection of important sample preconcentration methods is listed in Table 4.1.

Everaerts and coworkers developed an ITP-based system for CGE, where the transition from ITP to CGE was achieved by the mobility shift of DNA molecules from free solution (ITP part) to the sieving matrix (CGE part) [31] (Fig. 4.6). With the resulting intrinsic ITP preconcentration, large volume injections of up to 700 nL were permitted with accurate migration times in CGE, independent of injection plug length or sample ionic strength.

Foret et al. introduced ITP preconcentration strategies by applying either on-column transient ITP to inject relatively large volumes of samples and obtained a 50-fold concentration enhancement. They also developed a coupled-column system that provided a higher degree of freedom

Table 4.1 Selected electrophoretic sample preconcentration techniques used in CE.

Type of electrophoretic preconcentration techniques	Analyte charge	Resistance to sample salinity	Prevailing mode of coupling	EF
FASS (field-amplified sample stacking)	⊕ ⊖	−	Online	10^1-10^2
FESS (field-enhanced sample stacking)			online	
LVSS (large volume sample stacking)	⊕ ⊖	−	Online	10^2-10^3
FASI (field-amplified sample injection)	⊕ ⊖	−	Online	10^2-10^3
FESI (field-enhanced sample injection)			Online	
pH-driven techniques (dynamic pH junction, pH-mediated stacking)	⊕ ⊖	+	Online	10^1-10^3
Sweeping	⊕ ⊖ ○	O	Online	10^1-10^5
AFMC (analyte focusing by micelle collapse)	○	O	Online	10^1-10^2
MSS (micelle to solvent stacking)	⊕ ⊖	O	Online	10^1-10^2
ITP (isotachophoresis)	⊕ ⊖	+	Offline, inline	10^1-10^2
tITP (transient ITP)			Online	

Symbols: ⊕ / ⊖ = cations/anions, ○ = neutrals; + = very good, o = limited, − = unsuitable.
EF—enhancement factor.
Source: With permission from G. Jarvas, A. Guttman, N. Miekus, T. Baczek, S. Jeong, D.S. Chung, et al., Practical sample pretreatment techniques coupled with capillary electrophoresis for real samples in complex matrices, Trends in Analytical Chemistry 122 (2020) 115702 [30].

in the selection of separation conditions and injection of larger sample volumes. This latter arrangement resulted in a gain in detection level of at least a factor of 1000 [32]. Transient ITP is a convenient and important preconcentration method in CGE-based protein analysis using discontinuous buffer systems with appropriate leading and terminating electrolytes. Different options of transient ITP sample preconcentration methods are depicted in Fig. 4.7 [33]. In one of these approaches, the terminating electrolyte was replaced with the leading electrolyte after the focusing step and the separation was continued in zone electrophoresis mode (Panel A). Another interesting setup used only one background electrolyte, containing a low electrophoretic mobility co-ion, and the sample was supplemented with a salt of a highly mobile co-ion (Panel B). This arrangement resulted in transient ITP migration of the sample ions at the beginning of the

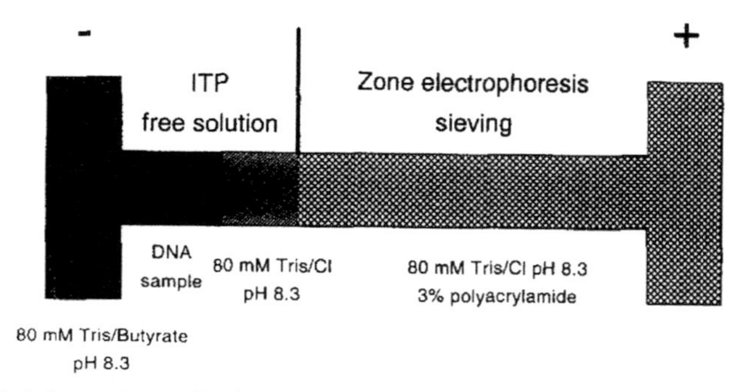

Figure 4.6 Schematics of the intrinsic isotachophoresis–CGE-based method. Preconcentration took place in free solution and separation takes place in the sieving matrix filled part of the same capillary. *With permission from M.J. van der Schans, J.L. Beckers, M.C. Molling, F.M. Everaerts, Intrinsic isotachophoretic preconcentration in capillary gel electrophoresis of DNA restriction fragments, Journal of Chromatography A 717 (1–2) (1995) 139–147 [31].*

separation and slowly changed to zone electrophoretic mode yielding detection limits of approximately 10^{-9} M [33]. A model was also developed to show the influence of sample injection time in stacking and nonstacking modes on migration time [34], correctly describing the experimental observation of increased migration time in stacking mode. Santiago's group modeled the dynamics of sample zone in ITP, especially for factors like electromigration, diffusion, buffer reactions, and nonlinear ionic strength effects [35]. Their perturbation analysis study allowed a rational approach for optimizing experimental parameters such as detector location, duration of the ITP process, and defining the right electrolyte composition to achieve high preconcentration ratio and sample accumulation rates. The same group successfully purified nucleic acids from whole blood samples in PCR-compatible form and separated from human blood lysate in only 3 minutes [36].

Recently, the ITP technology moved toward (micro)preparative purification of nucleic acids [37] in which case only small differences existed between the electrophoretic mobilities of short and large nucleic acids due to their similar surface charge densities. Kondratova et al. described the ITP method for concentration and isolation of DNA in agarose gel rods [38].

In addition to the utilization of electrophoresis principles, preconcentration techniques for dilute samples can utilize chromatographic methods as well. One of these approaches is solid-phase microextraction in conjunction with CE and tandem mass spectrometry to analyze protein complexes [39].

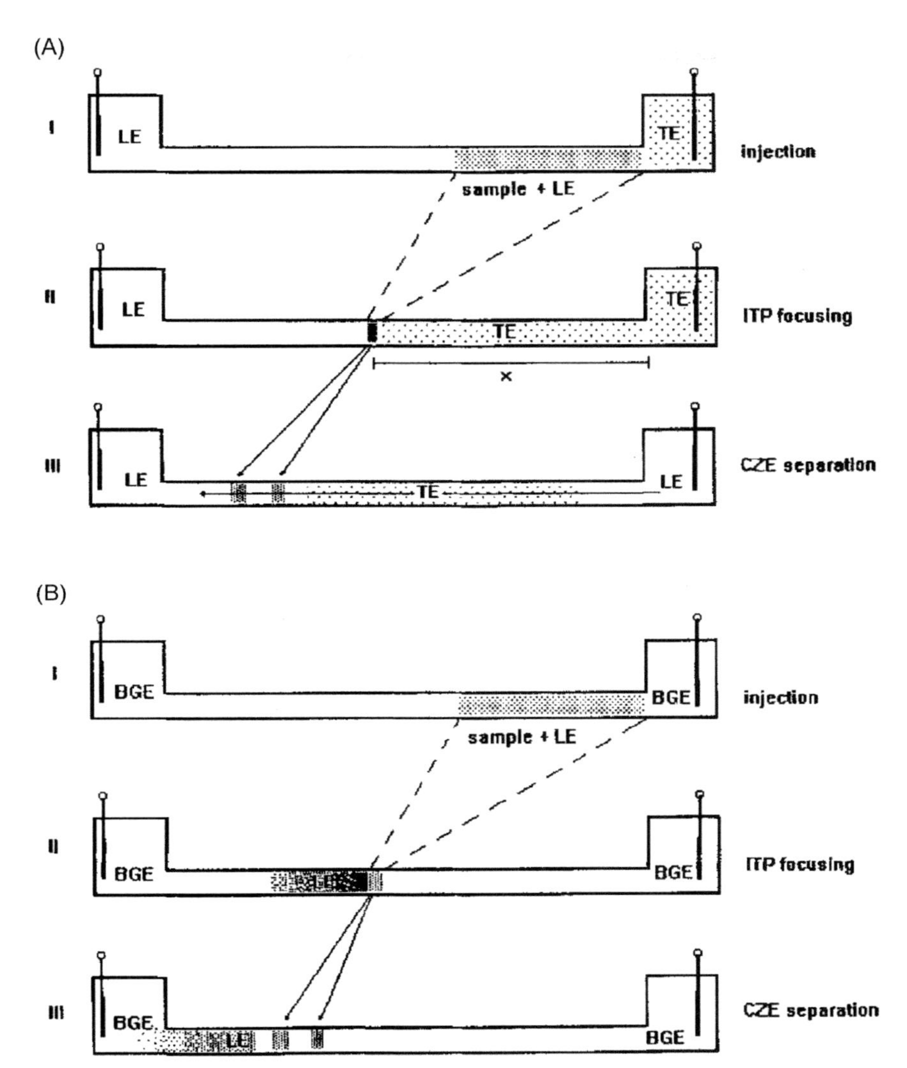

Figure 4.7 Different methods of transient ITP sample preconcentration in CE. (A) Replacement of the terminating electrolyte with the leading electrolyte and (B) sample contains high electrophoretic mobility co-ion as leading ions. *LE*, Leading electrolyte; *TE*, terminating electrolyte. *With permission from F. Foret, E. Szoko, B.L. Karger, Trace analysis of proteins by capillary zone electrophoresis with on-column transient isotachophoretic preconcentration, Electrophoresis 14 (5–6) (1993) 417–428 [33].*

Solid-phase extraction was also executed on HPLC-phase impregnated membranes for protein preconcentration [40]. While this method improved detection limits, it led to injection bias associated with the selectivity variation of the stationary phase used. Others attempted the use of reversed phase–based extraction on quartz wool and porous beads generating concentration detection limits for peptides in the mid-picomolar range [41]. An in-capillary solid-phase extraction sample cleanup setting was published in Ref. [42] that allowed preconcentration and eliminated the need to attach a precolumn prior to the separation capillary. As Fig. 4.8 shows, a 30-μm hole drilled in the capillary wall after the solid-phase extraction portion allowed the waste bypassing the separation part of the column after the preconcentration step. Similarly to the double T microchip injection method regularly utilized in

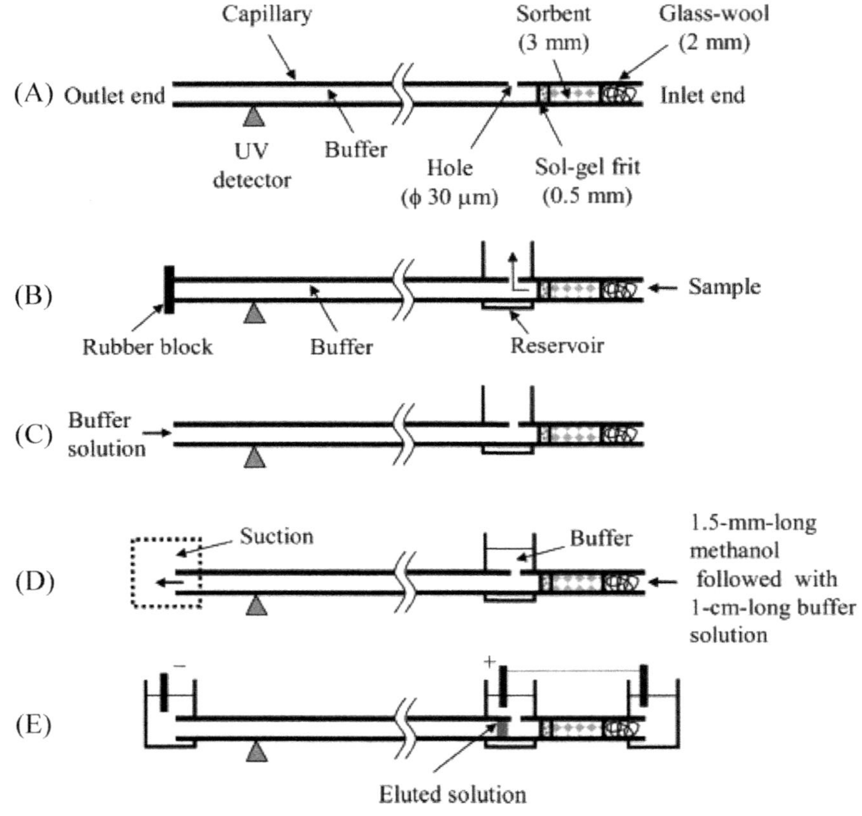

Figure 4.8 Schematics of in-capillary SPE-CE (A) and the procedure (B–E). (B) Sample loading, (C) refilling the capillary between the hole and the inlet end with buffer, (D) introducing methanol followed by buffer from the inlet end by a suction at outlet end, and (E) CE. *With permission from L.H. Zhang, X.Z. Wu, Capillary electrophoresis with in-capillary solid-phase extraction sample cleanup, Analytical Chemistry 79 (6) (2007) 2562–2569 [42].*

microfluidics formats, the electric field could be applied between the inlet of the capillary and the hole for the injection/preconcentration step and from the hole and the outlet of the capillary for the separation step.

A similar approach was reported by Wu and Umeda [43] forming a glass membrane at the injection end of the capillary by etching with HF and covered by cellulose acetate to enforce it. With this method, the sample components accumulated at the porous joint providing a preconcentration factor of several hundreds. The cellulose acetate-coated porous end was reportedly strong, stable, and applicable to CGE as well.

Inline sample pretreatment techniques based on liquid-phase microextraction (LPME) performed before sample injection and online sample preconcentration techniques during or after sample injection are discussed in a comprehensive review article of Jarvas et al. [30]. Emphasis was focused on the applicability to samples of high conductivity, commonly occurring for biological specimens. Inline LPME is a sample preconcentration technique, introduced by the Chung group, also capable of cleaning up sample matrices (including desalting) and sampling gas or solid surface substances [44]. Among various inline LPME techniques, single-drop microextraction (SDME), in-tube microextraction (ITME), and liquid extraction surface analysis (LESA) can be easily coupled with CE without any modification of existing instruments (Fig. 4.9). More importantly, due to the sample cleanup capability, these techniques can be directly applied to biological samples.

SDME uses an acceptor drop hanging on a syringe needle tip or capillary inlet end. Since the acceptor drop volume in SDME is less than a few microliters, the volume ratio between the sample donor and acceptor phases is very high, resulting in high enrichment factors (EFs). SDME can be performed in a direct immersion or headspace extraction modes. Although it represents a convenient and powerful sample pretreatment technique for CE, it often exhibits instability of the acceptor drop attached to the inlet end of the CE capillary. Careful counterbalancing of drop shrinkage due to surface tension should be performed by applying appropriate backpressures, especially for long extraction times. To overcome this disadvantage, ITME using a liquid plug inside the separation capillary as an acceptor phase has also been developed. For this method, it is not necessary to counterbalance the surface tension since no drop is attached to the capillary. Another advantage of this method is that all the extracted analytes are already inside the separation capillary. In liquid-phase microextraction, a liquid microjunction is formed on the sample

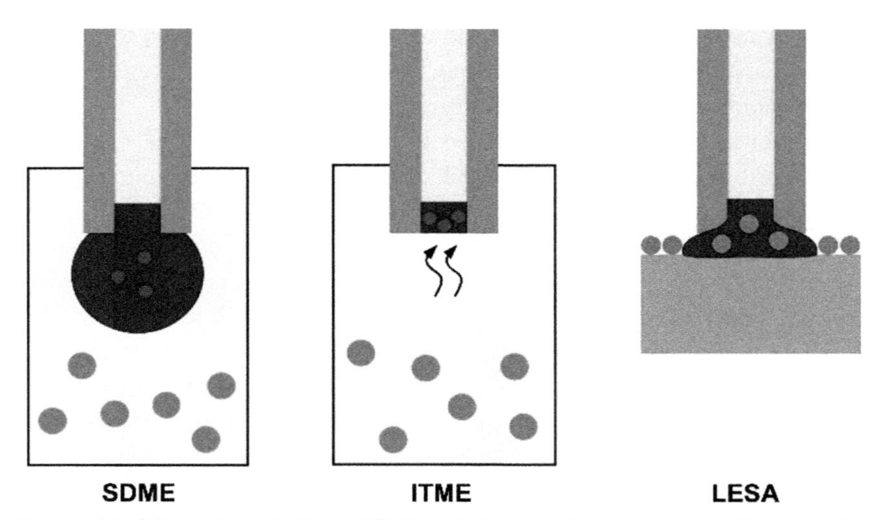

Figure 4.9 Schematics of three LPME techniques inline coupled with separation capillaries. *ITME*, In-tube microextraction; *LESA*, liquid extraction surface analysis; *SDME*, single-drop microextraction. *With permission from G. Jarvas, A. Guttman, N. Miekus, T. Baczek, S. Jeong, D.S. Chung, et al., Practical sample pretreatment techniques coupled with capillary electrophoresis for real samples in complex matrices, Trends in Analytical Chemistry 122 (2020) 115702 [30].*

surface, allowing direct sampling of the analytes without dilution and other complications [30].

Affinity-based preconcentration usually decreases sample complexity and provides improved detection sensitivity in CE analysis. Immunoaffinity capillary electrophoresis (IACE) is one of these techniques to isolate, concentrate, and analyze cell-free biomarkers and/or tissue or cell extracts from biological fluids. Isolation and concentration of analytes is accomplished through binding to one or more biorecognition affinity ligands immobilized to a solid support, while separation and analysis are achieved by high-resolution CE coupled to one or more detectors. Fig. 4.10 shows a schematic diagram of a unidirectional (A) and an orthogonal (B) design of online immunoaffinity analyte concentrator-microreactor (ACM) devices introduced by the Guzman lab [45]. The unidirectional design is operated without the presence of microvalves, whereas the orthogonal IACE design requires microvalves. This latter has four microvalves positioned at each entrance—exit port, which are crucial in controlling the fluid path through the transport or separation capillary. Biorecognition affinity capture ligands or affinity capture selectors are immobilized to a beaded or monolithic structure, positioned within the cavity

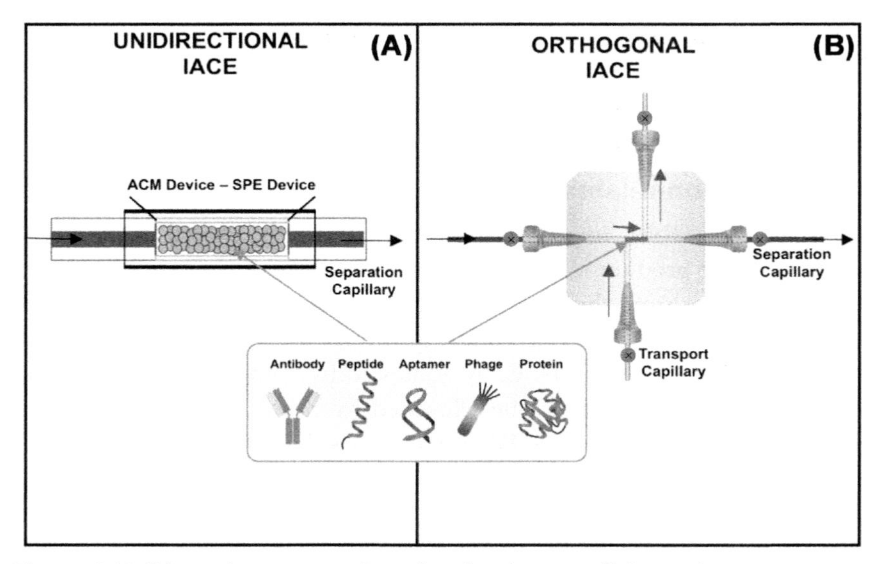

Figure 4.10 Schematic representation of online immunoaffinity analyte concentrator-microreactor (IACE) devices with unidirectional (A) and orthogonal (B) settings. *With permission from N.A. Guzman, D.E. Guzman, A two-dimensional affinity capture and separation mini-platform for the isolation, enrichment, and quantification of biomarkers and its potential use for liquid biopsy, Biomedicines 8 (8) (2020) 255 [45].*

of the ACM device, or immobilized directly to the inner surface of the cavity of the ACM. The affinity capture selectors can be antibodies, antibody fragments, lectins, aptamers, enzymes, phages, receptors, protein A, protein G, or a variety of substances having affinities for different kinds of molecules.

4.1.4 Effect of sample overloading

Sample overloading in CGE causes broader, asymmetric triangular peaks, thus decreases resolution [46]. Sample overload can affect the system efficiency by various mechanisms [47], one relates simply to the volume of sample injected into the system (V_S) relative to the total volume of the capillary column (V_C). The maximum number of theoretical plates (N_{max}) can be calculated by the following equation:

$$N_{max} = 12\left(V_C/V_S\right)^2 \quad (4.5)$$

Thus the theoretical plates can be increased by decreasing the injected volume or by increasing the length (i.e., the volume) of the capillary. Distortion of the electric field in the sample zone by the presence of the sample ions can also degrade the separation efficiency, referred to as electromigration dispersion, discussed earlier by Poppe and coworkers [46].

4.1.5 Injection-related artifacts

Reproducible DNA migration times in CGE are crucial for accurate assignment for product identification. Van der Schans et al. [48] observed DNA migration time shifts with changes in sample ionic strength and loss of resolution with increasing length of sample plug with pressure injection. The former one can be easily addressed by coinjecting of an internal standard. The latter one can be alleviated by using a 0.1 M Tris-acetate buffer preinjection before the DNA sample, as shown in Fig. 4.11.

Two injection-related artifacts were reported by Guttman and Schwartz [19] with CGE. The first was caused by consecutive injections from the same, low-volume (10–200 μL) aqueous sample, while the second one was triggered by using an oblique edge capillary end. The progressively smaller peak heights obtained with each injection of a DNA standard were alleviated by performing an intermediate electrokinetic injection of water prior to the sample injection (Fig. 4.12). The second problem was simply solved by perpendicularly polishing the end of the capillary after cutting to the right size for the analysis (Fig. 4.13).

4.2 Detection systems

Unlike in slab gel electrophoresis where the detection phase is usually decoupled from the separation phase (except for fluorescent scanning–based DNA sequencing), CGE utilizes online detectors, similar to that of in HPLC. The most frequently used detection techniques in CGE are optical based like UV/Vis absorbance, laser- or light-emitting diode (LED)-induced fluorescence, and refractive index. Less frequently used detection types include mass spectrometry, conductivity, and amperometric. Table 4.2 shows the molar (mol) and molarity (M) detection limits of these detection systems.

For optical detectors, one of the most important first steps is the proper removal of the outer polyimide coating layer of the fused silica capillaries in order to provide a transparent light path, referred to as the

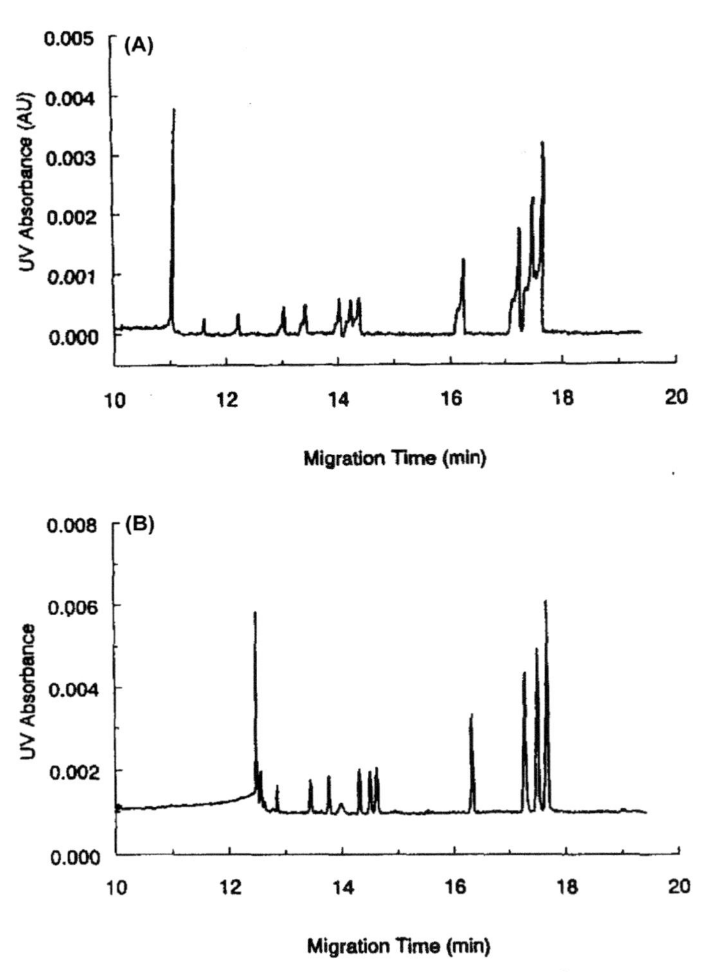

Figure 4.11 CGE separation of a φX-174 HAE III restriction fragment mixture dissolved in 20 mM NaCl without (A) and with (B) a presample injection of 0.1 M TRIS-acetate (pH 8.3). *With permission from M.J. van der Schans, J.K. Allen, B.J. Wanders, A. Guttman, Effects of sample matrix and injection plug on dsDNA migration in capillary gel electrophoresis, Journal of Chromatography A 680 (2) (1994) 511–516 [48].*

detection window. The most commonly used outer coating removal is by burning off the polyimide layer. If the inner capillary surface comprises any kind of covalent coating, burning may not be the optimal way for generating the window (i.e., damaging the inner coating), so removal of the polyimide is suggested by means of a razor blade under a microscope. There are several publications in the literature to address this issue in other

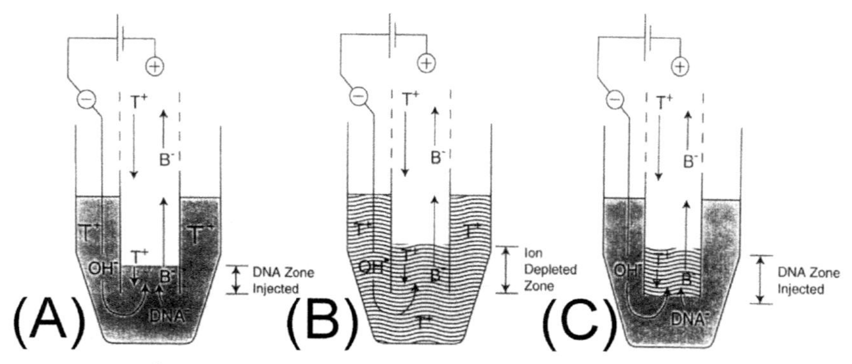

Figure 4.12 Schematic illustration of the electrokinetic sample introduction processes with intermediate electrokinetic injection of water. (A) Regular electrokinetic injection, (B) electrokinetic injection from water (water preinjection), and (C) electrokinetic sample injection after water preinjection. *With permission from A. Guttman, H. E. Schwartz, Artifacts related to sample introduction in capillary gel electrophoresis affecting separation performance and quantitation, Analytical Chemistry 67 (13) (1995) 2279–228 [19].*

ways, some of them are quick but unsophisticated [49], while others require simple devices to do so [50]. Once the detection window is made and a UV/Vis detection system is in place, the next step is to evaluate the sensitivity, noise, and linearity of the detector design, even considering effects such as refractive index changes during the analysis [51]. The general approach for UV detectors is the use of a lens to focus the light beam onto the inner cavity of the separation capillary. Extended optical path lengths (e.g., U- or Z-shaped) enable better detection limits. Another interesting way to increase path length was the use of rectangular cross section capillaries [52], which also exhibited excellent heat dissipation characteristics. Multireflection absorption cells also increase the effective optical path length and resulted in a 40-fold sensitivity enhancement [53]. Directing the light beam along the capillary axis was also attempted to increase the detection path length [54]. Additional selectivity was offered by the introduction of multiwavelength UV/Vis absorbance detection [55].

Instrumental and sample matrix factors influencing resolution, sensitivity, detectability, and quantitation are of high importance in CGE detection. In this respect, Demorest and Dubrow [56] studied the separation of oligonucleotides by CGE. The authors found that the substantial errors in the calculated percentage of individual oligonucleotides were caused by not correcting the peak areas for peak velocity. As shown in Fig. 4.14, at higher oligonucleotide concentrations, the detection response varied

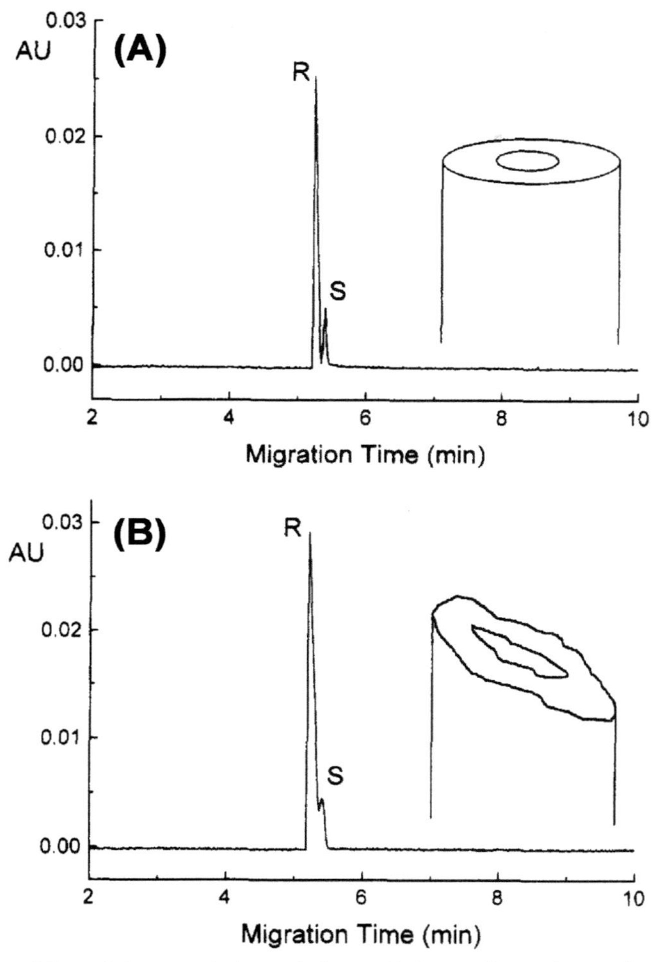

Figure 4.13 Effect of the actual physical shape of the capillary inlet on the resolution of naproxen enantiomers. (A) polished perpendicular and (B) oblique edge capillary end. *With permission from A. Guttman, H.E. Schwartz, Artifacts related to sample introduction in capillary gel electrophoresis affecting separation performance and quantitation, Analytical Chemistry 67 (13) (1995) 2279–228 [19].*

linearly with increasing sample concentration, however, at the lower concentration range the response varied anomalously, suggesting the importance of using internal standards for quantification.

The effect of electric field was investigated on the dispersion of oligonucleotides using multipoint detection [57]. A single gel-filled capillary was curved into loops, with several detection points aligned with a

Table 4.2 Detection systems used in capillary gel electrophoresis with their detection limits.

Detector	Approximate detection limits	
	Moles	**Molarity**
UV/Vis absorbance	10^{-13}–10^{-16}	10^{-5}–10^{-7}
Indirect absorbance	10^{-12}–10^{-15}	10^{-4}–10^{-6}
Fluorescence	10^{-15}–10^{-17}	10^{-7}–10^{-9}
Indirect fluorescence	10^{-14}–10^{-16}	10^{-6}–10^{-8}
Laser-induced fluorescence	10^{-18}–10^{-20}	10^{-13}–10^{-16}
Mass spectrometry	10^{-16}–10^{-17}	10^{-8}–10^{-10}
Amperometric	10^{-18}–10^{-19}	10^{-7}–10^{-10}
Conductivity	10^{-15}–10^{-16}	10^{-7}–10^{-9}
Refractive index	10^{-14}–10^{-16}	10^{-6}–10^{-8}
Radiometric	10^{-17}–10^{-19}	10^{-10}–10^{-12}

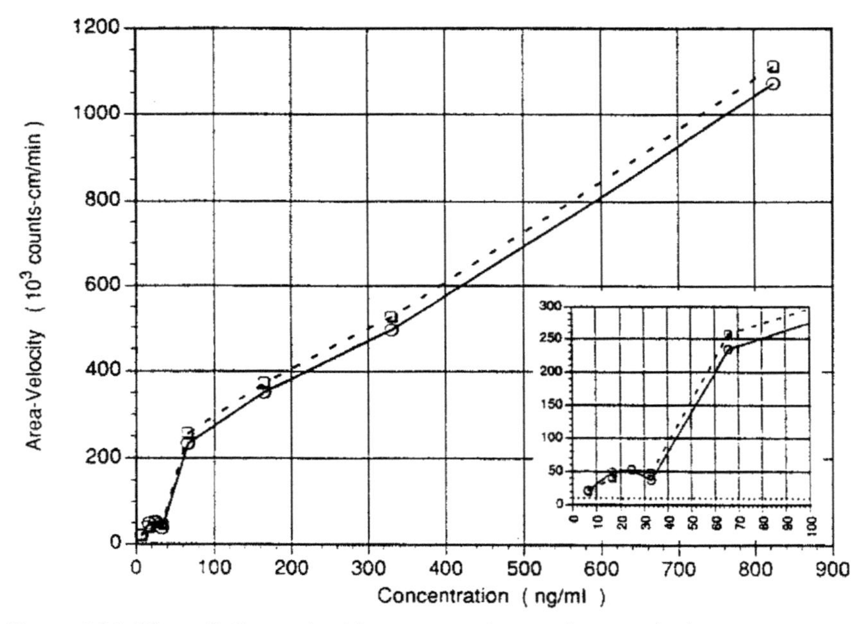

Figure 4.14 Effect of oligonucleotide concentration on the quantitative response, area times velocity at 260 nm. *With permission from D. Demorest, R. Dubrow, Factors influencing the resolution and quantitation of oligonucleotides separated by capillary electrophoresis on a gel-filled capillary, Journal of Chromatography 559 (1–2) (1991) 43–56 [56].*

common detector that allowed obtaining several electropherograms at different migration times during the electrophoresis run enabling the observation of changes in the spatial peak variance as a function of time.

Applying this method, the authors found that the diffusion coefficient of polydeoxyadenylic acids increased at higher voltages in polyacrylamide gel-filled capillaries.

4.2.1 UV absorbance and diode array detection

One of the most frequently used detection methods in CGE is UV absorbance. The small inner capillary diameter in CE only enables the use of very short optical path lengths, thus limited detection volumes resulting in restricted detection sensitivity. Light absorption detectors work with the principle of collecting as much light through the center of the capillary as possible. Given the fact that the usual inner diameter of a CE capillary is 25−100 μm, this represents a challenging task. Techniques using fiber optics and ball lenses can help to increase detection sensitivity [58]. Attempts for improvement also included the application of fiber optics to UV detection by Foret and Bocek [59] or on-column photodiode arrays [60]. The use of photodiode array detection enabled to obtain on-column absorbance spectra that could be used for the characterization of complex samples. The spectral range of such detectors is from 190 to 700 nm. Photodiode array detection enabled identification and precise characterization of separated DNA fragments [61], even with high UV background caused by the polyacrylamide gel within the capillary.

A Z-shaped detector cell configuration was also suggested to increase the optical path length and concomitantly increase absorbance. When compared to conventional on-column detection, a 3-mm cell could provide detection limits by more than one order of magnitude better [62]. The signal-to-noise ratio was improved by 6-fold for 3-mm Z-shaped flow cells and the signal increased up to 10-fold compared to regular on-column UV detection [63]. An important factor in all approaches was to optimize the aperture width for high sensitivity, while minimizing distortion of the migrating zones.

The UV transmittance of various gel-filled capillaries was compared by Macek et al. [64]. The authors found a 15% decrease at 260 nm with a $T = 6\%/C = 5\%$ cross-linked polyacrylamide gel. As Fig. 4.15 delineates, the intensity further dropped by 40% when the gel contained 20% polyethylene glycol as a stabilizer.

An interesting concept of capillary coupling was suggested by Schwarz and coworkers for the connection of a nonflow buffer-filled capillary to gel-filled capillaries as a low-background detection cell [65]. The UV detection sensitivity of the resulting capillaries was improved by at least

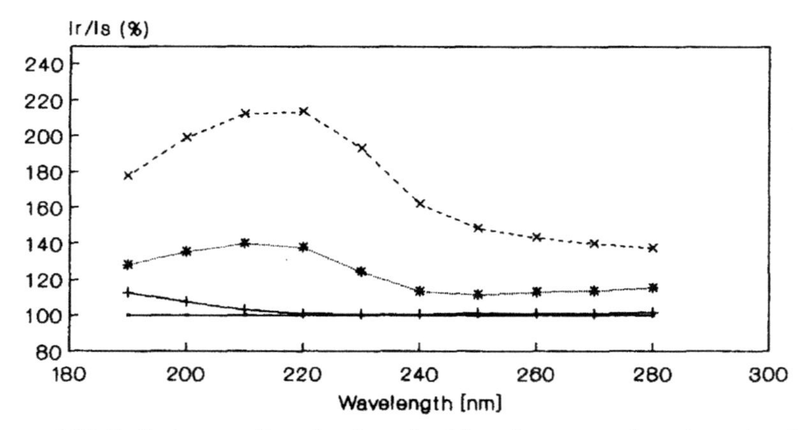

Figure 4.15 Optical properties of polyacrylamide gels measured as the ratio of the intensities between the reference and sample beam and related to water. •—Water; *—6%T/5%C gel; + —CGE buffer; x—6%T/5%C/20% polyethylene glycol gel. *With permission from J. Macek, U.R. Tjaden, J. Van der Greef, Resolution and concentration detection limit in capillary gel electrophoresis, Journal of Chromatography 545 (1) (1991) 177–182 [64].*

5× and was further enhanced by using a low-background running buffer. Fig. 4.16 depicts two of these approaches: Panel A shows the linearly coupled capillaries, while panel B shows a T-shaped capillary junction. Appropriate coating of the connected empty capillary part was crucial to maintain good resolution.

The cylindrical lens effect of the capillary walls can be eliminated with a paraxial approach using transformation matrices, which represented another option to increase detection limits. Detectors based on fluorescence and multipass absorption may benefit from this optical setup since the light remained collimated as passed through the detector [66].

Another approach to actually see electrophoretic separation on the fly was reported by using a charged coupled device (CCD) camera to image a 4-cm section of 0.2 mm i.d. capillary [67] that also allowed simultaneous monitoring of several narrower bore capillaries. A two-dimensional CCD camera-based detection system was reported by Pawliszyn's group in absorption imaging mode, one dimension representing the capillary length and the second dimension recorded the absorbance spectrum at the same time [68]. Another option this setup offered was simultaneous detection of multiple capillaries in an array format. Deep UV LED at 255 nm was also attempted for capillary absorbance detection, providing a compact detection source and compatible wavelength with a wide range of biological samples [69].

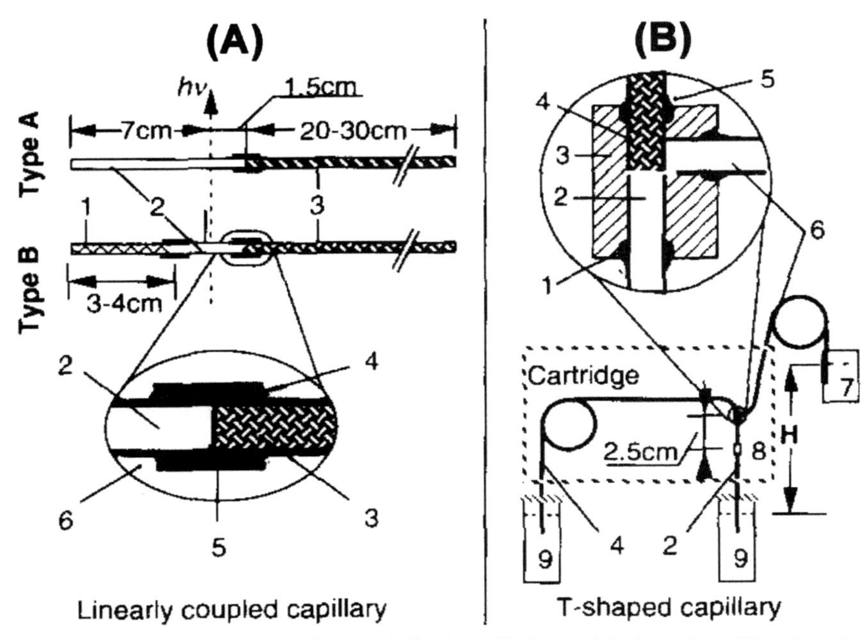

Figure 4.16 Structure of linearly coupled (A) capillaries and T-shaped capillary junction (B). *With permission from Y. Chen, J.-V. Hoeltje, U. Schwarz, Improving the UV detection sensitivity of condensed polyacrylamide gel-filled capillaries using non-flow buffer-filled capillaries as a detection cell, Journal of Chromatography A 700 (1–2) (1995) 35–42 [65].*

4.2.2 Laser-induced fluorescence detection

Fluorescence detection is another easily adaptable detection method for CGE. Similar to that of UV detectors, on-column detection can be readily accomplished, but in this instance by driving the excitation light onto the column and collecting the emission beam perpendicular to the excitation beam. Please note that laser-induced fluorescence (LIF) detection is not path length dependent, thus this approach can accommodate the use of very narrow bore capillaries [70]. Lasers, on the one hand while more amenable to focusing on small detection areas, on the other hand, they are limited as excitation sources due to their narrow spectral wavelength, in contrast to arc lamps, where the excitation wavelength can be varied by spectral filters before focusing onto the capillary [71]. The use of native fluorescence of solute molecules is another good option for LIF detection, not requiring any labeling steps.

For solute molecules not possessing intrinsic fluorescent characteristics, addition of a fluorophore tag is necessary that can be accomplished before (precapillary) or after (postcapillary) the CE separation. The precapillary mode has higher flexibility in terms of derivatization chemistry with no restrictions on reaction kinetics, for example, reaction time and temperature. In postcapillary mode fluorophore labeling takes place after the CE separation before the detection region. This approach is attractive in a way that less sample handling is required prior to analysis and the fluorescent tag is not changing the migration properties of the solute molecules. The drawback is that postcapillary detection requires rapid reaction kinetics. Postcapillary reactors were used to obtain fluorophore derivatized samples by Weinberger and coworkers [72]. The reaction took place in a gap-junction reactor chamber with 50 μm spacing between the capillaries. Another attempt utilized a postcolumn reactor employing the cathodic reservoir as a reaction chamber [73]. Swaile and Sepaniak evaluated pre- and on-column labeling of proteins with arylaminonaphthalene sulfonates [74].

One of the early LIF-based single wavelength detection systems for CGE was built by Drossman et al. [75] and the schematics of the settings is shown in Fig. 4.17. The entire system consisted of a plexiglass box

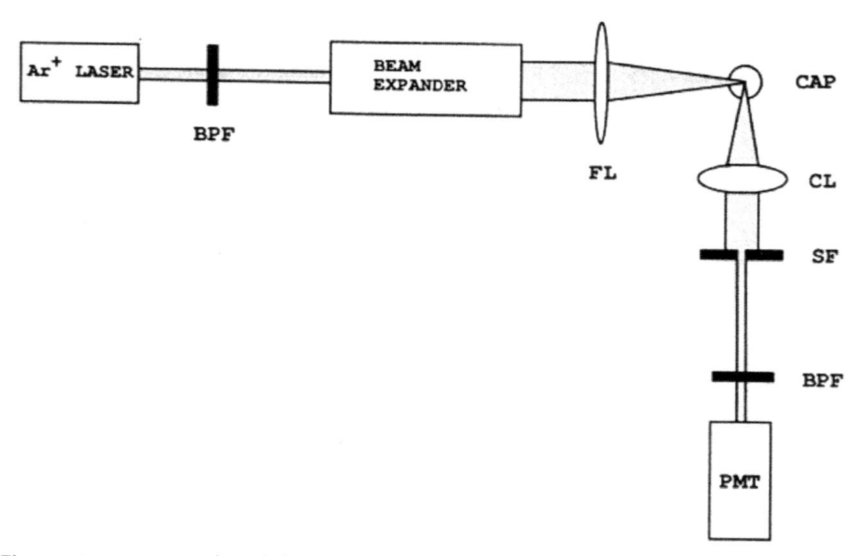

Figure 4.17 Laser-induced fluorescence detector for single wavelength detection of fluorescently labeled DNA fragments electrophoresing through a gel-filled capillary column. *BPF*, Bandpass filter; *CAP*, capillary; *CL*, collection lens; *FL*, focusing lens; *PMT*, photomultiplier tube; *SF*, spatial filter. *With permission from H. Drossman, J.A. Luckey, A.J. Kostichka, J. D'Cunha, L.M. Smith, High-speed separations of DNA sequencing reactions by capillary electrophoresis, Analytical Chemistry 62 (9) (1990) 900–903 [75].*

enclosing two buffer reservoirs, maintained at constant temperature with a heat control unit. Electric field was provided by a high-voltage power supply with a magnetic safety interlock and a control unit to adjust the applied potential. Injection of the samples was accomplished electrokinetically, directly from the sample vial.

To pursue higher detection limits of fluorescence detection, in one approach, the spectral background caused by the fused silica capillary was minimized by means of a sheath-flow cuvette fluorescence detector [76]. With a laser-based excitation source, the limit of detection (LOD) with the sheath-flow detection setting was at the mid zeptomol range for DNA sequencing by CGE [77]. An alternative approach was published by the Dovichi group using a sheath-flow cuvette [78], in which system the end of the separation column was inserted into a square quartz flow chamber, resulting in very sensitive measurements. Detection linearity for fluorescein isothiocyanate-labeled amino acids was over five orders of magnitude with the LOD of 1.7 zeptomol.

A low-cost, high-sensitivity LIF detection [79] employed a 0.75-mW He-Ne laser operating in the green (534.5 nm) region (Fig. 4.18), appropriate for tetramethylrhodamine (TMR)-labeled DNA fragments [79]. The detection limit of the system for the TMR was reportedly 500 ymol (~300 molecules) in capillary zone electrophoresis mode and 2 zmol (~1200 molecules) in CGE mode.

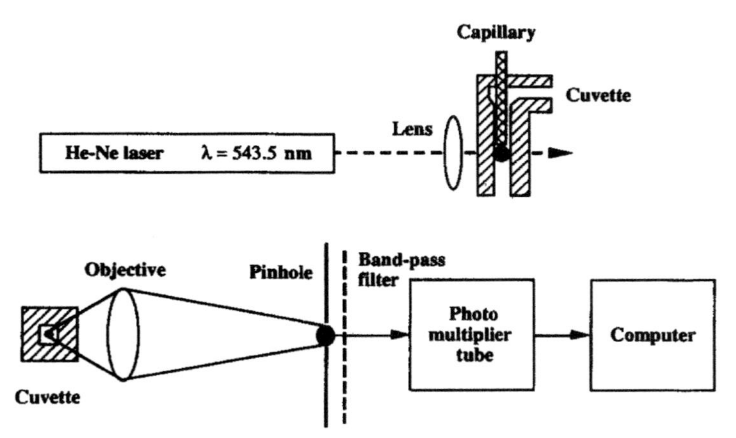

Figure 4.18 Optical scheme of the low-cost high-sensitivity LIF detection. *With permission from Chen, D.Y., H.P. Swerdlow, H.R. Harke, J.Z. Zhang, and N.J. Dovichi, Low-cost, high-sensitivity laser-induced fluorescence detection for DNA sequencing by capillary gel electrophoresis. Journal of Chromatography, 559 (1–2) (1991) 237–246. [79].*

For four color DNA sequencing, one of the first CGE apparatuses is shown in Fig. 4.19 [80]. This sensitive detector setting was capable of simultaneous measurements of fluorescence at four different wavelengths. The excitation light was provided by a 10-mW argon-ion laser and the emitted fluorescence signal was collected at right angles with a microscope objective, spatially filtered to remove scattered excitation light, and split into four parts of approximately equal intensity using beam splitters.

The performance of their settings was tested by analyzing the sequence of the bacteriophage vector M13mp19. The mixtures of the four DNA sequencing reactions were labeled with different fluorophores and separated in a 50-μm i.d. polyacrylamide gel-filled capillary (Fig. 4.20, upper panel). The correlation of the peaks obtained to the sequence of this particular template is depicted in the lower panel of Fig. 4.20.

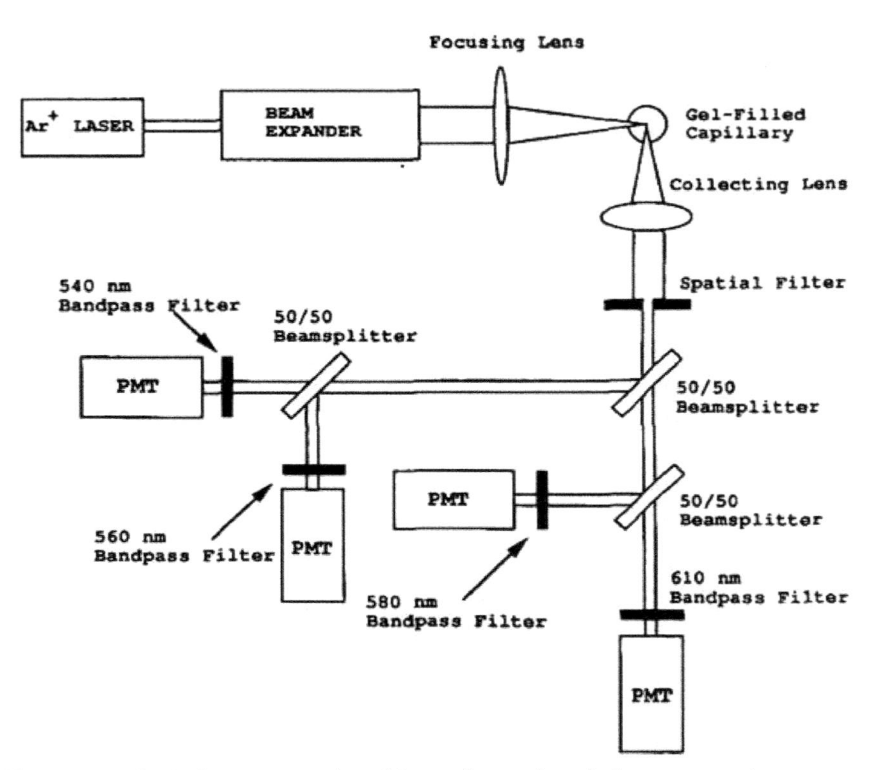

Figure 4.19 Optical system employed for multiwavelength fluorescence detection in capillary electrophoresis. *With permission from J.A. Luckey, H. Drossman, A.J. Kostichka, D.A. Mead, J. D'Cunha, T.B. Norris, et al., High speed DNA sequencing by capillary electrophoresis, Nucleic Acids Research 18 (15) (1990) 4417–4421 [80].*

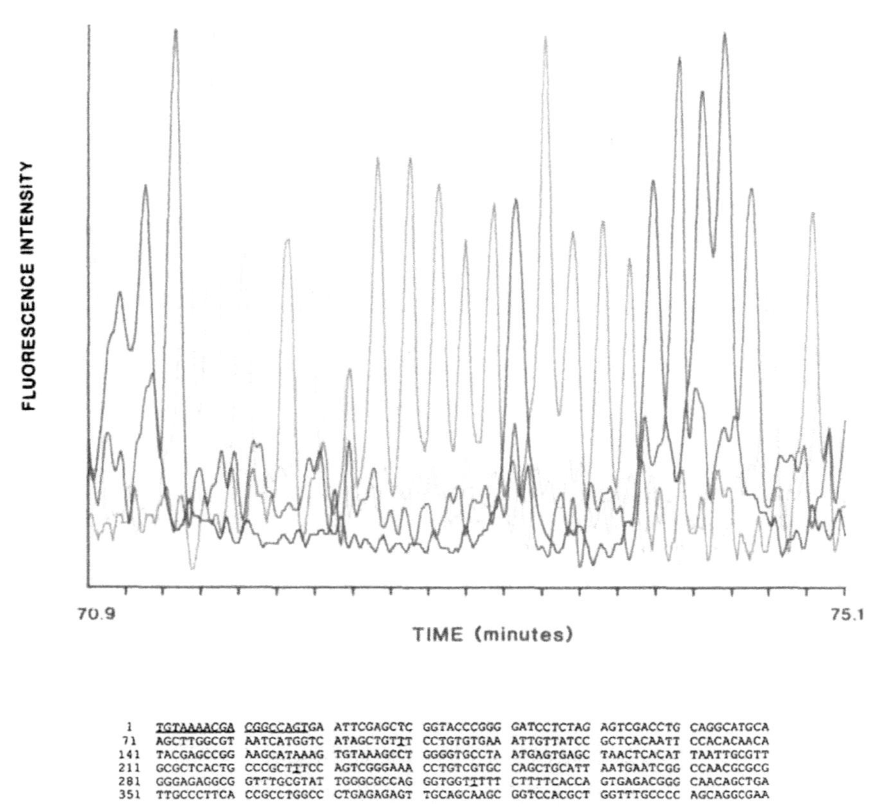

Figure 4.20 Representative portion of DNA sequence data obtained in CGE analysis of M13mpl9 (upper panel). The lower panel shows the corresponding sequence data. *With permission from J.A. Luckey, H. Drossman, A.J. Kostichka, D.A. Mead, J. D'Cunha, T. B. Norris, et al., High speed DNA sequencing by capillary electrophoresis, Nucleic Acids Research 18 (15) (1990) 4417–4421 [80].*

CCD detection was also attempted in fluorescent detection for CGE as it could provide multispectral imaging, thus simultaneously monitoring several different emission wavelengths. Such a multiwavelength fluorescence detection approach was introduced by Gesteland and coworkers [81] for DNA sequencing applications using a cryogenically cooled, low-noise two-dimensional CCD as a detector with a 94% duty cycle. Fig. 4.21 depicts a part of background subtracted fluorescent spectra in a 50-second time period of the separation of the M13mp19 sample. The selectivity advantage of multiwavelength detection improved the results of multicomponent analysis for bands C169 and T170, which otherwise electrophoretically co-migrated. Swerdlow

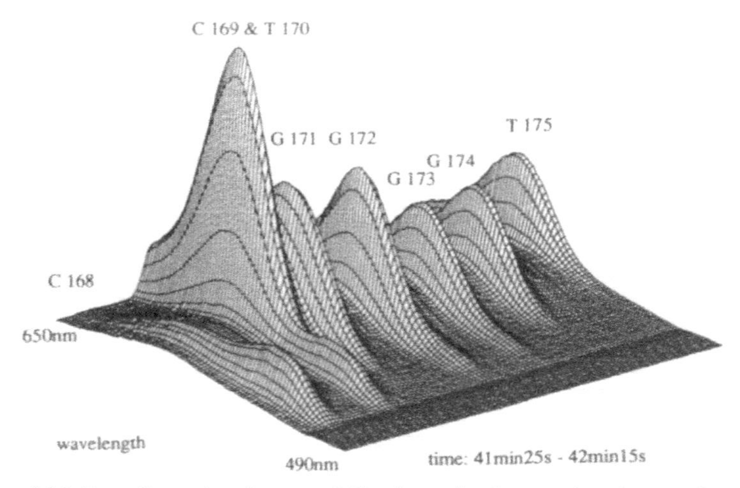

Figure 4.21 Two-dimensional raw CGE data (background subtracted) of the M13mpl9 sequencing reaction. Separation and detection conditions: $L = 42$ cm, $l = 35$ cm, $E = 190$ V/cm, 4%T/5%C polyacrylamide, laser excitation: 20 mW at 515 nm. *With permission from A.E. Karger, J.M. Harris, R.F. Gesteland, Multiwavelength fluorescence detection for DNA sequencing using capillary electrophoresis, Nucleic Acids Research 19 (18) (1991) 4955–4962 [81].*

et al. [77] compared three CE-based DNA sequencing methods using four, two, and one spectral-channel sequencing techniques. CCD detectors can also register the migrating sample zone in an axially illuminated capillary as was reported by Zare and coworkers [82]. They used snapshot acquiring mode to collect a series of wavelength and capillary position–dependent images and obtained detection limits for fluorescein isothiocyanate-labeled amino acids as low as 10^{-20} mol.

An epi-illumination fluorescence microscope was adapted to detect fluorophore-labeled species by Guzman and coworkers [83]. Their setup allowed detection of fluorescein derivatized amino acids at picomolar concentrations. The authors evaluated a 50-W mercury lamp with riboflavin and attained 500 amol detection limit [84]. Interestingly, fluorite microscope objective provided improved detection limits compared to glass objectives.

A laser-induced, confocal fluorescence detection setup was used by the Mathies group for high-sensitivity, on-column detection for a capillary array [85]. The linear array of capillaries was assembled on a translation stage and then scanned past by a laser excitation and confocal fluorescence detection setup. The small depth of focus of the excitation coupled with confocal detection through a spatial filter reduced the background scattering and reflections from the capillary walls. The high numerical aperture objective provided efficient collection of the fluorescence signal. Fig. 4.22 shows the schematics of the system. Laser light (1 mW, 488 nm at sample) was reflected by a long-pass dichroic beam splitter, passed through a 32 ×, N.A. 0.4 microscope objective, and brought to focus within a 100-μm i.d. capillary. The resulting fluorescence was collected by the objective, driven back through the beam splitter, and focused on a 400-μm diameter confocal pinhole. After the spatial filter, the emission was spectrally filtered and detected by a photomultiplier. A computer-controlled translation stage was used with 2.5-μm resolution. The output was amplified, digitized, and electronically stored.

The same group presented a DNA sequencing method utilizing capillary array electrophoresis with a two-color fluorescence detection and a two-dye labeling protocol. Since only two-dye-labeled primers were required, it was easier to select dyes with identical electrophoretic mobility shifts. The ratio of the signal in the two detection channels provided reliable identification of the sequencing fragments [86].

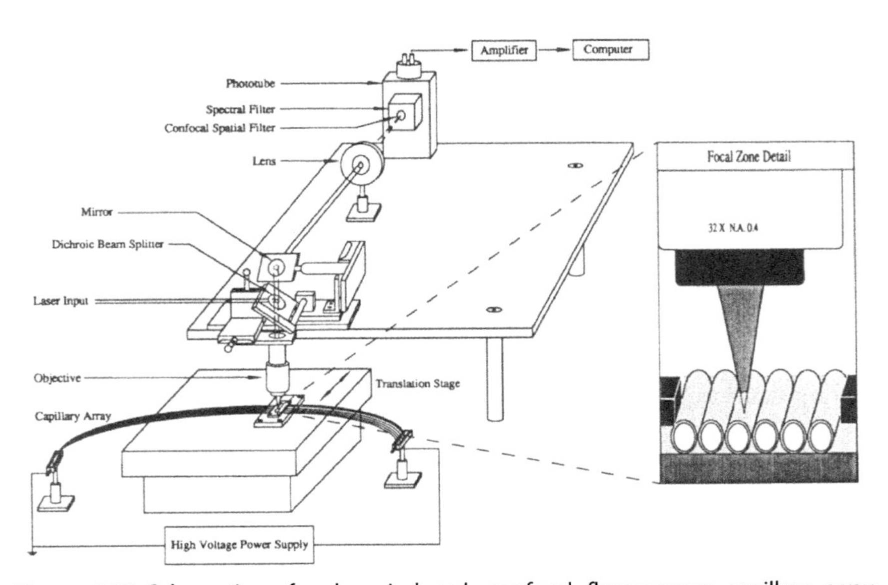

Figure 4.22 Schematics of a laser-induced, confocal fluorescence capillary array scanner. *With permission from X.C. Huang, M.A. Quesada, R.A. Mathies, Capillary array electrophoresis using laser-excited confocal fluorescence detection, Analytical Chemistry 64 (8) (1992) 967–972 [85].*

A laser-induced native fluorescence detection scheme for nucleic acids and DNA restriction fragments was developed by Milofsky and Yeung [87]. The 275.4 nm line from an argon ion laser or the 248 nm line from a waveguide KrF laser was used as an excitation source. Detection limits for guanosine and adenosine monophosphates were 1.5×10^{-8} and 5×10^{-8} M, respectively, up to three orders of magnitude linear range, better than that of with UV detection. The sensitivity for native fluorescence of DNA restriction fragments in gel-filled capillaries was close to that of UV absorption. The authors concluded that the decrease in performance in CGE-LIF analysis was caused by the high background fluorescence signal from the gel and quenching, calling for new gel formulations with lower background fluorescence or off-column detection. The same group evaluated several light excitation schemes for laser beam distribution to an array of capillaries [88]. The fluorescence signals from each capillary were simultaneously recorded at a high acquisition rate (0.6 frame/second) by a CCD camera allowing the use of a low energy output laser. Beale and Sudmeier [89] designed a LIF detector using epi-illumination and confocal optical detection in such a way that the detector scanned along the capillary length. The separation column was on a high precision a translational stage that moved the capillary through the probe beam.

Besides the use of common laser sources of He-Cd, Ar ion, or He-Ne lasers, in the mid-1990s interests were shifted to lower cost, smaller and longer lifetime semiconductor lasers and LEDs [90–96]. Advantages of LEDs are reduced size, easy operation, a wide range of available wavelengths (also in the UV region), low cost, and long lifetimes. Yeung and coworkers developed a diode laser-induced fluorescence (DIO-LIF) detection system for CE of amino acids. To optimize the detection performance the capillary position relative to the excitation laser beam was fixed as depicted in Fig. 4.23 [95]. Some of the drawbacks of LEDs are their low output energy and the spectral width of the output that in most instances should be filtered, further decreasing the energy output that hits the capillary. Other problems are reflections and scattering on the capillary surfaces. Coupling of LEDs with optical fibers was one way to somewhat increase the excitation power [97]. Application of UV LEDs is of a particularly high importance as UV lasers are very expensive [98].

An interesting development was reported by Nilsson et al. by applying real-time fluorescence imaging to view the electrophoretic process [99]. A large section of uncoated (outer polyimide coating) capillary was illuminated by a dye laser and the emitted light was collected by a CCD camera and processed by a computer to visualize real-time moving

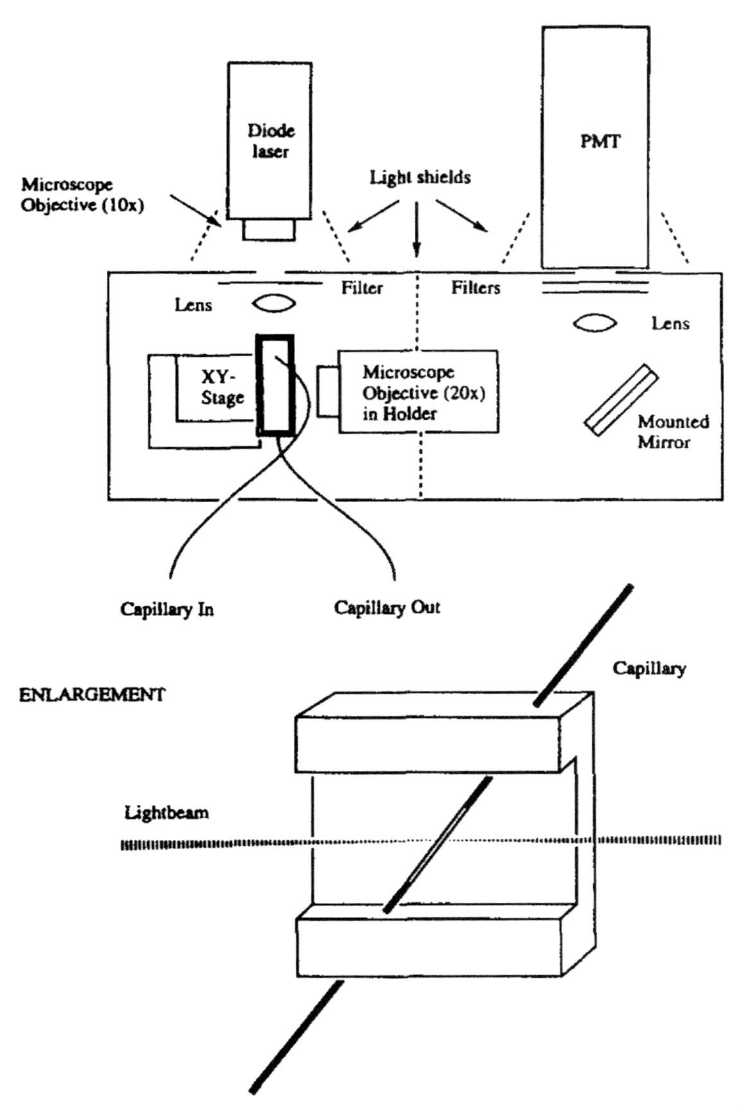

Figure 4.23 Illustration of the DIO-LIF detector for capillary electrophoresis. Enlarged the capillary position relative to the excitation laser beam. *With permission from A.J.G. Mank, E.S. Yeung, Diode laser-induced fluorescence detection in capillary electrophoresis after precolumn derivatization of amino-acids and small peptides, Journal of Chromatography A 708 (2) (1995) 309–321 [95].*

images. Another exciting attempt was the implementation of a single-molecule fluorescence burst detection for DNA analysis by CGE [100]. In their setting, a confocal fluorescence microscope was used to observe the fluorescence bursts from single DNA molecules labeled with thiazole orange. After the early attempts of using a scanning laser beam on a capillary array to read DNA sequences [101], more advanced scanning systems were developed that were reviewed in [102]. A good alternative for the scanning system was the introduction of imaging instruments enabling simultaneous monitoring of the separation progress in each capillary of the array, eliminating in this way the downtime of the system between scans. The use of a linear sheath-flow cuvette was one of these, where the capillary array was inserted into a rectangular cuvette and the buffer was pumped between the capillaries, drawing the effluent from each capillary as a sheath stream. A single laser beam shined through the sheath fluid providing the excitation source for the labeling dyes. Another approach was the use of square cross section capillaries for separation in conjunction with side illumination for the array to obtain simultaneous excitation without the requirement for a scanning detector [103]. From detection point of view, an interesting approach was to utilize the fluorescent lifetime in spite of multispectral imaging, reported by McGown and coworkers [104]. The fluorescent lifetime for a given set of dyes was apparently independent of the sieving matrix composition [105].

In order to reduce noise, infrared dyes were introduced for DNA sequencing with lifetimes ranging from 735 to 889 ps as reported by Soper and coworkers [106]. Others used diode laser-excited time-resolved fluorescence detection with rhodamine, oxazine, and cyanine dyes. Albeit, these dyes possessed similar excitation and emission spectra, their lifetime varied from 3.7 to 1.6 ns, which was adequate for proper sequence determination [107]. A new optical waveguide-based LIF detector was reported by Roeraade's group for multicapillary electrophoresis [108], utilizing the fact that the interior of the capillary can act as a liquid core waveguide. Illumination was arranged in a planar fashion and the capillary ends formed a two-dimensional array, which was imaged end-on with high light collection efficiency and excellent image quality. This setup was capable of four-color DNA sequencing in CGE in an array of 91 capillaries. A compact, 2D direct-reading fluorescence spectrograph was reported by Dovichi and coworkers with a sheath-flow-based detection cuvette supporting an array of 96 capillaries [109] (Fig. 4.24). Fluorescence from the capillary

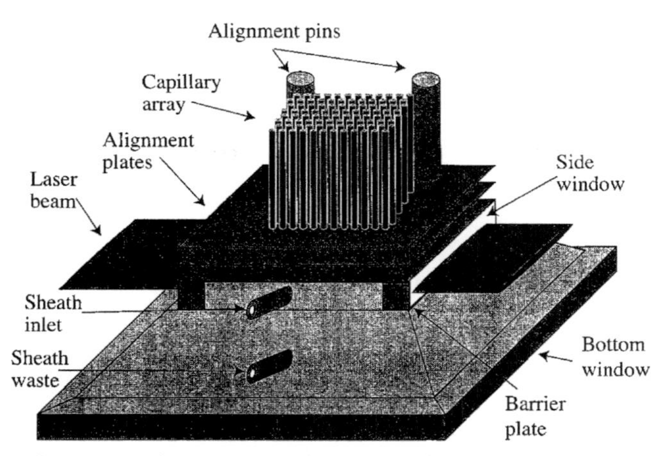

Figure 4.24 Illustration of the multicapillary sheath-flow cuvette for DNA sequencing fragment analysis by CGE. *With permission from J. Zhang, M. Yang, X. Puyang, Y. Fang, L.M. Cook, N.J. Dovichi, Two-dimensional direct-reading fluorescence spectrograph for DNA sequencing by capillary array electrophoresis, Analytical Chemistry 73 (6) (2001) 1234–1239 [109].*

columns was dispersed across the face of a CCD camera to simultaneously monitor DNA sequencing fragments as they migrated by.

Regarding the detection side, a cascaded avalanche photodiode photon counter was employed by Dada et al. [110] that increased the dynamic range for CE-LIF up to nine orders of magnitude. Their system apparently approached the fundamental limits as the dynamic range of CE detection is ultimately limited by molecular shot noise at low concentrations and by concentration-induced band broadening at high concentrations. The detector setting employed a cascade of four serially connected fiber optic beam splitters to generate four attenuated signals along with the primary signal and monitored the emitted light by a single-photon counting avalanche photodiode. The concentration detection limit for 5-carboxyl-TMR was 1 pM ranging up to 1 mM. This novel detection approach was tested on TMR-labeled glycosphingolipid GM1 (GM1-TMR) produced by single cells isolated from rat cerebellum.

Szarka and Guttman developed a smartphone-based real-time fluorescent imaging detection system for CGE-LIF analysis of APTS-labeled oligosaccharide samples [111]. The novel smartphone-based detection method for fluorescently labeled biomolecules greatly expanded the dynamic range and enabled retrospective correction for injections with unsuitable signal levels without the necessity to repeat the analysis (Fig. 4.25).

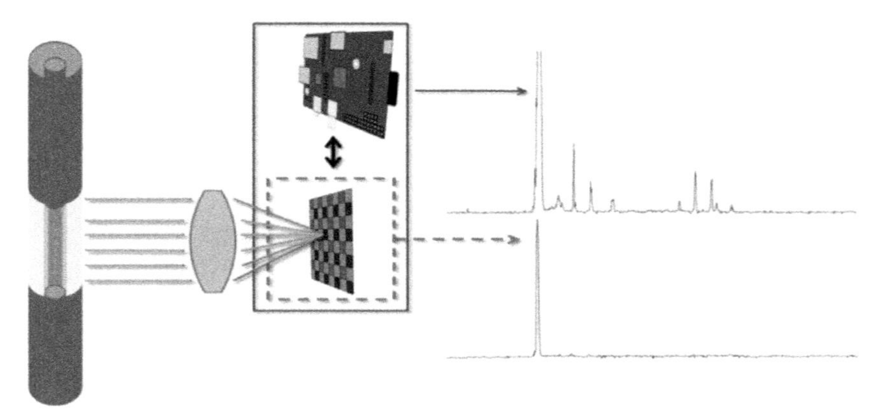

Figure 4.25 Scheme of the smartphone-based real-time fluorescent imaging detection system for CGE of labeled glycans. *With permission from M. Szarka, A. Guttman, Smartphone cortex controlled real-time image processing and reprocessing for concentration independent LED induced fluorescence detection in capillary electrophoresis, Analytical Chemistry 89 (20) (2017) 10673–10678 [111].*

4.2.3 Indirect detection methods

Indirect detection, where the analyte physically displaces a chromophore or fluorophore in the background electrolyte in light detection modes, offers a good alternative to avoid labeling of solute molecules lacking chromophore or fluorophore characteristics. In other words, indirect detection is considered as a universal mode of detection for solute molecules that cannot be detected otherwise or would require derivatization. This detection method can use UV/Vis absorbance, fluorescence, or amperometric detection setups. With the first two, the detection limit is mainly determined by the concentration of the visualization agent, in addition to other factors like dynamic reserve and displacement ratio [112]. An increase of the dynamic reserve improves LOD and can be optimized by decreasing the noise and intensifying the light source. Fig. 4.26 compares indirect and direct detection methods in CE [113].

Early demonstration on the feasibility of indirect detection methods in CE was reported by Hjerten et al. employing indirect UV absorbance detection of organic anions [114]. The feasibility to detect native DNA fragments by indirect fluorescent detection using noncross-linked separation matrices in CE was demonstrated later [115]. Indirect fluorometry was used to detect native DNA fragments by CGE as reported in [115]. An indirect thermooptical detection system was built by Ren et al. [116] using a He-Ne laser for pump and probe beams but with different power

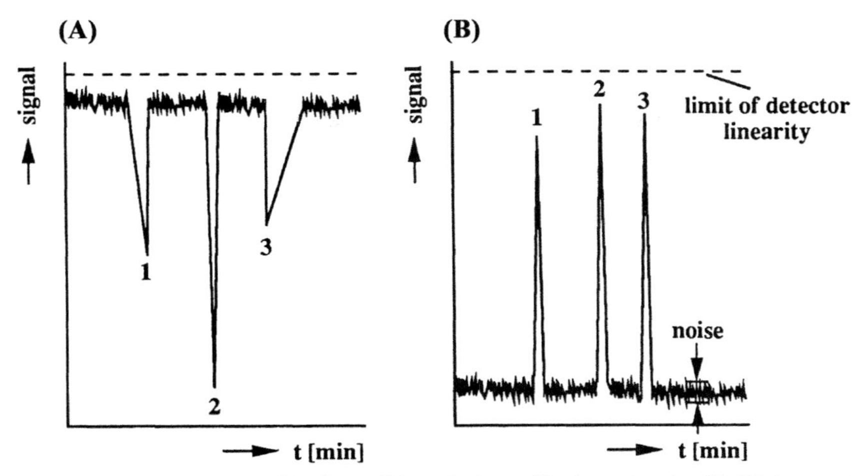

Figure 4.26 Comparison of indirect (A) and direct (B) detection in CE. *With permission from R. Kuhn, S. Hofstetter-Kuhn, Capillary Electrophoresis: Principles and Practice, Springer Laboratory, Berlin, 1993 [113].*

settings (20 mW and 2 mW, respectively). The background absorber was methylene blue that allowed 5×10^{-6} M detection limit for lysine. CE with indirect UV absorbance detection was used to estimate the effective charge and charge density of polyelectrolyte homopolymers of poly(diallyl dimethylammonium chloride), poly(acrylic acid), and poly(methacrylic acid) by Cottet and coworkers [117]. During the analysis, the charged polymer displaced an absorbing co-ion in the background electrolyte and directly correlated to the effective charge of the solute. In addition, their method easily distinguished statistical copolymers from diblock copolymers of similar chemical charge densities.

4.2.4 Electrochemical and conductivity detection

Electrochemical detection was first used in CE in the form of conductivity measurements by placing an electrode pair in the capillary to measure the current between the electrodes. Amperometric detection was introduced to CE by the Ewing group in 1987 [118], by applying a carbon fiber working electrode directly into the end of the capillary. Decoupling from the separation part with a small section covered by a porous glass capillary allowed the elimination of cross talks between the two electric circuits, resulting in sensitive amperometric detection. A pulsed amperometric detection scheme was applied for sensitive analysis of carbohydrates and glycopeptides [119] to monitor the desialylation process and used during

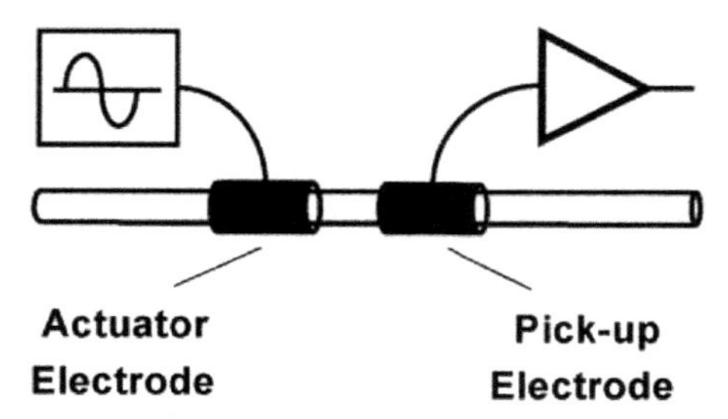

Figure 4.27 Illustration of a C^4D detection cell for capillary electrophoresis. *With permission from P. Kuban, P.C. Hauser, A review of the recent achievements in capacitively coupled contactless conductivity detection, Analytica Chimica Acta 607 (1) (2008) 15–29 [122].*

the characterization of oligosaccharides. An alternative approach to conventional multispectral imaging was reported by Kuhr and coworkers [120] by employing electrochemical detection. A set of four ferrocene derivatives were investigated with different scanning voltammetric signals distinguishable as they migrated out of the separation capillary.

Conductivity detection for CE can be operated in two modes: galvanic contact mode and contactless mode [121]. Galvanic contact can be performed with electrodes, which are placed in drilled holes through the capillary wall (on-column detection) or in an end-column wall jet arrangement (off-column detection). In case of the contactless method, the analytical signal is obtained with oscillometric techniques. Capacitively coupled contactless conductivity measurement (C^4D) is extensively reviewed in [122]. C^4D uses alternating current (AC) voltage to a galvanically isolated actuator electrode and measures the resulting AC current at the second pick-up electrode as illustrated in Fig. 4.27.

4.2.5 Miscellaneous detector systems

In 1989 Pentoney et al. described two types of miniaturized radioisotope detectors for CE [123]. One design utilized a commercially available semiconductor detector, the other was a scintillation-based detection system with LODs for ^{32}P nucleotide triphosphates of <1 fmol. The sensitivity of the detector was enhanced by flow programming, that is, the applied electric field was decreased to slow the migration rate and concomitantly increasing detection sensitivity that was found proportional to the residence time.

Thermooptical absorbance detection is a technique where the absorption of a so-called pump laser beam rises the local temperature at the detection point changing the refractive index, which is then measured by another laser (probe beam) [70,124]. The temperature rise is proportional to the thermal absorbance of the sample components at a given pump laser intensity. Fig. 4.28 shows the schematics of such a detector setup.

Another thermooptical type absorbance detection [125] utilized optical pumping with a frequency-doubled Ar-ion laser to induce UV light absorption. The generated heat changed the refractive index, which was probed with a laser diode or He-Ne laser utilizing a specially designed holographic optical system. This setting was capable of detecting nanomolar concentrations and subfemtomole masses. A new generation thermo-optical absorbance detection system was built with a pulsed KrF pump laser and a He-Ne laser probe beam [126], where the pump laser was focused into the capillary in a way that when the solute molecules migrated through the beam, absorption of the incident radiation generated a temperature rise in the surrounding medium that was monitored by the probe beam. Detection limits were found to be 10 times better than that of with on-column UV detection. A thermal lens detector was reported as a CE detection device that utilized a combination of a miniaturized

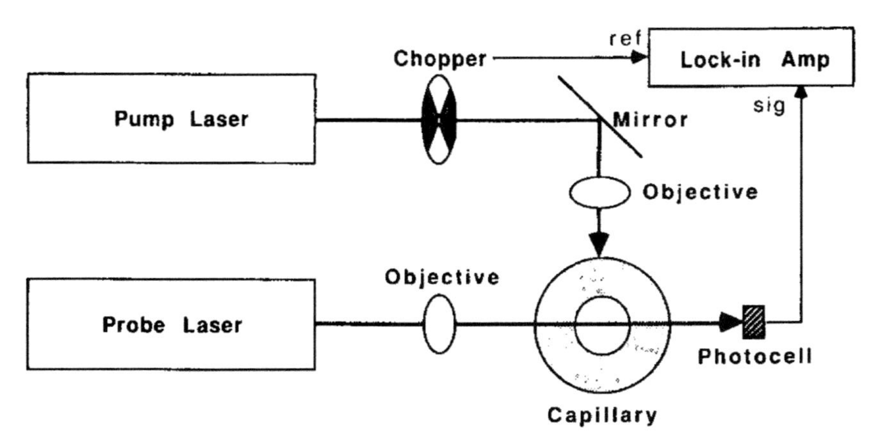

Figure 4.28 Thermooptical detector for capillary electrophoresis. Light from the pump laser is modulated with a mechanical chopper, reflected from a mirror, and focused with a microscope objective into the capillary. Light from the probe laser is focused with a microscope objective into the capillary. The transmitted beam is detected with a small-area photocell. The output of the photocell is sent to the signal channel (sig) of the lock-in amplifier (lock-in amp), whereas the reference channel (ref) is supplied by the chopper head. *With permission from M. Yu, N.J. Dovichi, Attomole amino acid determination: capillary-zone electrophoresis with laser-based thermooptical detection, Applied Spectroscopy 43 (2) (1989) 196–201 [124].*

fiber optic modified photothermal sensor based on a double-beam absorption scheme [127] using a tunable multiline argon ion laser as an excitation source and a 633 nm He-Ne laser for the probe beam source. To detect solute molecules possessing neither UV active nor fluorescent groups, a thermal lens microscopy approach offered a good alternative [128]. This detection method can also be used with a UV laser source [129] to obtain higher sensitivity.

Schlieren optics was utilized to evaluate the molecular sieving effect of linear polyacrylamide (LPA) solutions in real-time visualization [130] and to measure refractive index gradients in capillary columns [131]. Refractive index detectors were built based on off-axis capillary focusing [132] and with the utilization of holographic optical elements [133]. Sumitomo et al. synthesized a luminescent Eu^{3+} chelating reagent to covalently bind the 5-end of DNA through its dichroic functional group, while retaining the unique luminescent properties. This lanthanide chelate labeling reagent was applied for a DNA hybridization assay by time-resolved CGE with luminescence detection [134]. Soper et al. developed a near-IR time-correlated single-photon counting instrument for online fluorescence lifetime detection in CE. The lifetimes of the migrating components were detected using maximum likelihood estimators in the limit of low photocounts within the decay profile [135].

A photothermal approach was used for on-column detection of nonlabeled analytes with a lower detection limit of 1.8×10^{-7} M for riboflavin [136]. An online multichannel Raman spectroscopic detection system was designed and built by Chen and Morris with a CCD detector [137] that required voltage programming mediated reduction of analyte velocity in the detection zone to gain additional sensitivity (500 amol for methyl orange). Multipoint detection setting to interrogate the capillary at several points along the axis during the electrophoresis process provided an interesting new detector design as shown in Fig. 4.29. This detection mode allowed accurate mobility and relative peak area measurements as well as the observation of the unexpectedly discontinuous migration of φX DNA molecules [138].

A novel instrumental design was introduced comprising a planar integrated detection cell, where the illumination beam was guided through a path length of up to 1 cm [139]. The authors used antiresonant reflecting optical waveguides or TiO_2-SiO_2 Bragg layers to create a low-loss waveguide over the chosen path length. Haddad and coworkers reported on a model that considered stray and polychromatic light to predict on-column detection linearity [140] and compared the performance of four LEDs to

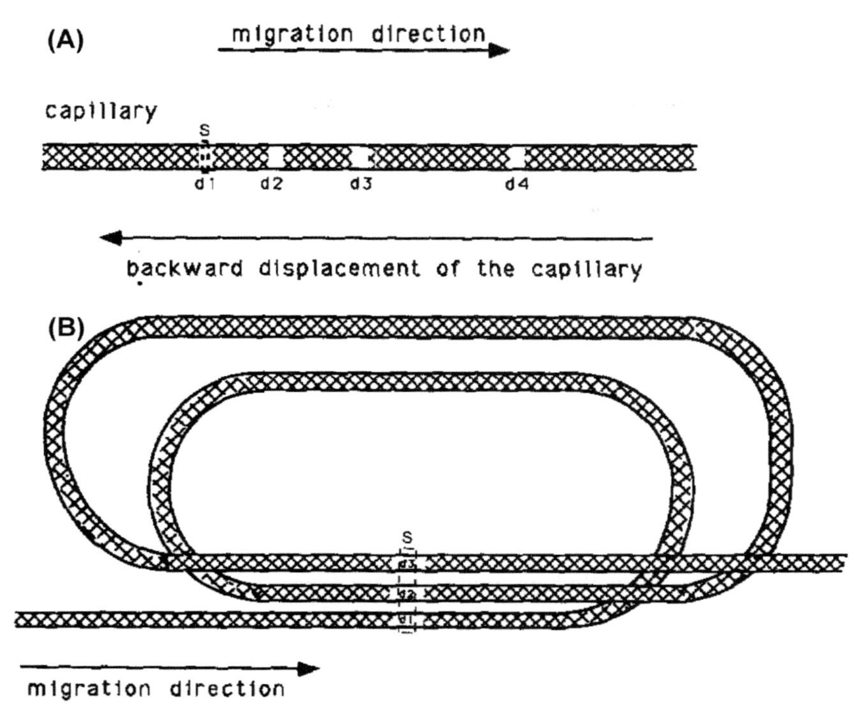

Figure 4.29 The principle of multipoint detection in capillary electrophoresis, where S represents the slit and d1, d2, d3, d4 represent the detection windows. (A) Method I: following a first detection at d1 the tubing is moved "backwards" for a second recording at d2, and so on. (B) Method II: the tubing is stationary. *With permission from T. Srichaiyo, S. Hjerten, Simple multipoint detection method for high-performance capillary electrophoresis, Journal of Chromatography 604 (1) (1992) 85–89 [138].*

mercury and tungsten light sources. They found that the performance of LEDs was close to or even better than that of the mercury light source in respect to linear range and noise. A double-beam absorbance detection setup with diode lasers and LEDs was also reported [141], where the diode laser and LED double-beam setting resulted in improved performance over conventional CE absorbance detection.

Miniaturized surface plasmon resonance detection was developed for CE by Zare's group [142]. The surface was functionalized by protein immobilization for selective protein detection at the level of 2 fmol with a dynamic range of three orders of magnitude. Chemiluminescence detection is another approach to attain very good detection limits. In this case, the solute molecule, or its derivatized form, is subject of a chemical reaction that produces light. The same group built an end-column chemiluminescence detector that reached the detection limit of 10^{-9} M for ATP [143]. CE was used with chemiluminescence detection with a novel double-on-column coaxial flow interface for the analysis of multiple PCR products for simultaneous qualitative and quantitative analysis of genetically modified organisms [144]. The amplified fragments were labeled with acridinium ester at the 5'-terminal through an amino modification and chemiluminescence was triggered by using the coaxial flow interface to mix the capillary outflow with a nitric acid/hydrogen peroxide solution. A similar detection scheme was applied by the same authors for the detection of peptides [145].

Recent developments in quantum dot (QD) chemistry allowed their use with biomolecular conjugation. In contrast to conventionally used organic dyes, QDs feature broader absorption spectra, higher photo stabilities, and better sensitivities. However, separation of QD conjugated biomolecules is not an easy task because of their size distribution and small difference in the charge to size ratio. Separation of QD-streptavidin/biotin/IgG mixtures was reported utilizing polyethylene oxide (PEO) additive to the background electrolyte [146]. The use of a polymeric additive into the CE run buffer improved the resolution of bioconjugated CdSe/ZnS QDs.

4.2.6 Coupling to mass spectrometers

Connecting CE with mass spectrometry (CE-MS) is a favorable method to obtain fast separation and structural information from biological materials [147]. Electrospray ionization (ESI) is the most common ionization mode for online CE-MS coupling in either sheath-flow mode [148] or sheathless modes [149] Fig. 4.30. In regard to mass analyzers, ion tap mass spectrometers were mostly used in the early days [150,151]. Fourier transform ion cyclotron resonance MS instruments were also attempted to couple with CE to provide sensitive and high-precision mass measurements [152–155].

In CE-MS interfacing, the sheath-flow approach (Fig. 4.30A) is more common than the sheathless form because of its simplicity, reliability, and easier implementation. However, while the sheath-flow interface type is more robust and relatively easy to use, the sheath liquid may decrease detection sensitivity because of sample dilution. As a matter of fact, sheathless interfaces are better suited for CE as they support low flow rates that is crucial to obtain better ionization efficiency, lower ion suppression, and higher sensitivity [156]. With sheathless ESI interface, the connection between CE and MS closes the electric circuit of the CE system; however, simultaneously providing adequate voltage for the electrospray is a challenge. Usual ways to address this issue are inserting a small electrode into a hole of the separation capillary, coating a small portion of the end of the column with a conductive material, or inserting the capillary end into a conducting sleeve or a microtee. Etching the capillary end by HF providing an electrical porous junction that can be incorporated in the capillary is another possibility. Similar to the porous junction, drilling a hole on the capillary can accommodate the same goal. Consequently, numerous interface designs have been published on sheathless interfaces using a liquid junction, a conductive capillary tip, or porous junction [157–159] (Fig. 4.30B–D). Again, the sheathless design may increase detection sensitivity by eliminating the mixing with the sheath liquid. One of the sheathless interface designs uses a conducting metal coating at the end of the capillary [160], grounding the high CE voltage and supporting the ESI spray voltage (Fig. 4.30C). The disadvantage of this interface design is the rapid degradation of the conductive coating.

Liquid junction [149] (Fig. 4.30B) offers a very efficient sheathless CE-MS interface, as in this case no metal coating is required. The setting includes a conductive wire (usually gold) at a gap between the separation and spray capillaries. Karger and coworkers designed a subatmospheric pressure interface that was a modified version of the liquid junction interface [161] using a replaceable micro-ESI tip enclosed in a subatmospheric chamber. This interface was also used to couple a microfabricated CE

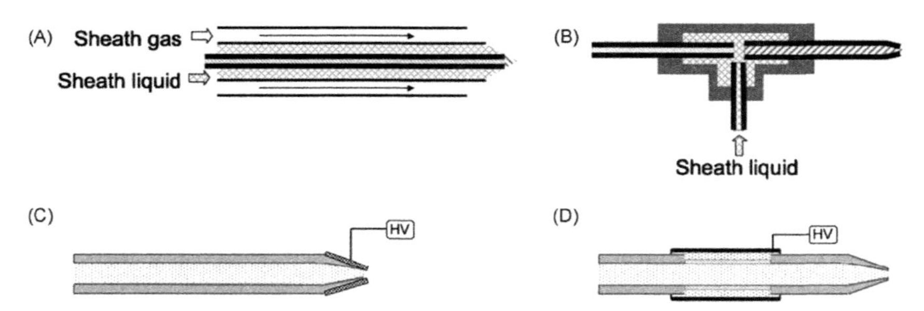

Figure 4.30 Most frequently used CE-MS interfaces. (A) Coaxial sheath-flow interface with sheath gas, (B) Liquid junction interface, (C) Conductive coating applied to the emitter tip and (D) Porous, etched capillary walls in metal sleeve. *With permission from E.J. Maxwell, D.Y. Chen, Twenty years of interface development for capillary electrophoresis-electrospray ionization-mass spectrometry, Analytica Chimica Acta 627 (1) (2008) 25–33 [147].*

device with ESI-MS [162]. Another attempt was the introduction of a split-flow CE/ESI-MS interface. In this setup, the electrical connection to the CE capillary outlet was obtained by diverting part of the CE buffer from the capillary through an opening near the capillary outlet with a contact of a sheath metal tube representing the CE ground and ESI electrodes. The advantage of this approach was the zero dead volume and/or bubble formation [163]. Another approach from the same group utilized a narrow platinum wire inserted into the CE capillary through a small hole near the outlet end. The design was successfully applied to CE-ESI-MS analysis of mixtures of peptides and proteins, attaining a detection limit of $\sim$4 fmol for myoglobin [164].

Smith and coworkers introduced one of the first electrospray interfaces for CE-MS in 1988 [165], which made electric contact with the background electrolyte via either a small needle or a thin metal film deposited on the surface of the capillary. The same group introduced online gel-filled capillary electrophoresis/mass spectrometry (CGE/MS). Relatively high concentrations of urea and other nonvolatile buffer components were used with online CGE-MS using a liquid junction—ion spray interface apparently without any problem for the mass spectrometer [166]. Foret et al. extensively studied both the theoretical and experimental aspects of moving ionic boundary formation inside the capillary tubing, especially the migration of sheath liquid counter-ions into the capillary [167] as these boundaries can alter the migration and resolution of the solute molecules being analyzed. Henion's group built a self-aligning liquid junction for CE-ESI-MS [168]. The sprayer needle and the junction between the CE capillary and the makeup flow were incorporated in a single unit allowing a simple way for precise alignment of the CE capillary and the sprayer needle. Numerous attempts were reported on the use of sheathless ESI interfaces, such as a tapered outlet of the analytical separation capillary to be positioned directly in front of the MS inlet [169]. One of the important ones was described by Janini [157] and further developed by Moini [164], which utilized etching of the outlet end of the capillary in a way that the wall thickness allowed electrons through but not buffer and solute molecules, thus provided good grounding for the CE separation and simultaneously the high voltage for the ESI.

Matching CE separation with MS detection is a challenge as the narrow time window of the peaks in high-speed CE separations is not compatible with the usually slower scan speed of the MS instruments. A regular MS/MS cycle usually takes a couple of seconds that is too slow for regular CE-based separations. One of the solutions to address this problem was field programming modulated by the MS feedback (scheme of the automated modulation system is depicted in Fig. 4.31). In other words, when MS/MS mode is required, the MS system sends a signal to the CE power supply that drops the separation voltage, concomitantly slowing down the separation speed, that is, providing more time for MS/MS analysis [170].

Coupling of the electrophoresis capillary with the MS instrument is also possible through a pressurized liquid junction interface. The Foret group [171] optimized such a nanoelectrospray interface between CE and MS for reliable proteomic analysis. In their setting, the electrode chamber at the injection side of the separation column and the spray liquid reservoir were pneumatically connected by a tube filled with pressurized nitrogen providing the exact counterbalance pressures at the inlet and outlet parts. The other important part of the system was the width of the gap between the separation and spray capillaries that was optimized at the 50–200 μm range.

Chen and coworkers described a novel CE-ESI-MS interface that decoupled the electrical and solution flow rate requirements of the separation and ionization processes [172]. The outlet end of the separation capillary was

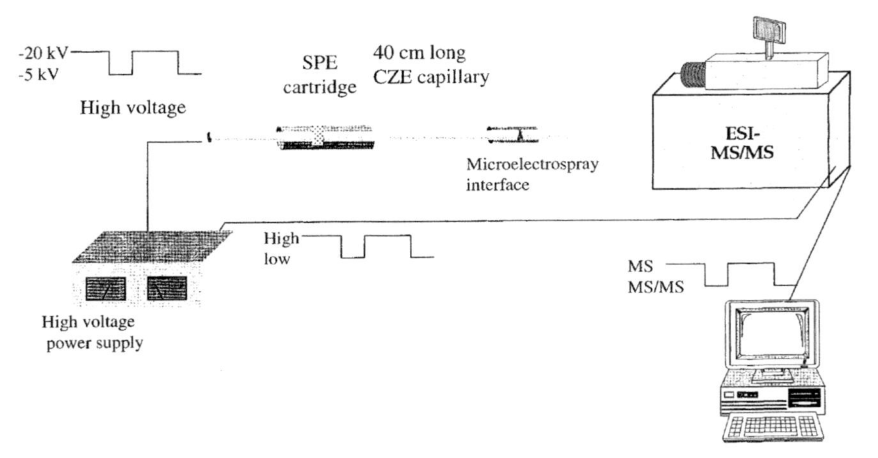

Figure 4.31 Scheme of the CE-MS/MS system coupled to the ESI-MS/MS system with automated field programming modulation of the analytical window and control of the mass spectrometer. *With permission from D. Figeys, G.L. Corthals, B. Gallis, D.R. Goodlett, A. Ducret, M.A. Corson, et al., Data-dependent modulation of solid-phase extraction capillary electrophoresis for the analysis of complex peptide and phosphopeptide mixtures by tandem mass spectrometry: application to endothelial nitric oxide synthase, Analytical Chemistry, 71 (13) (1999) 2279–2287 [170].*

inside a tapered and beveled stainless steel hollow needle that acted inside as an outlet buffer reservoir and outside as an ESI emitter. Additional advantage of the setting was the possibility of using capillary columns with any surface treatment. A modification buffer solution was introduced through a second capillary connected to the needle via a tee junction to improve the compatibility of the background electrolyte with the electrospray. The authors reported a 5-fold greater detection limit for amino acids compared to commercial sheath flow–based CE-MS interfaces.

The Dovichi group developed electroosmotically pumped sheath-flow nanoelectrospray interfaces for CE-MS coupling [173]. The separation capillary was connected through a cross-connector into a glass emitter. One side arm of the cross provided the fluidic contact with a sheath buffer reservoir that was connected to a power supply. The electric potential, applied to the sheath buffer, resulted in the generation of the electroosmotic flow in the emitter capable of pumping the sheath fluid at low nL/minute rates. Their modified interface was robust and produced a long lifetime without any sensitivity loss. Using this approach, enabled the single-shot, bottom-up proteomic analysis of 300 ng of *Xenopus laevis* fertilized egg proteome digest with a LPA-coated capillary to identify 1249 proteins and 4038 peptides in less than 2 hours with a peak capacity of ~330.

Tycova et al. developed a novel interface-free CE-nanospray ESI approach (Fig. 4.32) and demonstrated the capabilities in rapidly analyzing a cardioprotective drug (dexrazoxane) and its hydrolyzed form in blood

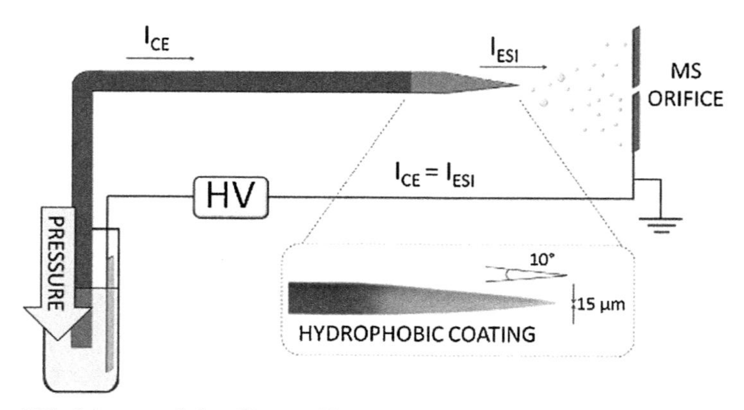

Figure 4.32 Scheme of the CE-nanoESI/MS system. *With permission from A. Tycova, M. Vido, P. Kovarikova, F. Foret, Interface-free capillary electrophoresis-mass spectrometry system with nanospray ionization-—analysis of dexrazoxane in blood plasma, Journal of Chromatography A 1466 (2016) 173–179 [174].*

plasma samples after deproteinization [174]. This interface-free system comprised a single piece of a narrow bore capillary acting as both electrophoretic separation column and nanospray emitter with low flow rates (30 nL/minute).

Neusüß and coworkers [175] introduced a two-dimensional capillary SDS gel electrophoresis method by utilizing an eight-port nanoliter valve containing four sample loops with increased distances between the separation dimensions as shown in Fig. 4.33. Successive co-injection of the solvent and a cationic surfactant simplified the decomplexation strategy.

CE can also be used in conjunction with MALDI-TOF-MS in online or offline modes. Matrix-assisted laser desorption/ionization time-of-flight mass spectrometry (MALDI-TOF-MS) has growing importance in proteomics research because it supports accurate mass determination with high detection sensitivity and easily automated. Caprioli's group utilized continuous depositing of the CE effluent onto a membrane, coated with the

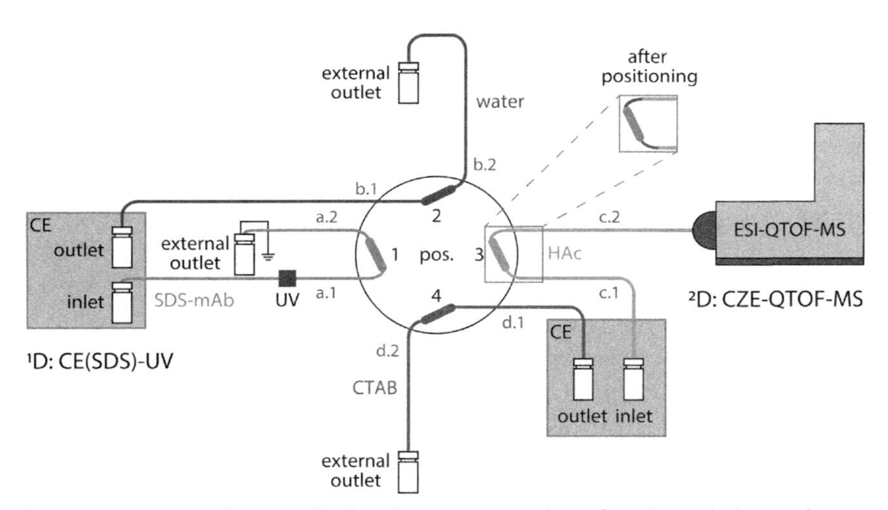

Figure 4.33 Setup of the CE(SDS)-CZE-MS system where first the gel electrophoresis analysis is performed (position 1). For each decomplexation zone, consisting of water (position 2) and the cationic surfactant CTAB (position 4), separate connections are utilized and positioned between both separation dimensions. The second dimension was a simple CZE-MS setup (position 3). For zone and sample transfer into the CZE-MS dimension, filled loops are switched to positon 3 via the multiposition electric microactuator. (*CZE*, capillary zone electrophoresis) *With permission from J. Römer, S. Kiessig, B. Moritz, C. Neusüß, Improved CE(SDS)-CZE-MS method utilizing an 8-port nanoliter valve, Electrophoresis 42 (4) (2021) 374–380 [175].*

MALDI matrix prior to MS analysis [176]. Foret and coworkers [177] built a sheath-flow interface for collecting protein fractions by transferring the pressure-mobilized effluent from cIEF into collection capillaries. The fractions were then analyzed by MALDI-TOF-MS. In another study, full-scan spectra were generated for a protein mixture with CE separation providing 3–4 second peak widths [178]. Since the speed of CE and MS measurements is different, mathematical correlation techniques were necessary to evaluate CE-ESI-TOF data to enhance the S/N ratio of the system [179].

4.3 Operation variables

4.3.1 Gel concentration

The relationship between the gel concentration and the electrophoretic mobility of the analytes is important in CGE. Chrambach and Rodbard introduced the so-called "Ferguson plot" [180], where the logarithmic electrophoretic mobility was plotted against the gel monomer concentration. Originally, this Ferguson plot was applied to estimate the size and free solution mobility of macromolecules. It is important to note that similar to SDS-PAGE, SDS-CGE also overestimates the molecular weights of, for example, glycoproteins, due to the facts that the glycosylation part does not bind SDS while the hydrodynamic volume of glycoproteins is larger than that of without the glycosylation part (decreased surface charge density). The Ferguson plot analysis can be used to correct this issue, although not widespread because of the time required to prepare and run all different concentration gels. In SDS-CGE, on the other hand, with replaceable sieving matrices an automated Ferguson analysis method was introduced by Werner et al. [181] for modified proteins (e.g., glycosylated). Their approach generated all data required for the Ferguson plot analysis using Eq. (4.6)

$$\mathrm{Log}(\mu) = \mathrm{Log}(\mu_o) - K_r(T) \tag{4.6}$$

where μ is the protein mobility, T is the sieving matrix concentration, μ_o is the free solution mobility, and K_r is the retardation coefficient. An illustration of this principle is delineated in Fig. 4.34 representing Ferguson plots for four hypothetical proteins. Proteins A, B, and C are

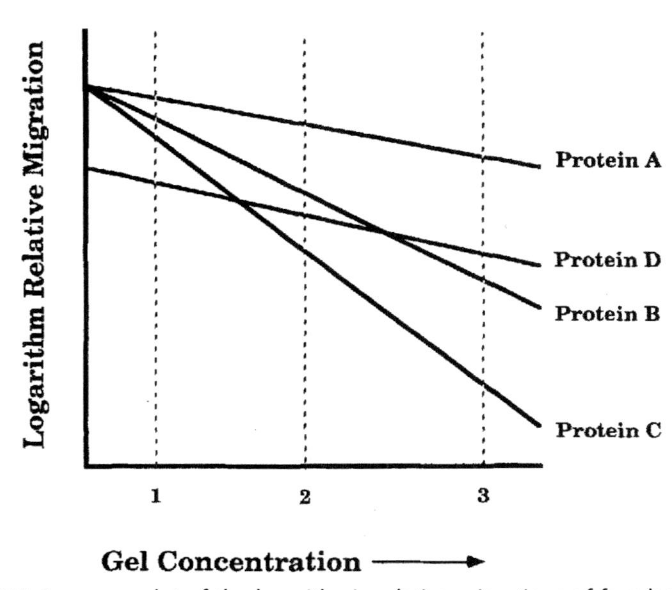

Figure 4.34 Ferguson plot of the logarithmic relative migrations of four hypothetical proteins against gel concentration. *With permission from W.E. Werner, D.M. Demorest, J.E. Wiktorowicz, Automated Ferguson analysis of glycoproteins by capillary electrophoresis using a replaceable sieving matrix, Electrophoresis 14 (8) (1993) 759–763 [181].*

nonglycosylated with molecular masses A < B < C, and protein D is a glycoprotein with a mass A = D. While the plots for proteins A, B, and C intercept at the *Y*-axis at one point, protein D intersects it at a point corresponding to a lower free solution mobility, indicating a lower charge-to-mass ratio, presumably due to the lack of SDS binding by the carbohydrate moieties, which otherwise increases the bulkiness of the complex. Please note that nonlinear Ferguson plots were obtained in borate cross-linked dextran gels in SDS-CGE of proteins (Chapter 2: Basic Principles of Capillary Gel Electrophoresis) and for DNA analysis in derivatized cellulose matrices (Chapter 3: Separation Matrix and Column Technology).

Separation of polystyrene carboxylates (PSC) up to 10 μm in diameter in 0.1%–0.9% solutions of noncross-linked polyacrylamide interestingly showed that the reduced mobility was not constant but increased in proportion to the load of PSC and inversely to the concentration of the sieving matrix, unless the background electrolyte (TBE) concentration was significantly increased (10-fold) [182]. The authors observed that both the Ferguson plots of PSC obtained at various PSC loads (Fig. 4.35) and the effects of the dissociating conditions (high ionic strength and detergent)

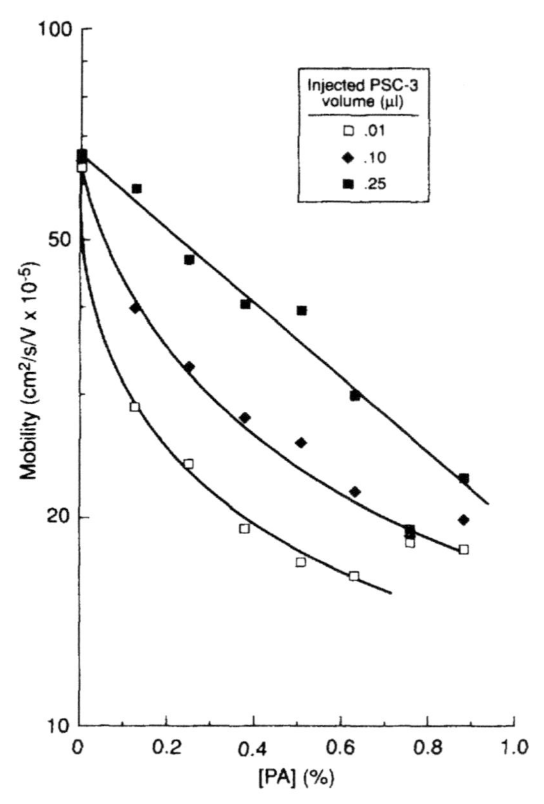

Figure 4.35 Ferguson plots of polystyrene carboxylate (PSC-3) at three different sample loads. *With permission from S.P. Radko, M.M. Garner, G. Caiafa, A. Chrambach, Molecular sieving of polystyrene carboxylate of a diameter up to 10 mm in solutions of uncrosslinked polyacrylamide of Mr 5 * 106 using capillary zone electrophoresis, Analytical Biochemistry 223 (1) (1994) 82–87 [182].*

signified that the variable mobilities were those of different PSC aggregation states. Thus only the fully associated or dissociated states of PSC provided constant mobilities independent of PSC load.

The mobility of single-stranded DNA as a function of cross-linker concentration in polyacrylamide CGE was investigated by Figeys and Dovichi [183]. Besides the usual notion that electrophoretic mobility decreases with increasing cross-linker concentration, the transition from normal mobility to the limiting mobility of biased reptation with stretching occurred for longer fragments as the cross-linker concentration was decreased. Rapid separations of SDS–protein complexes in the molecular mass range of 20,000–200,000 Da were also demonstrated resulting in linear Ferguson plots of log mobility versus % gel concentration, intercepting at 0% gel concentration representing the approximate free solution mobility in polyethylene oxide gels [184,185].

4.3.2 Separation voltage and temperature

Morris and coworkers [186] demonstrated the effect of analyte velocity modulation in CGE of nucleic acids by varying the separation voltage. Sinusoidal variation of the applied electric field strength improved resolution for both large and small size DNA fragments, while concomitantly decreased separation times. The authors suggested that the sinusoidal variation, similar to conventional pulsed field electrophoresis, probably caused reorientation of the fragments during CGE. The effect of electric field strength gradients on the separation of DNA restriction fragments was investigated by Guttman et al. [187]. It is known that the mobility of different size double-stranded DNA fragments is a function of the applied electric field strength, suggesting that the use of a nonuniform (i.e., time varying) electric field may alter the resolving power. Indeed, enhanced separation of DNA restriction fragments of up to 1,353 base pairs (bp) in size was achieved by applying a field strength gradient method as shown in Fig. 4.36. The shape of the gradient can be continuous or stepwise over time.

Mitnik et al. conducted a systematic study of dsDNA separation in hydroxypropylcellulose solution evaluating the concentration range from 0.1% to 1% and electric fields from 6 to 540 V/cm [188]. Their results using polymer concentrations of >0.4% and low field strengths showed good agreement with a recently proposed theory for gels with a pore size smaller than the persistence length of DNA. However, for more diluted solutions and high fields, the separation pattern was confirmed by the direct observation of the conformation of double-stranded DNA molecules in the polymer solution by fluorescence videomicroscopy.

Sequence-induced anomalous migration of dsDNA under native gel electrophoresis conditions is a well-known phenomenon that is greatly affected by the gel concentration and separation temperature. The problem is that this anomalous migration results in discrepancies between calculated and actual molecular sizes. Comparable to slab gel electrophoresis, Wenz reported strong dependence of the retardation effect, indicative of bent or curved DNA, as a function of polymer concentration and separation temperature in CGE [189].

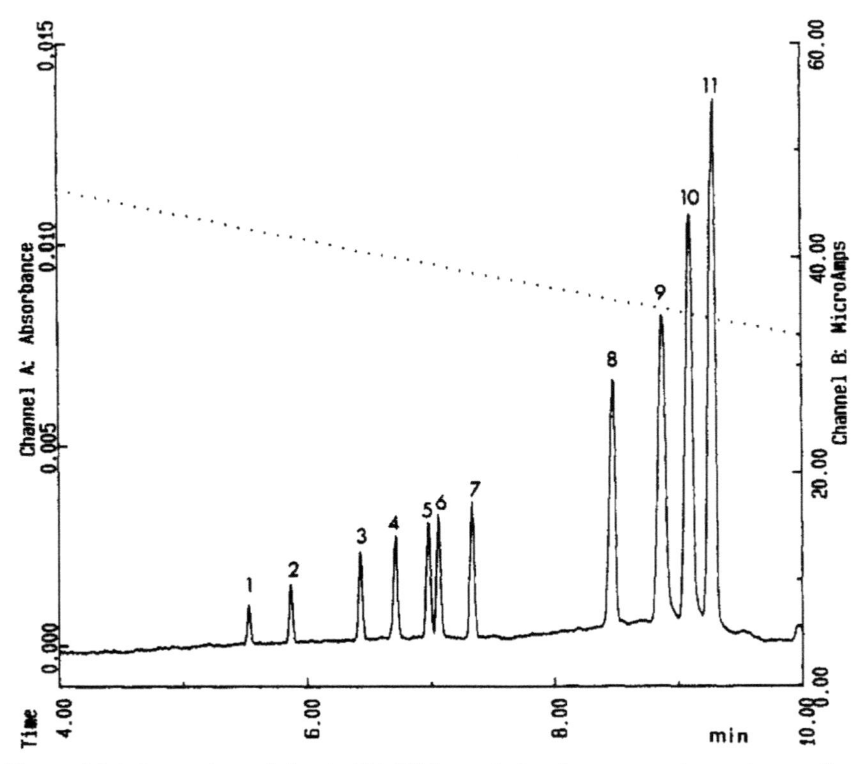

Figure 4.36 Separation of the ϕx174 DNA restriction fragment mixture by capillary gel electrophoresis with a decreasing voltage gradient. The dotted line represents the current output. *With permission from A. Guttman, B. Wanders, N. Cooke, Enhanced separation of DNA restriction fragments by capillary gel electrophoresis using field strength gradients, Analytical Chemistry 64 (20) (1992) 2348–2351 [187].*

The effect of temperature was studied in two different separation modes of high-performance CGE on the separation of DNA restriction fragments ranging up to 1.3 kbp [190]. In isoelectrostatic (use of constant applied electric field) and isorheic (use of constant current) modes, the migration properties and resolution of the dsDNA molecules were investigated as the function of capillary temperature between 20°C and 50°C. In the constant field separation mode, the migration time and resolution decreased with increasing temperature; however, in the constant current mode, an increase in the column temperature resulted a maximum in resolution, as shown in Fig. 4.37.

Temperature-mediated constant denaturant capillary electrophoresis (CDCE) was used to separate wild-type and different mutant N-ras exon

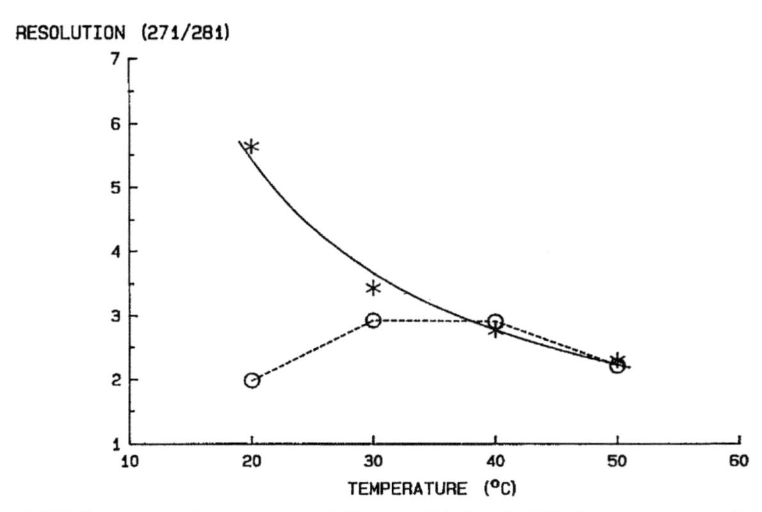

Figure 4.37 Resolution between the 271 and 281-bp dsDNA fragments as a function of the separation temperature using isoelectrostatic (*) and isorheic (○) separation modes. *With permission from A. Guttman, N. Cooke, Effect of temperature on the separation of DNA restriction fragments in capillary gel electrophoresis, Journal of Chromatography 559 (1–2) (1991) 285–294 [190].*

1 and 2 sequences by Thilly and coworkers [191]. Amplified DNA fragments of 151 bp (exon 1) and 150 bp (exon 2) were analyzed by CDCE mode with the denaturant zone temperature in the capillary corresponding to the melting temperature of the 111 bp (exon 1) and 110 bp (exon 2) low-melting domains. The denaturation and reannealing of wild-type and mutant fragments together created wild-type/mutant heteroduplexes, which were resolved from wild-type homoduplex.

Temperature-programmed CE was applied by Gelfi et al. for DNA point mutation analysis [192]. The method was based on the principle of temperature gradient gel electrophoresis, an alternative to denaturing gradient gel electrophoresis. Differential melting of mutant and wild-type PCR-amplified DNA fragments migrated along the separation space with a time programmed temperature gradient of typically 1°C–1.5°C/minute. Point mutants were fully resolved into a spectrum of four bands, with a dynamic range extending from 45°C up to 70°C.

The effect of separation temperature on peak efficiency in CGE was investigated using low-molecular-mass pharmaceutical compounds and carbohydrate molecules, as well as high-molecular-mass dsDNA and protein molecules [193]. It was observed that the simple reciprocal relationship between the

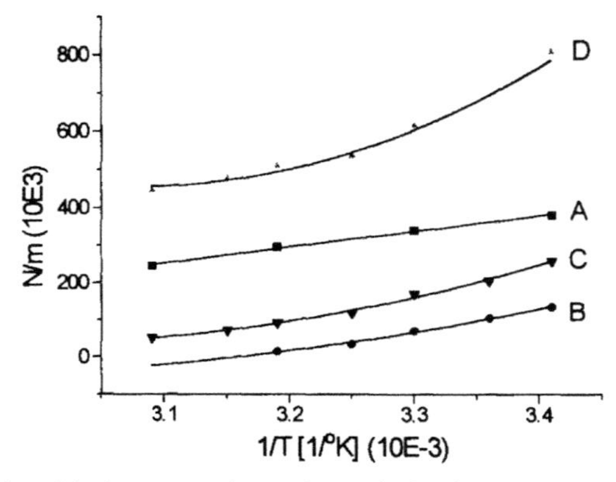

Figure 4.38 Relationship between the reciprocal absolute temperature and the theoretical plate numbers of (A) dsDNA-1,353-mer, (B) SDS–ovalbumin complex, (C) maltotriose-ANTS, and (D) R-propranolol. *With permission from A. Guttman, Effect of temperature on separation efficiency in capillary gel electrophoresis, Trends in Analytical Chemistry 15 (5) (1996) 194–198 [193].*

separation temperature and the theoretical plate numbers was distorted when the charge of the analyte changed with temperature. Also, the efficiency of the separations was susceptible to buffer pH changes and complexation by temperature variations. Fig. 4.38 shows the relationship between the reciprocal absolute temperature and the theoretical plate numbers for dsDNA fragments, SDS–protein complexes, as well as oligosaccharides and small drug molecules.

4.3.3 Capillary dimensions

Fluctuations of the lateral dimensions of a capillary influence migration time as well as peak width and concomitantly the resolution. The local electric resistance in a capillary tubing is inversely proportional to the actual capillary cross section, thus the intensity of the electric field may vary if the inner diameter of the capillary is not constant [194]. Slater and Mayer developed a theoretical framework for understanding such effects by examining a simple case where both the mobility and the diffusion coefficients were field-independent and concluded that optimal resolution was always obtained for perfectly flat walls.

The influence of column coiling on performance in CGE was studied by the Karger group [195]. While in open tubular columns, no significant effect of column coiling was observed on efficiency, with gel-filled columns the influence was significant, for example, a factor of 3 or more loss in plate count per coil. For relatively rigid gels of 9% T or higher LPA concentrations with 7 M urea, the loss of column efficiency was probably due to the inability of the polymer network to diffusional relaxation. A significant influence of coiling on column efficiency was also found with medium concentration gels of 3%–6% T LPA with no urea, where the polymer network structure was assumed to change under the influence of mechanical stress caused by coiling, resulting in anisotropy across the column. The change in separation performance of a DNA sequencing mixture in straight and coiled capillaries caused significant loss of resolution observable in the latter case, as shown in Fig. 2.4 (in Chapter 2: Basic principles of capillary gel electrophoresis).

4.3.4 Buffer systems

The effect of buffer system pH on the electrophoretic migration properties of single-stranded oligodeoxyribonucleotides in CGE was investigated by Guttman et al. [196]. Homooligodeoxyribonucleotides with equal chain lengths but with different monomer compositions were injected and the resulting electropherograms showed significant differences (Fig. 4.39) in relative migration when the pH of the gel-buffer and background electrolyte was varied from pH 6 to 8.

Application of an electrolyte step gradient during capillary gel electrophoretic separation of dsDNA fragments resulted in apparent peak shape improvement and decrease in the required analysis time [197]. Palm performed the separation of small DNA fragments on low-melting agarose gels at different Tris-acetate-EDTA (TAE) buffer concentrations [198]. He found that all DNA fragments maximally resolved at around 0.160–0.20 M TAE concentration (except the 634/800 bp fragment pair, where the resolution maxima was at 0.260 M). Fig. 4.40 shows the dependence of resolution of the different DNA fragments on the TAE buffer concentration.

4.3.5 Nonaqueous electrophoresis, organic modifiers

Organic solvents can be used as the background electrolyte (BGE) in nonaqueous capillary electrophoretic (NACE) separations if fulfill the following criteria: (1) suitable liquid range, (2) sufficient relative permittivity (ε), (3) relatively low viscosity (η), (4) good solvent properties, (5) not too high volatility, (6) chemical stability, (7) availability at reasonable cost and purity, and (8) compatibility with instrumental demands [199]. The most

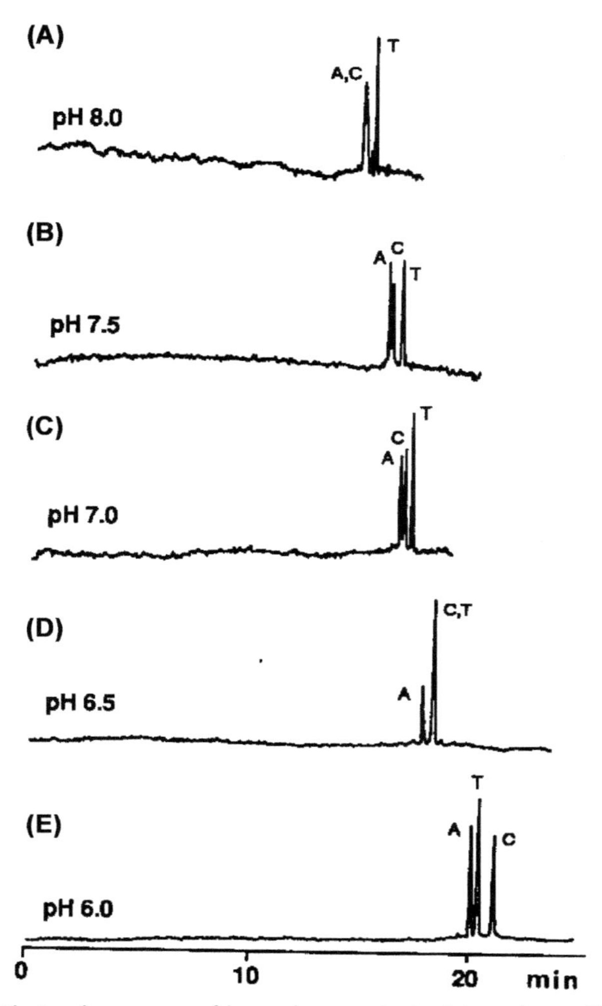

Figure 4.39 Electropherograms of homodecamer test mixtures by capillary gel electrophoresis at different pH values. All gels were prepared at the same pH as the running buffer: (A) pH 8.0; (B) pH 7.5; (C) pH 7.0; (D) pH 6.5; (E) pH 6.0. Peaks: A = p $(dA)_{10}$; C = $p(dC)_{10}$; T = $p(dT)_{10}$. *With permission from A. Guttman, A. Arai, K. Magyar, Influence of pH on the migration properties of oligonucleotides in capillary gel electrophoresis, Journal of Chromatography 608(1–2) (1992) 175–179 [196].*

relevant organic solvent parameter in NACE is relative permittivity because it describes the strength of the interactions between ions in the solvent. A preferable solvent has $\varepsilon > 30$, in which case the ionic dissociation of electrolytes can be considered to take place to a large or full

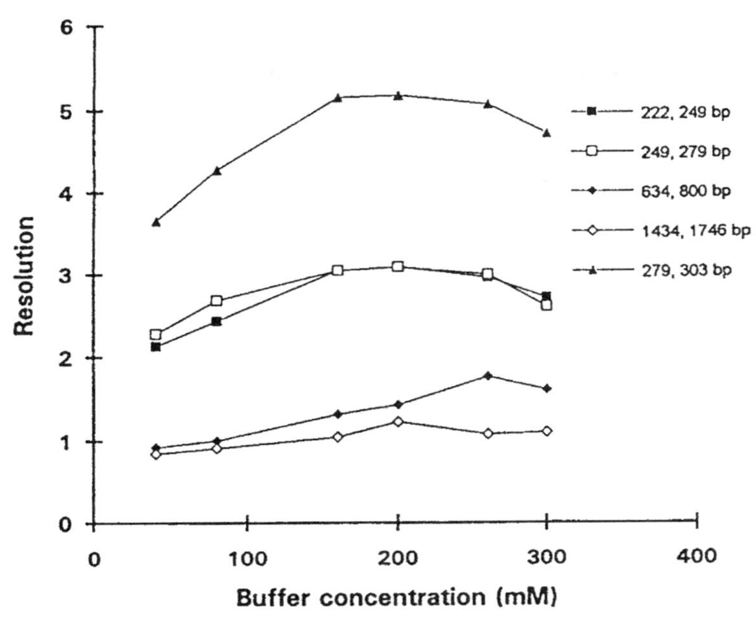

Figure 4.40 Resolution of DNA fragment pairs in low-melting agarose CGE as a function of TAE buffer concentration. *With permission from A.K. Palm, Capillary electrophoresis of DNA fragments with replaceable low-gelling agarose gels, Methods in Molecular Biology 162 (2001) 279–290 [198].*

extent. Although ethanol and 1-propanol with $10 \leq \varepsilon \leq 30$ can also be used as solvents, the dominant effect is extensive ion-pairing and ionic dissociation takes place to some extent only [200].

A strongly denaturing, nonaqueous CGE method was also developed for size separation of large RNA molecules [201]. The system addressed the analysis of mRNAs of up to 6000 nucleotides long. Comparing with a standard aqueous CGE method, the resolution of some components of an RNA ladder increased in the nonaqueous buffer system as shown in Fig. 3.31 (in Chapter 3: Separation matrix and capillary column technology). The authors optimized the polymer type, molecular weight, and polymer concentration on the resolution separation.

References

[1] S. Terabe, K. Otsuka, T. Ando, Band broadening in electrokinetic chromatography with micellar solutions and open-tubular capillaries, Analytical Chemistry 61 (3) (1989) 251–260.

[2] F.E.P. Mikkers, F.M. Everaerts, T.P.E.M. Verheggen, Concentration distributions in free zone electrophoresis, Journal of Chromatography 169 (1979) 1–10.

[3] H.F. Yin, S.R. Motsch, J.A. Lux, G. Schomburg, A miniature device for electrokinetic or hydrodynamic sample introduction from small volumes in capillary electrophoresis, HRC-Journal of High Resolution Chromatography 14 (4) (1991) 282–284.
[4] P.B. Hietpas, A.G. Ewing, DNA Separations from nanovials, Journal of Liquid Chromatography and Related Technologies 23 (1) (2000) 15–24.
[5] S.F.Y. Li, Capillary Electrophoresis: Principles, Practice and Application Journal of Chromatographic Library 52 (1992) pp. 48.
[6] J. Tehrani, R. Macomber, L. Day, Capillary electrophoresis—an integrated system with a unique split-flow sample introduction mechanism, HRC-Journal of High Resolution Chromatography 14 (1) (1991) 10–14.
[7] J.A. Lux, H.F. Yin, G. Schomburg, Construction, evaluation and analytical operation of a modular capillary electrophoresis instrument, Chromatographia 30 (1–2) (1990) 7–15.
[8] L. Nozal, L. Arce, B.M. Simonet, A. Rios, M. Valcarcel, New supported liquid membrane-capillary electrophoresis in-line arrangement for direct selective analysis of complex samples, Electrophoresis 27 (15) (2006) 3075–3085.
[9] Y. Rang, H. Zeng, H. Nakajima, S. Kato, K. Uchiyama, Quantitative on-line concentration for capillary electrophoresis with inkjet sample introduction technique, Journal of Separation Science 38 (15) (2015) 2722–2728.
[10] W.A. Williams, A. Hendrickson, A.F. Gillaspy, D.W. Dyer, L.A. Lewis, Oligonucleotide analysis by sequential injection before analysis (SIBA) capillary electrophoresis, Analytical Biochemistry 313 (1) (2003) 183–185.
[11] N.L. Kuehnbaum, A. Kormendi, P. Britz-McKibbin, Multisegment injection-capillary electrophoresis-mass spectrometry: a high-throughput platform for metabolomics with high data fidelity, Analytical Chemistry 85 (22) (2013) 10664–10669.
[12] Z. Kovács, M. Szarka, M. Szigeti, A. Guttman, Separation window dependent multiple injection (SWDMI) for large scale analysis of therapeutic antibody N-glycans, Journal of Pharmaceutical and Biomedical Analysis 128 (2016) 367–370.
[13] T. Sonehara, H. Kawazoe, T. Sakai, S. Ozawa, T. Anazawa, T. Irie, Ultra-slim laminated capillary array for high-speed DNA separation, Electrophoresis 27 (14) (2006) 2910–2916.
[14] X. Xu, G. Luo, X. Chu, B. Lin, Electrokinetic injection in capillary gel electrophoresis, Chinese Science Bulletin 40 (6) (1995) 482–487.
[15] X. Huang, M.J. Gordon, R.N. Zare, Bias in quantitative capillary zone electrophoresis caused by electrokinetic sample injection, Analytical Chemistry 60 (4) (1988) 375–377.
[16] K. Kleparnik, M. Garner, P. Bocek, Injection bias of DNA fragments in capillary electrophoresis with sieving, Journal of Chromatography A 698 (1–2) (1995) 375–383.
[17] J.R. Catai, E. Carrilho, Simplex optimization of electrokinetic injection of DNA in capillary electrophoresis using dilute polymer solution, Electrophoresis 24 (4) (2003) 648–654.
[18] B. Chen, G. Chen, M.G. Bartlett, Factors influencing the electrokinetic injection of oligonucleotides in capillary gel electrophoresis when using laser-induced fluorescence detection, Biomedical Chromatography: BMC 28 (3) (2014) 320–323.
[19] A. Guttman, H.E. Schwartz, Artifacts related to sample introduction in capillary gel electrophoresis affecting separation performance and quantitation, Analytical Chemistry 67 (13) (1995) 2279–2283.
[20] A. Allmendinger, L.H. Dieu, S. Fischer, R. Mueller, H.C. Mahler, J. Huwyler, High-throughput viscosity measurement using capillary electrophoresis

instrumentation and its application to protein formulation, Journal of Pharmaceutical and Biomedical Analysis 99 (2014) 51–58.
[21] M.C. Breadmore, Electrokinetic and hydrodynamic injection: making the right choice for capillary electrophoresis, Bioanalysis 1 (5) (2009) 889–894.
[22] S.E.Y. Li, Chapter 2, Sample injection methods, in: S.E.Y. Li (Ed.), Journal of Chromatography Library, Elsevier, 1992, pp. 31–54.
[23] R. Bernard, G. Loge, Poly(2-ethyl-2-oxazoline) as a sieving matrix for SDS-CE, Electrophoresis 30 (23) (2009) 4059–4062.
[24] M.R. Monton, S. Terabe, Field-enhanced sample injection for high-sensitivity analysis of peptides and proteins in capillary electrophoresis-mass spectrometry, Journal of Chromatography. A 1032 (1–2) (2004) 203–211.
[25] R.L. Chien, D.S. Burgi, Field-amplified polarity-switching sample injection in high-performance capillary electrophoresis, Journal of Chromatography 559 (1–2) (1991) 153–161.
[26] R.L. Chien, D.S. Burgi, Field amplified sample injection in high performance capillary electrophoresis, Journal of Chromatography A 559 (1991) 141–152.
[27] A. Vinther, F.M. Everaerts, H. Soeberg, Differences in absorbency levels as found in capillary zone electrophoresis under stacking conditions, HRC-Journal of High Resolution Chromatography 13 (9) (1990) 639–642.
[28] A. Vinther, X.H. Soeberg, Temperature elevations of the sample zone in free solution capillary electrophoresis under stacking conditions, Journal of Chromatography 559 (1991) 27–42.
[29] C.X. Zhang, M.M. Meagher, Sample stacking provides three orders of magnitude sensitivity enhancement in SDS capillary gel electrophoresis of adeno-associated virus capsid proteins, Analytical Chemistry 89 (6) (2017) 3285–3292.
[30] G. Jarvas, A. Guttman, N. Miekus, T. Baczek, S. Jeong, D.S. Chung, et al., Practical sample pretreatment techniques coupled with capillary electrophoresis for real samples in complex matrices, Trends in Analytical Chemistry 122 (2020) 115702.
[31] M.J. van der Schans, J.L. Beckers, M.C. Molling, F.M. Everaerts, Intrinsic isotachophoretic preconcentration in capillary gel electrophoresis of DNA restriction fragments, Journal of Chromatography A 717 (1–2) (1995) 139–147.
[32] F. Foret, E. Szoko, B.L. Karger, On-column transient and coupled column isotachophoretic preconcentration of protein samples in capillary zone electrophoresis, Journal of Chromatography 608 (1–2) (1992) 3–12.
[33] F. Foret, E. Szoko, B.L. Karger, Trace analysis of proteins by capillary zone electrophoresis with on-column transient isotachophoretic preconcentration, Electrophoresis 14 (5–6) (1993) 417–428.
[34] H.W. Zhang, X.G. Chen, Z.D. Hu, Influence of sample injection time of ions on migration time in capillary zone electrophoresis, Journal of Chromatography A 677 (1) (1994) 159–167.
[35] T.K. Khurana, J.G. Santiago, Sample zone dynamics in peak mode isotachophoresis, Analytical Chemistry 80 (16) (2008) 6300–6307.
[36] A. Persat, L.A. Marshall, J.G. Santiago, Purification of nucleic acids from whole blood using isotachophoresis, Analytical Chemistry 81 (22) (2009) 9507–9511.
[37] V. Datinská, I. Voráčová, U. Schlecht, J. Berka, F. Foret, Recent progress in nucleic acids isotachophoresis, Journal of Separation Science 41 (1) (2018) 236–247.
[38] V.N. Kondratova, O.I. Serd'uk, V.P. Shelepov, A.V. Lichtenstein, Concentration and isolation of DNA from biological fluids by agarose gel isotachophoresis, Biotechniques 39 (5) (2005) 695–699.
[39] W. Tong, A. Link, J.K. Eng, J.R. Yates 3rd, Identification of proteins in complexes by solid-phase microextraction/multistep elution/capillary electrophoresis/tandem mass spectrometry, Analytical Chemistry 71 (13) (1999) 2270–2278.

[40] E. Rohde, A.J. Tomlinson, D.H. Johnson, S. Naylor, Comparison of protein mixtures in aqueous humor by membrane preconcentration—capillary electrophoresis—mass spectrometry, Electrophoresis 19 (13) (1998) 2361—2370.
[41] C.J. Herring, J. Qin, An on-line preconcentrator and the evaluation of electrospray interfaces for the capillary electrophoresis/mass spectrometry of peptides, Rapid Communications in Mass Spectrometry: RCM 13 (1) (1999) 1—7.
[42] L.H. Zhang, X.Z. Wu, Capillary electrophoresis with in-capillary solid-phase extraction sample cleanup, Analytical Chemistry 79 (6) (2007) 2562—2569.
[43] X.Z. Wu, R. Umeda, In-capillary preconcentration of proteins for capillary electrophoresis using a cellulose acetate-coated porous joint, Analytical and Bioanalytical Chemistry 382 (3) (2005) 848—852.
[44] I.H. Sung, Y.W. Lee, D.S. Chung, Liquid extraction surface analysis in-line coupled with capillary electrophoresis for direct analysis of a solid surface sample, Analytica Chimica Acta 838 (2014) 45—50.
[45] N.A. Guzman, D.E. Guzman, A two-dimensional affinity capture and separation mini-platform for the isolation, enrichment, and quantification of biomarkers and its potential use for liquid biopsy, Biomedicines 8 (8) (2020) 255.
[46] X. Xu, W.T. Kok, H. Poppe, Change of pH in electrophoretic zones as a cause of peak deformation, Journal of Chromatography A 742 (1—2) (1996) 211—227.
[47] H.H. Lauer, D. Mcmanigill, Zone electrophoresis in open-tubular capillaries—recent advances, Trends in Analytical Chemistry 5 (1) (1986) 11—15.
[48] M.J. van der Schans, J.K. Allen, B.J. Wanders, A. Guttman, Effects of sample matrix and injection plug on dsDNA migration in capillary gel electrophoresis, Journal of Chromatography A 680 (2) (1994) 511—516.
[49] J.A. Lux, U. Hausig, G. Schomburg, Production of windows in fused-silica capillaries for in-column detection of UV-absorption or fluorescence in capillary electrophoresis or HPLC, HRC-Journal of High Resolution Chromatography 13 (5) (1990) 373—374.
[50] R.M. Mccormick, R.J. Zagursky, Polyimide stripping device for producing detection windows on fused-silica tubing used in capillary electrophoresis, Analytical Chemistry 63 (7) (1991) 750—752.
[51] G.J.M. Bruin, G. Stegeman, A.C. Vanasten, X. Xu, J.C. Kraak, H. Poppe, Optimization and evaluation of the performance of arrangements for UV detection in high-resolution separations using fused-silica capillaries, Journal of Chromatography 559 (1—2) (1991) 163—181.
[52] T. Tsuda, J.V. Sweedler, R.N. Zare, Rectangular capillaries for capillary zone electrophoresis, Analytical Chemistry 62 (19) (1990) 2149—2152.
[53] T.S. Wang, J.H. Aiken, C.W. Huie, R.A. Hartwick, Nanoliter-scale multireflection cell for absorption detection in capillary electrophoresis, Analytical Chemistry 63 (14) (1991) 1372—1376.
[54] J.A. Taylor, E.S. Yeung, Axial-beam absorbency detection for capillary electrophoresis, Journal of Chromatography 550 (1—2) (1991) 831—837.
[55] P. Gebauer, W. Thormann, Isotachophoresis in open-tubular fused-silica capillaries with on-column multiwavelength detection, Journal of Chromatography 545 (2) (1991) 299—305.
[56] D. Demorest, R. Dubrow, Factors influencing the resolution and quantitation of oligonucleotides separated by capillary electrophoresis on a gel-filled capillary, Journal of Chromatography 559 (1—2) (1991) 43—56.
[57] P. Sun, R.A. Hartwick, The effect of electric fields on the dispersion of oligonucleotides using a multi-point detection method in capillary gel electrophoresis, Journal of Liquid Chromatography 17 (9) (1994) 1861—1875.

[58] P. Lindberg, A. Hanning, T. Lindberg, J. Roeraade, Fiber-optic-based UV–visible absorbance detector for capillary electrophoresis, utilizing focusing optical elements, Journal of Chromatography A 809 (1) (1998) 181–189.
[59] F. Foret, P. Bocek, Use of optical fibers for the adaptation of UV detector LCD 254 for online detection in capillary columns, Chemické Listy 83 (2) (1989) 191–193.
[60] M.J. Sepaniak, D.F. Swaile, A.C. Powell, Instrumental developments in micellar electrokinetic capillary chromatography, Journal of Chromatography 480 (1989) 185–196.
[61] Y. Baba, R. Tomisaki, M. Tsuhako, Three-dimensional electropherogram for the separation of oligodeoxynucleotides and DNA restriction fragments using capillary gel electrophoresis with a photodiode array detector, Journal of Liquid Chromatography 18 (7) (1995) 1317–1324.
[62] S.E. Moring, R.T. Reel, R.E.J. Vansoest, Optical improvements of a Z-shaped cell for high-sensitivity UV absorbency detection in capillary electrophoresis, Analytical Chemistry 65 (23) (1993) 3454–3459.
[63] J.P. Chervet, R.E.J. Vansoest, M. Ursem, Z-shaped flow cell for UV-detection in capillary electrophoresis, Journal of Chromatography 543 (2) (1991) 439–449.
[64] J. Macek, U.R. Tjaden, J. Van der Greef, Resolution and concentration detection limit in capillary gel electrophoresis, Journal of Chromatography 545 (1) (1991) 177–182.
[65] Y. Chen, J.-V. Hoeltje, U. Schwarz, Improving the UV detection sensitivity of condensed polyacrylamide gel-filled capillaries using non-flow buffer-filled capillaries as a detection cell, Journal of Chromatography A 700 (1–2) (1995) 35–42.
[66] F. Maystre, A.E. Bruno, Laser-beam probing in capillary tubes, Analytical Chemistry 64 (22) (1992) 2885–2887.
[67] A. Palm, C. Lindh, S. Hjerten, J. Pawliszynn, Capillary zone electrophoresis in agarose gels using absorption imaging detection, Electrophoresis 17 (4) (1996) 766–770.
[68] J.Q. Wu, J. Pawliszyn, Absorption-spectra and multicapillary imaging detection for capillary isoelectric-focusing using a charge-coupled-device camera, Analyst 120 (5) (1995) 1567–1571.
[69] L. Krcmova, A. Stjernlof, S. Mehlen, P.C. Hauser, S. Abele, B. Paull, et al., Deep-UV-LEDs in photometric detection: a 255 nm LED on-capillary detector in capillary electrophoresis, Analyst 134 (12) (2009) 2394–2396.
[70] W.G. Kuhr, E.S. Yeung, Optimization of sensitivity and separation in capillary zone electrophoresis with indirect fluorescence detection, Analytical Chemistry 60 (23) (1988) 2642–2646.
[71] J.W. Jorgenson, K.D. Lukacs, Zone electrophoresis in open-tubular glass capillaries, Analytical Chemistry 53 (8) (1981) 1298–1302.
[72] M. Albin, R. Weinberger, E. Sapp, S. Moring, Fluorescence detection in capillary electrophoresis—evaluation of derivatizing reagents and techniques, Analytical Chemistry 63 (5) (1991) 417–422.
[73] D.J. Rose, Free-solution reactor for postcolumn fluorescence detection in capillary zone electrophoresis, Journal of Chromatography 540 (1–2) (1991) 343–353.
[74] D.F. Swaile, M.J. Sepaniak, Laser-based fluorometric detection schemes for the analysis of proteins by capillary zone electrophoresis, Journal of Liquid Chromatography 14 (5) (1991) 869–893.
[75] H. Drossman, J.A. Luckey, A.J. Kostichka, J. D'Cunha, L.M. Smith, High-speed separations of DNA sequencing reactions by capillary electrophoresis, Analytical Chemistry 62 (9) (1990) 900–903.
[76] Y.F. Cheng, S.L. Wu, D.Y. Chen, N.J. Dovichi, Interaction of capillary zone electrophoresis with a sheath flow cuvette detector, Analytical Chemistry 62 (5) (1990) 496–503.

[77] H. Swerdlow, J.Z. Zhang, D.Y. Chen, H.R. Harke, R. Grey, S. Wu, et al., Three DNA sequencing methods using capillary gel electrophoresis and laser-induced fluorescence, Analytical Chemistry 63 (24) (1991) 2835–2841.
[78] Y.F. Cheng, N.J. Dovichi, Subattomole amino acid analysis by capillary zone electrophoresis and laser-induced fluorescence, Science (New York, N.Y.) 242 (4878) (1988) 562–564.
[79] D.Y. Chen, H.P. Swerdlow, H.R. Harke, J.Z. Zhang, N.J. Dovichi, Low-cost, high-sensitivity laser-induced fluorescence detection for DNA sequencing by capillary gel electrophoresis, Journal of Chromatography 559 (1–2) (1991) 237–246.
[80] J.A. Luckey, H. Drossman, A.J. Kostichka, D.A. Mead, J. D'Cunha, T.B. Norris, et al., High speed DNA sequencing by capillary electrophoresis, Nucleic Acids Research 18 (15) (1990) 4417–4421.
[81] A.E. Karger, J.M. Harris, R.F. Gesteland, Multiwavelength fluorescence detection for DNA sequencing using capillary electrophoresis, Nucleic Acids Research 19 (18) (1991) 4955–4962.
[82] J.V. Sweedler, J.B. Shear, H.A. Fishman, R.N. Zare, R.H. Scheller, Fluorescence detection in capillary zone electrophoresis using a charge-coupled device with time-delayed integration, Analytical Chemistry 63 (5) (1991) 496–502.
[83] L. Hernandez, J. Escalona, N. Joshi, N. Guzman, Laser-induced fluorescence and fluorescence microscopy for capillary electrophoresis zone detection, Journal of Chromatography 559 (1–2) (1991) 183–196.
[84] L. Hernandez, R. Marquina, J. Escalona, N.A. Guzman, Detection and quantification of capillary electrophoresis zones by fluorescence microscopy, Journal of Chromatography 502 (2) (1990) 247–255.
[85] X.C. Huang, M.A. Quesada, R.A. Mathies, Capillary array electrophoresis using laser-excited confocal fluorescence detection, Analytical Chemistry 64 (8) (1992) 967–972.
[86] X.C. Huang, M.A. Quesada, R.A. Mathies, DNA sequencing using capillary array electrophoresis, Analytical Chemistry 64 (18) (1992) 2149–2154.
[87] R.E. Milofsky, E.S. Yeung, Native fluorescence detection of nucleic acids and DNA restriction fragments in capillary electrophoresis, Analytical Chemistry 65 (2) (1993) 153–157.
[88] K. Ueno, E.S. Yeung, Simultaneous monitoring of DNA fragments separated by electrophoresis in a multiplexed array of 100 capillaries, Analytical Chemistry 66 (9) (1994) 1424–1431.
[89] S.C. Beale, S.J. Sudmeier, Spatial-scanning laser fluorescence detection for capillary electrophoresis, Analytical Chemistry 67 (18) (1995) 3367–3371.
[90] M. Jansson, J. Roeraade, F. Laurell, Laser-induced fluorescence detection in capillary electrophoresis with blue-light from a frequency-doubled diode-laser, Analytical Chemistry 65 (20) (1993) 2766–2769.
[91] H. Kawazumi, J.M. Song, T. Inoue, T. Ogawa, Laser fluorometry using a visible semiconductor-laser and an avalanche photodiode for capillary electrophoresis, Analytical Sciences 11 (4) (1995) 587–590.
[92] F.T.A. Chen, A. Tusak, S. Pentoney, K. Konrad, C. Lew, E. Koh, et al., Semiconductor laser-induced fluorescence detection in capillary electrophoresis using a cyanine dye, Journal of Chromatography. A 652 (2) (1993) 355–360.
[93] T. Fuchigami, T. Imasaka, M. Shiga, Subatomole detection of amino-acids by capillary electrophoresis based on semiconductor-laser fluorescence detection, Analytica Chimica Acta 282 (1) (1993) 209–213.
[94] A.J.G. Mank, H. Lingeman, C. Gooijer, Semiconductor-laser-induced fluorescence detection in capillary electrophoresis using precolumn thiol derivatization, HRC-Journal of High Resolution Chromatography 17 (11) (1994) 797–798.

[95] A.J.G. Mank, E.S. Yeung, Diode laser-induced fluorescence detection in capillary electrophoresis after precolumn derivatization of amino-acids and small peptides, Journal of Chromatography A 708 (2) (1995) 309–321.
[96] T. Fuchigami, T. Imasaka, M. Shiga, Ultratrace analysis of biological substances by capillary electrophoresis/semiconductor laser fluorometry, in: Proc. SPIE-Int. Soc. Opt. Eng., Proceedings of Advances in Fluorescence Sensing Technology, 1993, vol. 1885, pp. 435–438.
[97] S. Zhao, H. Yuan, D. Xiao, Optical fiber light-emitting diode-induced fluorescence detection for capillary electrophoresis, Electrophoresis 27 (2) (2006) 461–467.
[98] S. Hapuarachchi, G.A. Janaway, C.A. Aspinwall, Capillary electrophoresis with a UV light-emitting diode source for chemical monitoring of native and derivatized fluorescent compounds, Electrophoresis 27 (20) (2006) 4052–4059.
[99] S. Nilsson, J. Johansson, M. Mecklenburg, S. Birnbaum, S. Svanberg, K.G. Wahlund, et al., Real-time fluorescence imaging of capillary electrophoresis, Journal of Capillary Electrophoresis 2 (1) (1995) 46–52.
[100] S. Takahashi, K. Murakami, T. Anazawa, H. Kambara, Multiple sheath-flow gel capillary-array electrophoresis for multicolor fluorescent DNA detection, Analytical Chemistry 66 (7) (1994) 1021–1026.
[101] R.J. Zagursky, R.M. McCormick, DNA sequencing separations in capillary gels on a modified commercial DNA sequencing instrument, Biotechniques 9 (1) (1990) 74–79.
[102] I. Kheterpal, R.A. Mathies, Capillary array electrophoresis DNA sequencing, Analytical Chemistry 71 (1) (1999) 31A–37A.
[103] S.X. Lu, E.S. Yeung, Side-entry excitation and detection of square capillary array electrophoresis for DNA sequencing, Journal of Chromatography A 853 (1–2) (1999) 359–369.
[104] H. He, B.K. Nunnally, L.-C. Li, L.B. McGown, On-the-fly fluorescence lifetime detection of dye-labeled DNA primers for multiplex analysis, Analytical Chemistry 70 (16) (1998) 3413–3418.
[105] L. Li, L.B. McGown, Effects of gel material on fluorescence lifetime detection of dyes and dye-labeled DNA primers in capillary electrophoresis, Journal of Chromatography A 841 (1) (1999) 95–103.
[106] J.H. Flanagan Jr., C.V. Owens, S.E. Romero, E. Waddell, S.H. Kahn, R.P. Hammer, et al., Near-infrared heavy-atom-modified fluorescent dyes for base-calling in DNA-sequencing applications using temporal discrimination, Analytical Chemistry 70 (13) (1998) 2676–2684.
[107] U. Lieberwirth, J. Arden-Jacob, K.H. Drexhage, D.P. Herten, R. Mueller, M. Neumann, et al., Multiplex dye DNA sequencing in capillary gel electrophoresis by diode laser-based time-resolved fluorescence detection, Analytical Chemistry 70 (22) (1998) 4771–4779.
[108] A. Hanning, J. Westberg, J. Roeraade, A liquid core waveguide fluorescence detector for multicapillary electrophoresis applied to DNA sequencing in a 91-capillary array, Electrophoresis 21 (15) (2000) 3290–3304.
[109] J. Zhang, M. Yang, X. Puyang, Y. Fang, L.M. Cook, N.J. Dovichi, Two-dimensional direct-reading fluorescence spectrograph for DNA sequencing by capillary array electrophoresis, Analytical Chemistry 73 (6) (2001) 1234–1239.
[110] O.O. Dada, D.C. Essaka, O. Hindsgaul, M.M. Palcic, J. Prendergast, R.L. Schnaar, et al., Nine orders of magnitude dynamic range: picomolar to millimolar concentration measurement in capillary electrophoresis with laser induced fluorescence detection employing cascaded avalanche photodiode photon counters, Analytical Chemistry 83 (7) (2011) 2748–2753.

[111] M. Szarka, A. Guttman, Smartphone cortex controlled real-time image processing and reprocessing for concentration independent LED induced fluorescence detection in capillary electrophoresis, Analytical Chemistry 89 (20) (2017) 10673–10678.
[112] T. Wang, R.A. Hartwick, Noise and detection limits of indirect absorption detection in capillary zone electrophoresis, Journal of Chromatography 607 (1) (1992) 119–125.
[113] R. Kuhn, S. Hofstetter-Kuhn, Capillary Electrophoresis: Principles and Practice, Springer Laboratory, Berlin, 1993.
[114] S. Hjerten, K. Elenbring, F. Kilar, J.L. Liao, A.J. Chen, C.J. Siebert, et al., Carrier-free zone electrophoresis, displacement electrophoresis and isoelectric focusing in a high-performance electrophoresis apparatus, Journal of Chromatography 403 (1987) 47–61.
[115] K.C. Chan, C.W. Whang, E.S. Yeung, Separation of DNA restriction fragments using capillary electrophoresis, Journal of Liquid Chromatography 16 (9–10) (1993) 1941–1962.
[116] J.C. Ren, B.C. Li, Y.Z. Deng, J.K. Cheng, Indirect thermo-optical detection for capillary electrophoresis, Talanta 42 (12) (1995) 1891–1895.
[117] N. Anik, M. Airiau, M.-P. Labeau, C.-T. Vuong, J. Reboul, P. Lacroix-Desmazes, et al., Determination of polymer effective charge by indirect UV detection in capillary electrophoresis: toward the characterization of macromolecular architectures, Macromolecules (Washington, DC) 42 (7) (2009) 2767–2774.
[118] R.A. Wallingford, A.G. Ewing, Capillary zone electrophoresis with electrochemical detection, Analytical Chemistry 59 (14) (1987) 1762–1766.
[119] P.L. Weber, S.M. Lunte, Capillary electrophoresis with pulsed amperometric detection of carbohydrates and glycopeptides, Electrophoresis 17 (2) (1996) 302–309.
[120] S.A. Brazill, P.H. Kim, W.G. Kuhr, Capillary gel electrophoresis with sinusoidal voltammetric detection: a strategy to allow four-\"color\" DNA sequencing, Analytical Chemistry 73 (20) (2001) 4882–4890.
[121] A.J. Zemann, Conductivity detection in capillary electrophoresis, Trends in Analytical Chemistry 20 (6) (2001) 346–354.
[122] P. Kuban, P.C. Hauser, A review of the recent achievements in capacitively coupled contactless conductivity detection, Analytica Chimica Acta 607 (1) (2008) 15–29.
[123] S.L. Pentoney Jr., R.N. Zare, J.F. Quint, On-line radioisotope detection for capillary electrophoresis, Analytical Chemistry 61 (15) (1989) 1642–1647.
[124] M. Yu, N.J. Dovichi, Attomole amino acid determination: capillary-zone electrophoresis with laser-based thermooptical detection, Applied Spectroscopy 43 (2) (1989) 196–201.
[125] B. Krattiger, A.E. Bruno, H.M. Widmer, R. Dandliker, Hologram-based thermooptic absorbency detection in capillary electrophoresis—separation of nucleosides and nucleotides, Analytical Chemistry 67 (1) (1995) 124–130.
[126] K.C. Waldron, J.J. Li, Investigation of a pulsed-laser thermo-optical absorbance detector for the determination of food preservatives separated by capillary electrophoresis, Journal of Chromatography B-Biomedical Applications 683 (1) (1996) 47–54.
[127] B.S. Seidel, W. Faubel, Miniaturized thermal lens device for capillary electrophoresis, Biomedical Chromatography 12 (3) (1998) 155–157.
[128] F. Kitagawa, Y. Akimoto, K. Otsuka, Label-free detection of amino acids using gold nanoparticles in electrokinetic chromatography-thermal lens microscopy, Journal of Chromatography. A 1216 (14) (2009) 2943–2946.
[129] F. Yu, A.A. Kachanov, S. Koulikov, A. Wainright, R.N. Zare, Ultraviolet thermal lensing detection of amino acids, Journal of Chromatography A 1216 (16) (2009) 3423–3430.

[130] T. Takagi, H. Kubota, S. Oishi, Application of schlieren optics to real-time monitoring of protein electrophoresis in crosslinker-free linear polyacrylamide solution, Electrophoresis 12 (6) (1991) 436–438.
[131] J. Pawliszyn, J.Q. Wu, Moving boundary capillary electrophoresis with concentration gradient detection, Journal of Chromatography 559 (1–2) (1991) 111–118.
[132] V.L. Kasyutich, I.I. Mahnach, Intelligent refractive index detector for capillary electrophoresis, Proc. SPIE-Int. Soc. Opt. Eng. 2208 (1995) 94–102.
[133] B. Krattiger, G.J.M. Bruin, A.E. Bruno, Hologram-based refractive-index detector for capillary electrophoresis—separation of metal-ions, Analytical Chemistry 66 (1) (1994) 1–8.
[134] K. Sumitomo, T. Ito, M. Sasaki, Y. Yamaguchi, Hybridization assay by time-resolved capillary gel electrophoresis with a lanthanide chelate, Chromatographia 67 (9/10) (2008) 715–721.
[135] S.A. Soper, B.L. Legendre Jr., D.C. Williams, Online fluorescence lifetime determinations in capillary electrophoresis, Analytical Chemistry 67 (23) (1995) 4358–4365.
[136] J.Q. Wu, T. Odake, T. Kitamori, T. Sawada, Ultrasensitive detection for capillary zone electrophoresis using laser-induced capillary vibration, Analytical Chemistry 63 (20) (1991) 2216–2218.
[137] C.Y. Chen, M.D. Morris, Online multichannel raman-spectroscopic detection system for capillary zone electrophoresis, Journal of Chromatography 540 (1–2) (1991) 355–363.
[138] T. Srichaiyo, S. Hjerten, Simple multipoint detection method for high-performance capillary electrophoresis, Journal of Chromatography 604 (1) (1992) 85–89.
[139] T. Delonge, H. Fouckhardt, Integrated optical-detection cell-based on bragg reflecting wave-guides, Journal of Chromatography A 716 (1–2) (1995) 135–139.
[140] M. Macka, P. Andersson, P.R. Haddad, Linearity evaluation in absorbance detection: the use of light-emitting diodes for on-capillary detection in capillary electrophoresis, Electrophoresis 17 (12) (1996) 1898–1905.
[141] W. Tong, E.S. Yeung, Simple double-beam absorption detection systems for capillary electrophoresis based on diode lasers and light-emitting diodes, Journal of Chromatography. A 718 (1) (1995) 177–185.
[142] R.J. Whelan, R.N. Zare, Surface plasmon resonance detection for capillary electrophoresis separations, Analytical Chemistry 75 (6) (2003) 1542–1547.
[143] R. Dadoo, A.G. Seto, L.A. Colon, R.N. Zare, End-column chemiluminescence detector for capillary electrophoresis, Analytical Chemistry 66 (2) (1994) 303–306.
[144] L. Guo, B. Qiu, Y. Chi, G. Chen, Using multiple PCR and CE with chemiluminescence detection for simultaneous qualitative and quantitative analysis of genetically modified organism, Electrophoresis 29 (18) (2008) 3801–3809.
[145] L. Guo, B. Qiu, Y. Jiang, Z. You, J.M. Lin, G. Chen, Capillary electrophoresis chemiluminescent detection system equipped with a two-step postcolumn flow interface for detection of some enkephalin-related peptides labeled with acridinium ester, Electrophoresis 29 (11) (2008) 2348–2355.
[146] G. Vicente, L.A. Colon, Separation of bioconjugated quantum dots using capillary electrophoresis, Analytical Chemistry 80 (6) (2008) 1988–1994.
[147] E.J. Maxwell, D.Y. Chen, Twenty years of interface development for capillary electrophoresis-electrospray ionization-mass spectrometry, Analytica Chimica Acta 627 (1) (2008) 25–33.
[148] M. Larsson, E.S.M. Lutz, Transient isotachophoresis for sensitivity enhancement in capillary electrophoresis-mass spectrometry for peptide analysis, Electrophoresis 21 (14) (2000) 2859–2865.

[149] J.C.M. Waterval, P. Bestebreurtje, H. Lingeman, C. Versluis, A.J.R. Heck, A. Bult, et al., Robust and cost-effective capillary electrophoresis-mass spectrometry interfaces suitable for combination with on-line analyte preconcentration, Electrophoresis 22 (13) (2001) 2701–2708.
[150] J.D. Henion, A.V. Mordehai, J.Y. Cai, Quantitative capillary electrophoresis ion spray mass-spectrometry on a benchtop ion-trap for the determination of isoquinoline alkaloids, Analytical Chemistry 66 (13) (1994) 2103–2109.
[151] R.S. Ramsey, S.A. Mcluckey, Capillary electrophoresis electrospray-ionization ion-trap mass-spectrometry using a sheathless interface, Journal of Microcolumn Separations 7 (5) (1995) 461–469.
[152] S.A. Hofstadler, J.H. Wahl, J.E. Bruce, R.D. Smith, Online capillary electrophoresis with fourier-transform ion-cyclotron resonance mass-spectrometry, Journal of the American Chemical Society 115 (15) (1993) 6983–6984.
[153] S.A. Hofstadler, J.H. Wahl, R. Bakhtiar, G.A. Anderson, J.E. Bruce, R.D. Smith, Capillary electrophoresis fourier-transform ion-cyclotron resonance mass-spectrometry with sustained off-resonance irradiation for the characterization of protein and peptide mixtures, Journal of the American Society for Mass Spectrometry 5 (10) (1994) 894–899.
[154] S.A. Hofstadler, F.D. Swanek, D.C. Gale, A.G. Ewing, R.D. Smith, Capillary electrophoresis electrospray-ionization Fourier-transform ion-cyclotron resonance mass-spectrometry for direct analysis of cellular proteins, Analytical Chemistry 67 (8) (1995) 1477–1480.
[155] J.H. Wahl, S.A. Hofstadler, R.D. Smith, Direct electrospray ion current monitoring detection and its use with online capillary electrophoresis mass-spectrometry, Analytical Chemistry 67 (2) (1995) 462–465.
[156] G. Jarvas, B. Fonslow, J.R. Yates 3rd, F. Foret, A. Guttman, Characterization of a porous nano-electrospray capillary emitter at ultra-low flow rates, Journal of Chromatographic Science 55 (1) (2017) 47–51.
[157] G.M. Janini, T.P. Conrads, K.L. Wilkens, H.J. Issaq, T.D. Veenstra, A sheathless nanoflow electrospray interface for on-line capillary electrophoresis mass spectrometry. Anal. Chem. 2003, 75, 1615–1619., Analytical Chemistry 75 (7) (2003) 1615–1619.
[158] J.H. Wahl, R.D. Smith, Comparison of buffer systems and interface designs for capillary electrophoresis-mass spectrometry, Journal of Capillary Electrophoresis 1 (1) (1994) 62–71.
[159] M.S. Kriger, K.D. Cook, R.S. Ramsey, Durable gold-coated fused-silica capillaries for use in electrospray mass-spectrometry, Analytical Chemistry 67 (2) (1995) 385–389.
[160] B.E. Chong, J. Kim, D.M. Lubman, J.M. Tiedje, S. Kathariou, Use of non-porous reversed-phase high-performance liquid chromatography for protein profiling and isolation of proteins induced by temperature variations for Siberian permafrost bacteria with identification by matrix-assisted laser desorption lionization time-of-flight mass spectrometry and capillary electrophoresis-electrospray ionization mass spectrometry, Journal of Chromatography B 748 (1) (2000) 167–177.
[161] B. Zhang, F. Foret, B.L. Karger, A microdevice with integrated liquid junction for facile peptide and protein analysis by capillary electrophoresis/electrospray mass spectrometry, Analytical Chemistry 72 (5) (2000) 1015–1022.
[162] B. Zhang, F. Foret, B.L. Karger, High-throughput microfabricated CE/ESI-MS: automated sampling from a microwell plate, Analytical Chemistry 73 (11) (2001) 2675–2681.
[163] M. Moini, Design and performance of a universal sheathless capillary electrophoresis to mass spectrometry interface using a split-flow technique, Analytical Chemistry 73 (14) (2001) 3497–3501.

[164] P. Cao, M. Moini, A novel sheathless interface for capillary electrophoresis/electrospray ionization mass spectrometry using an in-capillary electrode, Journal of the American Society for Mass Spectrometry 8 (5) (1997) 561–564.
[165] R.D. Smith, J.A. Olivares, N.T. Nguyen, H.R. Udseth, Capillary zone electrophoresis-mass spectrometry using an electrospray ionization interface, Analytical Chemistry 60 (5) (1988) 436–441.
[166] F. Garcia, J.D. Henion, Gel-filled capillary electrophoresis mass-spectrometry using a liquid junction ion spray interface, Analytical Chemistry 64 (9) (1992) 985–990.
[167] F. Foret, T.J. Thompson, P. Vouros, B.L. Karger, P. Gebauer, P. Bocek, Liquid sheath effects on the separation of proteins in capillary electrophoresis electrospray mass-spectrometry, Analytical Chemistry 66 (24) (1994) 4450–4458.
[168] T. Wachs, R.L. Sheppard, J. Henion, Design and applications of a self-aligning liquid junction-electrospray interface for capillary electrophoresis-mass spectrometry, Journal of Chromatography B, Biomedical Applications 685 (2) (1996) 335–342.
[169] M. Mazereeuw, A.J.P. Hofte, U.R. Tjaden, J. vander Greef, A novel sheathless and electrodeless microelectrospray interface for the on-line coupling of capillary zone electrophoresis to mass spectrometry, Rapid Communications in Mass Spectrometry 11 (9) (1997) 981–986.
[170] D. Figeys, G.L. Corthals, B. Gallis, D.R. Goodlett, A. Ducret, M.A. Corson, et al., Data-dependent modulation of solid-phase extraction capillary electrophoresis for the analysis of complex peptide and phosphopeptide mixtures by tandem mass spectrometry: application to endothelial nitric oxide synthase, Analytical Chemistry 71 (13) (1999) 2279–2287.
[171] P. Kusy, K. Kleparnik, Z. Aturki, S. Fanali, F. Foret, Optimization of a pressurized liquid junction nanoelectrospray interface between CE and MS for reliable proteomic analysis, Electrophoresis 28 (12) (2007) 1964–1969.
[172] E.J. Maxwell, X. Zhong, H. Zhang, N. van Zeijl, D.D. Chen, Decoupling CE and ESI for a more robust interface with MS, Electrophoresis 31 (7) (2010) 1130–1137.
[173] L. Sun, G. Zhu, Z. Zhang, S. Mou, N.J. Dovichi, Third-generation electrokinetically pumped sheath-flow nanospray interface with improved stability and sensitivity for automated capillary zone electrophoresis-mass spectrometry analysis of complex proteome digests, Journal of Proteome Research 14 (5) (2015) 2312–2321.
[174] A. Tycova, M. Vido, P. Kovarikova, F. Foret, Interface-free capillary electrophoresis-mass spectrometry system with nanospray ionization—analysis of dexrazoxane in blood plasma, Journal of Chromatography A 1466 (2016) 173–179.
[175] J. Römer, S. Kiessig, B. Moritz, C. Neusüß, Improved CE(SDS)-CZE-MS method utilizing an 8-port nanoliter valve, Electrophoresis 42 (4) (2021) 374–380.
[176] H.Y. Zhang, R.M. Caprioli, Capillary electrophoresis combined with matrix-assisted laser desorption/ionization mass spectrometry; continuous sample deposition on a matrix-precoated membrane target, Journal of Mass Spectrometry 31 (9) (1996) 1039–1046.
[177] F. Foret, O. Muller, J. Thorne, W. Gotzinger, B.L. Karger, Analysis of protein-fractions by micropreparative capillary isoelectric-focusing and matrix-assisted laser-desorption time-of-flight mass-spectrometry, Journal of Chromatography A 716 (1–2) (1995) 157–166.
[178] J.F. Banks, T. Dresch, Detection of fast capillary electrophoresis peptide and protein separations using electrospray ionization with a time-of-flight mass spectrometer, Analytical Chemistry 68 (9) (1996) 1480–1485.
[179] D.C. Muddiman, A.L. Rockwood, Q. Gao, J.C. Severs, H.R. Udseth, R.D. Smith, et al., Application of sequential paired covariance to capillary electrophoresis electrospray-ionization time-of-flight mass-spectrometry—unraveling the signal from the noise in the electropherogram, Analytical Chemistry 67 (23) (1995) 4371–4375.

[180] A. Chrambach, D. Rodbard, Polyacrylamide gel electrophoresis, Science (New York, N.Y.) 172 (3982) (1971) 440–451.
[181] W.E. Werner, D.M. Demorest, J.E. Wiktorowicz, Automated Ferguson analysis of glycoproteins by capillary electrophoresis using a replaceable sieving matrix, Electrophoresis 14 (8) (1993) 759–763.
[182] S.P. Radko, M.M. Garner, G. Caiafa, A. Chrambach, Molecular sieving of polystyrene carboxylate of a diameter up to 10 mm in solutions of uncrosslinked polyacrylamide of Mr 5 * 10^6 using capillary zone electrophoresis, Analytical Biochemistry 223 (1) (1994) 82–87.
[183] D. Figeys, N.J. Dovichi, Mobility of single-stranded DNA as a function of cross-linker concentration in polyacrylamide capillary gel electrophoresis, Journal of Chromatography 645 (2) (1993) 311–317.
[184] A. Guttman, J.A. Nolan, N. Cooke, Capillary sodium dodecyl sulfate gel electrophoresis of proteins, Journal of Chromatography 632 (1–2) (1993) 171–175.
[185] A. Guttman, P. Shieh, J. Lindahl, N. Cooke, Capillary sodium dodecyl sulfate gel electrophoresis of proteins. II. On the Ferguson method in polyethylene oxide gels, Journal of Chromatography A 676 (1) (1994) 227–231.
[186] T. Demana, M. Lanan, M.D. Morris, Improved separation of nucleic acids with analyte velocity modulation capillary electrophoresis, Analytical Chemistry 63 (23) (1991) 2795–2797.
[187] A. Guttman, B. Wanders, N. Cooke, Enhanced separation of DNA restriction fragments by capillary gel electrophoresis using field strength gradients, Analytical Chemistry 64 (20) (1992) 2348–2351.
[188] L. Mitnik, L. Salome, J.L. Viovy, C. Heller, Systematic study of field and concentration effects in capillary electrophoresis of DNA in polymer solutions, Journal of Chromatography A 710 (2) (1995) 309–321.
[189] H.M. Wenz, Capillary electrophoresis as a technique to analyze sequence-induced anomalously migrating DNA fragments, Nucleic Acids Research 22 (19) (1994) 4002–4008.
[190] A. Guttman, N. Cooke, Effect of temperature on the separation of DNA restriction fragments in capillary gel electrophoresis, Journal of Chromatography 559 (1–2) (1991) 285–294.
[191] R. Kumar, J.S. Hanekamp, J. Louhelainen, K. Burvall, A. Onfelt, K. Hemminki, et al., Separation of transforming amino acid-substituting mutations in codons 12, 13 and 61 of the N-ras gene by constant denaturant capillary electrophoresis (CDCE), Carcinogenesis 16 (11) (1995) 2667–2673.
[192] C. Gelfi, L. Cremonesi, M. Ferrari, P.G. Righetti, Temperature-programmed capillary electrophoresis for detection of DNA point mutation, Biotechniques 21 (5) (1996) 926–932.
[193] A. Guttman, Effect of temperature on separation efficiency in capillary gel electrophoresis, Trends in Analytical Chemistry 15 (5) (1996) 194–198.
[194] G.W. Slater, P. Mayer, Electrophoretic resolution vs fluctuations of the lateral dimensions of a capillary, Electrophoresis 16 (5) (1995) 771–779.
[195] S. Wicar, M. Vilenchik, A. Belenkii, A.S. Cohen, B.L. Karger, Influence of coiling on performance in capillary electrophoresis using open tubular and polymer network columns, Journal of Microcolumn Separations 4 (4) (1993) 339–348.
[196] A. Guttman, A. Arai, K. Magyar, Influence of pH on the migration properties of oligonucleotides in capillary gel electrophoresis, Journal of Chromatography 608 (1–2) (1992) 175–179.
[197] A. Guttman, E. Szoko, Capillary gel electrophoretic separation of DNA restriction fragments in a discontinuous buffer system, Journal of Chromatography A 744 (1–2) (1996) 321–324.

[198] A.K. Palm, Capillary electrophoresis of DNA fragments with replaceable low-gelling agarose gels, Methods in Molecular Biology 162 (2001) 279–290.
[199] S.P. Porras, M.L. Riekkola, E. Kenndler, The principles of migration and dispersion in capillary zone electrophoresis in nonaqueous solvents, Electrophoresis 24 (10) (2003) 1485–1498.
[200] M.L. Riekkola, Recent advances in nonaqueous capillary electrophoresis, Electrophoresis 23 (22–23) (2002) 3865–3883.
[201] T. Lu, L.J. Klein, S. Ha, R.R. Rustandi, High-resolution capillary electrophoresis separation of large RNA under non-aqueous conditions, Journal of Chromatography. A 1618 (2020) 460875.

CHAPTER FIVE

Applications

5.1 Capillary gel electrophoresis of DNA

The advent of capillary gel or polymer network-based electrophoresis methods have greatly improved our capabilities to separate and analyze biologically important polymers, such as oligonucleotides, even down to the single-nucleotide resolution level. Denaturing capillary gel electrophoresis (CGE) is utilized mainly in size separation of relatively short single-stranded oligonucleotides and in DNA sequencing, for this latter to alleviate hairpin region-mediated structural abnormalities that may affect base calling. Urea and formamide are the most commonly used denaturants. Nondenaturing gels or polymer networks, on the other hand, are used when separation should be based on the size and/or shape of nucleic acids, such as dsDNA fragment analysis, as well as to reveal secondary structure differences (e.g., single-nucleotide or conformational polymorphisms) [1]. Capillary electrophoresis of DNA in sieving polymers for molecular diagnostics applications, such as screening for inherited human genetic defects, quantitative gene dosage, microbiology/virology, therapeutic antisense DNA separation, aptamer screening, and forensic analysis, is also summarized [2,3].

5.1.1 Analysis of single-stranded oligonucleotides

For the millions of synthetic oligonucleotides produced annually, purity analysis in a high-resolution rapid and automated fashion is of high importance. CGE also proved to be effective between phosphorylated and dephosphorylated forms [4], oligonucleotide deprotection products [5], and separating synthetic branched DNA molecules [6]. It is important to note that one of the main differences between CGE and slab gel electrophoresis is the detection mode that is space based in slab gel systems and time based in CGE. In other words, in CGE all solute molecules are traveling through the entire effective column length until get detected.

One of the most important applications of CGE analysis of DNA is sequencing [7,8] based on the first high-resolution separations of single-stranded

Capillary Gel Electrophoresis.
DOI: https://doi.org/10.1016/B978-0-444-52234-4.00001-5

oligonucleotides by Guttman et al. [9]. Picomole amounts of a mixture of polydeoxyadenylic acids, $p(dA)_{10-30,40-60}$ (Fig. 5.1), were baseline resolved in less than 11 minutes by high-performance CGE using a cross-linked (7.5% T/3.3% C) polyacrylamide gel. These publications pointed out the high potential of CGE-based biopolymer analysis, and the possibility of fully automated DNA sequencing.

In comparison to other separation methods, Warren and Vella [10] used CGE and anion-exchange high-performance liquid chromatography (HPLC) in the analysis of synthetic oligodeoxyribonucleotides based on sample load, sample preparation method (e.g., heating), electric field strength/gradient and solute size. Fig. 5.2 depicts the effect of electric field in CGE (Fig. 5.2A) and gradient elution in anion-exchange HPLC (Fig. 5.2B) on the separation of a mixture of short oligonucleotides (40–60-mers). As one can observe, CGE showed significantly better separation performance.

In a similar study, Gelfi et al. [11] assessed the purity of synthetic oligodeoxyribonucleotides by comparing reversed-phase high-performance liquid

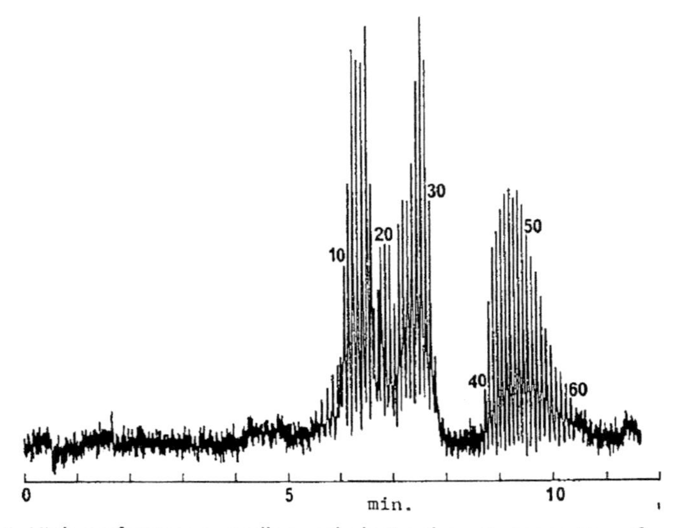

Figure 5.1 High-performance capillary gel electrophoresis separation of a polydeoxyadenylic acid mixture, $(dA)_{10-30,\ 40-60}$. Capillary: 270 mm × 0.075 mm i.d.; running buffer: 0.1 M Tris/0.25 M borate/7 M urea (pH 8.3), gel composition: 7.5% T/3.3% C acrylamide/bisacrylamide, and applied electric field: 400 V/cm. *With permission from A. Guttman, A. Paulus, A.S. Cohen, B.L. Karger, H. Rodriguez, W.S. Hancock, High performance, capillary gel electrophoresis: high resolution and micropreparative applications, in: Electrophoresis'88, Copenhagen VCH, Weinheim, Germany, 1988 [9].*

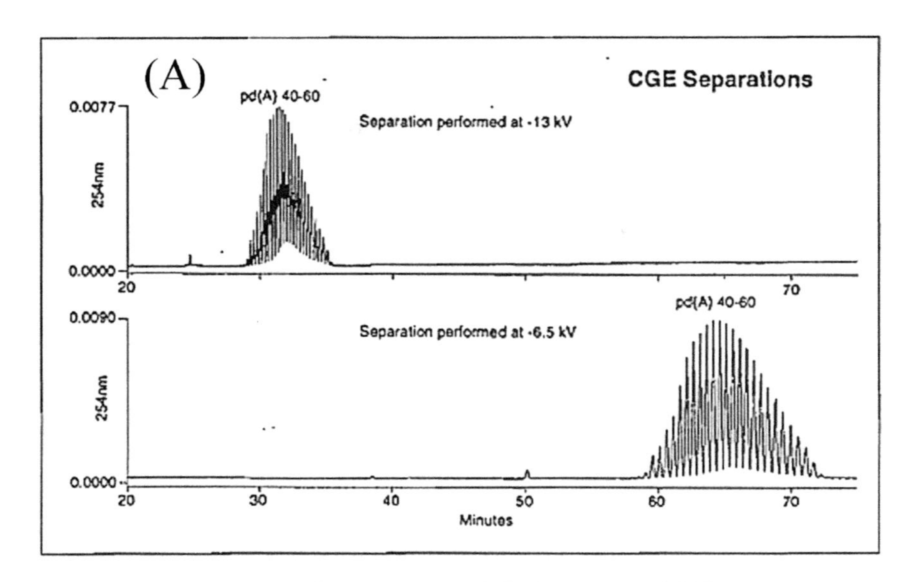

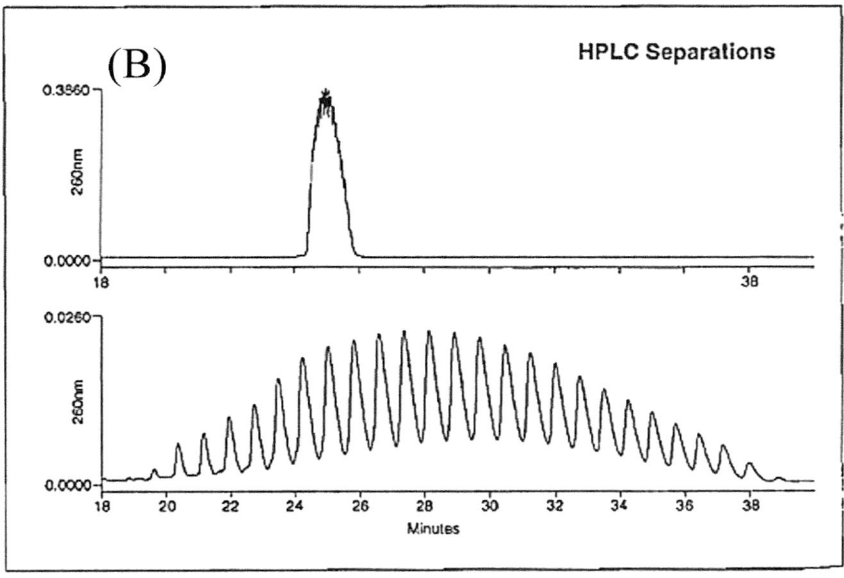

Figure 5.2 Comparison of capillary gel electrophoresis and high-performance liquid chromatography analysis of oligonucleotides. (A) Effect of the applied field strength on the separation of pd(A)$_{40-60}$ oligonucleotides in capillary gel electrophoresis. (B) Effect of solvent gradient on oligonucleotide separations by anion-exchange high-performance liquid chromatography. *With permission from W.J. Warren, G. Vella, Analysis of synthetic oligodeoxyribonucleotides by capillary gel electrophoresis and anion-exchange HPLC. Biotechniques 14 (4) (1993) 598–606 [10].*

chromatography (RP-HPLC), polyacrylamide slab gel electrophoresis (PAGE) and CGE (Fig. 5.3). This latter utilized highly concentrated (18%T) entangled polymer networks to baseline resolve all shorter fragments and provided

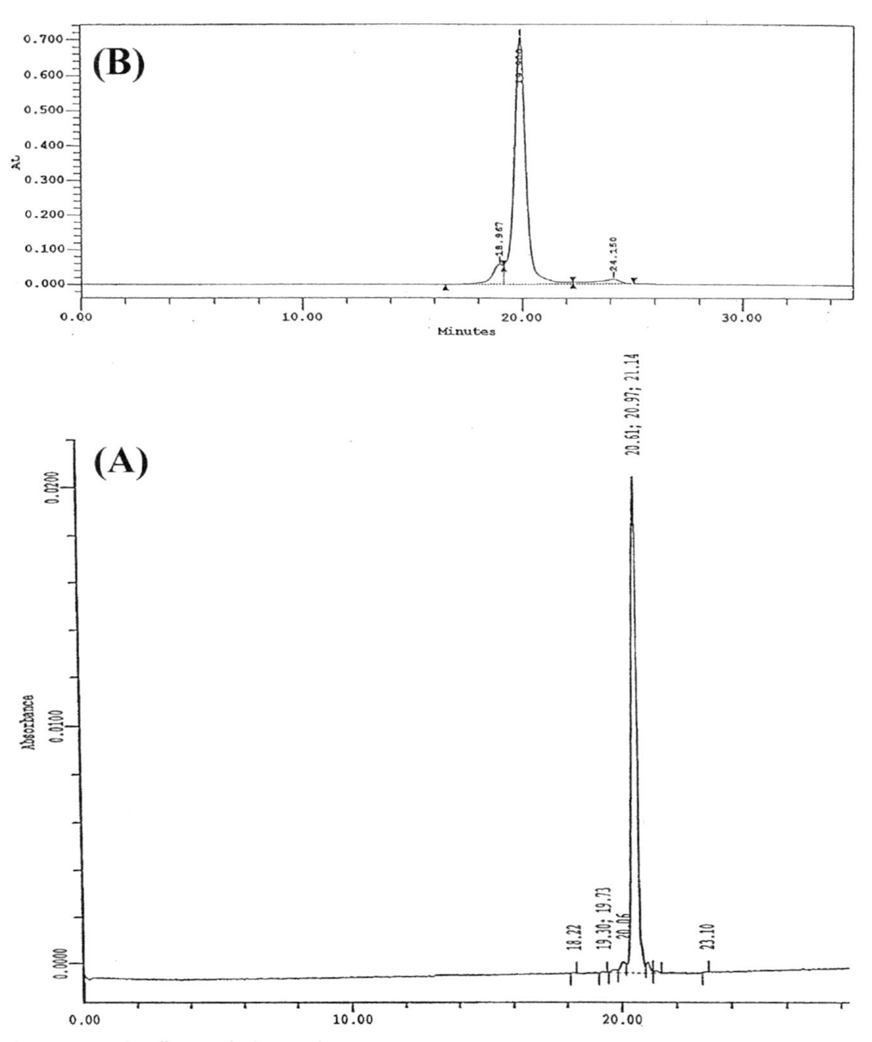

Figure 5.3 Capillary gel electrophoresis (A) and strong anion-exchange high-performance liquid chromatography (B) analysis of a 3′-terminal phosphorothioate oligonucleotide. *With permission from V.T. Ravikumar, D.C. Capaldi, W.F. Lima, E. Lesnik, B. Turney, D.L. Cole, Antisense phosphorothioate oligodeoxyribonucleotide targeted against ICAM-1: synthetic and biological characterization of a process-related impurity formed during oligonucleotide synthesis. Bioorganic & Medicinal Chemistry 11 (21) (2003) 4673–4679 [12].*

precise evaluation for the amount of impurities. Cole and coworkers synthesized a 3′-terminal phosphorothioate oligonucleotide and analyzed with CGE and strong anion-exchange HPLC as depicted in Fig. 5.3 [12].

High-resolution analysis of single-stranded oligosaccharides down to the single base resolution level has always been a high importance both for low base number oligonucleotide analysis (e.g., primers) and for large DNA sequencing fragments. Early attempts utilized similar cross-linked polyacrylamide gel compositions as in PAGE to attain this resolution level. Using CGE in 6% T/5% C polyacrylamide gel composition [13] resulted in very high resolution for a mixture of polydeoxythymidilic acids, pd $(T)_{20-160}$, with theoretical plate numbers as high as 30 million (Fig. 5.4), reaching the level of sequencing grade separation.

Cohen et al. [14] achieved fast separation of a larger size crude oligodeoxynucleotide probe (70-mer) and a slab gel-purified 99-mer in less than 8 minutes (Fig. 5.5) demonstrating the ability of the method to rapidly characterize freshly generated primers and probes. In this latter instance, 5% T/3.3% C polyacrylamide gel composition was used. The lack of adequate stability of cross-linked polyacrylamide gels within microbore columns initiated a rapid development to find novel, more capillary friendly sieving matrices [15]. One of the solutions was the development of dynamically cross-linked polymer solutions as sieving matrices for the electrophoretic separation of poly-ionic biopolymers with similar surface charge densities [16].

The effects of separation parameters on the analysis of nucleic acids were evaluated on resolution and separation efficiency by Macek et al. [17]. They found that the best resolution was achieved with long columns and optimized electric field strengths. Paulus and Ohms got excellent separation of oligonucleotides of 20–50 bases following a similar optimization strategy [18]. The influence of sample size on resolution was also discussed. The importance of instrumental and sample matrix effects on resolution, sensitivity, and detectability in gel-filled capillaries was evaluated with oligodeoxynucleotide samples [19] and possible errors were found in sample quantitation if peak areas were not corrected for their velocity, that is, the time they needed to pass the detection zone. Righetti's group used histidine-based isoelectric buffer at pH = pI conditions with linear polyacrylamide (LPA) to separate oligonucleotides [20]. By the increase in the applied electric field strength with such isoelectric buffers, the Joule heating was significantly reduced with concomitant increase in the separation performance [20].

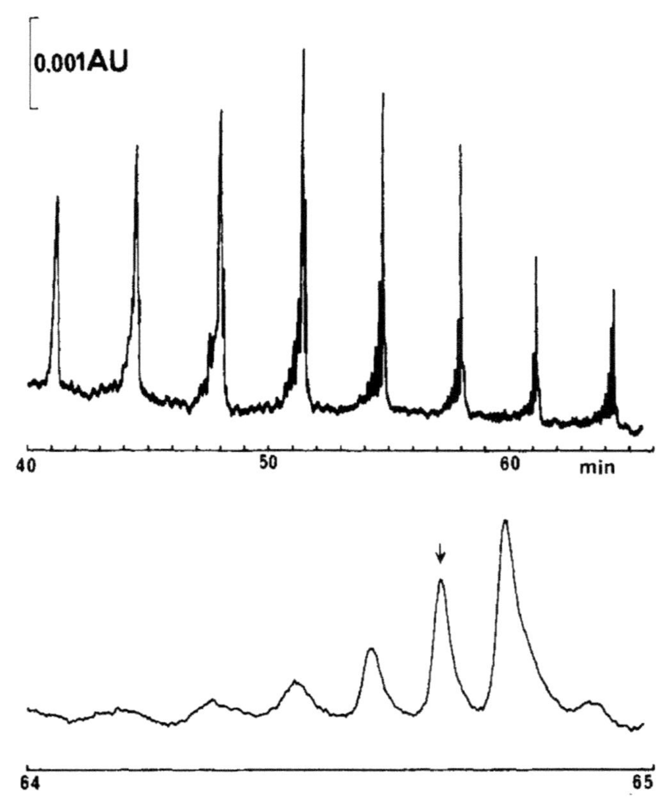

Figure 5.4 High-performance capillary gel electrophoresis separation of a polydeoxythymidylic acid mixture ($pd(T)_{20-160}$). Capillary dimensions, 1000 mm × 0.075 mm i.d. (effective length: 800 mm); applied electric field: 200 V/cm, 8.2 μA. The arrow points to the peak with 30 million plates. *With permission from A. Guttman, A.S. Cohen, D.N. Heiger, B.L. Karger, Analytical and micropreparative ultrahigh resolution of oligonucleotides by polyacrylamide gel high-performance capillary electrophoresis. Analytical Chemistry 62 (2) (1990) 137–141 [13].*

The electrophoretic mobility of single-stranded DNA in denaturing polyacrylamide gel-filled capillaries was analyzed as the function of the applied electric field by Luckey and Smith [21]. The resultant mobility plots showed complex functions of the fragment size and electric field. Fig. 5.6 illustrates the two regimes of mobility behavior for five different electric field strengths between 100 and 400 V/cm. The lower size fragments followed the Ogston equation, while the higher ones the $1/N$ dependency, thus deviated from the Ogston behavior. Rodbard and

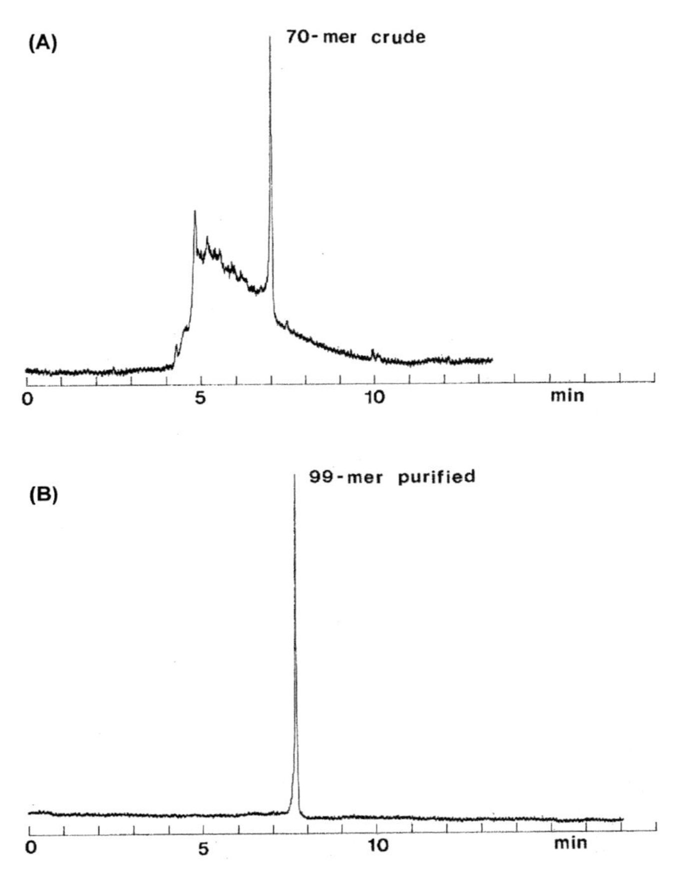

Figure 5.5 Capillary gel electrophoresis analysis of a crude 70-mer (A) and a slab gel purified 99-mer (B) oligodeoxynucleotide in cross-linked 5% T/3.3% C polyacrylamide. *With permission from A.S. Cohen, D.R. Najarian, A. Paulus, A. Guttman, J.A. Smith, B.L. Karger, Rapid separation and purification of oligonucleotides by high-performance capillary gel electrophoresis. Proceedings of the National Academy of Sciences of the United States of America 85 (24) (1988) 9660–9663 [14].*

Chrambach also studied the relationship between molecular geometry and electrophoretic mobility as summarized in [22].

To accurately model the variations in the electrophoretic mobility of nucleic acids observed with electric field strength used in CGE, the extended Ogston sieving theory was modified with the reptation model, where stretching of the migrating DNA molecules points in the direction of the electric field, as shown in Fig. 5.7 [21].

In this modified Ogston model, the applied electric field strength in conjunction with the gel matrix effects considered to cause distortion of

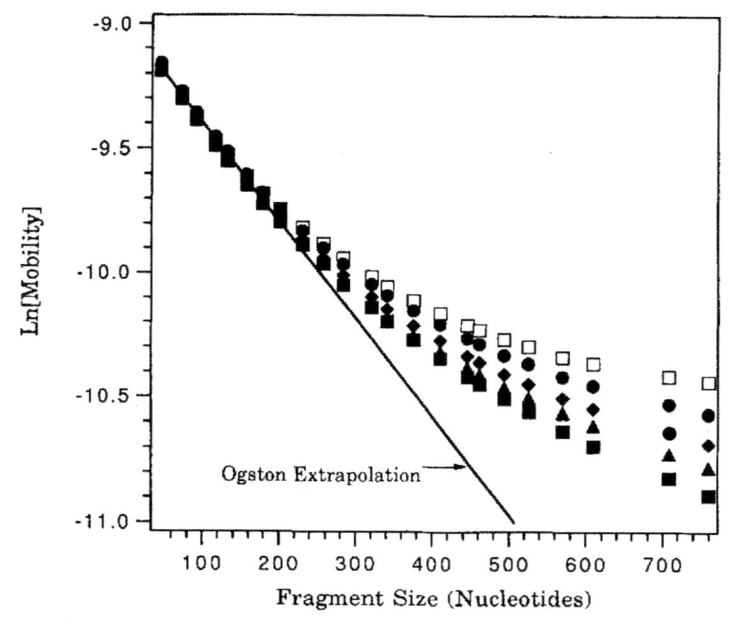

Figure 5.6 Illustration of two regimes of mobility behavior by plotting the electrophoretic mobilities as the function of DNA fragment size at various electric field strengths: (■) 50, (▲) 100, (♦) 150, (•) 200, and (□) 400 V/cm. In the reptation regime, the mobilities follow a reciprocal dependency with the DNA fragment size and thus deviated from the Ogston behavior (Ogston extrapolation line). *With permission from J.A. Luckey, L.M. Smith, A model for the mobility of single-stranded DNA in capillary gel electrophoresis. Electrophoresis 14 (5–6) (1993) 492–501 [21].*

the shape of the DNA, changing the effective size of the migrating molecule. The stretched DNA had smaller cross-section than that of the gel pore and thus could migrate through as it were a smaller molecule. The electrophoretic mobility depended on the applied electric field and the size of the fragment. The extended Ogston model Eq. (5.1) accurately predicted the mobilities of DNA fragments in all three mobility regimes, providing a single model for all of the observed behaviors. Considering that molecular distortions was only responsible for the deviation from the Ogston model, with the extended Ogston equation, a simple relationship was obtained between this relative deformation, the applied electric field, and the size of the DNA. Mobility curves generated by this Eq. (5.1) [21] showed excellent agreement with the experimental data.

$$\ln \mu = \ln \mu^0 - \alpha\lambda Cn\left[1+[a_1 E]^{1/4} n^{3/2}\right]^{-1} \tag{5.1}$$

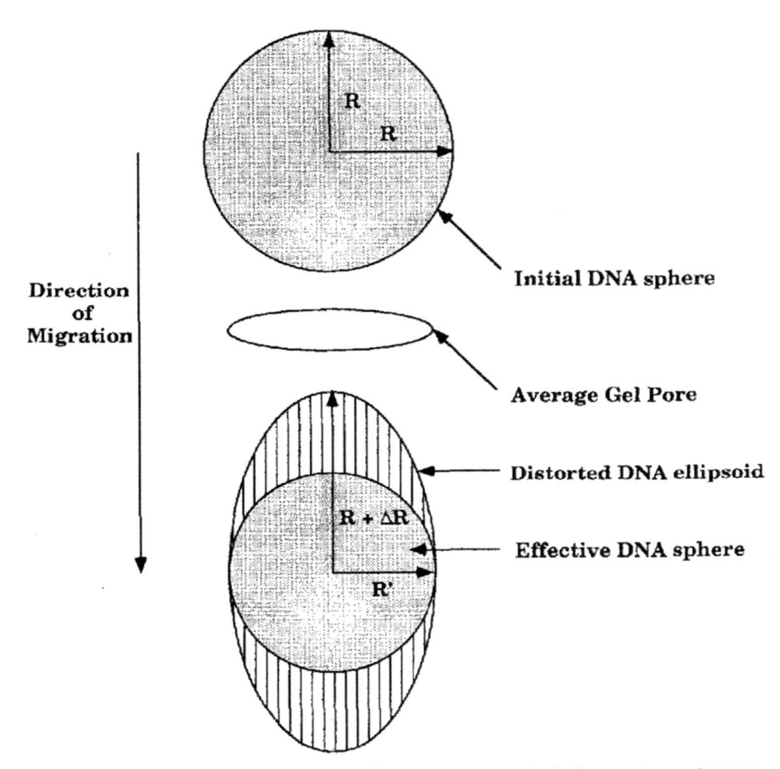

Figure 5.7 Schematic representation of the presumed deformation of DNA molecules in the presence of high electric fields in polyacrylamide gels. An initial random coil with a radius of R is pulled through a relatively rigid gel pore as a result of the applied electric field. This caused the DNA to deform to a spherical ellipsoid with a decreased radius of *R*′. This distorted molecule supposed to be functionally equivalent to a DNA sphere of radius *R* as shown. *With permission from J.A. Luckey, L.M. Smith, A model for the mobility of single-stranded DNA in capillary gel electrophoresis. Electrophoresis 14 (5–6) (1993) 492–501 [21].*

where μ is the electrophoretic mobility; μ^0 is the free-solution electrophoretic mobility at zero gel concentration; α is the constant of proportionality; C is the gel concentration; n is the chain length expressed in nucleotide units; E is the electric field strength, and λ and a_1 are the constants.

The gel in the capillary also supports diffusion-limited electromigration. Therefore peaks can be sharpened by applying a gradient of different system parameters especially gel concentration. The general idea came from slab gel electrophoresis, in which case, only the smaller solute molecules are traveling through the gel concentration gradient, the larger ones left behind, and at the end of the run, they are distributed over the gel according to their size. To exploit the same phenomenon in CGE, the effect of viscosity gradients was investigated by Guillouzic et al. [23] by using continuous and stepwise gradients. The authors demonstrated that viscosity gradients in CE did not result in improvement in resolution as all molecules traveled through the entire gradient (Fig. 5.8), that is, the

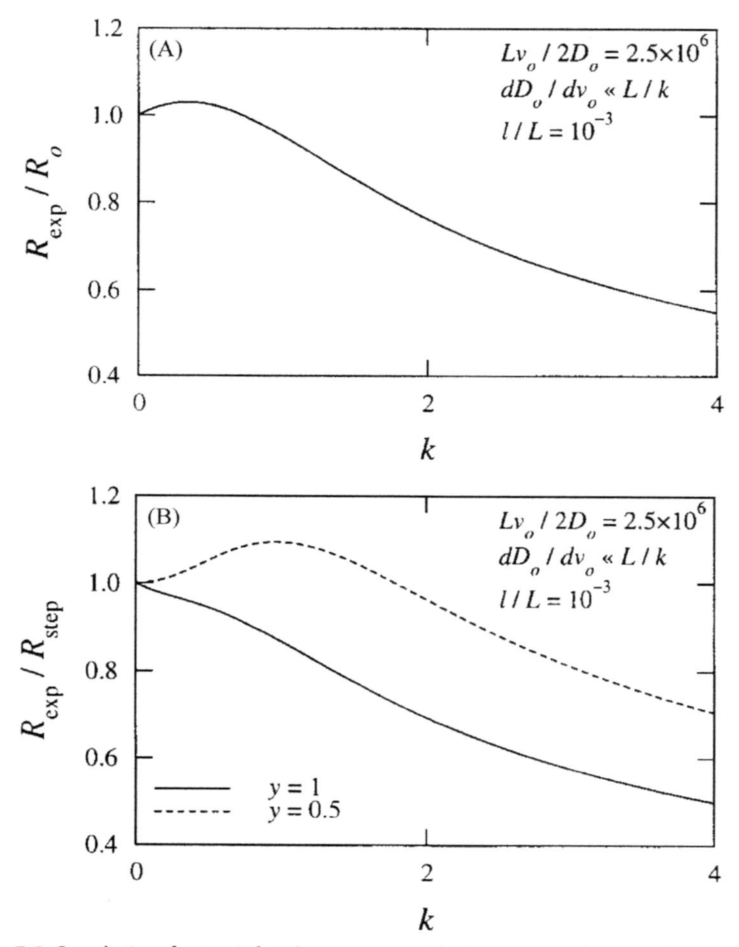

Figure 5.8 Resolution factor R for the exponential viscosity gradient with a nonnegligible injection width. R is divided (A) by its uniform viscosity limit R_o and (B) by the resolution factor for the step gradient with the same total variation in viscosity. *With permission from S. Guillouzic, L.C. McCormick, G.W. Slater, Electrophoresis in the presence of gradients: I. Viscosity gradients. Electrophoresis 23 (12) (2002) 1822–1832 [23].*

performance of electrophoretic separation in narrow bore capillaries cannot be improved in that way.

5.1.1.1 Effect of the nucleotide sequence on CGE separation

The influence of the primary sequence (base composition) on the electrophoretic migration properties of single-stranded oligodeoxyribonucleotides in capillary polyacrylamide gel electrophoresis was investigated using homo- and heterooligomers under denaturing and nondenaturing conditions [24]. Heterooligodeoxyribonucleotides with equal chain lengths but with different base composition exhibited significant differences in their electrophoretic mobilities. The migration properties of heterooligomers were found to be highly dependent on their base composition. Eq. (5.2) was derived to predict the relative migration times for both in denaturing and nondenaturing polyacrylamide CGE.

$$t'(A_aT_tC_cG_g) = \frac{a}{n}t'(A_n) + \frac{t}{n}t'(T_n) + \frac{c}{n}t'(C_n) + \frac{g}{n}t'(G_n) \qquad (5.2)$$

where t' is the relative migration time normalized to the internal standard, n is the oligonucleotide chain length ($n = a + t + c + g$) and a, t, c, and g are the numbers of the individual bases in the oligonucleotide. The parameters $t'(A_n)$, $t'(T_n)$, $t'(C_n)$, and $t'(G_n)$ correspond to the relative migration times of the same chain length homooligomers of adenylic, thymidylic, cytidylic, and guanylic acids, respectively. Using Eq. (5.2), Fig. 5.9 shows the successful prediction of the separation of human K-ras oncogene probes differing only by one nucleotide in the middle of the oligomer.

Simultaneous determination of the migration coefficient of each base in heterogeneous oligo-DNA samples by capillary gel electrophoresis was attempted by Satow et al. [25]. The authors investigated the migration time of the solute molecules in a gel-filled capillary using the Gauss least-squares method as observation function. Their approach also assumed that the migration time of the oligonucleotides were dependent on their base composition and chain length. With their suggested method, the authors could accurately estimate the migration time of oligo-DNAs of any known sequence without secondary structures. In addition, the deviation of the actual migration time from the calculated one could suggest the presence of secondary structures, such as hairpins, even in the presence of 7 M urea. It was also suggested that the results of the analysis may be able to predict chain length reversal in DNA sequencing by CGE.

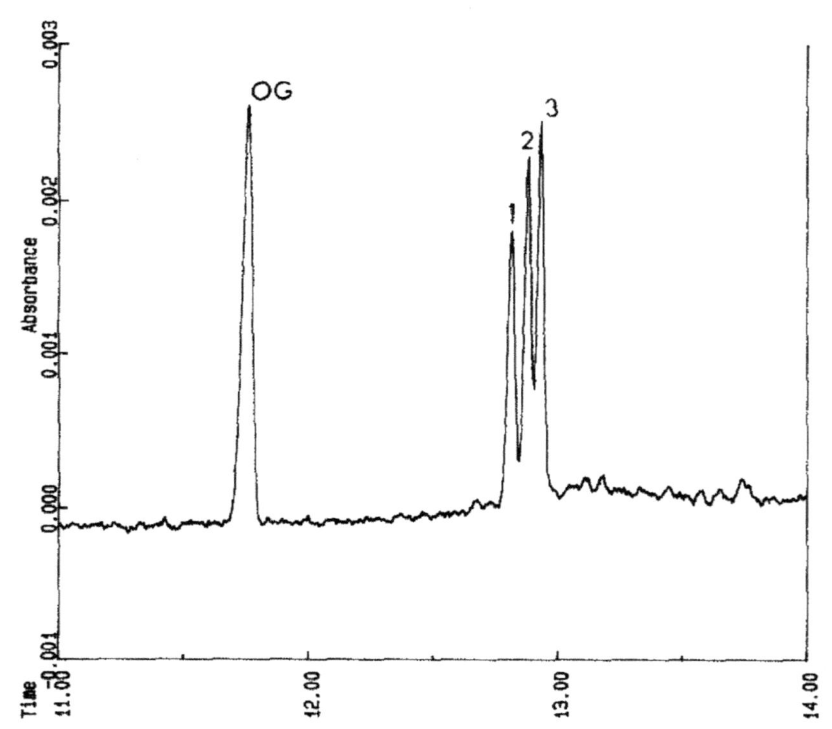

Figure 5.9 Nondenaturing capillary polyacrylamide gel electrophoresis separation of human K-ras oncogene probes. Peaks: 1 = dGTTGGAGCT-G-GTGGCGTAG: 2 = dGTTGGA GCT-C-GTGGCGTAG; and 3 = dGTTGGAGCT-T-GTGGCGTAG. *With permission from A. Guttman, R.J. Nelson, N. Cooke, Prediction of migration behavior of oligonucleotides in capillary gel electrophoresis, Journal of Chromatography 593 (1–2) (1992) 297–303 [24].*

Another prediction scheme for CGE migration times using base-specific migration coefficients was developed by Bischoff and coworkers [26] who chemically synthesized oligodeoxyribonucleotides and evaluated their migration properties in three different polyacrylamide-based gel matrices (10% T LPA gel, as well as 5% T/5% C and 3% T/3% C cross-linked polyacrylamide gels). Computer-aided prediction of migration times was subsequently evaluated to confirm the size and base composition of oligonucleotides with adequate accuracy. Interestingly, the CGE migration of oligonucleotides at pH 3.5 was mainly dependent on the charge per base ratio that significantly altered selectivity, complementing the analyses under commonly used basic pH conditions.

Konrad and Pentoney [27] investigated the increased electrophoretic mobility of single-stranded DNA molecules in respect to their intramolecular base pairing at the 3′ end to address one of the common problems in DNA sequencing, that is, peak compressions of comigrating multiple oligomers differing in length, therefore making sequence determination complicated. Incorporation of formamide into the gel formulations, application of different separation field strengths, and external heating of the capillary were examined for their ability to reduce peak compression, as shown in Fig. 5.10. The developed method was suitable to separate

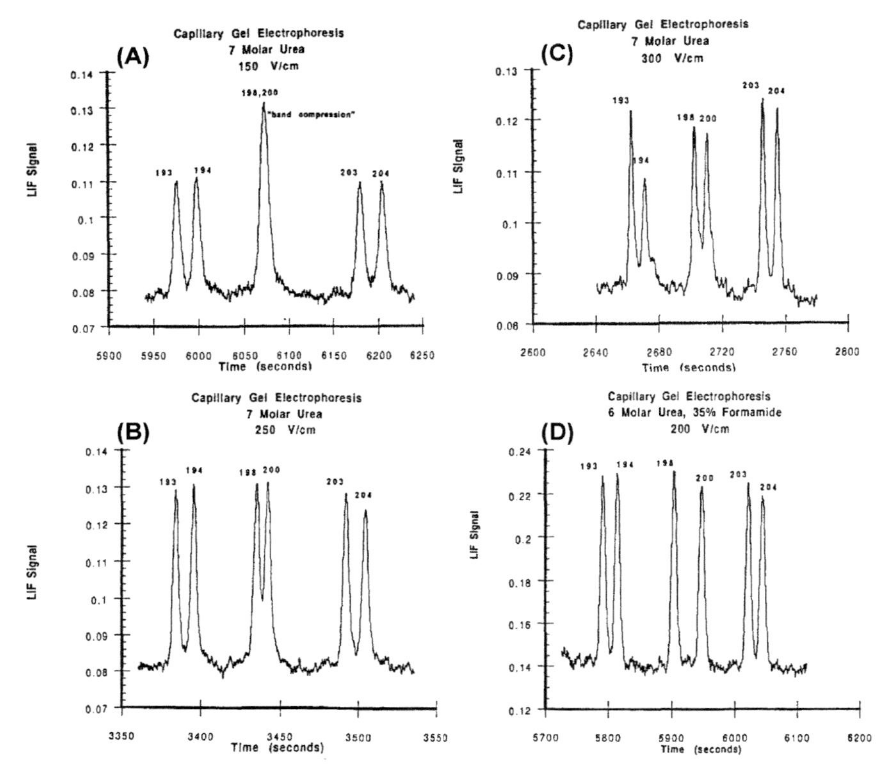

Figure 5.10 Capillary gel electrophoresis separation of fluorescently labeled DNA fragments terminated with ddA from Bacteriophage M13mp18 DNA. The separation of the 198 and 200 nucleotide polymers improved with increasing field strength (A–C). The best separation was achieved in formamide containing gels (D), however with increased separation time. *With permission from K.D. Konrad, S.L. Pentoney Jr., Contribution of secondary structure to DNA mobility in capillary gels, Electrophoresis 14 (5–6) (1993) 502–508 [27].*

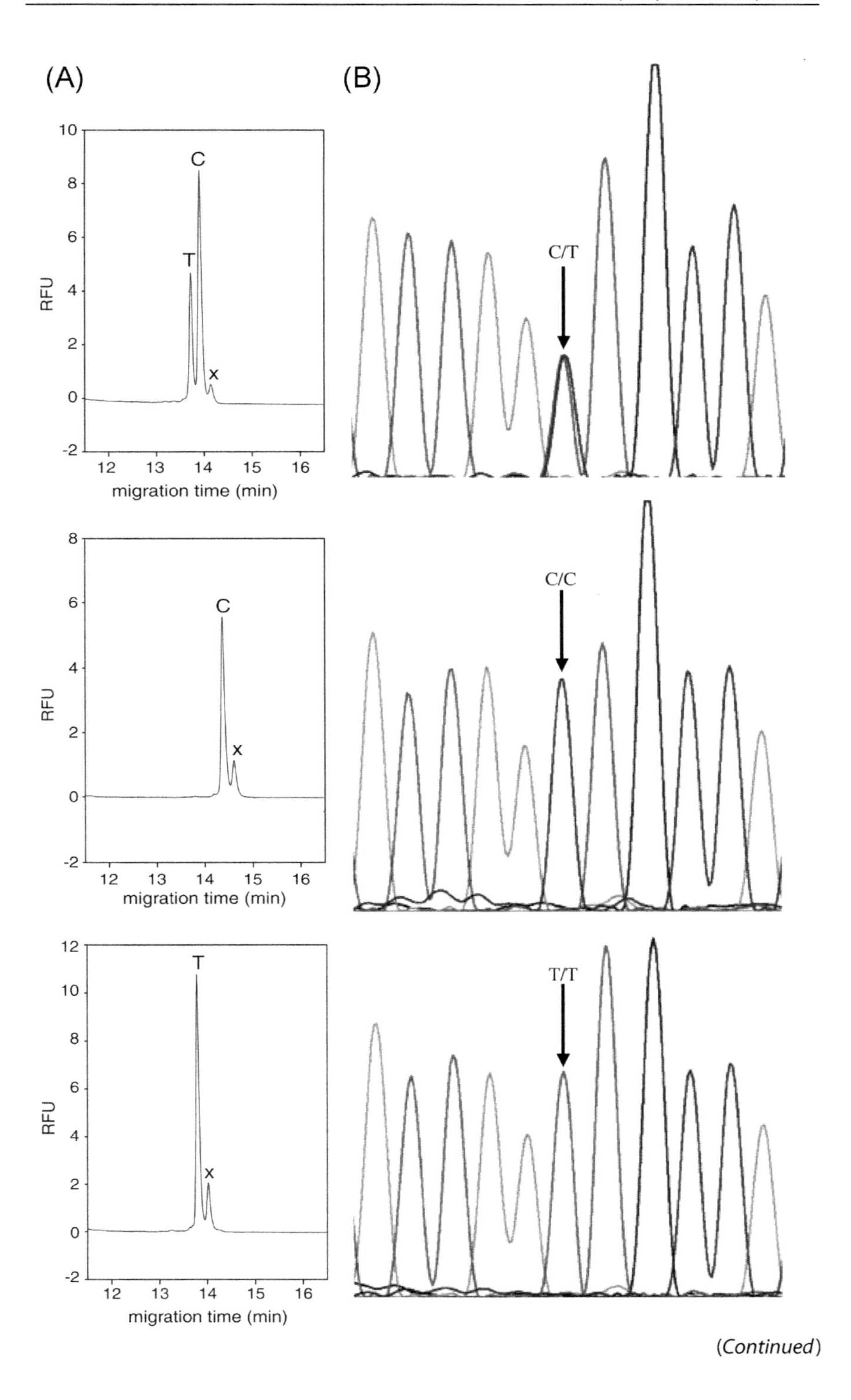

(*Continued*)

synthetic oligonucleotides of the same length but differing in composition by one base, or of the same base composition but different sequence.

5.1.1.2 Single-stranded conformation polymorphism

Single-stranded conformation polymorphism (SSCP) is one of the important CGE-based DNA mutation detection techniques, where a single-stranded DNA molecule is analyzed under nondenaturing conditions to reveal sequence-specific conformation differences. SSCP can detect minor sequence variations between closely related DNA fragments as even small differences may cause conformational secondary structure changes resulting in electrophoretic mobility shifts in CGE. Mitchelson and coworkers applied entangled polymer solutions to fractionate various ssDNA samples with minor mutations [28]. The metastable isomers originating from bi- or trimolecular self-annealing interactions between ssDNAs at complementary regions were detected by their approach with increased discrimination power [29]. A CGE method was developed by Cheng et al. [30] for screening of the 452C > T single-nucleotide polymorphism (SNP) in the γ-glutamyl hydrolase gene. This gene plays a role in methotrexate metabolism, a chemotherapeutic for the treatment of arthritis and leukemia. The DNA samples were analyzed after polymerase chain reaction (PCR) amplification using SSCP-CGE with 1.5% w/v hydroxypropylmethyl cellulose matrix and the genotyping results showed good agreement with DNA sequencing data as shown in Fig. 5.11. Alternative CGE methods were also developed to detect specific mutations or just for general mutation screening as reported in [31,32]. In the study of Kuypers et al. [33] the PCR-amplified p53 tumor suppressor gene, known to be frequently mutated in malignant cells, was the subject of the analysis. Two single-stranded DNA fragments of 372 bases in length differing by only one nucleotide were successfully separated, proving that CGE was a powerful method for the detection of point mutations in DNA sequences. The introduction of Pluronic polymers as sieving matrices opened up new opportunities for SSCP analysis with improved resolution. However,

◀ **Figure 5.11** Single-stranded conformation polymorphism-capillary gel electrophoresis analysis of homozygous and heterozygous genotypes of 452C > T in the γ-glutamyl hydrolase gene (A). DNA sequencing of 452C > T fragments (B). *With permission from H.L. Cheng, S.S. Chiou, Y.M. Liao, Y.L. Chen, S.M. Wu, Genotyping of single nucleotide polymorphism in gamma-glutamyl hydrolase gene by capillary electrophoresis. Electrophoresis 32 (15) (2011) 2021–2027 [30].*

optimizing Pluronic-based SSCP-CGE represented a challenge, because of its physical properties in solutions varied with temperature, especially near the gelation point, where their viscoelasticity sharply changed from that of a Newtonian fluid to a hydrogel. Hwang et al. [34] successfully separated two single-base-pair-differing DNA fragments (wild-type RB1 and its g.2162C > T variation) by using a commercial genetic analyzer with Pluronic F108 solution as sieving matrix. The resolution was significantly enhanced by increasing the separation temperature from 19 to 30°C.

5.1.1.3 DNA sequencing by capillary gel electrophoresis

One of the biggest achievements of the utilization of CGE at the turn of the new millennium was the completion of the draft DNA sequence of the human genome [35,36]. DNA sequencing defines the type and order of nucleotides in the genome. Before the introduction of second- and higher-generation sequencing instrumentations, the Sanger dideoxynucleotide technique was the method of choice, in which DNA polymerase was used to synthesize oligonucleotides starting at a common primer and building up a complementary strand to the target sequence. The sequencing reactions were terminated by the incorporation of one of the four dideoxynucleotides, which were labeled by different fluorophores to be detected at different wavelengths. The obtained so-called sequencing ladder was subject to size separation by CGE. The labeled fragments were migrating through the detection zone, where the individual nucleotides were determined by their spectral properties (i.e., fluorescent emission signal). With the advent of replaceable gel matrices [37] and high sensitivity laser-induced fluorescence (LIF) detection systems [38], automated DNA sequencers were rapidly developed by several companies making possible to decipher the human genome years before originally planned [36]. Consequently, supporting single base separation of DNA sequencing fragments was one of the fastest growing areas during the HUGO project, especially aiming to reduce analysis time to increase throughput. The sample preparation and detection/interpretation parts of CGE-based sequencing methods were just the adaptations of techniques earlier used in slab gel electrophoresis. The four dyes (Fig. 5.12) for LIF and an intensified diode-array detector were crucial for the identification of the spectral characteristics of the dyes on the individual sequencing fragments.

Karger's group [15,39] was among the first ones who demonstrated the potential of capillary polyacrylamide gel electrophoresis in conjunction

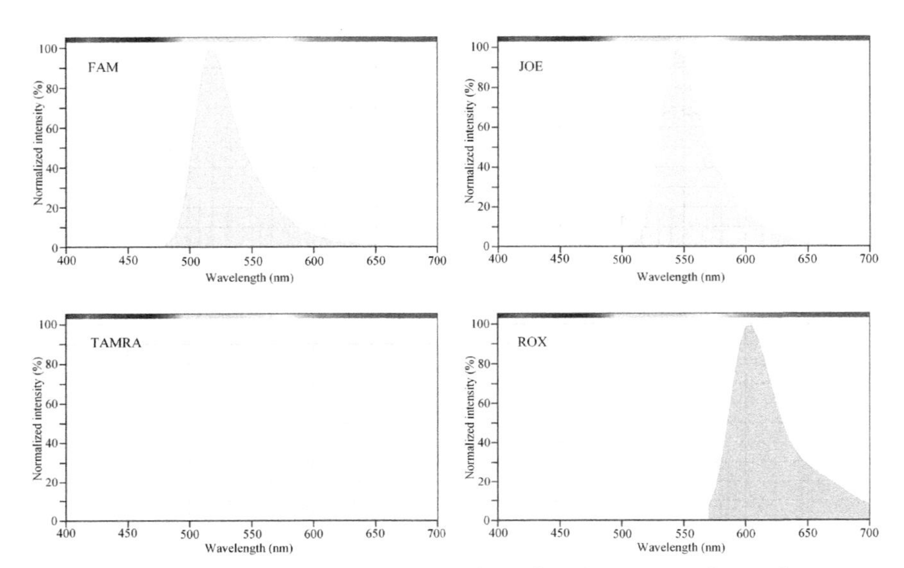

Figure 5.12 Normalized emission spectra of the four conventionally used sequencing dyes (FAM, JOE, TAMRA, and ROX).

with LIF detection for high temperature DNA sequencing as depicted in Fig. 5.13. Please note that in all four electropherograms, baseline resolution was observed with the length difference of a single-nucleotide residue. The approach not only featured high resolution but also rapid separation time, high detection sensitivity, the possibility of multiple injections onto a single column, and the potential for automation. Baseline resolution of DNA fragments by single base differences in length was also demonstrated up to hundreds of bases with the estimated detection limit in the subattomole range. The authors envisioned that ultimately, single column operation could satisfy general laboratory sequencing needs but for higher throughput requirements, such as for the Human Genome Project, a capillary array was required to analyze multiple samples simultaneously, possibly employing four different fluorescent dyes in one separation.

The same group [40] demonstrated the importance of high temperature during gel electrophoresis to achieve the rapid separation of single-stranded DNA molecules. Choosing the appropriate electrophoresis parameters (e.g., separation voltage and temperature) for high resolution of DNA sequencing fragments plays an important role to obtain rapid and high read length separations. Sequencing speed of more than 1000 bases

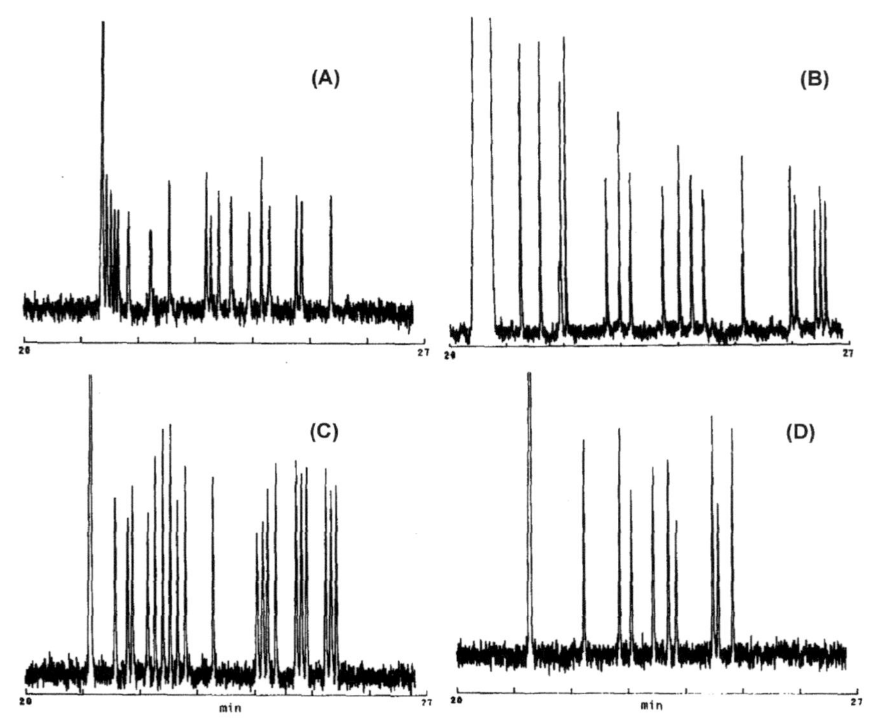

Figure 5.13 Electropherograms of chain-termination sequencing reaction products. Template, TEM80.3. Primer, JOE-PRMI8.1. (A) dA reaction: primer was extended in the presence of ddATP. (B) dC reaction: primer was extended in the presence of ddCTP. (C) dG reaction: primer was extended in the presence of ddGTP. (D) dT reaction: primer was extended in the presence of ddTTP. *With permission from A.S. Cohen, D.R. Najarian, B.L. Karger, Separation and analysis of DNA sequence reaction products by capillary gel electrophoresis. Journal of Chromatography 516 (1) (1990) 49–60 [39].*

per hour was successfully achieved by increasing the separation temperatures up to 80°C (Fig. 5.14) [41,42]. CGE-based DNA sequencing at high temperature has multiple benefits, such as speeding up the separation, increasing accuracy, and read length in addition to providing better denaturation [43]. The effect of small temperature changes during DNA sequencing by CGE was investigated by the Dovichi group [44]. They used Monte Carlo simulation to demonstrate the effect of temperature on flow initiation within the capillary and concluded that even small alterations in capillary temperatures can make remarkable changes in separation efficiency. An important implication of the application of higher temperature is the change in the activation energy requirement of the separating

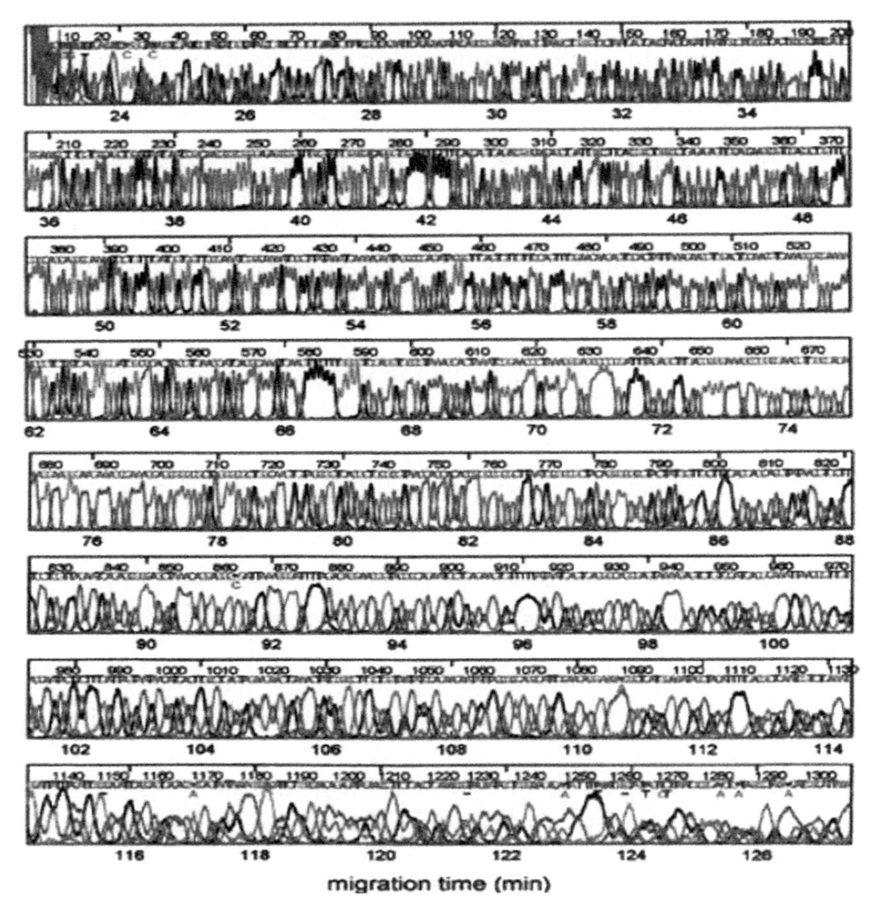

Figure 5.14 Long read length capillary gel electrophoresis-based DNA sequencing at 70°C using high-molecular-mass polymer matrices, optimized applied electric fields, and base calling software. *With permission from B.L. Karger, A. Guttman, DNA sequencing by CE, Electrophoresis 30 (Suppl. 1) (2009) S196–S202 [41].*

DNA molecules [45]. The activation energy associated with the electrophoretic migration of a solute molecule can be accessed through recording the electrophoretic mobilities at various temperatures (Arrhenius plots). In CGE, the activation energy is considered as the energy needed by the solute molecules to overcome the obstacles represented by the sieving matrix, and any deformation of the solute or the polymer network is expected to cause a decrease in the activation energy.

Formamide-modified polyacrylamide gels were applied in CGE-based DNA sequencing to avoid the compression effects occasionally found during the separation of DNA sequencing fragments, especially in G/C-rich regions. The denaturing capacity of the sieving gel was improved by increasing the concentration of formamide above 10% in urea/polyacrylamide sequencing gels and therefore minimized compressions [46]. By the mid-2000s, it was well understood that high-resolution CGE separation of large DNA sequencing fragments can be achieved in dilute but high-molecular-weight polymers [47]. Also, the use of denaturing agents was necessary to alleviate secondary structure related problems, such as compressions, ensuring in this way uniform fragment spacing during the sequencing run. Therefore denaturing chemicals, such as urea, formamide, or dimethyl sulfoxide, were used at high separation temperatures of 70°C−75°C [48].

In an interesting application, extremely long capillaries were used to separate long DNA fragments of cytosine-reaction products by CGE with LIF detection [49]. The authors obtained high-resolution separation of DNA fragments using a 200-cm-long capillary filled with 3% T/5% C polyacrylamide gel and by applying 170 V/cm electric field strength. With this method, sequencing read lengths of up to 680 bases with >0.5 resolution for adjacent peaks were obtained in about 10 hours. They also achieved high resolution of fragments with lower base numbers.

DNA sequencing samples may contain high amounts of salts, the template, and some proteins, such as the polymerase enzyme, among other impurities. The high salt content can cause injection bias as the low-molecular-mass salt components are competing with the DNA fragments to enter the capillary column during electrokinetic injection. One of the conventional purification methods for DNA sequencing fragments is ethanol precipitation to remove the protein and salt content originating from the cycle sequencing reaction. However, that was rather labor intensive and also unreliable in regard to the remaining amount of salt in the samples. Combination of ultrafiltration membranes and spin columns resulted in significant and consequent decrease in salt concentration and thus helped to attain high sequencing read lengths. While, spin-column technology [50] addressed the issues, it was not easily automatable as required centrifugation.

Swerdlow's group developed a readily automatable online sample cleanup method [51] allowing direct injection of unpurified dye-primer sequencing reaction products without pretreatment. According to their method, pH-mediated on-column preconcentration of DNA sequencing fragments was simply achieved by electrokinetic injection of hydroxide ions; thus the neutralization reaction with the cationic buffer components resulted in lower conductivity and consequent field focusing. This stacking method apparently eliminated the problem with sample impurities. The same group has also addressed certain problems for CGE-based DNA sequencing [52]. They emphasized that even solving the problem of bubble formation during the polymerization of the sieving matrix, bubbles can also be formed during the electrophoresis process. These bubbles usually form near the injection end of the capillary and in most instances seriously impair separation performance. Cutting off a few millimeters of the capillary at the injection end usually eliminated the problem. An automated sample preparation technique was reported in multiplexed high-throughput DNA sequencing by Tan and Yeung [53] that utilized size-exclusion chromatography. After purification the denaturation and stacking injection steps for CGE were made simultaneously at a cross assembly set at c.70°C. Read out of 460 bases was accomplished with 98% accuracy with this high-throughput DNA sequencing method.

Smith's group obtained high-speed CGE separations of fluorescently labeled DNA sequencing fragments generated in enzymatic reactions with attomole detection levels. They also proposed that the application of CGE technology to automated DNA sequencing may lead to the development of a second generation automated sequencer, capable of efficient and cost-effective analysis at the genomic scale [54]. Comparison of conventional slab gel electrophoresis and CGE for the separation of a fluorescein-labeled C-reaction of M13mp19 DNA is depicted in Fig. 5.15. The conventional slab gel approach (Fig. 5.15A) applied an electric potential of 1200–1500 V on a 40-cm long, 0.4-mm-thick gel with a 26-cm distance to the detector from the injection spot. In CGE analysis (Fig. 5.15B), 10 kV electric potential was applied on a 50-cm long, 50-μm i.d. gel-filled capillary (25 cm to the detector window). Fig. 5.15A and B shows the same region of data, from the primer peak to base 329. Both data sets were obtained on 4% denaturing polyacrylamide gels. Higher resolution and 5-fold reduction in separation time were obtained with CGE compared to conventional slab gel electrophoresis.

Low zeptomole detection limits were obtained for the sequencing of tetramethylrhodamine-labeled DNA fragments by CGE using a green helium-neon laser (543.5 nm) in conjunction with a sheath-flow cuvette detection settings, even during a 70 bases/hour sequencing speed [55,56].

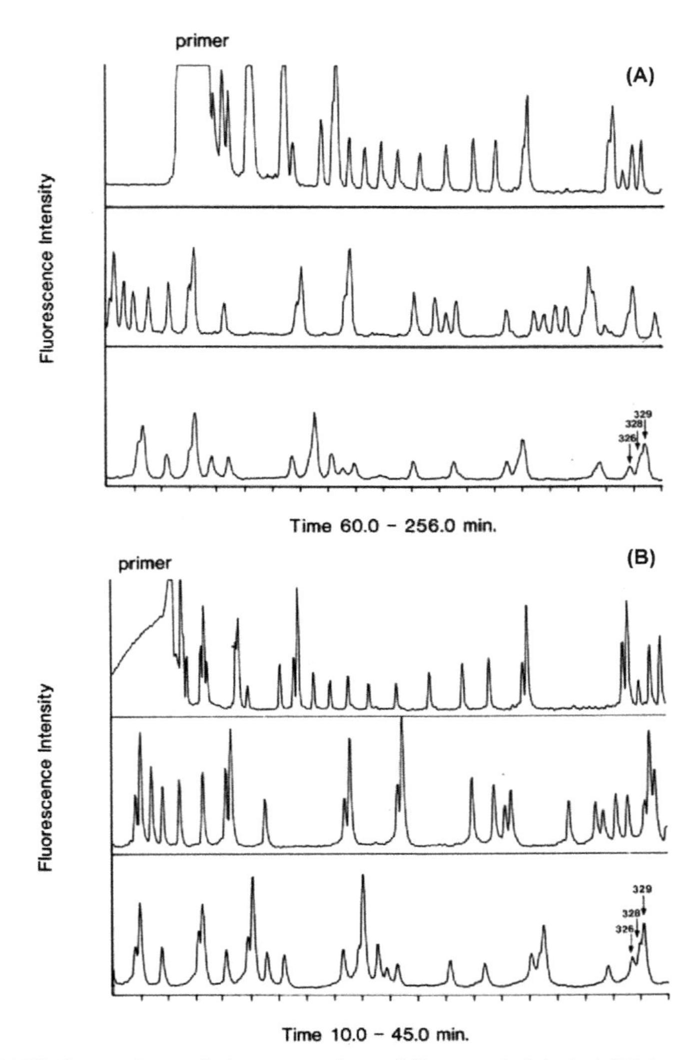

Figure 5.15 Comparison of the separation of fluorescein-labeled DNA sequencing fragments by conventional slab gel electrophoresis (A) and capillary gel electrophoresis (B). *With permission from H. Drossman, J.A. Luckey, A.J. Kostichka, J. D'Cunha, L.M. Smith, High-speed separations of DNA sequencing reactions by capillary electrophoresis. Analytical Chemistry 62 (9) (1990) 900–903 [54].*

The same authors applied CGE to southern blotting allowing online detection of DNA hybridization in solution [57]. Swerdlow et al. [58] introduced a postcolumn LIF detector that utilized a sheath-flow cuvette and minimized in this way any background signal due to light scatter from the gel and capillary, attaining a mass detection limit of 6000 fluorescein-labeled DNA fragment molecules. Better detection limits were obtained for highly diluted neat dye solution. Actually, their results for DNA sequencing were unprecedented at that time [59,60]. The same group also attempted the utilization of peak-height encoded DNA sequencing strategy using CGE-LIF technique [61]. Pentoney et al. [62] utilized the same method in CGE with a single fluorophore labeling to DNA sequence determination. The Tabor and Richardson [63] strategy was utilized for enzymatic chain-termination sequencing of DNA using relative peak intensities. The modified Tabor and Richardson method utilized two reactions, each containing complementary mixtures of only three dideoxynucleotide triphosphates (ddNTP) in the concentration ratios of 4:2:1 and the DNA sequence was detected by reading the relative peak heights and by assigning the missing ddNTP to gaps between the peaks in the electropherograms (Fig. 5.16), with the main advantage of detection simplicity [62].

A two-color peak-height encoded DNA sequencing method for CGE with LIF detection was introduced in [64]. The method was further developed to high-speed and high-accuracy DNA sequencing by CGE, that is, two-color/peak-height encoded sequencing, at 40°C [65]. TAMRA- and ROX-labeled primers were extended in the presence of ddGTP and ddTTP and the amounts of dideoxynucleotides were adjusted to produce a 3:1 peak-height ratio for easier sequence reading. Another approach was the use of a two-color confocal fluorescence detection setup with a two-dye labeling Sanger protocol by capillary array electrophoresis. The DNA fragments were separated in narrow bore capillaries and read by means of a binary coding scheme as each fragment set was labeled with a characteristic ratio of two-dye-labeled primers [66]. Using a 25 capillary array, this approach reached DNA sequencing rate of almost 25 kb/hour [67] in 1993.

In another attempt, four-color DNA cycle sequencing was performed on an M13mp18 template using dye-labeled primers by several groups [68,69] in 5%T noncross-linked polyacrylamide gel containing 7 M urea. The Dovichi group used high temperature that reduced secondary structure formation and also increased the analysis speed allowing separation of fragments up to 640 bases in less than 2 hours. A multiplexed, four-decay fluorescence detection scheme was evaluated by McGown and coworkers [70] for DNA sequencing. This four-decay fluorescence detection method used fluorescence lifetime instead of color to discriminate among the four

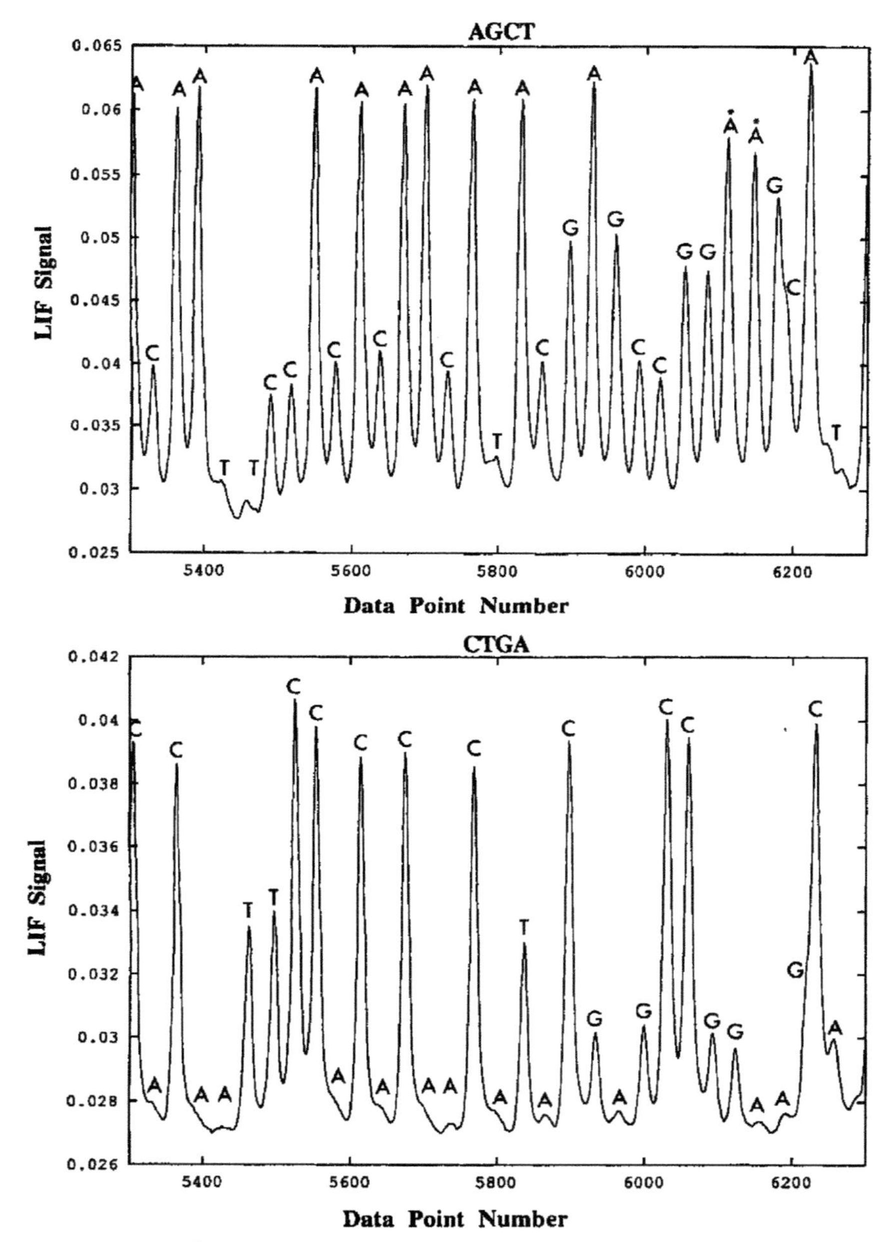

Figure 5.16 Capillary gel electrophoresis based DNA sequencing showing a portion of the separation of fluorescently labeled fragments generated enzymatically from M13mp18 template applying three dideoxy chain terminators in the concentration ratios (upper trace) 4A:2G:1C and (lower trace) 4C:2T:1G. *With permission from S.L. Pentoney Jr., K.D. Konrad, W. Kaye, A single-fluor approach to DNA sequence determination using high performance capillary electrophoresis, Electrophoresis 13 (8) (1992) 467–474 [62].*

bases. NBD-aminohexanoic acid, rhodamine green, and Bodipy 505/515 were identified to improve the resolution of their individual lifetimes in mixtures in batch measurements when conjugated to DNA. A good comparison of one-, two- and four-spectral-channel sequencing was given by Swerdlow et al. [71].

Since the mass-to-charge ratio of oligonucleotides is apparently independent of the size of the fragment, a sieving matrix is required to separate them by electrophoretic methods. It has been a long history of attempts to generate high-performance and low-viscosity sieving matrices that allow easy replacement of the separation medium within the capillary between each runs to avoid cross-contamination. One of the alternatives is to use a large uncharged tag that changes the mass-to-charge ratio and thus enables separation of DNA fragments in free-solution mode. Hence, this innovative approach is called end-labeled free-solution electrophoresis [72]. Barron and coworkers introduced such a free-solution conjugate electrophoresis method by utilizing a 516-amino-acid-long drag-tag for rapid DNA sequencing as depicted in Fig. 5.17 [73]. Using a four-color LIF detector, up to 265-base read lenght was obtained in free-solution sequencing in 30 minutes. The same group even turned this approach upside down, that is, attached a constant size DNA fragment to different length uncharged polymers for characterization. The fluorescently labeled DNA enabled polydispersity determination [74] as such polymers represented different hydrodynamic drags and the balance of electric field-mediated migration and friction with the various contour length polymers was measured.

5.1.2 dsDNA fragments and PCR products

High-resolution analysis of double-stranded DNA (dsDNA) molecules is of high importance in genotyping and mutation analysis in molecular biology laboratories. Actually, the same general system can be used for genotyping as for sequencing including PCR amplification followed by fragment analysis using single capillary or capillary array electrophoresis, but with different gel compositions. One of the major classes of genotyping markers is short tandem repeats (STR) varying in repeat numbers among individuals. Single nucleotide polymorphism (SNP) is another form of mutation, in which case single-base differences are identified. While STRs mostly used in paternity and forensic investigations [75], SNPs can be important disease markers [76]. Other methods to identify

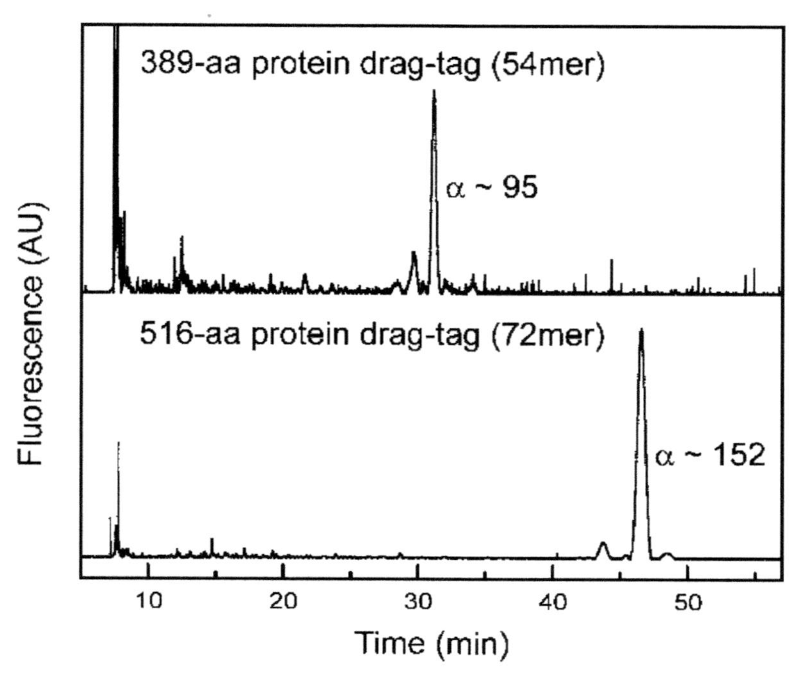

Figure 5.17 Electrophoretic analysis of the purity of protein drag-tags; the peaks at 7 minutes is "free" unconjugated DNA (30-mer oligomer), and the second peak is the DNA-drag-tag conjugate [54-mer (390-aa) and 72-mer (516-aa) proteins conjugated to DNA]. *With permission from J.C. Albrecht, J.S. Lin, A.E. Barron, A 265-base DNA sequencing read by capillary electrophoresis with no separation matrix, Analytical Chemistry 83 (2) (2011) 509–515 [73].*

mutations utilized the mismatched region in the sequence between the wild-type and target sequence, cleaved and separated the resulting fragments to identify the mutation location [77]. PCR amplification can be multiplexed and the resulting fragments analyzed by CGE analysis to detect multiple mutations in a single analysis step [78].

One of the early applications with the first commercially available CE instrument built by Brownlee and coworkers [79] (see instrument picture in Chapter 1) was dsDNA analysis with UV absorbance and fluorescence detectors using 50–100-μm-i.d. fused-silica capillary tubings. They achieved 15 mg/mL sensitivity (signal-to-noise ratio of 3) for ethidium bromide (EB)-stained herring sperm dsDNA by fluorescence detection. The authors also obtained rapid separations for restriction fragments, as well as for whole phage, viral, and plasmid DNAs, decades before the age of viral and gene therapy. Fig. 5.18 depicts the CGE separation of the

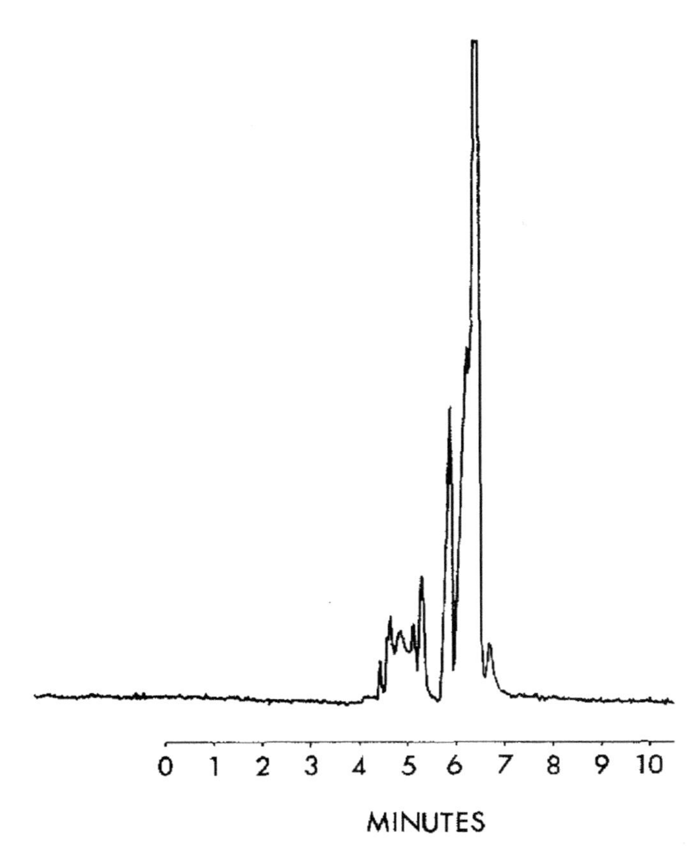

Figure 5.18 Capillary gel electrophoresis electropherogram of the Hae III fragments of the φX DNA using 3%T/5%C polyacrylamide gel-filled fused-silica capillary column (20 cm × 100 μm i.d.) with fluorescence detection. *With permission from T.J. Kasper, M. Melera, P. Gozel, R.G. Brownlee, Separation and detection of DNA by capillary electrophoresis. Journal of Chromatography 458 (1988) 303–312 [79].*

Hae III fragments of the φX DNA using a cross-linked polyacrylamide (3%T/5%C) gel-filled capillary and EB staining.

5.1.2.1 Genotyping and mutation analysis

Genotyping represents a similarly difficult separation challenge as DNA sequencing. Precise sizing of multiplexed STR loci was accomplished by the Mathies group using energy-transfer fluorescent primers by high-resolution capillary array electrophoresis [80]. This method established the feasibility of high-resolution and large-scale STR profiling. Analysis of a polyadenine tract, the $(A)_{10}$ repeat, within the cysteine rich domain of the transforming growth factor-beta type II receptor gene in colorectal cancer was also evaluated [81]. Optimized conditions enabled the determination of one nucleotide difference in 8–32 nucleotides. Stellwagen et al. [82] have studied DNA conformation and structure using CGE and found that counter-ions bound preferentially to DNA oligomers with A-tracts, especially in A_nT_n sequence motif. Genetic profiling of grape plant variants and clones were analyzed by using dynamic size-sieving capillary electrophoresis in conjunction with random amplified polymorphic DNA analysis [83]. Compared to slab gel electrophoresis with EB staining, CGE with LIF detection provided superior separation efficiency and detection limits in revealing polymorphic differences. Several companies have commercialized special systems for SNP genotyping [84,85].

Analysis of PCR-amplified DNA fragments by CGE-LIF was also reported from three different genetic loci [apolipoprotein B (apoB), variable number of tandem repeat (VNTR) locus D 1S80 and mitochondrial DNA] [86]. The developed separation system could be useful in analyzing PCR-amplified DNA fragments from loci that differed in number of base repeats and in size. DNA point mutations associated with certain diseases were also revealed by PCR in conjunction with CGE [33]. One of the innovative approaches to utilize CGE for oligonucleotide mutation analysis was the single base extension with a labeled dideoxynucleotide from a primer located at a putative mutation site [87]. The labeled short single-stranded primer extension products were analyzed by capillary electrophoresis using 4% LPA gel with 7 M urea denaturing sieving medium and LIF detection. The method was capable to readily distinguish the primer extension product from the primer.

Constant denaturant capillary electrophoresis (CDCE) was introduced by Karger and coworkers as a high-resolution approach to mutation analysis [88]. Using a zone of constant temperature and denaturant concentration, a simple, rapid, and reproducible system was implemented to separate mutant from wild-type DNA sequences with high resolution. A 206 bp DNA sequence portion of the human mitochondrial genome had a contiguous low-melting domain (112 bp) and a high melting domain (94 bp), in which case the separation of heteroduplexes from wild-type homoduplexes was not possible with conventional CGE. The sequences differed by a single bp at position 30 , the GC designated sequence had a GC base pair, and the AT designated sequence had an AT base pair. The separation parameters for CDCE were 36°C column temperature and denaturant buffer with 3.3 M urea plus 20% (v/v) formamide. At this temperature, the GC and AT homoduplexes, as well as the GT and AC heteroduplexes, were baseline separated as shown in Fig. 5.19. For a typical 100-bp fragment, point mutation-containing heteroduplexes were separated from wild-type homoduplexes in approximately 30 minutes, rendering this approach applicable to low frequency mutation identification, mutational spectrometry, and genetic screening of pooled samples for the detection of rare variants. CDCE allowed high-resolution separation of single-base variations, occurring in an approximately 100 bp isomelting DNA sequence, based on their differential melting temperatures. By coupling CDCE for highly efficient enrichment of mutants with high-fidelity PCR, Thilly and coworkers [89] have developed an analytical method for detecting point mutations at frequencies as low as 10^{-6} in

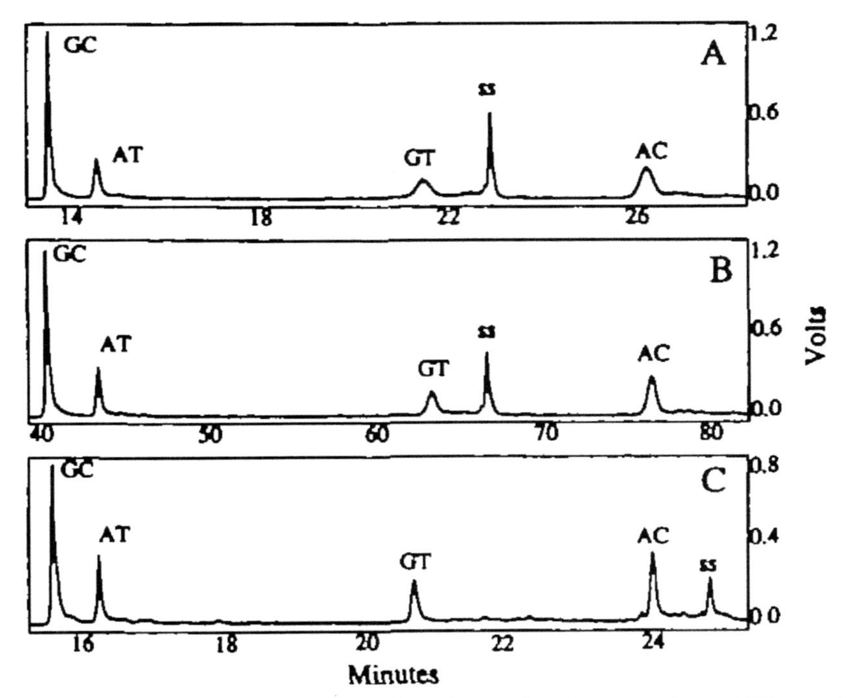

Figure 5.19 Constant denaturant capillary electrophoresis separation of GC and AT homoduplexes, as well as the GT and AC heteroduplexes at various electric field strengths and temperatures. (A) 3.3 M urea, 20% formamide, 36°C, 250 V/cm. (B) 3.3 M urea, 20% formamide, 36°C, 83 V/cm. (C) 0 M urea, 0% formamide, 63°C, 125 V/cm. *With permission from K. Khrapko, J.S. Hanekamp, W.G. Thilly, A. Belenkii, F. Foret, B.L. Karger, Constant denaturant capillary electrophoresis (CDCE): a high resolution approach to mutational analysis. Nucleic Acids Research 22 (3) (1994) 364–369 [88].*

human genomic DNA. Their instrument setup is shown in Fig. 5.20. The approach had promising implications in human genetic analysis, including studies of mitochondrial mutations, measurement of mutational spectra and population screening for disease-associated SNPs. The same group used CDCE for pooled blood samples to identify SNPs in Scnn1a and Scnn1b genes [90] and introduced a two point LIF detection method to improve mutation identification [91].

An efficient high-throughput method was reported for genotyping clonal CRISPR/Cas9-mediated mutants from crude genomic DNA, straight from cell cultures. The workflow of the developed method to genotype clonal CRISPR/Cas9-mediated mutants included the extraction of crude genomic DNA from cells with the help of a direct lysis buffer and fluorescent PCR coupled to multicapillary (24 ×) gel electrophoresis using POP-7 polymer as sieving matrix as shown in Fig. 5.21 [92].

A highly sensitive CGE-LIF method was applied for the detection of the most prevalent mutation of medium-chain acyl-coenzyme A dehydrogenase (MCAD) deficiency at position 329 of lysine to glutamic acid substitution in [93]. Thiazole Orange was used as fluorescent intercalating dye with a low cross-linked polyacrylamide gel (3%T/0.5%C) sieving matrix. The DNA fragment possessing the mutation site was PCR

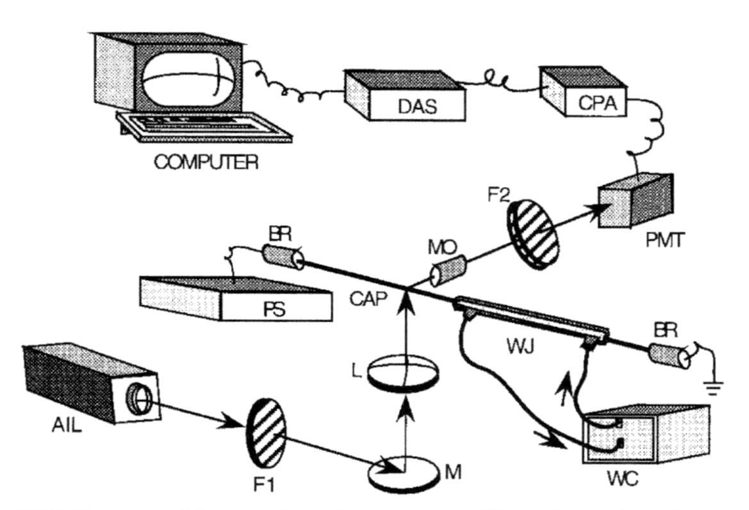

Figure 5.20 Diagram of the constant denaturant capillary electrophoresis apparatus. *With permission from X.C. Li-Sucholeiki, K. Khrapko, P.C. Andre, L.A. Marcelino, B.L. Karger, W.G. Thilly, Applications of constant denaturant capillary electrophoresis/high-fidelity polymerase chain reaction to human genetic analysis, Electrophoresis 20 (1999) 1224–1232 [89].*

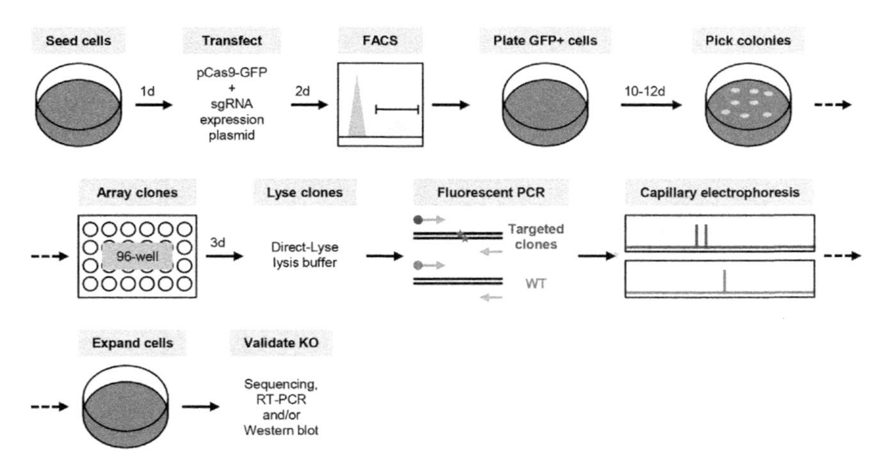

Figure 5.21 Workflow of high-throughput genotyping via fluorescent polymerase chain reaction coupled to multicapillary gel electrophoresis. *With permission from M. K. Ramlee, T.D. Yan, A.M.S. Cheung, C.T.H. Chuah, S. Li, High-throughput genotyping of CRISPR/Cas9-mediated mutants using fluorescent PCR-capillary gel electrophoresis. Scientific Reports 5 (2015) 15587 [92].*

amplified with two sets of allele-specific oligonucleotide primers, followed by CGE-LIF analysis. The mutant allele produced a 175 bp DNA fragment, whereas the normal allele generated a 202 bp DNA fragment. These two PCR products were baseline separated with CGE-LIF, thus suitable for rapid diagnosis of MCAD deficiency. Others performed restriction fragment length polymorphism (RFLP) analysis of apolipoprotein E (apoE) [94] by applying CGE for the separation of three common alleles. The E4 isoform was associated with heart disease through increased levels of total cholesterol and beta-lipoprotein. Homozygous E4 alleles produced single 72-bp DNA fragments, homozygous E3 alleles single 91-bp fragments, and heterozygous E3/E4 alleles both the 72- and 91-bp fragments. This apoE genotyping method can be effectively used for the genetic diagnosis of heart diseases. Rapid size-selective separation and identification of plasmid DNA (pBR322 and pBR328) were performed by the Sepaniak group with CGE using the water-soluble polymer methyl cellulose (100,000 MW) as sieving matrix, shown in Fig. 5.22 [95]. The two plasmid samples were distinguished based on differences in the CGE migration patterns of their HinfI enzyme digests. EB was used as intercalation dye for the DNA fragments and laser fluorimetry as detection system. High separation efficiency of 10^6 plates/m was achieved with good fragment resolution of the different restriction digest molecules. In addition,

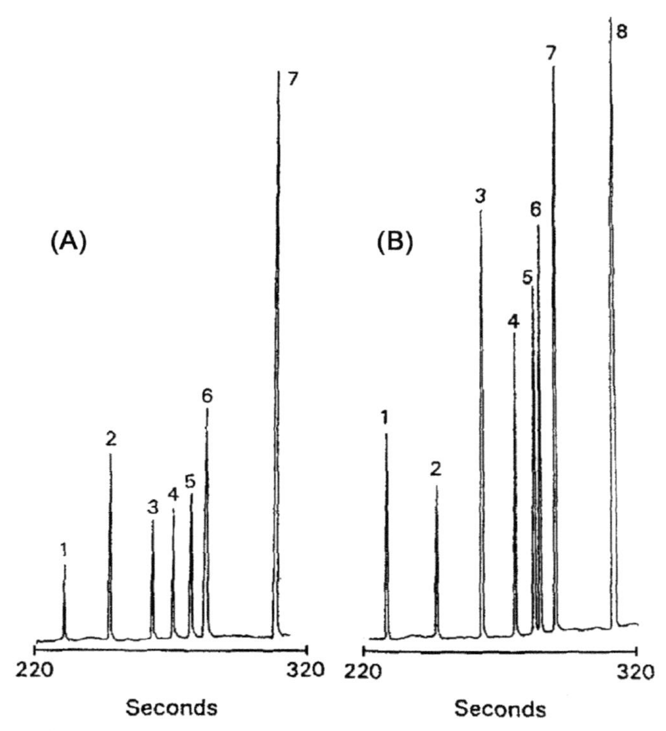

Figure 5.22 Electropherograms of the pBR322 (A) and pBR328 (B) HinfI digest fragments at −500 V/cm applied electric field. Identified fragment sizes: (A) pBR322—(1) l54 bp; (2) 220 bp, 221 bp; (3) 298 bp; (4) 344 bp; (5) 396 bp; (6) 506 bp, 517 bp; and (7) 1631 bp. (B) pBR328—(1) 145 bp, 154 bp; (2) 230 bp; (3) 320 bp; (4) 360 bp; (5) 440 bp; (6) 520 bp; (7) 660 bp; and (8) 1780 bp. *With permission from B.K. Clark, C.L. Nickles, K.C. Morton, J. Kovac, M.J. Sepaniak, Rapid separation of DNA restriction digests using size selective capillary electrophoresis with application to DNA fingerprinting. Journal of Microcolumn Separations 6 (5) (1994) 503–513 [95].*

unique migration profiles of the digest were obtained within 120 seconds under optimized conditions.

5.1.2.2 DNA fragment analysis

In the early 1990s with the exponential increase of the use of PCR technology, CGE became a popular tool for the analysis of PCR products. Another application started gaining popularity those years was RFLP analysis. For dsDNA fragments, the addition of an intercalator agent not only made pre-PCR fluorophore labeling unnecessary but also enhanced resolution as shown in Fig. 5.23 [96]. Intercalating dyes, such as TOTO and

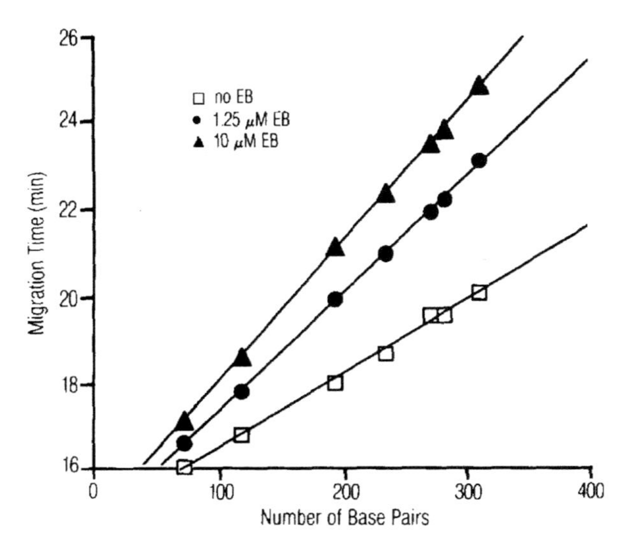

Figure 5.23 Effect of ethidium bromide concentration on the resolution of DNA restriction fragments. *With permission from K.J. Ulfelder, H.E. Schwartz, J.M. Hall, F.J. Sunzeri, Restriction fragment length polymorphism analysis of ERBB2 oncogene by capillary electrophoresis. Analytical Biochemistry 200 (2) (1992) 260–267 [96].*

YOYO, enabled visualization of femtogram per nanoliter quantities of dsDNA fragments.

Kuhr's group [97] studied the separation of dsDNA and single-stranded DNA restriction fragments in capillary electrophoresis with polymer solutions under alkaline conditions (pH 11) in epoxy-coated capillaries and achieved theoretical plate numbers exceeding several millions (Table 5.1). At pH 12, single-stranded DNA molecules were still well separated in entangled hydroxyethyl cellulose (HEC) solutions; however, the resolution significantly decreased in dilute polymer solutions. Heller [98,99] thoroughly studied the separation of dsDNA and single-stranded DNA in linear poly-*N*,*N*-dimethylacrylamide (PDMA) matrix and found significant differences between the experimental and by earlier scaling law predicted data [100]. Analysis of gamma radiation-induced damage to plasmid DNA was evaluated by using dynamic size-sieving CGE [101]. On-column sample preconcentration and separation of ΦX 174 RF DNA-*HaeIII* digest fragments were demonstrated in an open tubular capillary system utilizing electroosmotic flow with polyethylene oxide sieving matrix [102]. This polymer also featured wall coating capability, especially in low ionic strength buffer systems.

Table 5.1 Dependence of separation efficiency of DNA fragments on buffer pH and ionic strength.

Theoretical plate numbers (plates/m) at addition of

DNA fragments (bp)	32 mM	
	NaOH pH 11	NaCl pH 8.2
72	3,110,000	1,067,000
118	2,226,000	867,000
194	2,300,000	800,000
234	2,533,000	602,000
271	3,667,000	a
281	3,400,000	a
310	2,600,000	867,000
603	2,067,000	400,000
872	1,900,000	230,000
1078	1,967,000	183,000
1353	1,010,000	90,000

Note: Sodium chloride and sodium hydroxide were added to 0.5% hydroxyethyl cellulose in TBE buffer to obtain the same pH and ionic strength, respectively. (a) not detectable due to comigration
Source: With permission from Y. Liu, W.G. Kuhr, Separation of double- and single-stranded DNA restriction fragments: capillary electrophoresis with polymer solutions under alkaline conditions, Analytical Chemistry 71 (9) (1999) 1668–1673 [97].

A high-performance CGE system with a polysiloxane-coated capillary and hydroxypropylmethyl cellulose (HPMC) sieving matrix was used for the analysis of coamplified DNA sequences (HIV-1 and HLA-DQ-a) by Schwartz et al. [103]. Addition of EB to the sieving matrix resulted in better peak resolution but longer migration times due to the positive charge of this intercalator dye. Other groups used methylcellulose sieving medium for the analysis of large dsDNA restriction fragments up to 23,000 bp [104]. By coating the inner walls of a fused-silica capillary with LPA and filling with 0.5% methylcellulose, even the very large λDNA/HindIII digest fragments (up to 23 kBp) were well separated as depicted in Fig. 5.24. The resolution and quantitation of this CGE method was demonstrated as complementary to slab gel electrophoresis for applications, such as plasmid integrity confirmation [105].

Ulfelder et al. [96] analyzed the restriction fragment length polymorphism of the ERBB2 oncogene by CGE with EB intercalating dye containing sieving buffer. The addition of this intercalator dye to the buffer system in conjunction with field amplified injection led to increased sample detectability. The RFLP samples were analyzed for homo- or heterozygosity as shown in Fig. 5.25.

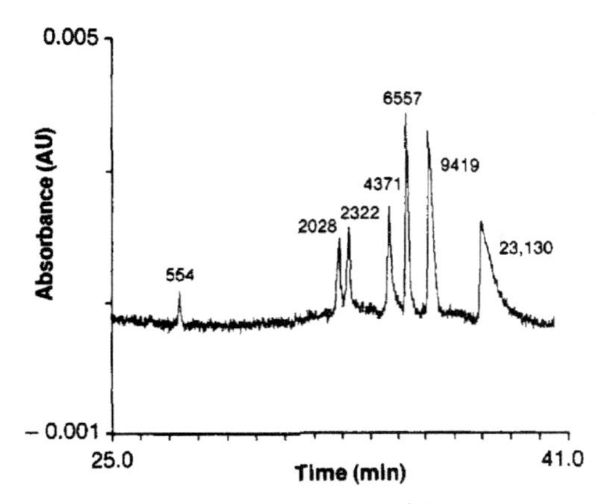

Figure 5.24 Capillary gel electrophoresis analysis of the λDNA/HindIII restriction fragment digest mixture. Separation was performed at 10 kV in a linear polyacrylamide-coated capillary filled with 50 mM Tris-borate (pH 8.0), 2.5 mM EDTA background electrolyte containing 0.5% methylcellulose (MW ~88,000). *With permission from M. Strege, A. Lagu, Separation of DNA restriction fragments by capillary electrophoresis using coated fused silica capillaries. Analytical Chemistry 63 (13) (1991) 1233–1236 [104].*

Separation of DNA restriction fragments in diluted solutions of LPA, HEC, and polyvinylalcohol (PVA) were investigated by Schomburg and coworkers [106]. Interestingly, PVA solutions became inhomogeneous after a few separations resulting in deterioration of separation power. However, good separations of DNA restriction fragments were obtained using LPA or HEC solutions after surface pretreatment of the fused-silica capillaries either by PVA rinsing or coating. Hydroxylic polymers, capable of forming suitable networks, were strongly adsorbed to fused-silica surfaces and suppressed the electroosmotic flow. Rossomando et al. [107] generated reverse transcriptase (RT) PCR products from the RNA of poliovirus and analyzed the products by slab gel electrophoresis on 4% agarose gels, compared to CGE using LPA matrix. The authors also quantified the amount of the RT-PCR product using a standard curve.

Heteroduplex DNA polymorphism analysis utilizes the conformational polymorphisms that alters the electrophoretic mobility of the fragments and thus can be employed to detect a nonrestrictable loci [108]. The addition of EB and glycerol to the sieving matrix resulted in improved peak resolution and good reproducibility. Reannealed PCR products were used directly for mutation screening with full automation, as depicted in Fig. 5.26.

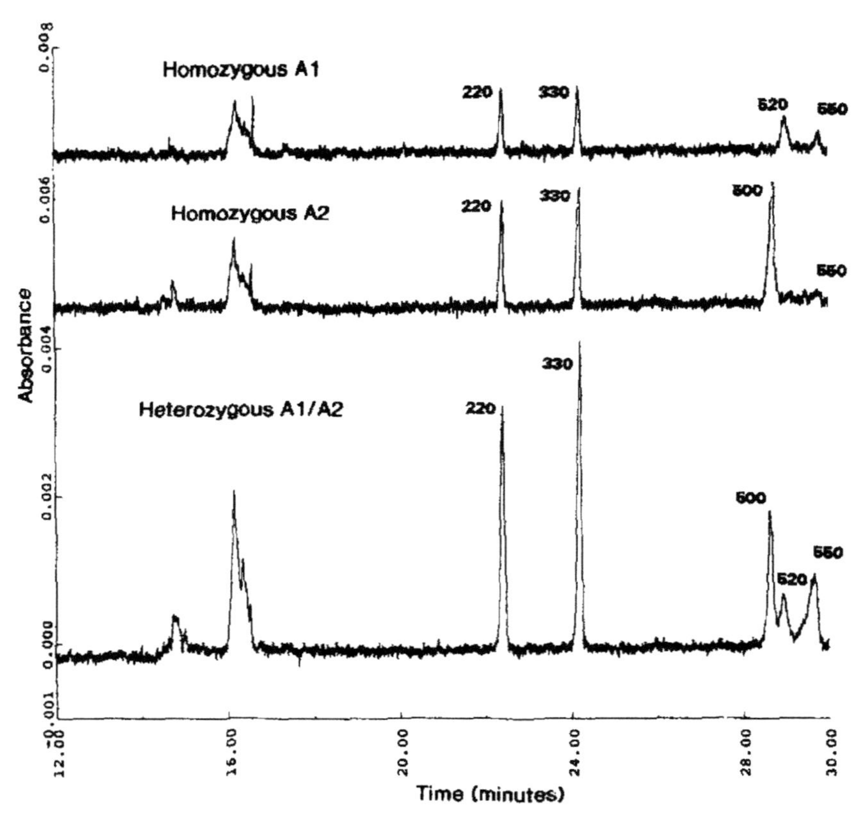

Figure 5.25 Capillary gel electrophoresis-laser-induced fluorescence separation of polymerase chain reaction-derived restriction fragment length polymorphism samples demonstrating *MboI* polymorphism. Upper trace: homozygous for allele A1 (520-bp fragment); middle trace: homozygous for allele A2 (500-bp fragment); and lower trace: heterozygous for A1 and A2. *With permission from K.J. Ulfelder, H.E. Schwartz, J.M. Hall, F.J. Sunzeri, Restriction fragment length polymorphism analysis of ERBB2 oncogene by capillary electrophoresis, Analytical Biochemistry 200 (2) (1992) 260–267 [96].*

Marino et al. performed the analysis of PCR products and DNA restriction fragments from soybean and human (ΦX174/HinfI) samples in less than 70 minutes using CGE with UV detection at 260 nm [109]. The effective length of the capillary was 40 cm, filled with a cross-linked polyacrylamide gel of 3% T/3% C. The 72–1353-bp DNA molecular mass marker set was used as internal standard. The applied method was rapid, sensitive, and reproducible for the analysis of DNA samples under 1000 bp.

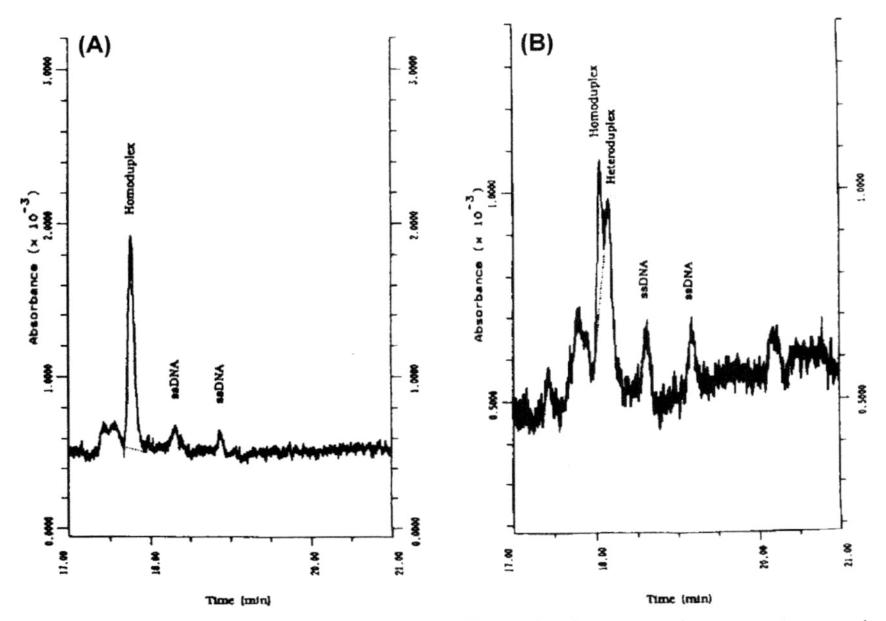

Figure 5.26 Capillary gel electrophoresis analysis of polymerase chain reaction product mixtures from S_{Fi} (S strain) and F_{It} (F strain) homoduplex (A) and heteroduplex (B). *With permission from J. Cheng, T. Kasuga, K.R. Mitchelson, E.R.T. Lightly, N.D. Watson, W.J. Martin, et al., Polymerase chain reaction heteroduplex polymorphism analysis by entangled solution capillary electrophoresis, Journal of Chromatography A 677 (1) (1994) 169–177 [108].*

In Kennedy disease, the PCR-amplified products exhibit two DNA fragments in heterozygous female carriers. One DNA fragment has a size range between 468 and 495 bp in the normal polymorphic population. The fragment corresponding to the pathological state is 573 bp. Chiari and coworkers applied CGE with a novel polymer networks (consisting of 8% polyacryloylamino ethoxyethanol with 0% crosslinker) to separate these fragments and detected by UV absorbance at 254 nm (Fig. 5.27) [110]. The use of this sieving matrix shortened the analysis time by 20-fold compared to standard polyacrylamide, while maintaining the same resolving power, probably due to the larger pore size of this new separation medium.

Herbert and coworkers applied dilute solutions of HEC as a replaceable sieving matrix for rapid analysis of DNA fragmentation in whole digests of apoptotic rat thymocytes [111]. They used LIF detection with the highly sensitive nucleic acid stain of YO-PRO-1. The CGE method

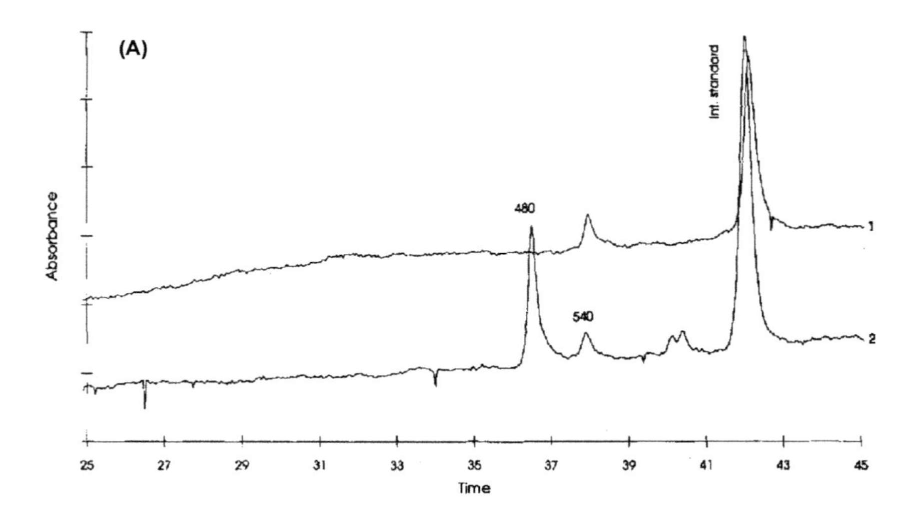

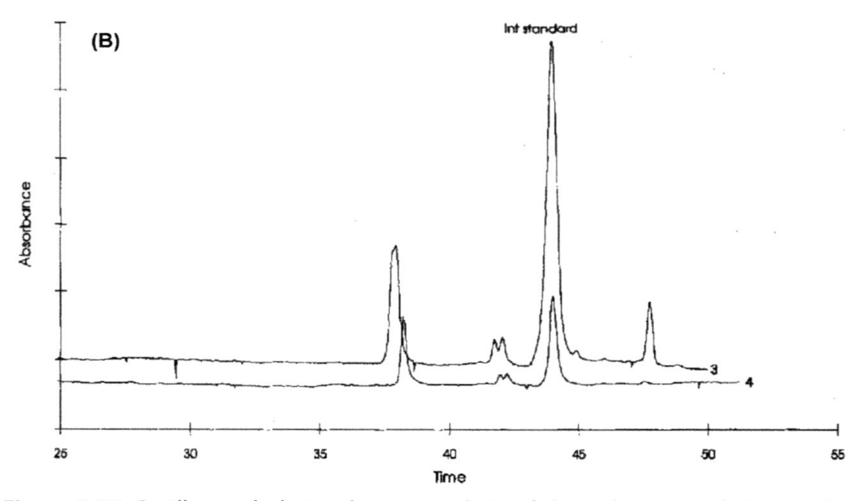

Figure 5.27 Capillary gel electrophoresis analysis of the polymerase chain reaction-amplified 5′ region of the first exon of the androgen receptor gene in four members of a family with Kennedy's disease. Traces: A1: affected male; A2: obligate daughter carrier; B3: normal female (daughter of 2 and 4); and B4: normal unrelated male (spouse of 2). Internal standard: 650 bp fragment. *With permission from M. Nesi, P.G. Righetti, M.C. Patrosso, A. Ferlini, M. Chiari, Capillary electrophoresis of polymerase chain reaction-amplified products in polymer networks: the case of Kennedy's disease. Electrophoresis 15 (5) (1994) 644–646 [110].*

was very sensitive and could be automated requiring 1000–2000-fold fewer cells compared to traditional slab gel electrophoresis. Gelfi et al. [112] performed the analysis of CAG triplet polymorphism in families carrying androgen insensitivity syndrome using PCR-amplified samples and CGE applying 6% LPA as a dynamic sieving matrix. DNA fragments of 139 and 160 bp were baseline separated and visualized by UV absorbance detection at 254 nm. Please note that even fragments of 136 and 139 bp were somewhat resolved providing some diagnostic value. Mathies and coworkers [113] performed genetic typing of the STR polymorphism of HUMTHO1 by capillary array electrophoresis with energy-transfer fluorescent dye-labeled PCR primers. The absorbance and fluorescence emission spectra of fluorescently labeled THO1 primers are depicted in Fig. 5.28. A replaceable sieving matrix of 0.8% HEC was used with 9-aminoacridine resolution enhancer.

Barron et al. utilized dilute and semidilute uncross-linked HEC solutions as sieving matrices for CGE of DNA restriction fragments. They investigated the effects of HEC molecular weight and concentration on resolution, as well as the relation between the former parameters and the polymer entanglement threshold concentration. The distance between

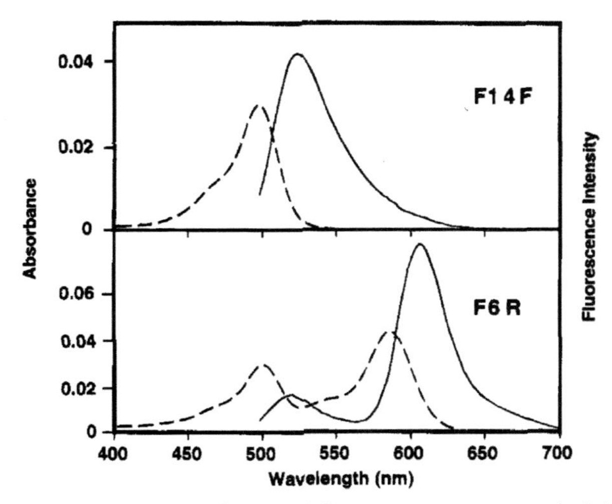

Figure 5.28 Absorption (dashed line) and fluorescence emission (solid line) spectra of the 9-aminoacridine-labeled THOI primers F14F and F6R measured in 1 × TBE. *With permission from Y. Wang, J. Ju, B.A. Carpenter, J.M. Atherton, G.F. Sensabaugh, R. A. Mathies, Rapid sizing of short tandem repeat alleles using capillary array electrophoresis and energy-transfer fluorescent primers. Analytical Chemistry 67 (7) (1995) 1197–1203 [113].*

entanglement points was dependent on HEC concentration only. The authors suggested that information about the entanglement threshold could help to predict the range of HEC concentrations of given molecular weights, which will separate certain dsDNA fragments. Based on these assumptions, excellent separation of ΦX174/*Hae*III DNA restriction fragments (72−1353 bp) were observed by capillary electrophoresis even in ultra-dilute in HEC solutions of 0.025% HEC, as depicted in Fig. 5.29 [114].

A mixture of low- and high-molecular-weight HEC polymers was applied as sieving matrix for the separation of DNA restriction fragments by capillary electrophoresis [115] and the separation was strongly influenced by the average HEC molecular weight as well as by its concentration in the electrophoresis buffer. The addition of a very small amount of high-molecular-weight HEC to a solution of low-molecular-weight HEC (mixed polymer solution: 0.30% HEC 27,000/0.025% HEC 105,000) lead to a significant improvement in the separation of larger DNA fragments (>603 bp), while retaining good resolution of the smaller DNA fragments. The lower viscosity of this mixed polymer sieving matrix also reduced capillary loading time.

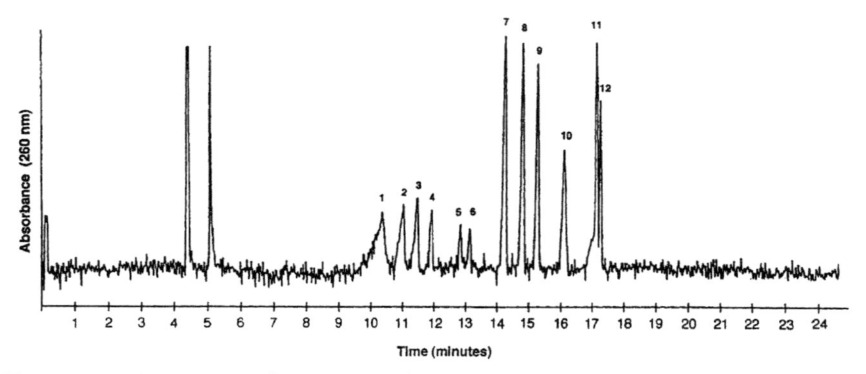

Figure 5.29 Separation of a mixture of λ-HindIII and ΦX174-HaeIII restriction fragments in 0.025% (w/w) hydroxyethyl cellulose (90,000−105,000 g/mol) with capillary electrophoresis. The far left peak corresponds to a neutral marker (mesityl oxide) and the second peak at left is an impurity present in the DNA sample. Peak identification: 1 = 23,130 bp, 2 = 9416 bp, 3 = 6557 bp, 4 = 4361 bp, 5 = 2322 bp, 6 = 2027 bp, 7 = 1353 bp, 8 = 1078 bp, 9 = 872 bp, 10 = 603 bp, 11 = 310 + 281 + 271 bp, and 12 = 234 + 194 + 118 + 72 bp. *With permission from A.E. Barron, W.M. Sunada, H.W. Blanch, Capillary electrophoresis of DNA in uncrosslinked polymer solutions: evidence for a new mechanism of DNA separation. Biotechnology and Bioengineering 52 (2) (1996) 259−270 [114].*

PCR-amplified HIV-1 DNA template was analyzed in background DNA from human peripheral blood mononuclear cells by an automated CGE-LIF system [116]. The electrophoresis gel buffer contained a DNA-fluorescence stain (LIFluor) enabling detection of multiple targets in the same tube by a single injection with high precision. Alternative sieving matrices were used for the analysis of forensic DNA samples [117] and polymorphic spacers of ribosomal DNA [118], both amplified by PCR prior to CGE analysis. Intercalating dyes other than EB, such as POPO-3, YOYO-3, and YOYO-1, were successfully used for dsDNA detection [119]. Single-molecule detection of DNA fragments was reported by Anazawa et al. [120] without using PCR amplification. Direct quantitation of the DNA molecules was possible by simple counting. Yeung's group utilized the solution of different molecular weight polyethylene oxide (PEO) sieving matrices for STR genotyping CGE [121]. By comparing the separation performance under denaturing and nondenaturing conditions, the latter one resulted in faster separations at elevated temperature conditions (Fig. 5.30).

Monitoring of ligation reactions during the production of long DNA probes was also accomplished by CGE with LIF detection [122]. The separation gel comprised LPA with the molecular mass of >2 million. The approach of the authors shed light into the nature of impurities and enabled easy optimization of the reaction conditions to obtain high-quality DNA probes and sensitively detect percentages of transgenic maize lower than 1%.

Holland and coworkers applied phospholipid-based highly viscous nanogels in a narrow bore capillary for the separation of DNA biomarkers of *Streptococcus pyogenes* strains [123]. The nanogel was composed of dimyristoyl-*sn*-glycero-3-phosphocholine (DMPC) and 1,2-dihexanoyl-*sn*-glycero-3-phosphocholine (DHPC), forming self-assembling structures, such as nanodisks and wormlike micelles. The 2.5% phospholipid ([DMPC]/[DHPC] = 2.5) containing nanogel effectively separated and sized dsDNA fragments up to 1500 bp with the resolution of 1%. Excellent separation was obtained in the range of 200−1500 bp allowing to distinguish invasive strains of *S. pyogenes* and *Aspergillus* species by utilizing the differences in gene sequences of collagen-like proteins in these organisms. Using a reversible stacking gel of a small plug of 10% phospholipid concentration before the 2.5% phospholipid containing nanogel filled capillary, a 12-fold increase was obtained in separation efficiency, as depicted in Fig. 5.31.

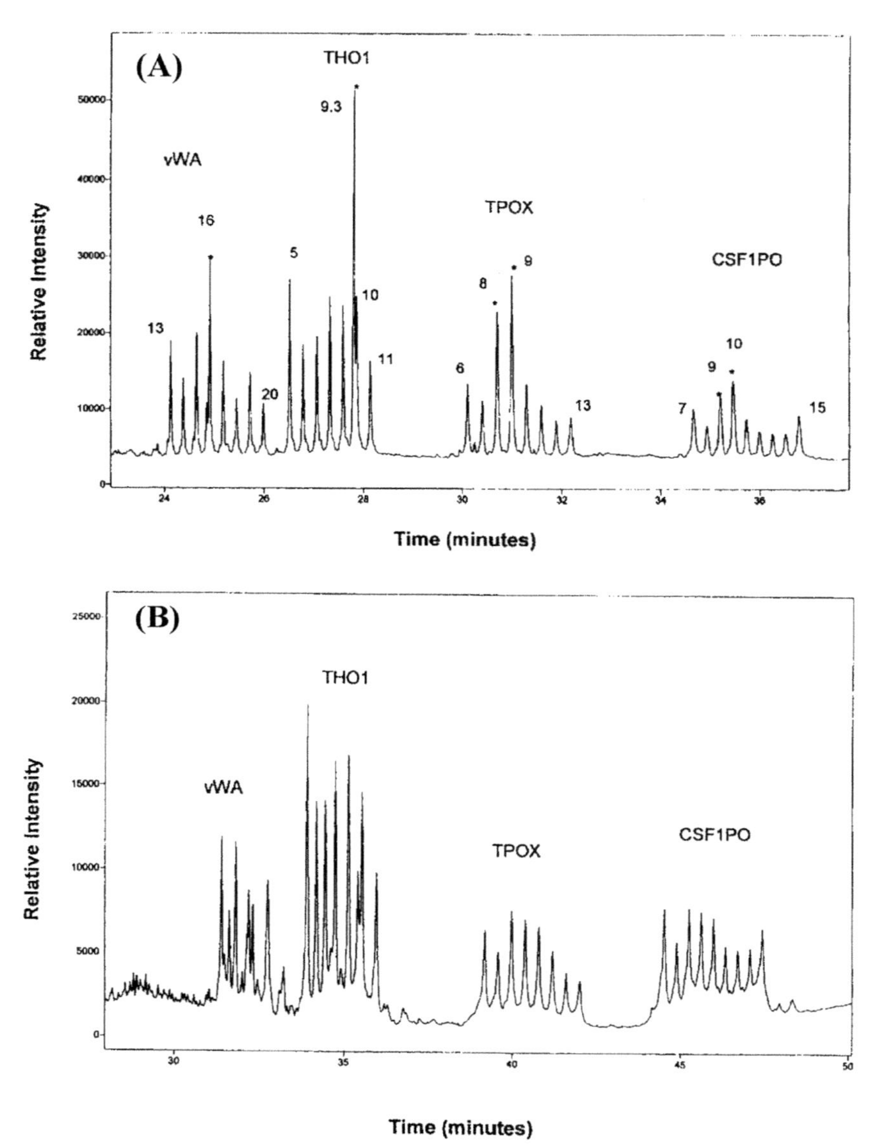

Figure 5.30 (A) Electropherogram of the coinjected mixture of CTTv ladder and individual DNA samples under denaturing condition. Separation matrix: 1.6% 8,000,000 M_r polyethylene oxide and 1.5% 600,000 M_r polyethylene oxide dissolved in 1 × TBE with 3.5 M urea. Denaturing: 95°C for 1 min. (B) Electropherogram of the same sample as in (A) obtained by using a urea-free matrix (under nondenaturing condition) at room temperature. Separation matrix: 1.9% 8,000,000 M_r polyethylene oxide and 1.8% 600,000 M_r polyethylene oxide in 1 × TBE. *With permission from N.Y. Zhang, E.S. Yeung, Simultaneous separation and genetic typing of four short tandem repeat loci by capillary electrophoresis. Journal of Chromatography A 768 (1) (1997) 135–141 [121].*

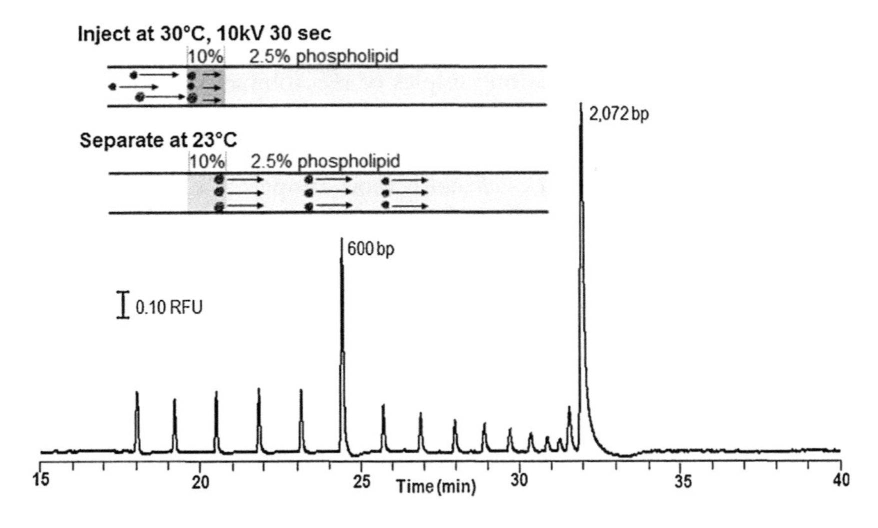

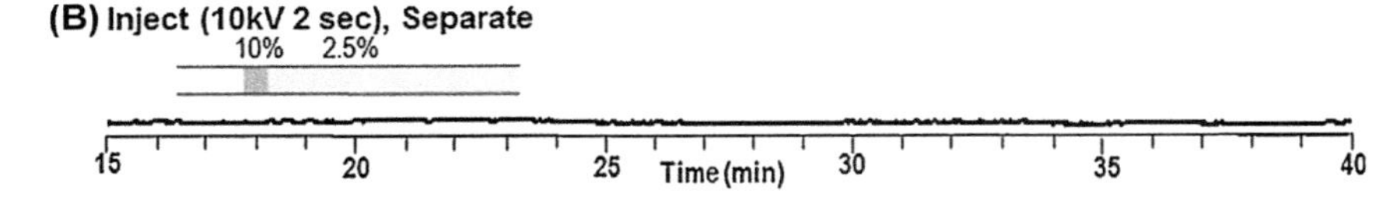

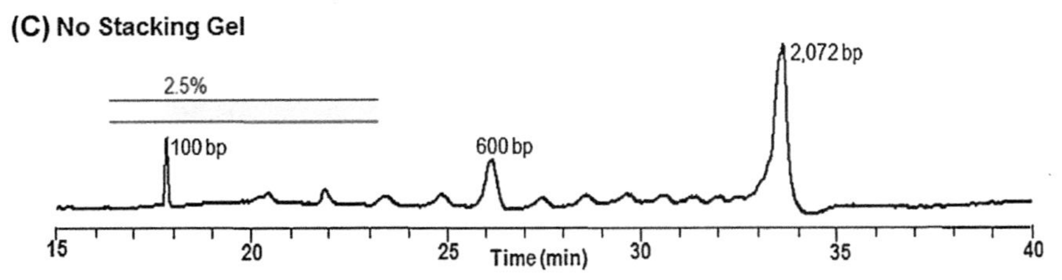

Figure 5.31 (A) discontinuous phospholipid nanogels in which the DNA is stacked in a small plug of 10% phospholipid and separated in the 2.5% phospholipid, (B) discontinuous phospholipid nanogels in which the 10 kV 2 s injection is too short to stack DNA, and (C) a continuous nanogel using a 10 kV 30 s injection without a stacking plug. *With permission from B.C. Durney, B.A. Bachert, H.S. Sloane, S. Lukomski, J.P. Landers, L.A. Holland, Reversible phospholipid nanogels for deoxyribonucleic acid fragment size determinations up to 1500 base pairs and integrated sample stacking, Analytica Chimica Acta 880 (2015) 136–144.* [123].

5.1.3 Large chromosomal DNA and pulsed-field electrophoresis

CGE of large DNA molecules up to 48.5 kb in length in different diluted polyacrylamide solutions provided evidence for polymer concentration and DNA length-dependent stretching and orientation of these species, suggesting effective separation with the use of polymer concentrations of about 0.6% [124]. Plasmid DNA in the size range of 3000–20,000 bp were prepared from cultivation samples of recombinant *Escherichia coli* and analyzed by CGE [125] to evaluate the genetic stability during production, as shown in Fig. 5.32. With the developed method the plasmid DNA from recombinant *E. coli* cultivation samples was analyzed within 30 minutes and also used to verify product purity after the downstream process.

Large DNA fragments size-dependently elongate under constant electric fields, resulting in a biased reptation mode, which has a negative effect on resolution. To alleviate this problem, pulsed-field capillary electrophoresis in ultra-dilute solutions of hydrophilic polymers was used for the separation of long dsDNA fragments above a million base pairs [126]. HEC (MW: 438,000) was used for the size range of 8–50 kbp and a mixture of HEC and polyethylene oxide for larger fragments as shown in Fig. 5.33. The larger DNA fragments up to 1.6 million bp were separated in less

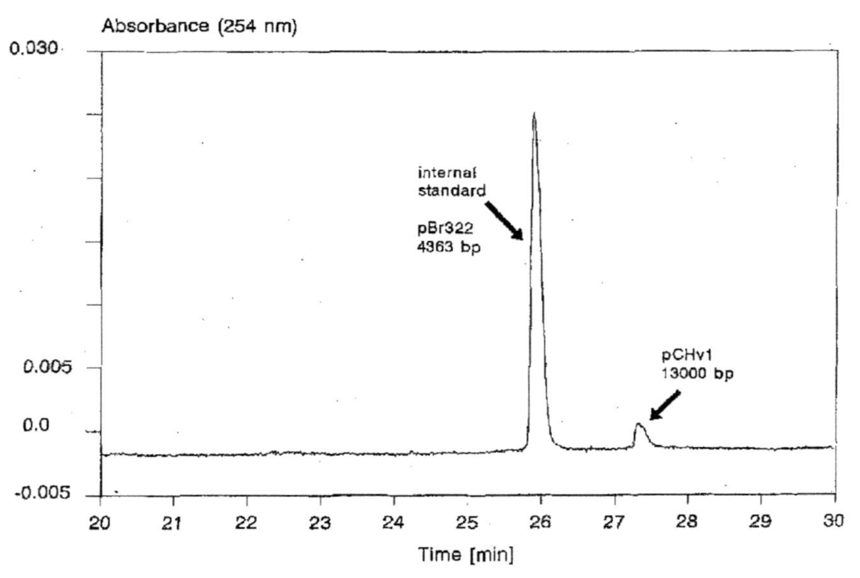

Figure 5.32 Capillary gel electrophoresis separation of the plasmid DNA of an *Escherichia coli* strain containing the plasmid pCHvl mixed with pBr322 DNA as internal standard. *With permission from K. Hebenbrock, K. Schuegerl, R. Freitag, Analysis of plasmid-DNA and cell protein of recombinant Escherichia coli using capillary gel electrophoresis, Electrophoresis 14 (8) (1993) 753–758 [125].*

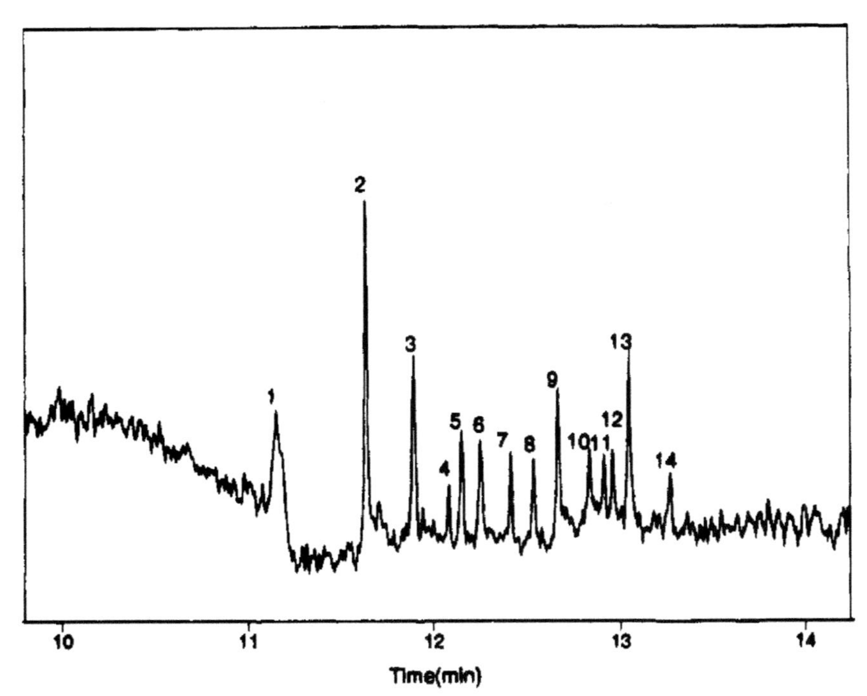

Figure 5.33 Capillary gel electrophoresis separation of megabase size dsDNA fragments in the mixture of 0.00375% hydroxyethyl cellulose and 0.0020% polyethylene oxide. Pulsed electric field: 100 V/cm + 14 Hz square wave, 250% modulation. Peaks: (1) 0.21, (2) 0.28, (3) 0.35, (4) 0.44, (5) 0.55, (6) 0.60, (7) 0.68, (8) 0.75, (9) 0.79, (10) 0.83, (11) 0.94, (12) 0.97, (13) 1.10, 1.12, and (14) 1.6 Mbp. *With permission from Y. Kim, M.D. Morris, Rapid pulsed field capillary electrophoretic separation of megabase nucleic acids. Analytical Chemistry 67 (5) (1995) 784–786 [126].*

than 13 minutes by applying rapid pulsed electric fields to the polymer matrix filled capillary. At higher dc fields and in shorter capillaries, chromosomal size DNA separation time could be reduced to as short as 3–5 minutes.

Pulsed-field CGE in diluted methylcellulose solutions was used to separate multikilobase length DNA fragments from a λDNA/HindIII digest [127]. Field inversion increased resolution for fragments longer than about 500 bp as shown in Fig. 5.34. Intermolecular hydrogen bonding was suggested as the cause of apparent cellulose fiber entanglement at concentrations below the calculated entanglement limit.

Highly efficient separations of large DNA fragments were achieved with the use of entangled polyacrylamide solution (MW: 5–6,000,000)

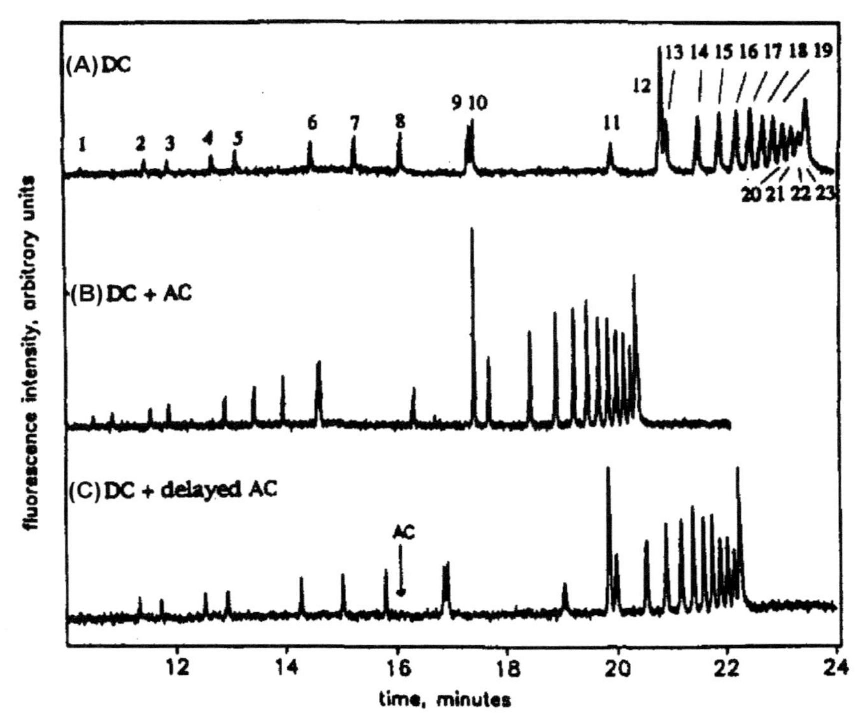

Figure 5.34 Pulsed-field capillary gel electrophoresis separation of a 1 kb DNA ladder (A) 180 V/cm dc only; (B) pulsed field, 180 V/cm dc + 500-Hz sinusoidal wave modulation, 130% modulation; (C) same conditions as in (B), but the modulation started 16 min after injection. Peaks: (1) 75, (2) 134, (3) 154, (4) 201, (5) 220, (6) 298. (7) 344, (8) 396, (9) 506, (10) 517, (11) 1018, (12) 1636, (13) 2016, (14) 3054, (15) 4072, (16) 5090, (17) 6108, (18) 7128, (19) 8144, (20) 9162, (21) 10,180, (22) 11,198, and (23) 12,216 bp. *With permission from Y. Kim, M.D. Morris, Pulsed field capillary electrophoresis of multikilobase length nucleic acids in dilute methyl cellulose solutions. Analytical Chemistry 66 (19) (1994) 3081–3085 [127].*

under pulsed-field separation conditions [128] with sinusoidal field and field-inversion pulsing regimes as Fig. 5.35 depicts. Even λDNA concatemers with sizes of 48.5–1000 kb were separated at 30 Hz in 0.4% high-molecular-weight LPA in c.3 hours with this method.

Field-inversion capillary gel electrophoresis (FICGE) was applied for the separation of 0.1–23 kbp DNA fragments in a low-melting, low-gelling agarose gel. The amplitude of the voltage pulses, the pulse times, and gel concentrations influenced the separation factor similarly to that of in polyacrylamide gels. Very high resolutions were obtained by continuously altering the pulse times and/or voltages according to a program, tailor-made for the size of the DNA molecules to be separated (programmed FICGE). The advantages of using nontoxic agarose gels were easy preparation and UV-transparency. The resolution was equivalent or even better compared to polyacrylamide gels with minimal risk of bubble formation. In addition, these gels were replaceable, therefore very convenient for automated analyses [129].

Heiger et al. analyzed DNA restriction fragment mixtures in low cross-linked polyacrylamide gel–filled capillary columns [130] and obtained great separation of dsDNA fragments up to 12 kbp using pulsed electric field. A 20% increase in peak separation was achieved in pulsed field at 100 Hz as shown in Fig. 5.36. Reducing the electric field strength and pulsing further enhanced the separation. A similar but different approach was reported by Demana et al. [131] who used velocity modulation in gel-filled capillary columns for DNA analysis. They found that in addition to improve resolution, velocity modulation also shortened migration times.

The migration pattern of dsDNA was studied under pulsed-field conditions in narrow bore capillaries using HEC matrix [132]. The migration behavior of dsDNA molecules proceeded by a sigmoidal migration regime including the Ogston sieving, the transition, and reptation regimes. Interestingly, when higher molecular mass HEC was used as sieving matrix the transition regime disappeared. The mobility of the dsDNA molecules decreased with the increase of pulse frequency. The mobility difference, on the other hand, between the adjacent dsDNA fragments, especially the longer ones increased with elevating pulse frequency, thus favoring for the separation of long DNA fragments. The mobility of dsDNA increased with the increase of the temperature in the capillary due to the decrease in the sieving polymer viscosity. Variable frequency pulsed-field CGE of nucleic acid restriction fragments up to 1300 bp was reported by others [133], in which case the frequency of the sinusoid wave, superimposed on the DC field, was randomly varied throughout the duration of the analysis to enhance separation. DNA fragments of various sizes exhibited optimum conditions for separation by numerous frequency changes and resolution of long fragments showed strong modulation frequency dependence. Modulations with center frequencies near 100 Hz resulted higher resolution of the 603–1353-bp fragments.

A very interesting CGE formulation was developed [134] utilizing star-shaped PDMA polymers with the intention to reduce the viscosity of

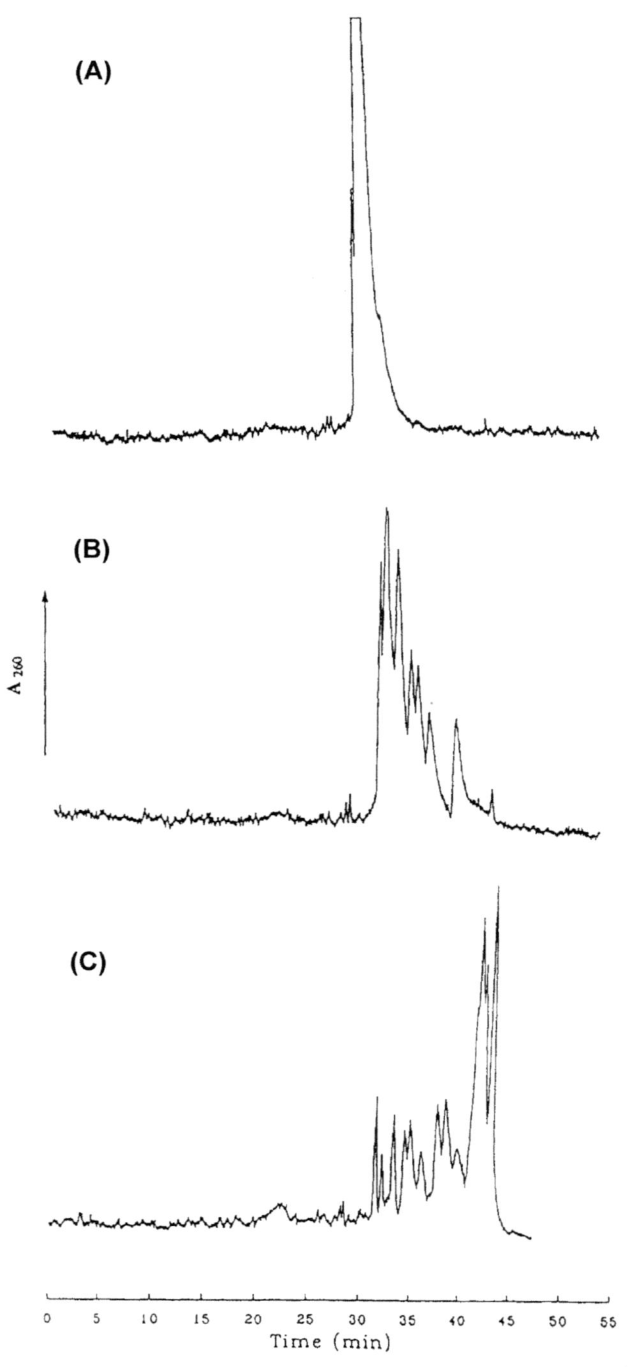

(*Continued*)

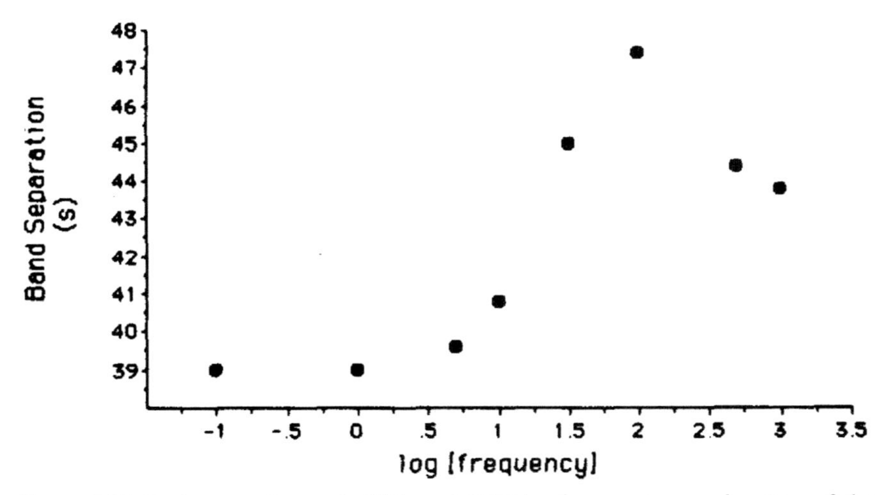

Figure 5.36 Peak separations of 4363- and 7253-bp fragments as a function of the logarithmic frequency of unidirectional pulse waveform. *With permission from D.N. Heiger, A.S. Cohen, B.L. Karger, Separation of DNA restriction fragments by high performance capillary electrophoresis with low and zero crosslinked polyacrylamide using continuous and pulsed electric fields. Journal of Chromatography 516 (1) (1990) 33–48 [130].*

the separation matrix. The authors obtained very high resolution between the 271- and 281-bp fragments of the φX 174 restriction digest mixture with 5% star-shaped polymer S-III (Fig. 5.37) solution as a sieving matrix. In general, the star-like polymers apparently offered higher separation performance compared to their linear counterparts, probably due to frequent bumping of the solute molecules to the sieving polymer. To further increase resolving power, square-wave pulsed electric field was applied for large dsDNA fragment analysis using HEC polymer (MW: 1,300,000) [135]. Evaluation of the sieving polymer concentration, the strength, frequency, and modulation depth of the pulse field showed that small DNA fragments were better separated with low polymer concentration (~0.1%) and modulation depth, while larger fragments better separated with high polymer concentration (>0.5%) and modulation depth.

◀ **Figure 5.35** Resolution dependence of the mixed digest of λDNA (8.3–48.5 kb) on the pulsing frequency of the applied electric field. (A) $f = 1$ Hz; (B) $f = 5$ Hz, and (C) $f = 15$ Hz. *With permission from J. Sudor, M.V. Novotny, Separation of large DNA fragments by capillary electrophoresis under pulsed-field conditions. Analytical Chemistry 66 (15) (1994) 2446–2450 [128].*

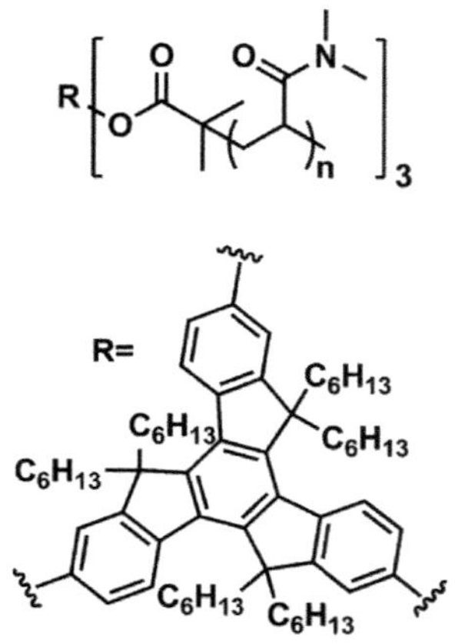

Figure 5.37 Star-shaped poly-*N*,*N*-dimethylacrylamide polymer S-III used in capillary gel electrophoresis of DNA restriction fragments. *With permission from F. Gao, C. Tie, X.-X. Zhang, Z. Niu, X. He, Y. Ma, Star-shaped polymers for DNA sequencing by capillary electrophoresis, Journal of Chromatography A 1218 (20) (2011) 3037–3041 [134].*

5.1.3.1 RNA analysis

High-resolution separation of ribonucleic acids by CGE is also of high importance and was demonstrated as early as the 1990s. A mixture of polyriboadenylic acids ranging from 12 to 18 units was successfully separated in 5 minutes [136] using a polyacrylamide gel-filled capillary. Capillary electrophoresis of RNA in dilute and semidilute polymer solutions of HEC and the dependence of solute mobility on its chain length was consistent with separation by transient entanglement mechanism [137]. Another application employed 1% polyvinylpirrolidone (PVP, 1.3 MDa) in 1 × TBE buffer with 4 M urea and 0.5 μM EB separation medium for automated and quantitative RNA screening by CGE, using commercially available instrumentation and reagents. The approach did not require coated capillaries and enabled high-throughput and large-scale analysis of RNA samples (Fig. 5.38) [138].

Short RNA (LMM RNA, 70–135 bases) fingerprinting was reported by Katsivela and Hoefle [139] using CGE with 0.5% HPMC (MW ~86,000) entangled polymer network as sieving additive. The authors reported the short lifetime of 30–50 runs of the DB-1 type capillary coating that was originally developed for gas chromatography applications, that is, not for aqueous systems. They also utilized cross-linked polyacrylamide gels to analyze the same samples but attained adequate resolving power only up to 79 bases.

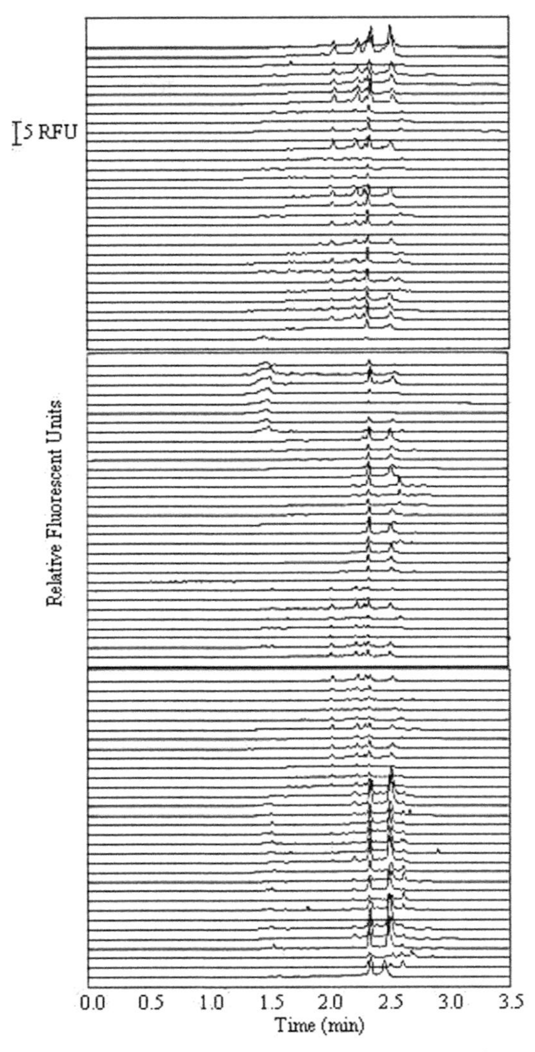

Figure 5.38 Automated high-throughput large-scale RNA analysis by capillary gel electrophoresis. Typical batch processing profiles of a 96-well sample plate. Total RNA samples from rice (traces 1–76 from top), *Arabidopsis* (traces 77–95), and yeast (trace 96); 6 μL each in a 96-well plate. Conditions: 50 μm i.d. capillary, $\ell = 10$ cm ($L = 30$ cm); sieving medium 1% polyvinylpirrolidone (MW = 1.3 MDa), 4 M urea, 1 × TBE, 0.5 μM ethidium bromide; $E = 500$ V/cm; 25°C. RNA samples were diluted in deionized water and denatured at 65°C for 5 min prior to analysis. The sample tray was stored at 4°C in the capillary electrophoresis instrument during processing. Injection: vacuum (5 s at 3.44 kPa). The separation matrix was replaced after each run, 2 min at 551 kPa. *With permission from J. Khandurina, H.S. Chang, B. Wanders, A. Guttman, Automated high throughput RNA analysis by capillary electrophoresis, Biotechniques 32 (2002) 1226–1228 [138].*

CGE supports quantitative mRNA analysis to study gene expression. Even low-abundance mRNA transcripts can be investigated using quantitative competitive reverse transcription PCR (QC-RT-PCR) coupled with CGE for rapid product analysis [140]. Archived formalin fixed and paraffin-embedded (FFPE) tissues can also be analyzed by RT-PCR and CGE [141]. RNA extraction was performed from single 6–8 μm formalin fixed and paraffin-embedded histological human tissue sections [142]. The first step of the RNA extraction was to eliminate the paraffin with solubilization in an organic solvent, such as xylene. After deparaffinization, the tissue proteins were removed by digestion with a proteolytic enzyme or a mixture of enzymes. Following the protein removal, RNA extraction and precipitation were performed with isopropanol using glycogen as carrier because of the low concentration. Markkanen and coworkers developed a novel RNA extraction method from laser-capture microdissected normal and cancer-associated stroma FFPE tissues by applying an additional focused ultrasonication step [143]. This efficient RNA extraction workflow is depicted in Fig. 5.39.

Gene expression was monitored by CGE in individual mammalian cells after RT-PCR [144] using HPMC sieving matrix (MW 10,000) to obtain size-based separation in conjunction with EB staining to yield excellent sensitivity. In another similar effort, RNA was directly sampled and separated at the single-cell level (without extraction) by CGE [145] using similar separation conditions as above but adding polyvinylpyrrolidone to the separation gel buffer system to eliminate electroosmotic flow and mannitol to enhance the separation. CGE-LIF was used for direct determination of the amount of RNA species present in cardiac tissue by Goldsmith et al. [146]. Gene expression was detected using a fluorescently labeled riboprobe, specific for a given RNA species and validated by analyzing levels of 28S RNA. The method was also applied for the determination of the amount of discoidin domain receptor 2 mRNA in cardiac tissues. Separations of the free and complexed probes were accomplished in HPMC (MW 10,000)-filled capillaries with LIF detection.

Fluorophore-tagged antisense oligonucleotides were directly hybridized with miRNA in cetyltrimethylammonium bromide (CTAB) containing buffer and analyzed by CGE with 2% polyethylene oxide (MW: 8,000,000) matrix in the presence of electroosmotic flow with 7 M urea [147]. Urea was added to prevent self-complementary binding of the probe, while the complex structure of miRNA/DNA remained. Hybridization in the presence of CTAB allowed annealing at 50°C, a temperature that was significantly less

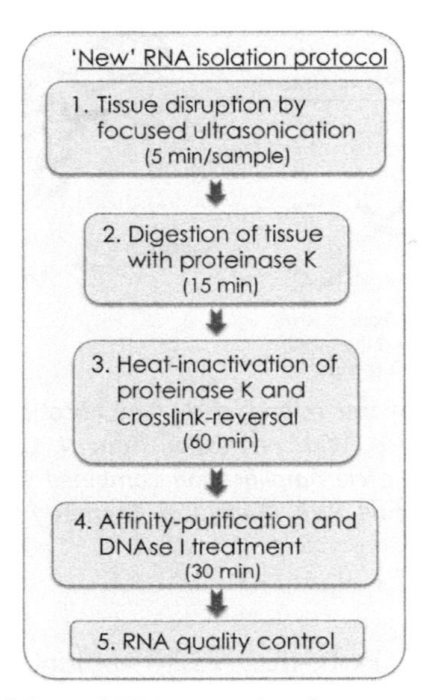

Figure 5.39 Isolation workflow of RNA extraction from normal and cancer-associated stroma formalin fixed and paraffin-embedded tissue with focused ultrasonication. *With permission from P. Amini, J. Ettlin, L. Opitz, E. Clementi, A. Malbon, E. Markkanen, An optimised protocol for isolation of RNA from small sections of laser-capture microdissected FFPE tissue amenable for next-generation sequencing, BMC Molecular Biology 18 (1) (2017) 22 [143].*

than that of the computer-calculated melting temperature of 66.4°C. CGE provided excellent specificity and showed only negligible effects of intrinsic interferences, such as with human total RNA, primary miRNA, or precursor miRNA.

Another small RNA analysis method was reported using rolling circle amplification (RCA) technique [148] without ligation, as shown in Fig. 5.40. The circular probes were designed for small RNA targets as templates and the products were analyzed by CGE using 4% PVP (MW: 1,300,000). The method allowed specific and sensitive detection of small RNAs with the size of about 200 bases in biological samples. Simultaneous detection of multiple miRNAs was developed by utilizing denaturing CGE with LIF detection [149]. The advantage of the separation method was the combination with tandem adenosine-tailed DNA bridge-assisted splinted ligation, resulting in the detection of five Epstein – Barr virus mRNAs in a single run, in addition to the mRNA isomers. The linear detection range of the analysis was three orders of magnitude (1.0 nM to 1.0 pM) with 2.5 zmol detection limit.

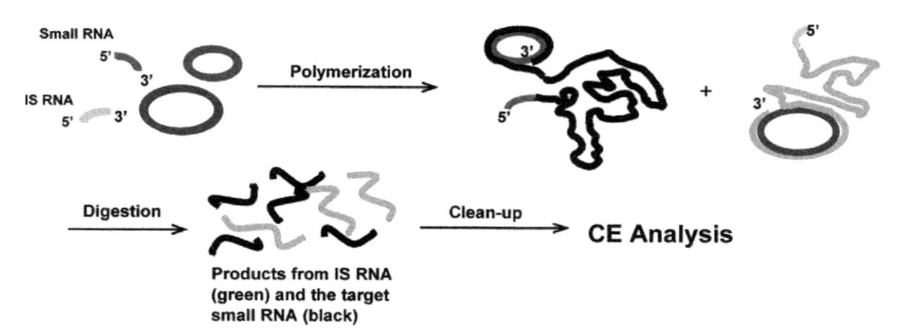

Figure 5.40 Schematics of the rolling circle amplification capillary electrophoresis assay for small RNA analysis. *With permission from N. Li, C. Jablonowski, H. Jin, W. Zhong, Stand-alone rolling circle amplification combined with capillary electrophoresis for specific detection of small RNA, Analytical Chemistry 81 (12) (2009) 4906–4913 [148].*

The method was fast, PCR-free, easily multiplexed, and cost-effective, making it applicable to large-scale screening of iso-miRNAs.

An interesting method was introduced for the identification of rapidly growing mycobacteria (RGM) by amplification of the 16S–23S rRNA internal transcribed spacer (ITS) followed by separation of the amplified fragments by CGE [150]. Nineteen ATCC mycobacterium strains and 178 clinical RGM isolates (12 species) were studied. CGE-based PCR fragment analysis was performed using a commercial 48-capillary DNA analyzer with 50 cm POP-7 gel-filled capillaries. The authors found that the sizes of the ITS fragments ranged from 332 to 534 bp and their separation by CGE allowed the identification of clinically relevant RGMs, for example, the important *Mycobacterium abscessus* and *Mycobacterium massiliense* species. Dou and coworkers performed the analysis of small interfering RNA (siRNA) by capillary electrophoresis in HEC with different molecular weight (MW: 90,000; 250,000; 720,000; and 1,300,000) solutions [151] and studied the migration mechanism of dsRNA by mobility and resolution length (RL) plots (Fig. 5.41). They concluded that the concentration of high MW HEC sieving polymer rather affected the RL of small dsRNA fragments, while the concentration of low-MW HEC mostly affected the RL of large dsRNA fragments.

Selective and high-resolution separation of siRNA mixtures were performed with CGE [152] by applying PVA-coated capillary and a solution of 27% (w/v) polyethylene glycol (MW: 35,000) in 200 mM Bis–Tris with 20% (w/v) acetonitrile as sieving media. Others designed a flexible,

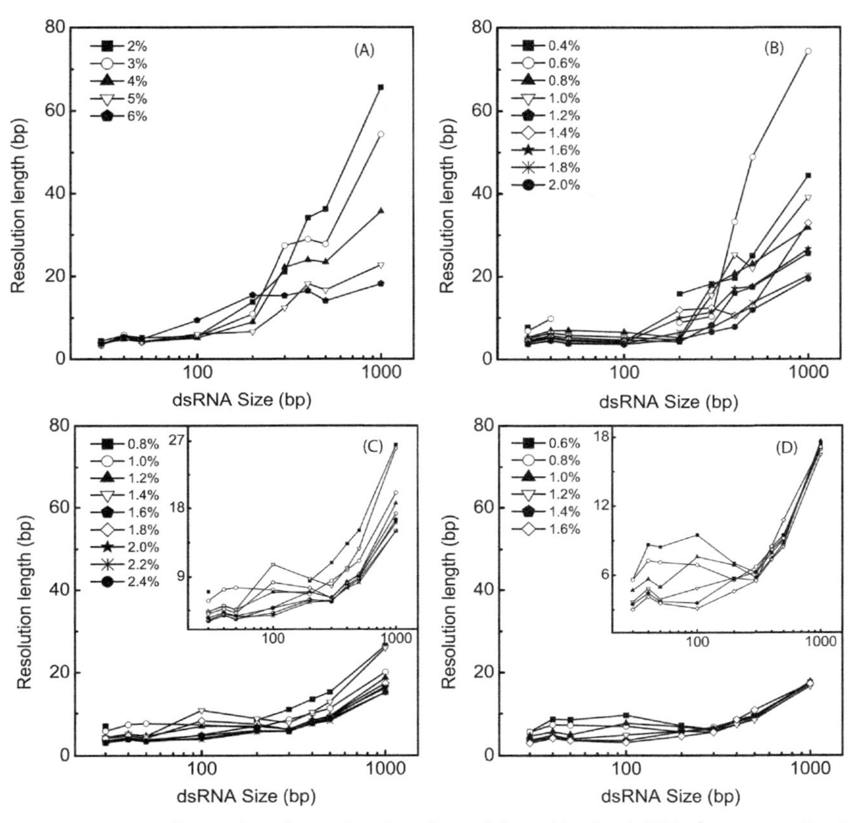

Figure 5.41 Resolution length in the function of logarithmic dsRNA fragment size in hydroxyethyl cellulose solution with MW: (A) 90k, (B) 250k, (C) 720k, and (D) 1300k. *With permission from C. Liu, Y. Yamaguchi, X. Zhu, Z. Li, Y. Ni, X. Dou, Analysis of small interfering RNA by capillary electrophoresis in hydroxyethylcellulose solutions, Electrophoresis 36 (14) (2015) 1651–1657 [151].*

low-viscosity sieving medium for capillary gel electrophoretic separation of mRNA, based on high mass linear PVP (MW: 1,300,000) and glycerol (range: 10%–20%) [153]. The sieving medium was optimized with a central composite face-centered factorial design, which model allowed to predict suitable sieving media compositions for their response(s) of interest, for purity and stability analysis of polynucleotides with the size range of 100–1000 bases.

A CGE method was introduced to analyze PCR fragments with 5′-labeled primers for the diagnosis of SARS-CoV2 [154]. The group isolated the RNAs from the collected samples, amplified two fragments of the SARS-CoV2, and synthetized the cDNA. Then, 5′-end fluorochrome labeling reaction was performed to one of the primers for each pair. The amplicons were separated by a commercial CGE-based sequencer.

A nonaqueous CGE method was developed for separating large RNA molecules by size or length under strongly denaturing conditions [155]. The authors applied formamide-based separation buffers containing 100 mM MES (pH 6.0) and 1 mM EDTA with 0.16% w/v PEO (>5 MDa) high-molecular-weight polymers for the analysis of mRNA vaccines. The proposed method can be used for stability-indicating determination of mRNA purity. DeDionisio and Gryaznov [156] reported on the analysis of RNA hydrolysis in the presence of RNase H and monitored of the digestion products by CGE.

5.1.4 Antisense DNA and modified oligonucleotides

Antisense therapeutics, such as phosphodiester oligonucleotides and their analogs, represent a promising class of molecules for treating viral infections and various cancers. In the early 1990s, Cohen et al. [157] applied CGE with noncross-linked polyacrylamide gel (13%) in the presence of 8.3 M urea to antisense DNA analysis and demonstrated its separation potential for ssDNA analogs known as phosphorothioates, depicted in Fig. 5.42. Base composition analysis of synthetic oligodeoxyribonucleotides by CGE was reported by Dinh et al. [158]. The separation of deoxyribonucleotide monophosphates was accomplished at constant voltage (isoelectrostatic) and constant temperature (isothermal) modes.

A comparison of chromatography (RP-HPLC and ion-exchange HPLC) with electrophoretic separations (slab gel and CGE) for the analysis of antisense phosphorothioate oligomers revealed that electrophoretic separations were more effective, especially for oligomers longer than 20 bases [159]. Later, a comparative quantitative analysis of antisense phosphorothioate oligonucleotides was reported in biological fluids using anion-exchange chromatography and CGE. Although CGE was recognized as an effective method for the analysis of normal phosphodiester deoxyoligonucleotides (DNA), it was found that when the same electrophoretic principles were applied to the analysis of phosphorothioate DNA, peak broadening occurred due to the phosphorothioate moiety. It necessitated the optimization of gel concentration, buffer additives, and pH for CGE to be effective in precise and accurate analysis of phosphorothioate oligonucleotides [160].

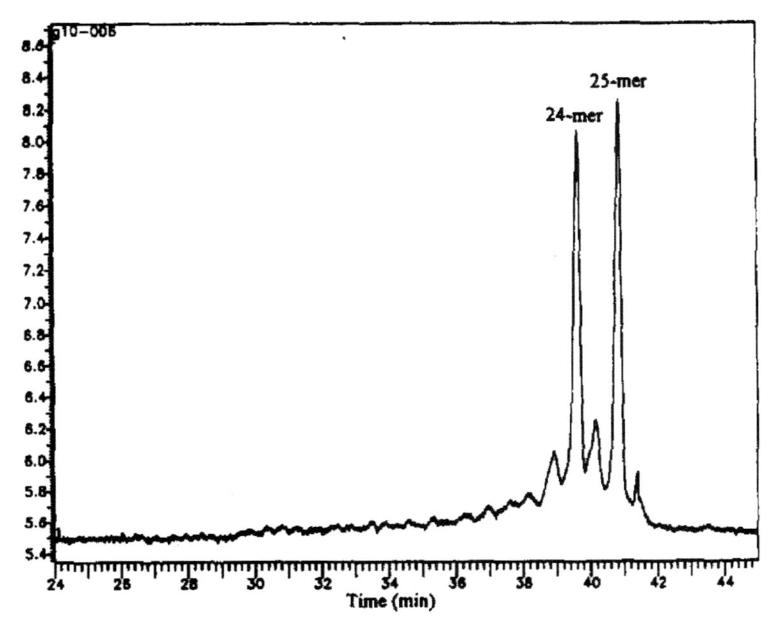

Figure 5.42 High-performance capillary gel electrophoresis separation of a mixture of 24- and 25-mers of synthetic oligodeoxyribonucleotides. Effective length of the capillary column was 28 cm; conditions: 13% T/0% C, 8.3 M urea, 0.2 M TBE (pH 8.3); applied electric field 400 V/cm. *With permission from A.S. Cohen, M. Vilenchik, J.L. Dudley, M.W. Gemborys, A.J. Bourque, High-performance liquid chromatography and capillary gel electrophoresis as applied to antisense DNA. Journal of Chromatography 638 (2) (1993) 293–301 [157].*

Binding of peptide nucleic acids (PNAs, Fig. 5.43) to their complementary oligonucleotides was studied by CGE [161], capable to resolve the free and bound species enabling to measure the relative binding kinetics and the stoichiometry of binding. It was found that the binding kinetics depended on the relative sequence orientation of the target oligonucleotides, while the stoichiometry of the binding was 1:1 for the heteroduplex.

A 20-mer phosphorothioate oligodeoxynucleotide was used to treat mice bearing A549 (human lung carcinoma) tumors to evaluate its malignant cell growth inhibition effect [162]. Recovery of the molecules from the tumor tissue following CGE analysis revealed that, 24 hours after the final dose of the treatment, predominantly intact, full-length 20-mer material was present together with some apparent exonuclease degradation products (e.g., n-1- and n-2-mers). These results clearly demonstrated that

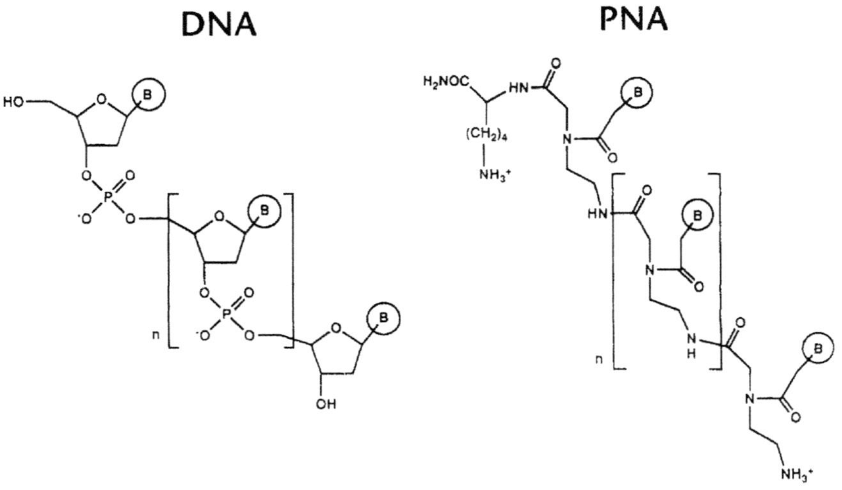

DNA PNA

Figure 5.43 Structures of a DNA and a peptide nucleic acid with B = purine or pyrimidine base (A, C, T, or G). *With permission from D.J. Rose, Characterization of antisense binding properties of peptide nucleic acids by capillary gel electrophoresis. Analytical Chemistry 65 (24) (1993) 3545–3549 [161].*

the unmetabolized intact phosphorothioate oligodeoxynucleotide reached at and consequently accumulated in the tumor tissue. A CGE method was developed for the accurate quantitation of a 21-mer phosphorothioate oligonucleotide and its degradation products in an intravitreal formulation [163]. Uncertainties of the electrokinetic injection mode necessitated the use of an external reference standard in the sample matrix similar to that of the drug product. The use of such internal standard improved analysis accuracy and precision. The detection limit of this CGE-based rapid monitoring of phosphorothioates was as low as 2 ng/mL (Fig. 5.44) with the use of various on-the fly staining (EB) and covalent labeling techniques (labeled primer), as reported by Vilenchik et al. [164]. The same group also introduced a CGE-LIF method that enabled complete sequence determination of antisense DNA analogs of unknown sequences [165]. Bruin et al. [166] applied 10% LPA gel to check the stability of chemically modified antisense oligonucleotides in model cellular environments against various nucleases.

5.1.5 Biomedical and forensic applications

Major biomedical applications on size-sieving capillary electrophoresis of DNA are discussed in details for diagnostics [167–169], DNA typing

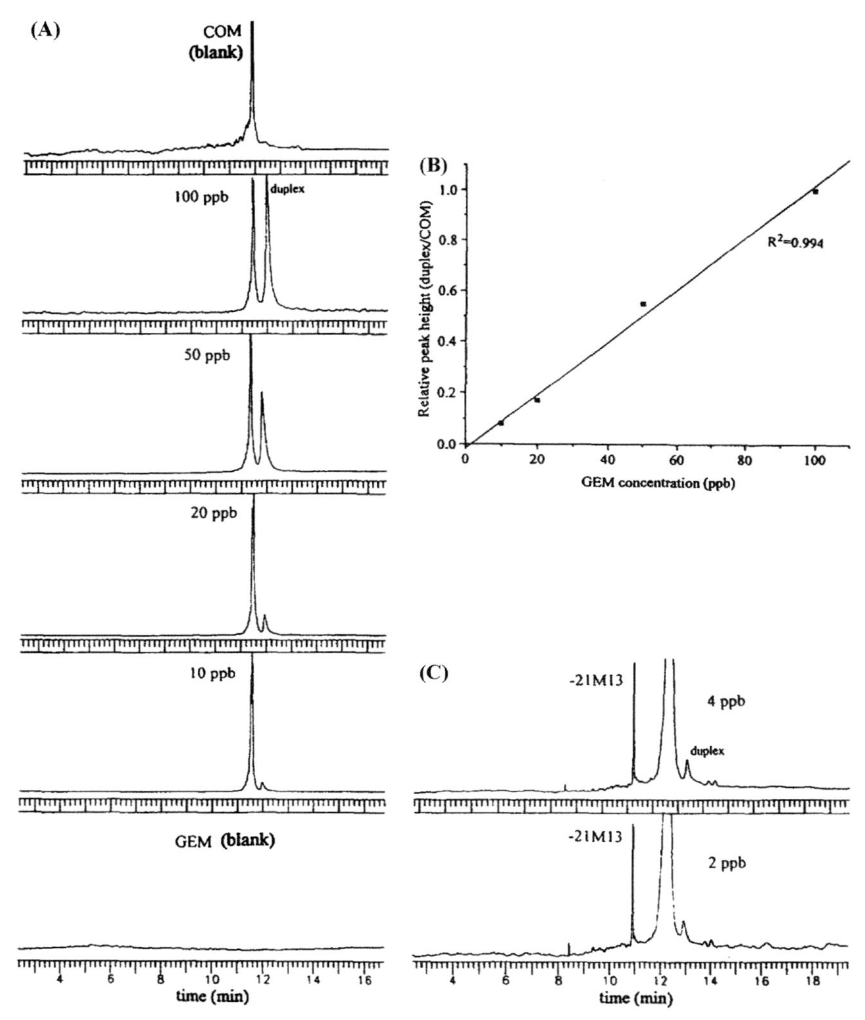

Figure 5.44 Monitoring and analysis of antisense DNA by high-performance capillary gel electrophoresis. (A) Low detection limits achieved by capillary gel electrophoresis with laser-induced fluorescence detection; (B) relative peak-height duplex/complementary probe molecule (COM) in the function of targeted phosphorothioate DNA (GEM) concentration; and (C) analysis under optimal conditions with the use of primer −21M13 labeled with fluorescent dye as an internal standard. *With permission from M. Vilenchik, A. Belenky, A.S. Cohen, Monitoring and analysis of antisense DNA by high-performance capillary gel electrophoresis. Journal of Chromatography A 663 (1) (1994) 105–113 [164].*

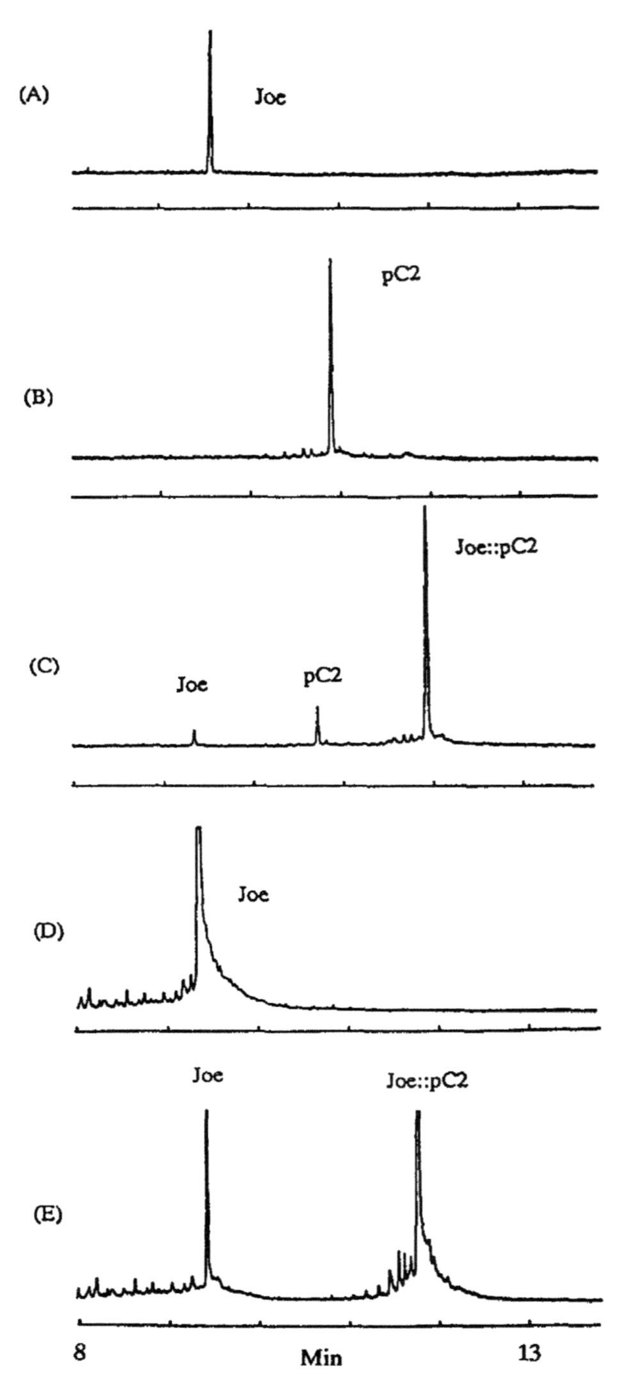

(Continued)

[170–172], antisense DNA analysis [20,173–175], linear and supercoiled DNA separation [176–178], SSCP study [179–181], DNA sequencing [40,68,182–185], and DNA restriction fragment analysis [106,186,187]. Comprehensive reviews on the specific application of capillary electrophoresis in molecular diagnostics are given in Refs. [188,189].

Southern blotting can be a useful tool in the diagnosis of genetic diseases (e.g., Fagile X syndrome) [190] or in forensics [191]. The feasibility of using gel-filled capillary columns to perform Southern blotting was demonstrated with online detection and identification of DNA molecules by precolumn hybridization and CGE [57]. Parameters affecting the hybridization of DNA molecules in solution were also studied. Fig. 5.45 shows the use of a fluorescently tagged oligonucleotide as hybridization probe in solution with complementary DNA molecules and analyzed by UV and LIF detection in CGE. The hybridized species were clearly identified by both UV and LIF detection. The effects of probe concentration and annealing temperature were also reported.

McCord et al. introduced CGE in the field of forensic DNA testing [117], emphasizing that for the routine application of CGE to forensic DNA analysis, a number of requirements must be met. In the analysis of DNA amplified by PCRs, these requirements included high-resolution and consistent results, as well as reproducible runs. Genetic markers of interest in this field consist of DNA fragments of >200 bp with repeat units as small as 2 bp.

A normalized extraction protocol was developed and used in conjunction with fast PCR reactions (42–51 minutes utilizing KAPA2GTM Fast Multiplex PCR Kit) and capillary array electrophoresis (24–28 minutes) using POP-6 Polymer and a 22 cm effective capillary length for forensic analysis [192]. The new extraction procedure resulted in a 37% reduction in processing time for a 96-well platform to generate high-quality STR profiles from buccal swabs and Buccal DNA Collectors, with 95%–99% and 88%–91% pass rates, respectively. Other forensic applications of

◀ **Figure 5.45** DNA identification by hybridization with a fluorophore-tagged oligonucleotide probe using capillary gel electrophoresis. (A) Joe-labeled 17-mer; (B) pC2; (C) equal amounts of Joe-labeled primer and pC2 annealed prior to electrophoresis (UV absorption); (D) the same sample as (A) analyzed by CGE-LIF; and (E) the same sample as (C) analyzed by CGE-LIF. *With permission from J.W. Chen, A.S. Cohen, B.L. Karger, Identification of DNA molecules by pre-column hybridization using capillary electrophoresis, Journal of Chromatography 559 (1–2) (1991) 295–305 [57].*

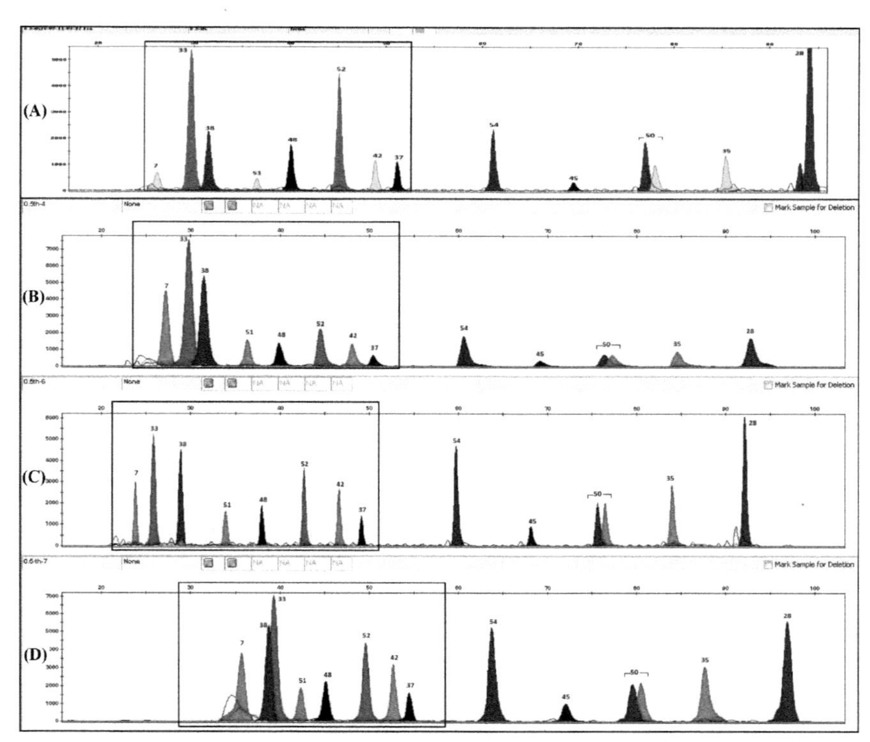

Figure 5.46 Comparison of SNaP shot profiles of 13th multiplex assay using (A) POP-4; (B) POP-4, (C) POP-6, and (D) POP-7 polymers. Boxes show the most mobility-affected single-nucleotide polymorphisms. *With permission from W. Goodwin, S. Alimat, Analysis of four PCR/SNaPshot multiplex assays analyzing 52 SNPforID markers. Electrophoresis 38 (7) (2017)1007–1015 [193].*

DNA analysis with CGE identified commonly used SNPs and STRs, as shown in Fig. 5.46 [193,194].

In the biomedical side, CGE was utilized for PCR products analysis of cystic fibrosis (CF) mutation AF508 from homozygous and heterozygous individuals with base line separation of two complementary single-stranded (95-base) DNA fragments [195]. DNA fragments differing in size by only one base (105 and 106) in the PCR products of CF mutation at 3905 insT from heterozygous individuals were also separated within 45 minutes. The VNTR alleles of apoB can serve as a risk factor marker for coronary heart disease if contains larger repeat numbers. Thus rapid CGE assay, using entangled polymer solutions, was applied to the analysis of PCR-amplified apoB VNTR loci for diagnostic purposes. Excellent resolution of two alleles differing by one or two 16-bp repeat sections in the size range up to 600 bp was obtained, while alleles differing in length of 2 or 4 repeat units in the size range of 600 to 1000 bp were also distinguishable [196]. An apoE genotyping method was developed for the diagnosis of Alzheimer's disease (AD) [197]. The one-step multiplex PCR from whole blood without DNA purification enabled to minimize the risk of DNA loss or contamination. The amplified PCR products with different DNA lengths, such as the apoE genotypes related to AD (i.e., 112-, 253-, 308-, 444-, and 514-bp DNA), were identified within 2 minutes by an extended voltage programming (VP)-based method utilizing the sieving gel matrix of 1.75% PVP (MW: 1,000,000). The detection range of the method was 6.4–62.0 pM and 5–540 times faster and ca100–100,000 times more sensitive than previous Alzheimer's diagnosis methods (Table 5.2). Yamaguchi and coworkers successfully identified PCR products of different species of anaerobic bacteria, such as *Porphyromonas gingivalis*, *Treponema denticola*, and *Tannerella forsythia*, with CGE within 8.5 minutes in 0.5% PEO (MW: 4,000,000) containing sieving matrix [198]. The method can be used to identify anaerobic bacteria species causing the inflammatory disease periodontitis.

Inverse-shifting PCR (IS-PCR) and short-end injection were used in the CGE-based diagnosis of hemophilia A (HA) with high specificity (Fig. 5.47). The method allowed rapid genotyping of intron 22 inversion of the factor VIII gene (F8), which causes 40%–50% of severe bleeding disorder of HA patients [199]. In IS-PCR, three specific primers were used to amplify a 512-bp amplicon for wild-type and 584 bp amplicon for patients with intron 22 inversion. Short-end injection CGE was performed using 1 × TBE buffer containing 0.3% (w/v) PEO (MW: 80,000). The F8 gene of HA patients and carriers was detected within 5 minutes. Others developed a VP-based CGE method with LIF detection for fast and highly sensitive analysis of DNA molecules related to angiotensin-converting enzyme insertion/deletion polymorphism [200]. These DNA molecules influence the predisposition to such ailments as cardiovascular disease, high blood pressure, myocardial infarction, and AD. The amplified products of the angiotensin-converting enzyme insertion/deletion polymorphism (190- and 490-bp DNA) were baseline separated within 3.2 minutes by applying optimum VP conditions and utilizing a sieving gel matrix of 2.0% w/v polyvinylpyrrolidone (PVP, MW = 1,000,000).

A CGE method was described to detect circular DNA ligation formation products to predict ligated DNA competent cell transformation efficiency, an important biomedical application [201]. The analysis was based

Table 5.2 Comparison of total analysis time, detection time, and LODs for various Alzheimer's disease detection methods.

Method	Analytes	Total analysis time[a]	Detection time[b]	LOD[c] (nM)
CE-MS	β-Secretase inhibitors	–	~1 h	5.0
Surface plasmon resonance	Tau protein	≥15 h	~40 min	7.0
Electrochemical immunosensor	apoE4	–	–	0.3
Magnetosandwich immunoassay	apoE	1 day	–	68.0
Competitive phage ELISA	13-(*E*,*E*)-HODE	2 day	~18 h	9.2–25.9
Open sandwich ELISA	13-(*E*,*E*)-HODE	1 day	~18 h	2.2
Nested-PCR/immunoblotting	apoE genotypes (ε2, ε3, ε4)	≥8 h	–	–
PCR-GoldMag LFA	apoE genotypes (ε2, ε34, ε4)	~9 h	~1.5 h	–
CE-UV	Amyloid peptides	≥6 h	~10 min	300–500
MALDI-TOF-MS	Sr, GA, DA	≥12 h	–	3.0–8.0
Direct multiplexed PCR/CE-LIF	ApoE genotypes (ε2, ε34, ε4)	<47 min	<2 min	$(6.4-62.0) \times 10^{-3}$

Note: CE-MS, Capillary electrophoresis–mass spectrometry; *CE-UV*, capillary electrophoresis with UV detection; *ELISA*, enzyme-linked immunosorbent assay; *DA*, dopamine hydrochloride; *13-(E,E)-HODE*, 13(*R*,*S*)-hydroxy-9(*E*),11(*E*)-octadecadienoic acid; *GA*, glutamic acid; *LOD*, limit of detection; *MALDI-TOF-MS*, matrix-assisted laser desorption ionization–time-of-flight mass spectrometry; *PCR-GoldMag LFA*, PCR-GoldMag lateral flow assays; *Sr*, serotonin.
[a]Total analysis time: total analysis time from whole blood to data acquisition.
[b]Detection time: total time from sample injection to detection.
[c]LODs were determined at signal-to-noise ratio $(S/N) = 3$.
Source: With permission from N. Woo, S.K. Kim, Y. Sun, S.H. Kang, Enhanced capillary electrophoretic screening of Alzheimer based on direct apolipoprotein E genotyping and one-step multiplex PCR, Journal of Chromatography B 1072 (2018) 290–299 [197].

on the relative migration differences between the same size linear and circular DNA molecules in the presence of an intercalating dye. Fig. 5.48 shows the electropherogram of pICD1LS supercoiled (dark line) and linear fragments (sizes on the peaks) with circular forms, supercoiled and open (solid line).

Kleparnik et al. utilized a high-pH buffer system (pH >12) containing LPA sieving matrix, which resulted in denaturation of DNA molecules and accommodated rapid separation even in uncoated capillaries. The

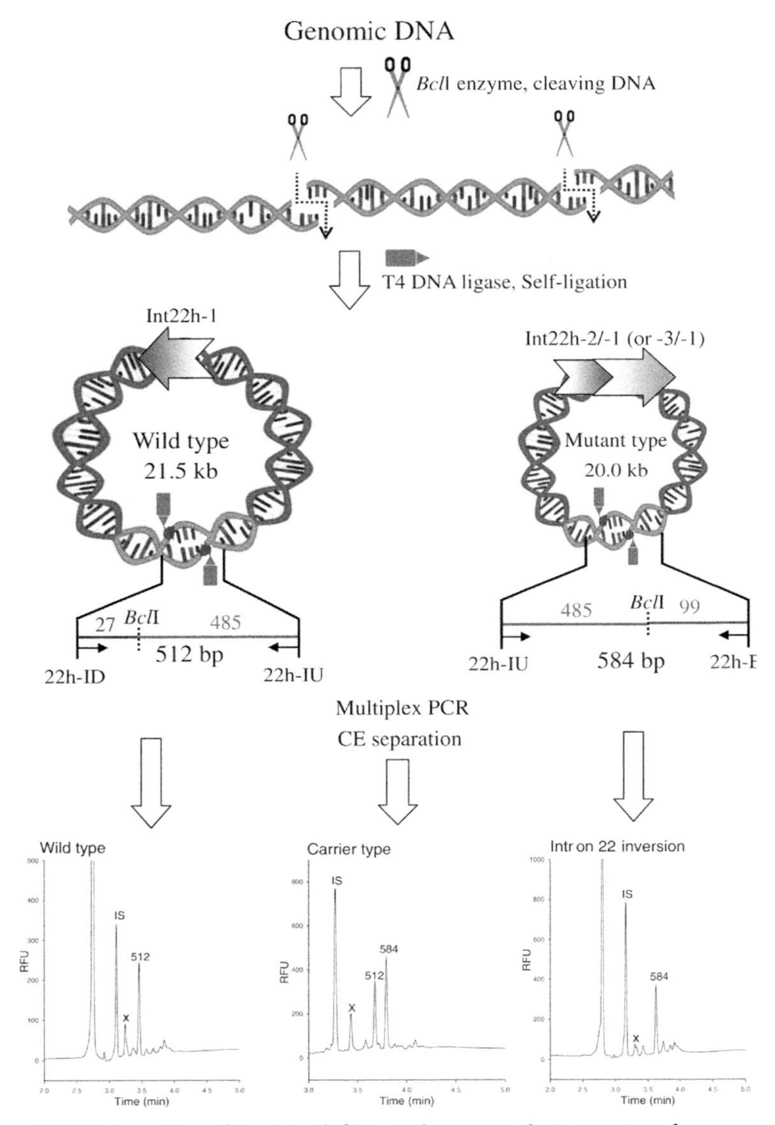

Figure 5.47 Schematics of inverse-shifting polymerase chain reaction for genotyping the intron 22 inversion in F8. Polymerase chain reaction amplicons from the relevant circular double-stranded DNA were separated by short-end injection capillary gel electrophoresis. IS internal standard (KRIT1), 512 wild-type allele, 584 intron 22 inversion, X nonspecific product. *With permission from T.Y. Pan, C.C. Wang, C.J. Shih, H.F. Wu, S.S. Chiou, and S.M. Wu, Genotyping of intron 22 inversion of factor VIII gene for diagnosis of hemophilia A by inverse-shifting polymerase chain reaction and capillary electrophoresis, Analytical and Bioanalytical Chemistry 406 (22) (2014) 5447–5454 [199].*

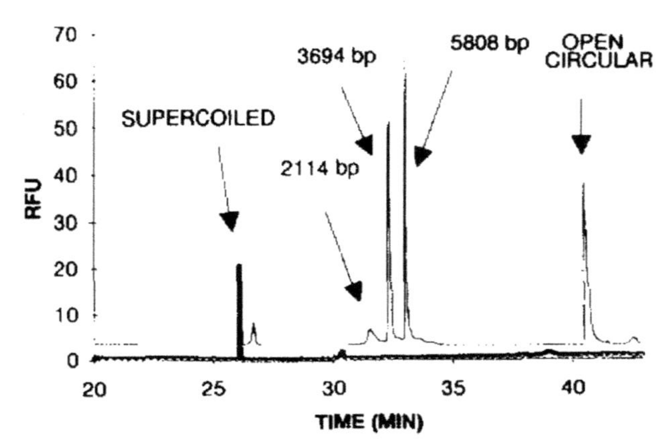

Figure 5.48 Capillary gel electrophoresis analysis of pICD1LS phage display vector. *With permission from B.C. Courtney, K.C. Williams, Q.A. Bing, J.J. Schlager, Capillary gel electrophoresis as a method to determine ligation efficiency. Analytical Biochemistry 228 (2) (1995) 281–286 [201].*

developed method can be applied as a fast and reliable procedure for diagnostic DNA sequencing [202]. Qualitative analysis of single-cell multiplex RT-PCR products was accomplished by CGE with LIF detection using replaceable HPMC matrix and EB labeling [203], and applied in cancer biomarker discovery. Others reported of the use of CGE-based mutation detection of the breast cancer susceptibility genes of BRCA1 and BRCA2 [204,205]. Heteroduplex analysis by capillary array electrophoresis provided rapid mutation detection in large multiexon genes in a validated fashion [206]. All samples were amplified by mutiplex PCR using fluorescently labeled primers and mixed with reference sequences to generate heteroduplexes for CGE analysis to scan variations in the BRCA1, BRCA2, MLH1, MSH2, and MSH6 genes. The highly homologous spinal muscular atrophy genes (SMN1 and SMN2) are hard to differentiate and quantify representing a challenge for the diagnosis of SMA patients and carriers. Point mutation or deletion of SMN1 or conversion of SMN1 to SMN2 was reported as directly responsible for SMA in patients or carriers. A fluorescence-based conformation-sensitive CGE method was developed to quantitatively analyze PCR products covering the variable position of the SMN1/SMN2 genes [207].

For the food and beverage industry, exogenous DNA was used as target to detect genetically modified organisms by applying CGE (2.5% PVP) coupled with electrochemiluminescent detection system to identify roundup ready soy [208]. Four pairs of primers were used in a multiple PCR reaction. One of the four amplified sequences was endogenous, while the other three were exogenous. The results revealed that the method can accurately identify roundup ready soy.

5.2 Capillary gel electrophoresis of proteins

CGE provides unique selectivity in protein analysis. In a sieving media, protein molecules migrate according to their hydrodynamic volume to charge ratio. The most frequently used ones are replaceable gels of LPA, polyethylene oxide, polyethylene glycol, dextran, pullulan, and alkylated cellulose derivatives. Due to the replaceability of the sieving matrix, the process can be fully automated. Capillaries utilized for the CGE of proteins usually possess some kind of an inner surface coating to eliminate electroosmotic flow and prevent solute adsorption.

Native proteins were separated by CGE by Wu and Regnier [209] using LPA gel matrices. The authors found that polyacrylamide gels in the concentration range of 3.5%–5% did not exhibit any size-based separations for native proteins with MW 20,000–47,000 and suggested that the separation was solely based on the charge to hydrodynamic ratios of the proteins. CGE analysis of rat testis H1 histone variants and their phosphorylated modifications was performed by Lindner et al. [210] They studied the effect of buffer pH, HPMC sieving polymer and buffer concentrations on separation performance. Under optimized conditions and with the use of uncoated fused-silica capillary column, eight H1 histone subfractions, including two H1 Deg histones and H1t and their phosphorylated modifications, were baseline separated as shown in Fig. 5.49. The separation of H1 histones provided an important new alternative technique to HPLC and traditional slab gel electrophoresis.

High concentration polyacrylamide gels with low cross-linking ratio were successfully applied to separate the individual polyamino acid oligomers either with UV absorbance detection at 220 nm or with the more sensitive LIF detection of 3-(4-carboxybenzoyl)-2-quinolinecarboxaldehyde (CBQCA) tagged forms [211]. A linear relationship was shown between the absolute electrophoretic mobility and $q/M_r^{2/3}$ (q: net negative charge of the peptides; M_r: molecular mass) with a good regression coefficient of $r^2 = 0.993$ [212].

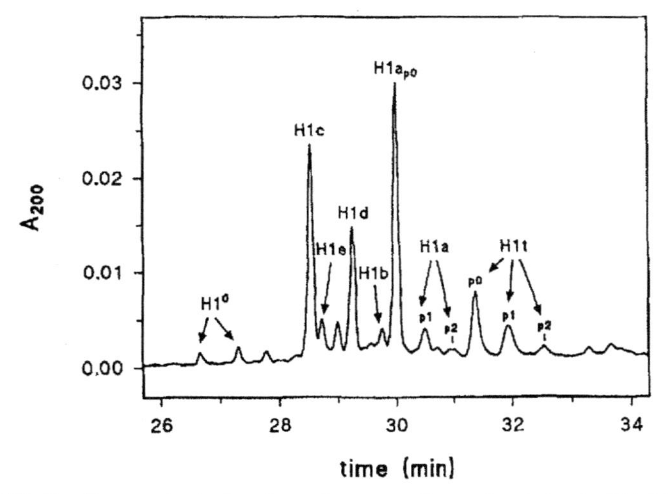

Figure 5.49 Capillary gel electrophoresis separation of rat testis H1 histones in 0.03% hydroxypropylmethyl cellulose sieving material. *With permission from H. Lindner, M. Wurm, A. Dirschlmayer, B. Sarg, and W. Helliger, Application of high-performance capillary electrophoresis to the analysis of H1 histones, Electrophoresis 14 (5–6) (1993) 480–485 [210].*

Preconcentration techniques for dilute samples can utilize either chromatographic or electrophoresis principles. The former one is the use of solid-phase extraction, while the latter is based on isotachophoretic stacking. Solid-phase extraction was also executed on silica-based C2 HPLC phase membranes for protein preconcentration, which improved detection limits but might introduced injection bias associated with the selectivity variation of the stationary phase used [213]. Others used reversed-phase-based extraction on quartz wool and porous beads generating concentration detection limits for peptides in the mid-picomolar range [214]. Transient isotachophoresis is a convenient and important preconcentration method in CE-based protein analysis [215].

5.2.1 Capillary SDS gel electrophoresis

The most frequently used CGE method for peptides and proteins is sodium dodecyl sulfate capillary gel electrophoresis (SDS-CGE), in which case the surfactant covered species assumable have the same surface charge density; therefore their separation is practically based on their sizes [216]. Capillary sodium dodecyl sulfate (SDS) gel electrophoresis is a rapid automated separation and characterization technique for protein molecules and contemplated as a modern instrumental approach to SDS-PAGE [217]. In SDS-CGE mode, the disulfide bonds of proteins are broken and the denatured molecules are covered by SDS micelles rendering them more or less identical mass-to-charge ratios. Size separation of SDS–protein complexes can be readily attained in narrow bore fused-silica capillaries filled with cross-linked gels [218–220] or noncross-linked polymer networks [221]. Fig. 5.50 depicts a typical applications of the technique for the analysis of a standard protein test mixture ranging from 14.2 to 205 kDa.

Karger and coworkers demonstrated for the first time that cross-linked polyacrylamide gel-filled capillaries could be applied for high-performance size separation of proteins [218]. They reported excellent separation power of this system by obtaining baseline separation of the A and B chains of insulin in less than 8 minutes using a 7.5% T/3.3% C polyacrylamide gel containing 8 M urea and 0.1% SDS. The same group polymerized various concentration cross-linked polyacrylamide gels (10%, 7%, and 5% T with 3.3% C) in 75 μm inner diameter fused-silica capillaries of 10 or 20 cm length and modified a commercial UV detector in house to hold the capillary in the optical path. Standard proteins of alpha-lactalbumin, beta-lactoglobulin (β-Lg),

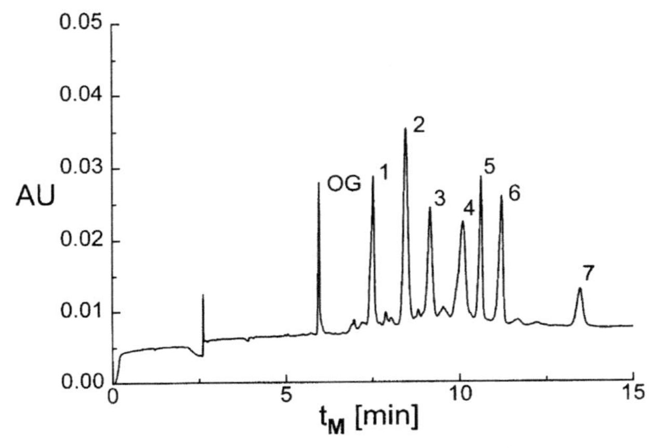

Figure 5.50 Capillary sodium dodecyl sulfate gel electrophoresis trace of a protein test mixture. (1) β-Lactalbumin, MW 14,200; (2) carbonic anhydrase, MW 29,000; (3) ovalbumin, MW 45,000; (4) bovine serum albumin, MW 66,000; (5) phosphorylase B, MW 97,400; (6) β-galactosidase, MW 116,000; (7) myosin, MW 205,000; *OG*, tracking dye Orange-G. Conditions: $E = 300$ V/cm, injection: 100 ng protein mix, detection: 214 nm. *With permission from A. Guttman, P. Shieh, J. Lindahl, and N. Cooke, Capillary sodium dodecyl sulfate gel electrophoresis of proteins. II. On the Ferguson method in polyethylene oxide gels. Journal of Chromatography A 676 (1) (1994) 227–231 [221].*

trysinogen, and pepsin (Fig. 5.51), as well as myoglobin fragments, were separated with high efficiencies (200,000 plates/m) and linear Ferguson plots of log mobility versus gel concentration were observed.

A detailed comparative study on the separation of proteins by SDS-PAGE and capillary SDS gel electrophoresis was published by Guttman and Nolan [222]. Fig. 5.52 depicts the electrophoresis patterns and Fig. 5.53 delineates the standard calibration curves for the SDS-PAGE slab and SDS-CGE experiments exhibiting the results of 65 biologically important and well characterized proteins, respectively. The differences in deviations of the individual proteins from the linear regression line in both cases were probably caused by the use of different polymeric sieving matrices in SDS-PAGE and SDS-CGE and the post-translational modifications of the samples.

High-performance capillary electrophoresis using replaceable PEO polymer network in the presence of SDS was shown as a viable alternative to SDS-PAGE by Benedek and Thiede [223]. Details, such as the effects of the actual size of the sieving polymers used and their concentration on the separation performance, were studied. The MW 100,000 PEO was found to provide good resolution and thus selected for further work.

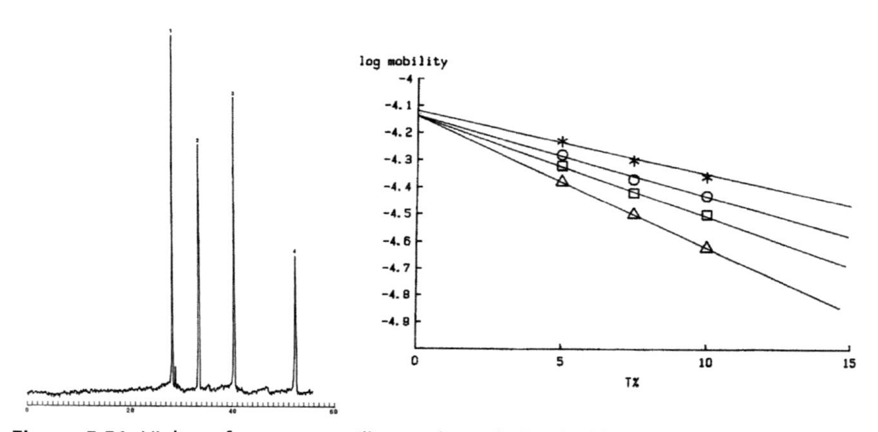

Figure 5.51 High-performance capillary sodium dodecyl sulfate polyacrylamide slab gel electrophoresis separation of proteins (left panel) and the Ferguson plot of log mobility versus percent T for the four proteins (right panel). Conditions: 400 V/cm, 24 μA, 27°C; $T = 10\%$, $C = 3.3\%$. Buffer, 90 mM Tris-NaH_2PO_4 (pH = 8.6), 8 M urea, 0.1% sodium dodecyl sulfate. Samples: 1 = α-lactalbumin; 2 = β-lactoglobulin; 3 = trypsinogen; 4 = pepsin. *With permission from A.S. Cohen, B.L. Karger, High-performance sodium dodecyl sulfate polyacrylamide gel capillary electrophoresis of peptides and proteins, Journal of Chromatography 397 (1987) 409–417 [218].*

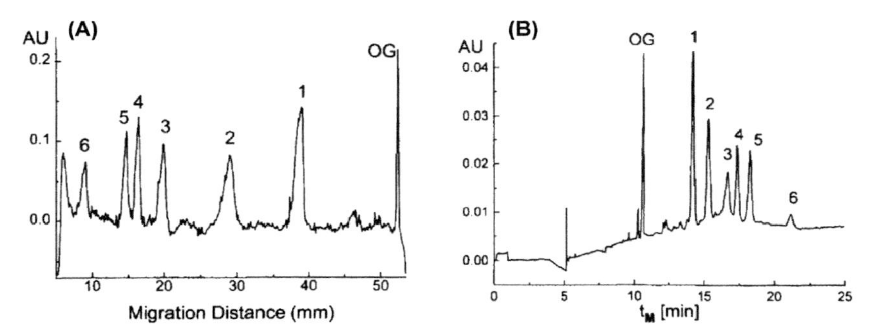

Figure 5.52 Separation patterns of a six protein test mixture by sodium dodecyl sulfate polyacrylamide slab gel electrophoresis (A) and capillary sodium dodecyl sulfate gel electrophoresis (B). *With permission from A. Guttman, J. Nolan, Comparison of the separation of proteins by sodium dodecyl sulfate-slab gel electrophoresis and capillary sodium dodecyl sulfate-gel electrophoresis, Analytical Biochemistry 221 (2) (1994) 285–289 [222].*

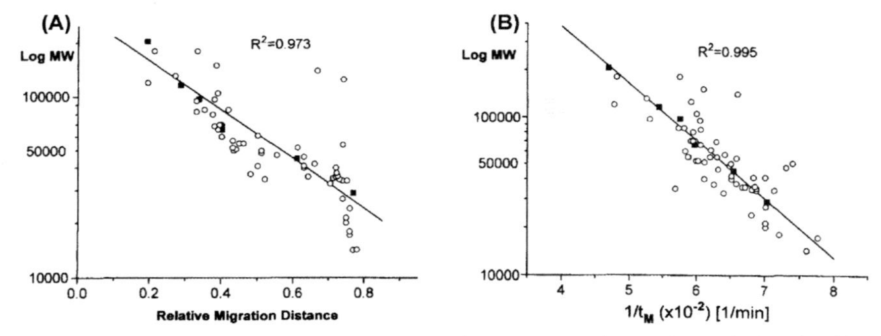

Figure 5.53 Molecular weight estimation of 65 proteins by sodium dodecyl sulfate polyacrylamide slab gel electrophoresis (A) and sodium dodecyl sulfate capillary gel electrophoresis (B). *With permission from A. Guttman, J. Nolan, Comparison of the separation of proteins by sodium dodecyl sulfate-slab gel electrophoresis and capillary sodium dodecyl sulfate-gel electrophoresis, Analytical Biochemistry 221 (2) (1994) 285–289 [222].*

Capillary SDS gel electrophoresis proved to be a very important bioanalytical tool for rapid molecular weight estimation [224] and purity check of recombinant proteins in the modern biotechnology industry [225]. SDS-CGE is widely used to separate and quantify monoclonal antibody (mAb) fragments and impurities [226–228]. The typical peaks expected in mAb samples are the light chain (LC), the heavy chain (HC), and the nonglycosylated heavy chain (ngHC) under reduced condition separations. Hunt and Nashabeh [229] explored a biotechnology perspective on the use of capillary electrophoresis-SDS sieving of therapeutic recombinant proteins. They used precolumn fluorophore labeling of a recombinant mAb to obtain low nanomolar detection limits. Their assay illustrated the advantages of enhanced precision and robustness, speed, ease of use, and on-line detection in monitoring bulk manufacture of protein pharmaceuticals. SDS-CGE with LIF detection was recently applied for the purity analysis of adeno-associated virus (AAV) capsid proteins [230,231]. A comprehensive review on CGE of proteins was published by Guttman in 1996 [217]. Since then, the field moved toward biomedical, biotechnology, and gene therapy applications [232].

Kilar and coworkers [233] successfully studied outer membrane proteins, lipopolysaccharides, hemolysin, and the in vivo and in vitro virulence of wild-type *Proteus penneri* 357 and its two isogenic mutant variants (a transposon and a spontaneous mutant) using capillary electrophoresis with dynamic sieving. CGE was found to be suitable for comparative analysis of bacterial protein patterns of genetic variants also providing valuable insights in connection with bacteriological virulence. Other interesting applications of capillary SDS gel electrophoresis were reported for the protein characterization of bacterial lysates [234] and profiling of human serum proteins [235]. Dextran and PEG as UV-transparent sieving matrices were successfully used for the molecular mass-based analysis of SDS–protein complexes with 214-nm UV detection by Ganzler et al. [236]. The low viscosities of these separation matrices allowed routine replacement in the column and hundreds of injections in a single capillary with migration time reproducibilities of >0.5% relative standard deviation (RSD).

The influence of temperature on the sieving effect of different polymer matrixes in capillary SDS gel electrophoresis of proteins was studied by Guttman et al. [237]. Dextran, polyethylene oxide, and LPA sieving polymers (Fig. 5.54) were used in CGE to study the effect of temperature on the separation of SDS–protein complexes in the molecular mass range of 14,400–97,400 Da.

The migration properties and band broadening of proteins were studied as a function of column temperature ranging from 20°C and 50°C. An equation has been derived that relate the effect of temperature on the mobility of the solute:

$$\ln \mu = \ln(Q/A) - E_a/RT \tag{5.3}$$

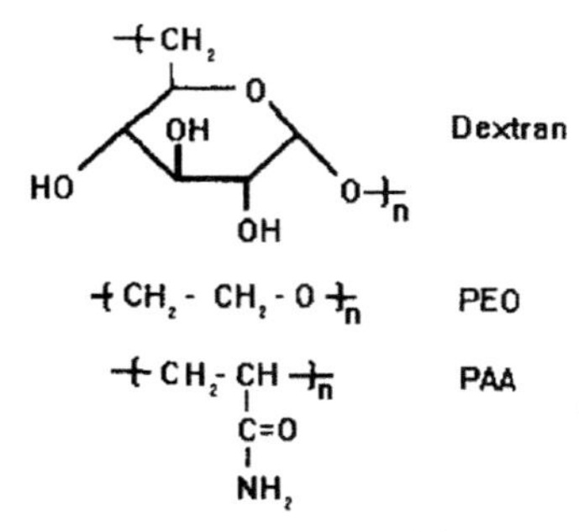

Figure 5.54 Structures of the monomer units of frequently used polymers in capillary sodium dodecyl sulfate gel electrophoresis. Dextran, polyethylene oxide, and polyacrylamide. *With permission from A. Guttman, J. Horvath, N. Cooke, Influence of temperature on the sieving effect of different polymer matrixes in capillary SDS gel electrophoresis of proteins. Analytical Chemistry 65 (3) (1993) 199–203 [237].*

where μ is the electrophoretic mobility of the solute, Q is the charge, A is a constant (characteristic on the polymer network), E_a is the sum of the activation energies for the viscous media (E_{av}) and the actual conformation (E_{ac}), R is the universal gas constant, and T is the absolute temperature. Using isoelectrostatic separation mode, the migration time of the SDS–protein molecules decreased in both instances as temperature increased within the examined time interval; however, peak efficiency increased with elevated temperature in the dextran-based column but decreased when PEO sieving material was used, as shown in Fig. 5.55.

The effect of operational variables on the separation of proteins by capillary SDS gel electrophoresis was investigated [238], and a general migration velocity Eq. (5.4) was derived that was supported by the experimental data.

$$\nu \sim C\,E\,\mathrm{MW}^k \exp(1/T) \tag{5.4}$$

where ν is the migration velocity, C is a combined constant including friction, charge, and viscosity, E is the applied electric field strength, MW is the molecular mass of the solute, k is the exponent that represents information about the apparent shape of the polyionic sample during CGE, and T is the absolute temperature. The average value of k was found as -0.2 in these experiments with the use of polyetylene oxide gels. This equation gave a better understanding of the shape of the SDS–protein complexes during their separation in CGE as discussed later. Results were presented regarding the effects of different operational variables, such as gel concentration, electric

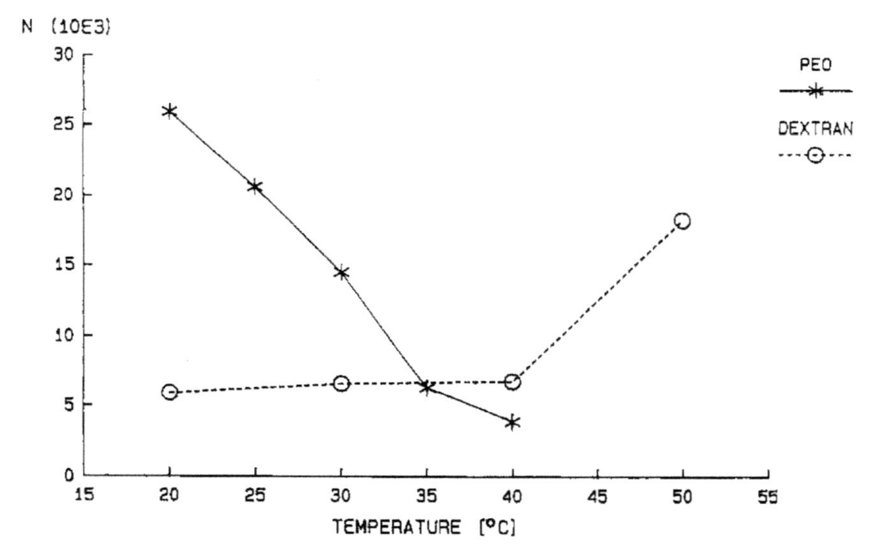

Figure 5.55 Relationship between the column temperature and the theoretical plate numbers of the ovalbumin peak using polyethylene oxide (*) and dextran (o) filled capillary columns in SDS-CGE. *With permission from A. Guttman, J. Horvath, N. Cooke, Influence of temperature on the sieving effect of different polymer matrixes in capillary SDS gel electrophoresis of proteins. Analytical Chemistry 65 (3) (1993) 199–203 [237].*

field strength, solute molecular mass, and running temperature, on the electrophoretic migration properties in different-sized gels.

SDS-CGE using a borate cross-linked dextran separation matrix is widely employed today for rapid consistency analysis of therapeutic proteins in manufacturing and release testing [232]. Transient borate cross-linking of the semirigid dextran polymer chains leads to a high-resolution separation gel for SDS–protein complexes [220]. To understand the migration and separation basis, Guttman and Filep [239] explored various gel formulations of dextran monomer and borate crosslinker concentrations. Ferguson plots were determined for the intact and subunit forms of an mAb drug. The resulting nonlinear concave curves pointed to nonclassical sieving behavior as shown in Fig. 5.56. The study revealed that by modulating the dextran (monomer) and borate (crosslinker) concentration ratios of the sieving matrix, one can optimize the separation for specific biopharmaceutical modalities by following Fig. 5.56C.

Effects of column temperature, internal column diameter, and column length on the performance of SDS polyacrylamide gel-filled capillary electrophoresis were reported in [240]. It was found that the increase in

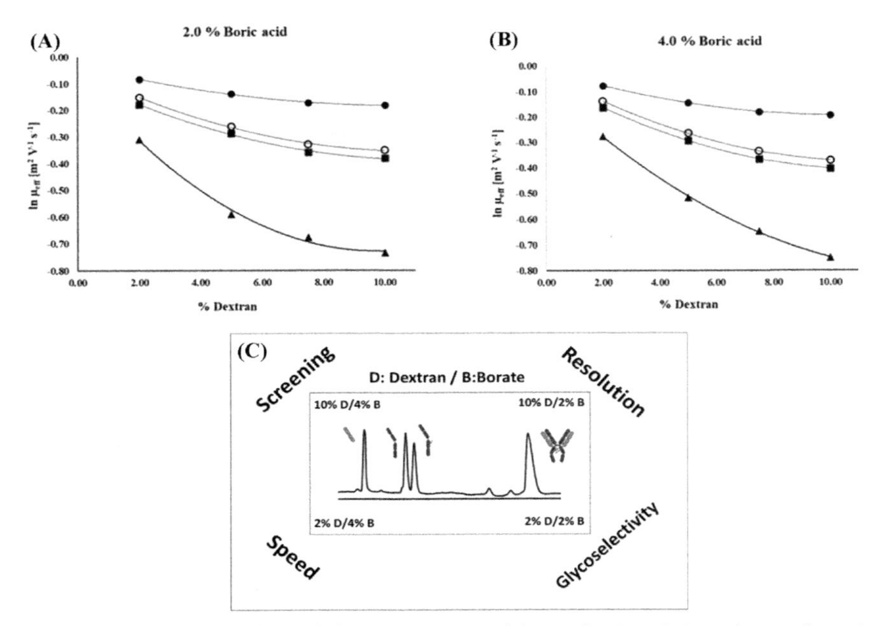

Figure 5.56 Ferguson plots of the EOF corrected logarithmic relative electrophoretic mobilities of the intact monoclonal antibody and its subunits as the function of monomer (dextran) concentration with 2.0% (A) and 4.0% (B) boric acid crosslinker. C: optimization scheme. Symbols (A and B): ● light chain, ○ deglycosylated heavy chain, ■ heavy chain, and ▲ intact mAb. *With permission from A. Guttman, C. Filep, B.L. Karger, Fundamentals of capillary electrophoretic migration and separation of SDS proteins in borate cross-linked dextran gels, Analytical Chemistry 93 (26) (2021) 9267–9276 [220].*

column temperature resulted in an exponential decrease in peak efficiency and best separation performance was obtained between 21°C and 24°C. An increase in the internal column diameter resulted in linear peak migration time decrease. Size-based separation of recombinant bovine growth hormone and its aggregates was attempted by the same group using nonacrylamide polymer-based SDS-CGE with excellent migration time reproducibility over 140 runs but a gradual loss of efficiency [241]. Shieh et al. [242] investigated the run-to-run and batch-to-batch reproducibility and stability of polyethylene oxide filled capillary columns and found that the migration time RSD in the course of 190 runs was in between 0.351% and 0.453% (Table 5.3).

Linear, noncross-linked polyacrylamide gels were utilized to baseline separate a four protein test mixture according to their molecular weights

Table 5.3 Run-to-run and batch-to-batch reproducibility study of SDS-CGE in polyethylene oxide filled capillary columns.

Run[a]	OG	α-LAC	CA	OVA	BSA	PHOS	β-GAL	MYO
lst	5.94	7.50	8.44	9.15	10.12	10.60	11.19	13.40
2nd	5.94	7.50	8.43	9.14	10.12	10.59	11.18	13.40
3rd	5.93	7.49	8.42	9.13	10.10	10.58	11.17	13.38
4th	5.90	7.44	8.38	9.07	10.05	10.53	11.12	13.32
5th	5.90	7.45	8.39	9.08	10.04	10.53	11.12	13.31
6th	5.93	7.47	8.42	9.11	10.07	10.55	11.14	13.34
7th	5.94	7.50	8.45	9.16	10.14	10.61	11.21	13.43
8th	5.89	7.43	8.38	9.08	10.04	10.52	11.11	13.31
9th	5.88	7.42	8.36	9.06	10.01	10.50	11.09	13.29
191st	5.93	7.45	8.44	9.08	10.08	10,58	11.20	13.45
194th	5.90	7.44	8.42	9.07	10.06	10.57	11.18	13.43
195th	5.90	7.44	8.40	9.06	10.04	10.56	11.15	13.41
199th	5.88	7.45	8.40	9.08	10.05	10.56	11.16	13.42
200th	5.87	7.43	8.39	9.06	10.03	10.55	11.15	13.41
Average	5.90	7.46	8.41	9.09	10.06	10.56	11.15	13.39
SD	0.024	0.030	0.030	0.041	0.042	0.038	0.043	0.061
RSD (%)	0.400	0.398	0.351	0.448	0.413	0.358	0.388	0.453

Note: BSA, Bovine serum albumin; *CA*, carbonic anhydrase; *β-GAL*, β-galactosidase; *α-LAC*, α-lactalbumin; *MYO*, myosin; *OG*, Orange-G; *OVA*, ovalbumin; *PHOS*, phosphorylase; *SDS-CGE*, sodium dodecyl sulfate capillary gel electrophoresis.

[a]Injections 1–9: protein standard; injections 10–190: fetal calf serum; injections 191–200; protein standard.

Source: With permission from P.C.H. Shieh, D. Hoang, A. Guttman, N. Cooke, Capillary sodium dodecyl sulfate gel electrophoresis of proteins. I. Reproducibility and stability. Journal of Chromatography A 676 (1) (1994) 219–226 [242].

by SDS-CGE covering 17,800–77,000-Da MW range and other mixtures with good resolving power [243]. The Regnier group applied noncross-linked LPA gels to obtain size-based separations of SDS–protein complexes in the MW rage of 14–205 kDa [244]. Baseline separation of model proteins that varied by 10% in molecular mass were achieved. Deyl et al. used CGE in 35 cm length (effective to the detector) and 75 μm i.d. untreated capillaries filled with 4% noncross-linked polyacrylamide to separate collagen type I polypeptide α-chains and their chain polymers (β11, β12, γ) even higher than γ-chains. Separations were run in 50 mM Tris-glycine buffer (pH 8.8) with the sampling port at the anodic end of the capillary with UV detection at 220 nm [245]. Guttman and Nolan reported on rapid separations of proteins in the molecular mass range of 20–200 kDa by SDS-CGE and demonstrated with excellent linearity as well as intra- and interday migration time reproducibility. Monomer-dimer forms of the recombinant human ciliary neurotrophic factor were analyzed by CE-SDS under reducing and nonreducing conditions [246]. Wiktorowicz and coworkers used a commercially available sieving matrix in capillary electrophoresis that separated proteins by as little as 4% size (Fig. 5.57) [247]. The method was applicable to both native and denatured proteins [248].

In contrast to LPA gels, dextrans are low UV absorbing polysaccharides and thus advantageous for CGE-based protein analysis. In protein separations, Karim et al. [249] revealed an interesting relationship between the molecular weight of the dextran sieving polymer as well as the migration time and efficiency. The authors attained very rapid separation and enhanced resolution using narrow molecular weight distribution dextran polymers. The same group also reported on the separation mechanism for high-molecular-weight proteins (MW >10,000) that was considered to be free draining and different from just sieving only [250]. Polyethylene oxide matrices received considerable attention by exploring its use in nucleic acid [251], protein [221,223,238,242,252], and complex

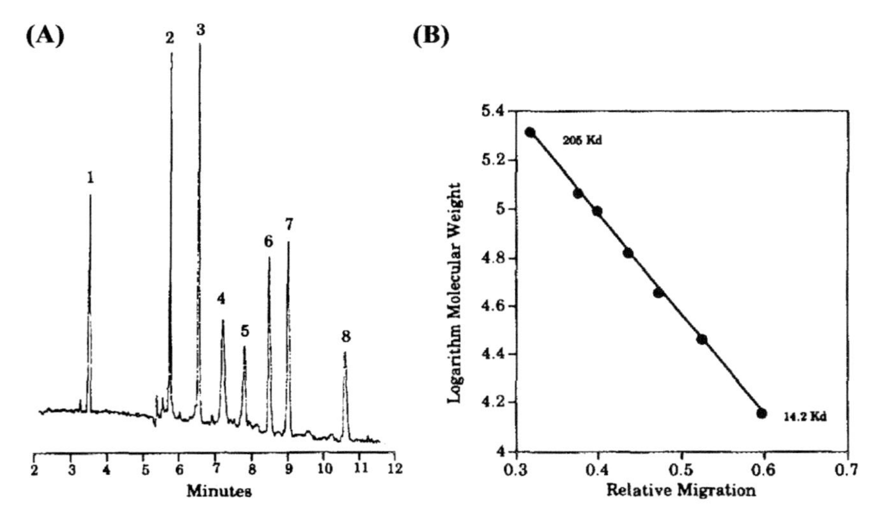

Figure 5.57 (A) Capillary sodium dodecyl sulfate gel electrophoresis of protein mixture [(1) α-lactalbumin, (2) carbonic anhydrase, (3) ovalbumin, (4) bovine serum albumin, (5) phosphorylase b, (6) β-galactosidase, (7) myosin] under denaturing conditions. (B) Logarithmic molecular weight plotted as a function relative migration. *With permission from W.E. Werner, D.M. Demorest, J. Stevens, J.E. Wiktorowicz, Size-dependent separation of proteins denatured in SDS by capillary electrophoresis using a replaceable sieving matrix. Analytical Biochemistry 212 (1) (1993) 253–258 [247].*

carbohydrate [253] analysis. Other research groups reported the separation and comparative analysis of apolipoproteins by capillary zone and capillary SDS gel electrophoresis using polyacrylamide matrices [254,255].

The use of linear poly[*N*-(acryloylamino)ethoxyethanol] was suggested by Chiari et al. as a 500 times more hydrolysis resistant and more hydrophilic polymer than that of polyacrylamide [256]. They also reported about the use of PVA (MW: 133,000) as an efficient sieving matrix for SDS–protein complexes in the concentration range of 4%–6%. Decreasing the inner diameter of the separation capillary from 75 to 25 μm, the apparent entanglement threshold shifted to extremely diluted PVA solutions and adequate sieving was still obtained below 1% PVA. The authors suggested that the residual, free silanols present (even in a coated capillary) acted as nucleation sites for H-bond formation and aggregation of free PVA molecules [257]. Addition of modifiers incorporated into the separation matrix in CGE offered a possible complexation equilibrium and thus additional selectivity. Similar effect can be obtained if a modifier is added to the background electrolyte.

5.2.2 Ultrafast separations

An ultrafast protein analysis method, using SDS-mediated gel electrophoresis was developed by Benedek and Guttman [224], where a mixture of standard proteins with molecular masses ranging from 14,200 to 94,700 Da were separated within 3 minutes without compromising efficiency or resolution as shown in Fig. 5.58. The analysis was performed at 888 V/cm in a 7-cm effective (27 cm total) length 50 μm i.d. capillary.

By applying a similar length separation distance, rapid separation of SDS–protein complexes is described in [258]. Standard proteins with the molecular mass range of 29,000–97,400 were baseline resolved in <2 minutes. Rapid analysis of the light and heavy chains of human immunoglobulin G (IgG) was reported, and the results were compared with applications using longer separation distances, revealing that rapid and efficient analysis with adequate resolution was obtained by the use of short separation distances (Fig. 5.59). Schultz and coworkers performed microchannel-based SDS-CGE of six standard proteins in the molecular mass range from 9 to 116 kDa and demonstrated its speed and high resolution [259]. The six proteins were separated within 35 s with adequate separation efficiency. The separation channel was 40-μm deep, 100-μm wide at the top, 20-μm wide at the bottom, and 4.5 cm in length.

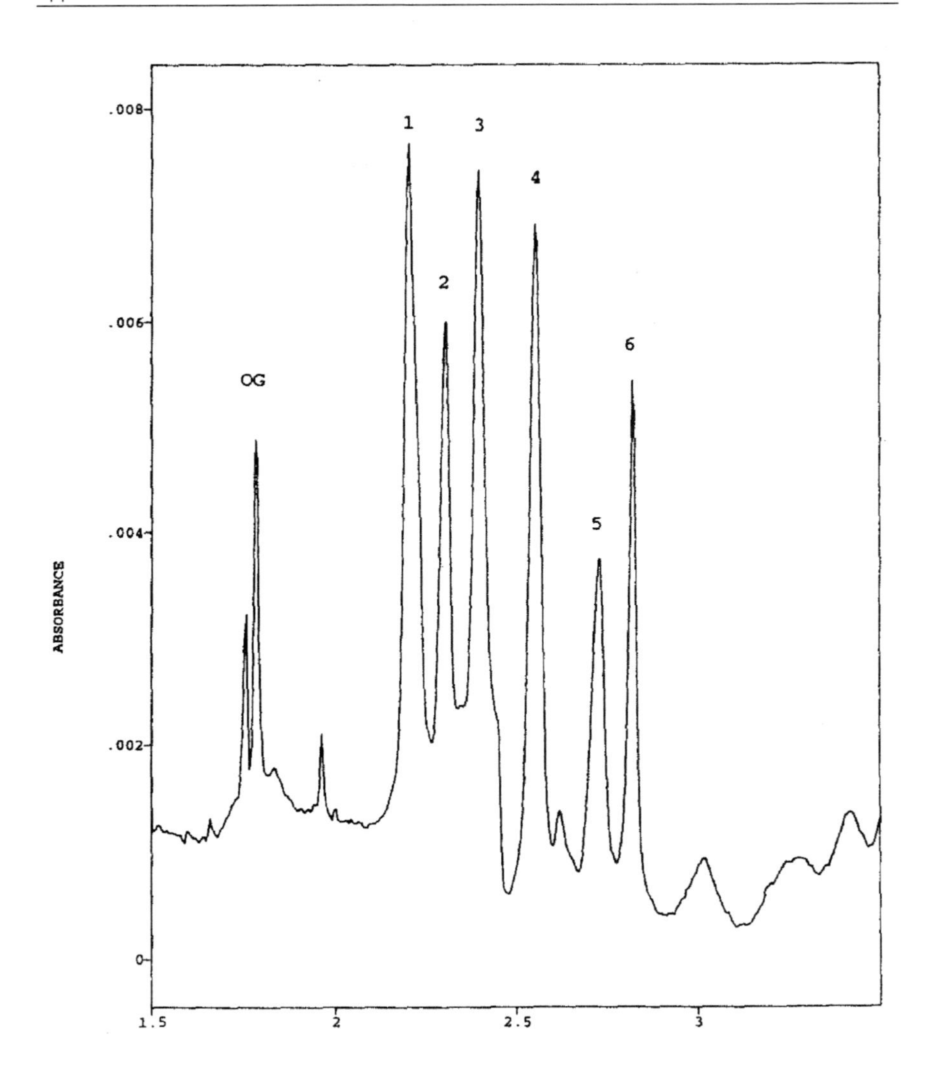

Figure 5.58 Ultrafast sodium dodecyl sulfate capillary gel electrophoresis separation of a standard protein mixture at high field strength (888 V/cm) using sodium dodecyl sulfate capillary gel electrophoresis. *With permission from K. Benedek, A. Guttman, Ultra-fast high-performance capillary sodium dodecyl sulfate gel electrophoresis of proteins, Journal of Chromatography A 680 (2) (1994) 375–381 [224].*

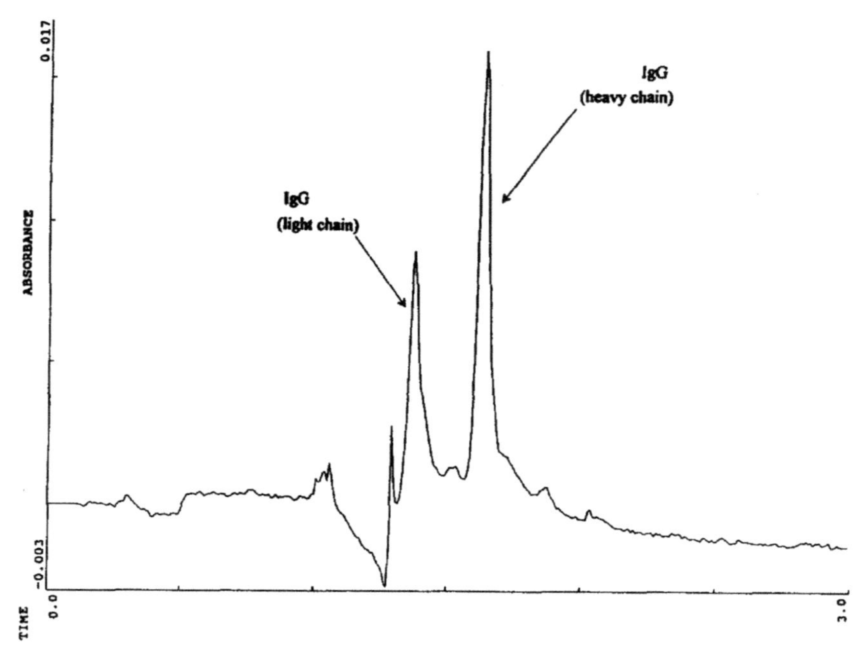

Figure 5.59 Rapid separation of sodium dodecyl sulfate–protein complexes of the light and heavy chains of human IgG in the short, 7-cm end of the 27 cm × 50 μm i.d. capillary. *With permission from R. Lausch, T. Scheper, O.W. Reif, J. Schloesser, J. Fleischer, R. Freitag, Rapid capillary gel electrophoresis of proteins, Journal of Chromatography 654 (1) (1993) 190–195 [258].*

5.2.3 Capillary gel isoelectric focusing of proteins

Capillary isoelectric focusing (cIEF) separates proteins or peptides according to their isoelectric point (pI) [260]. In a usual process, the sample is mixed with ampholytes and filled into a narrow bore fused-silica capillary with one end in a high-pH solution and the other end in a low-pH solution. When the electric field is applied a pH gradient is formed and the solute molecules are focused according to their pI values (Fig. 5.60) [261]. Detection can be accomplished in a one- or two-step process as well as by imaging the entire separation channel or connected to mass spectrometry (MS) [262]. The one-step cIEF method usually utilizes polymer buffer additives, such as cellulose derivatives to reduce, but not eliminate EOF, allowing proteins simultaneously focusing and migrating toward the detection window [263]. In the two-step process, the solute molecules are first focused and then mobilized by either chemical or hydrodynamic means [264,265]. This latter process requires EOF free capillary columns.

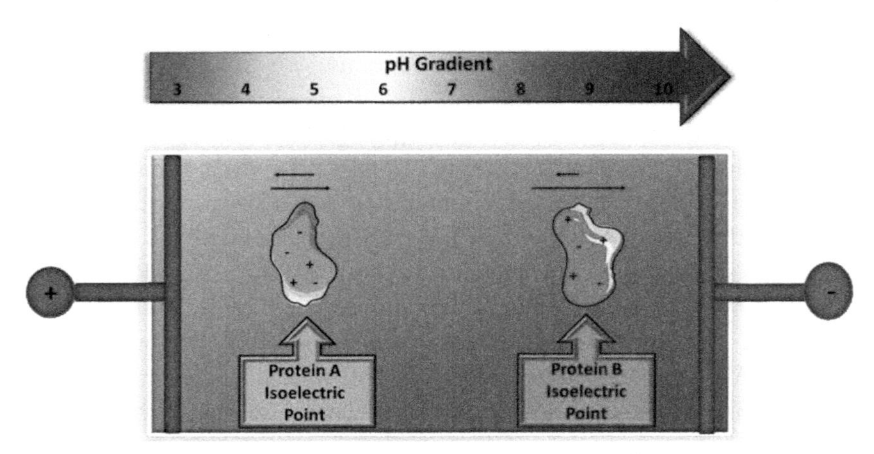

Figure 5.60 Principle of isoelectric focusing. Proteins with different isoelectric points migrate with the application of high electric field in the presence of a pH gradient until the net charge of a protein is zero. *With permission from M.R. Pergande, S.M. Cologna, Isoelectric point separations of peptides and proteins, Proteomes 5 (1) (2017) 4 [261].*

Real-time imaging of cIEF analysis was reported by several groups either using CCD camera-based detection after laser illumination [266,267] or spatial scanning LIF detection systems [268]. Zarabadi et al. [269] applied whole column UV imaging detection cIEF to analyze salivary α-amylase collected from different glands. The developed imaging cIEF method separated amylase isozymes in a rapid and automated way within 6 minutes. Imaged capillary isoelectric focusing (iCIEF) separation combined with downstream MS detection was also reported [262].

cIEF is an important technique in the emerging field of the analysis of biotherapeutics [270]. Grossbach [271] has described microelectrofocusing of proteins in capillaries as early as 1972. He used acrylamide based anticonvective medium for focusing proteins in the nanogram range and generated sharp focused zones. Others demonstrated the high resolution of this method by the separation of hemoglobin variants F and A where the pI difference was only 0.05 pH units and the resolution of the isoforms of an anticarcinoembryonic antigen mAb [272]. Differentiation between biofilm-positive and -negative yeast strains is important to choose an adequate therapeutic strategy due to higher resistance of the biofilm-positive strains to antifungals. Horka et al. [273] separated and identified similar strains of Candida cells and/or their lysates, based on cIEF with UV detection in acidic pH gradient. The same authors developed a fast and a simple lysis technique for the outer cell membrane and found characteristic fingerprints in the lysate electropherograms.

Rapid determination of full and empty adeno-associated virus capsid ratio was reported by using cIEF [274][275]. AAV-based pharmaceutical products should possess the therapeutic protein-coding transgene along with a promoter and a poly A sequence; however, manufacturers cannot guarantee the provision of only full capsids. Partially filled and empty capsids may also be part of the product affecting its efficacy and safety. The authors developed an automated cIEF method applicable in the biopharmaceutical industry for rapid determination of the full and empty capsid ratio of AAVs as shown in Fig. 5.61.

cIEF method was utilized for the separation and charge characterization of the heterogeneity of high-molecular-weight glutenin subunits (HMW-GS) in common wheat using LPA- and PVA-coated capillaries [276]. The method was able to distinguish 8–12 isoforms of HMW-GS with pI values in the range of 4.72–6.98 from the various wheat cultivars. Another cIEF method was developed for the charge variant characterization of therapeutic monoclonal antibodies (mAb) and reported in [277]. They applied a mixture of 5% Pharmalyte 8–10.5 and 1% Pharmalyte

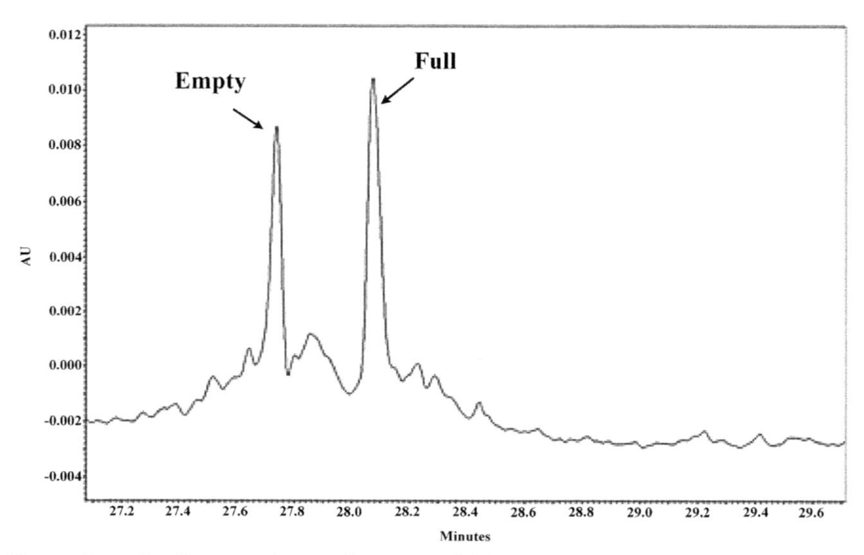

Figure 5.61 Capillary isoelectric focusing of full and empty adeno-associated virus serotype 5. *With permission from T. Li, T. Gao, H. Chen, P. Pekker, A. Menyhart, A. Guttman, Rapid determination of full and empty adeno-associated virus capsid ratio by capillary isoelectric focusing, Current Molecular Medicine 20 (10) (2020) 814–820 [274].*

3–10 as ampholyte solution, because this improved the resolution of the heterogeneous peaks. The acidic and basic charge variants of the mAb were baseline separated under optimized conditions. The developed approach showed good specificity and linearity within the concentration range of 0.03–0.20 mg/mL for the mAb test item.

A rapid, high-resolution cIEF was attempted with whole column imaging detection for the glycoform analysis of recombinant human erythropoietin (rhEPO) [278]. The glycoforms of rhEPO from three different sources were baseline separated within 5 minutes. The developed method was capable to detect rhEPO misuse in sport with high sensitivity imaging detection.

Salas-Solano et al. conducted an interlaboratory study to test the robustness of iCIEF for the analysis of monoclonal antibodies [279]. The participants in this study used an iCIEF instrument equipped with an UV detector at 280 nm. Fig. 5.62 shows a typically obtained electropherogram of a recombinant anti-α1-antitrypsin mouse mAb. Seven charge variants were baseline separated and the authors found that the pI data for the different charge variants were in very good agreement between the laboratories with RSD values of less than 0.8%. The RSD of the peak area values

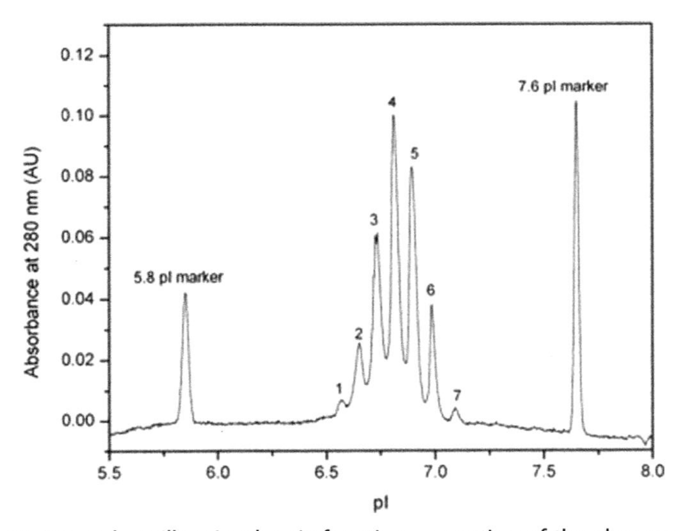

Figure 5.62 Imaged capillary isoelectric focusing separation of the charge variants of anti-α1-antitrypsin mouse mAb. *With permission from O. Salas-Solano, B. Kennel, S.S. Park, K. Roby, Z. Sosic, B. Boumajny, et al., Robustness of iCIEF methodology for the analysis of monoclonal antibodies: an interlaboratory study. Journal of Separation Science 35 (22) (2012) 3124–3129 [279].*

were less than 11% across different laboratories, although the different labs used different lots of ampholytes and multiple instruments. They concluded that the robust iCIEF method was appropriate to support the process development of monoclonal antibodies.

Protein therapeutics are usually produced in different heterogeneous forms during bioprocessing, including posttranslational modification-mediated charge variants, which should be analyzed at various stages of the manufacturing course. Isoelectric focusing (IEF) is one of the most frequently used methods to analyze protein charge heterogeneities. Dovichi and coworkers developed a rapid and reproducible system by coupling cIEF to a high-resolution MS for host cell protein analysis of a recombinant human monoclonal antibody [280]. Current profiles for the mobilization during capillary isoelectric focusing with different sheath flow buffers is shown in Fig. 5.63. The use of immobilized trypsin reduced the digestion time to 10 min, with a total analysis time of 4h. Host cell protein impurities were also analyzed in a recombinant mAb product.

5.2.4 Fluorescent labeling of proteins for SDS-CGE

Although most proteins have native fluorescence characteristics due to residues, such as tryptophan, the enhanced fluorescence properties of

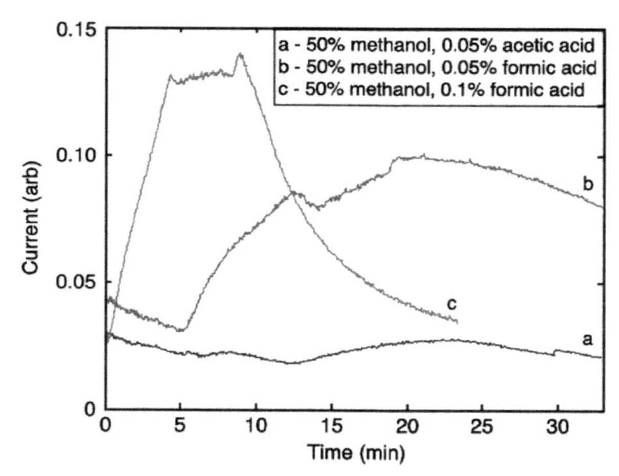

Figure 5.63 Current profiles for the mobilization during capillary isoelectric focusing with different sheath flow buffers: (a) 50% methanol and 0.05% acetic acid; (b) 50% methanol and 0.05% formic acid; and (c) 50% methanol and 0.1% formic acid. *With permission from Zhu G., Sun L., Wojcik R., Kernaghan D., McGivney IV J.B., Dovichi N.J., A rapid cIEF–ESI–MS/MS method for host cell protein analysis of a recombinant human monoclonal antibody. Talanta 98 (2012) 253–256 [280].*

fluorophore labels with the flexibility in choosing the location are more suitable for a number of applications [281]. Fluorophore labeling methods can either be through covalent linkage, noncovalent complexation or via genetic engineering, such as addition of green fluorescent protein [281,282]. Covalent modification of peptides and proteins can be performed via the amine, carboxyl, and thiol functional groups [283], but the most common strategies for fluorescent labeling of biomolecules is through the amine or thiol groups. In case of multiamine group possessing molecules, such as proteins, multiple labeling can occur resulting in band broadening as the different derivatization rate molecules are migrating slightly differently. Noncovalent labeling is another way to increase the detection limit for protein separations in CE [284]. Noncovalent dyes tend to form multiple labeled products, but do not require a chemical reaction prior to analysis. For example, protein–fluorophore interactions occur through physical mechanisms, including hydrophobic interactions, electrostatic interactions, and hydrogen bonding.

In several instances, the unconjugated dye has very low fluorescence signal compared to the conjugated product [285], which is preferred as in this case no purification is required before CGE analysis (Fig. 5.64). However, some of the noncovalent staining dyes have very high fluorescence in the presence of SDS micelles, increasing in this way the background level. NanoOrange, on the other hand, can be used with up to 0.05% SDS with no such issues [286]. The third category is genetically engineered tagging that is very specific as the labeled protein is generated by a fusion between the protein of interest and an autofluorescent protein tag [287,288]. The frequently used fluorescence dyes applied in CE with LIF detection of peptides and proteins are summarized in [283] and the most important ones are listed in Table 5.4 with their excitation and emission wavelengths.

Precolumn derivatization of proteins was applied using fluorescamine, naphthalene-2,3-dicarboxaldehyde (NDA), and *o*-phthaldialdehyde to enhance absorption and/or fluorescence detection in SDS-CGE [289]. Compared with underivatized proteins, absorption sensitivity increased by more than 20-fold at 280 nm. Fig. 5.65 compares the separation of unlabeled (left panel) and NDA-labeled (right panel) myoglobin and conalbumin by SDS-CGE, showing a significant increase in UV absorbance after derivatization. Fluorescent detection provided attomole detection limits and increased sensitivity with increasing protein molecular mass. Important to note that although the precolumn labeling step somewhat

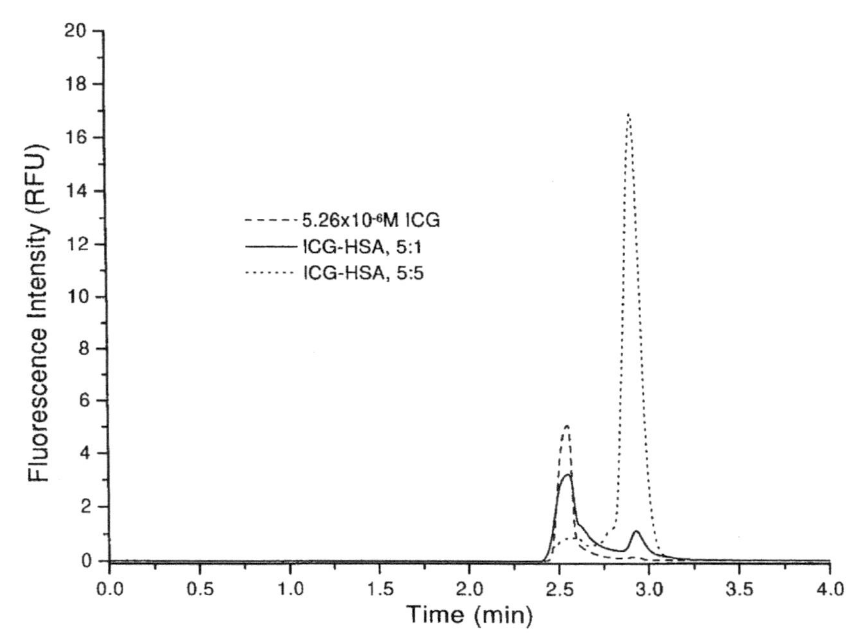

Figure 5.64 Electropherogram of free indocyanine green (*ICG*) and ICG–human serum albumin conjugates. *With permission from C. Colyer, Noncovalent labeling of proteins in capillary electrophoresis with laser-induced fluorescence detection. Cell Biochemistry and Biophysics 33 (3) (2000) 323–337 [285].*

decreased separation efficiency, still did not result in multiple peaks usually caused by heterogeneous labeling approaches. Fluorescence detection with the use of CBQCA yielded detection limits in the zeptomol range [290]. Postcolumn derivatization of proteins [291] with *o*-phthaldialdehyde also resulted in excellent detection limits, ranging from femtomoles to attomoles.

Dies-Masa and coworkers [292] found that *N*-arylaminonaphthalene sulfonate dyes provide good fluorescence emission when associated with some proteins, while their emission in aqueous buffers was very small. Others reported nanomolar detection limits for human serum albumin with iodocyanine green, a different type of noncovalent fluorescent label as shown in Fig. 5.66 [293]. Picomolar detection limits for proteins in SDS-CGE were obtained by using Sypro Red, a noncovalent reagent, however, only when the amount of SDS in the running buffer was below the critical micelle concentration [294] and the protein samples were labeled before injection.

Table 5.4 Fluorescence dyes applied in the capillary electrophoretic separation of peptides and proteins.

Dye name	Structure/type	Excitation max (nm)	Emission max (nm)
OPA	*ortho*-Phthalaldehyde	340	455
NDA	Naphthalene-2,3-dicarboxaldehyde	419	493
FC	4-Phenylspiro[furan-2(3*H*), 1′-phthalan]-3,3′-dione	382	480
FQ	3-(2-Furoyl)quinoline-2-carboxaldehyde	486	600
CBQCA	3-(4-Carboxybenzoyl)-2-quinolinecarboxaldehyde	465	550
6-AQC	6-Aminoquinolyl-*N*-hydroxysuccinimidyl carbamate	248	398
FITC	Fluorescein isothiocyanate	491	516
NBD-F	4-Fluoro-7-nitrobenzofurazan	465	535
NBD-Cl	4-Chloro-7-nitrobenzofurazan	337	512
Chromeo 503	Pyrylium dye	503	600
Cy5	Sulfoindocyanine succinimidyl ester	649	666
ICG	Indocyanine green	789	814
NanoOrange	Merocyanine dye	470	570
Sypro Red	Merocyanine dye	300, 550	630
ANS	l-Anilinonaphthalene-8-sulfonate	350	505
Alexa Fluor 488	Sulfonated fluorescein derivative	490	525
Dylight 488	Sulfonated fluorescein derivative	493	518

Pre- or on-column labeling of proteins with 3-(2-furoyl)quinoline-2-carboxaldehyde (FQ) and their separation in a replaceable polymer matrix of pullulan or hydroxypropylcellulose was reported by the Dovichi group [295]. Interestingly, the separation was not influenced by possible multiple labeling of the sample species. However, since the FQ labeling procedure uses highly poisonous potassium-cyanide to catalyze the labeling reaction, due to its safety concerns it is getting not preferred. The same group reported rapid and efficient capillary SDS gel electrophoresis separation of proteins by characterizing HT29 human colon cancer adenocarcinoma cells [296]. The cells were lysed inside a capillary, followed by protein denaturation with SDS, fluorophore labeled with 3-(2-furoyl)-quinoline-2-carboxyaldehide and separated using an 8% pollulan sieving matrix. Typical resolution was found to be around 30 protein components of a single HT29 cell, similar to the peak capacity of SDS-PAGE. Fluorescent

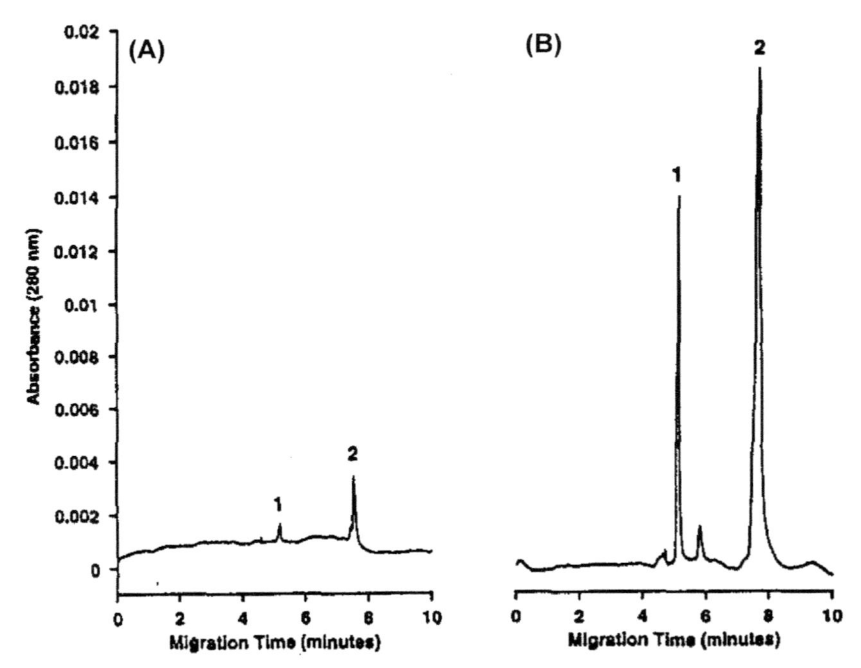

Figure 5.65 Sodium dodecyl sulfate capillary gel electrophoresis separation of unlabeled (A) and naphthalene-2,3-dicarboxaldehyde-labeled (B) proteins. Peaks: 1 = myoglobin (90 femtomoles); 2 = conalbumin (48 femtomoles). *With permission from E.L. Gump, C.A. Monnig, Pre-column derivatization of proteins to enhance detection sensitivity for sodium dodecyl sulfate non-gel sieving capillary electrophoresis, Journal of Chromatography A 715 (1) (1995) 167–177 [289].*

detection provided high sensitivity ranging from 10^{-10}–10^{-11} M. Single-cell level analysis was completed in 45 minutes.

A CGE assay was developed by Feng and Arriaga [297] utilizing LIF detection for carbonyl-modified proteins in mitochondrial fractions collected from rat skeletal muscle. Carbonylation is apparently a good marker for oxidative damage and an important modification to study in aging. Alexa 488 hydrazide was used for specific carbonyl labeling and FQ for protein labeling. The sensitivity of the method was at the femtomole level of carbonyls in proteins with the molecular masses of 26–30 kDa. In another approach, a pseudo SDS dye was synthesized by bonding a fluorescein headgroup to an alkyl chain with the goal to replace some of the SDS molecules in the protein SDS complexes and to act as a fluorescent tag during SDS-CGE [298]. Good detection limit of 0.13 ng/mL was attained for bovine serum albumin (BSA) with the dynamic range of five orders of magnitude.

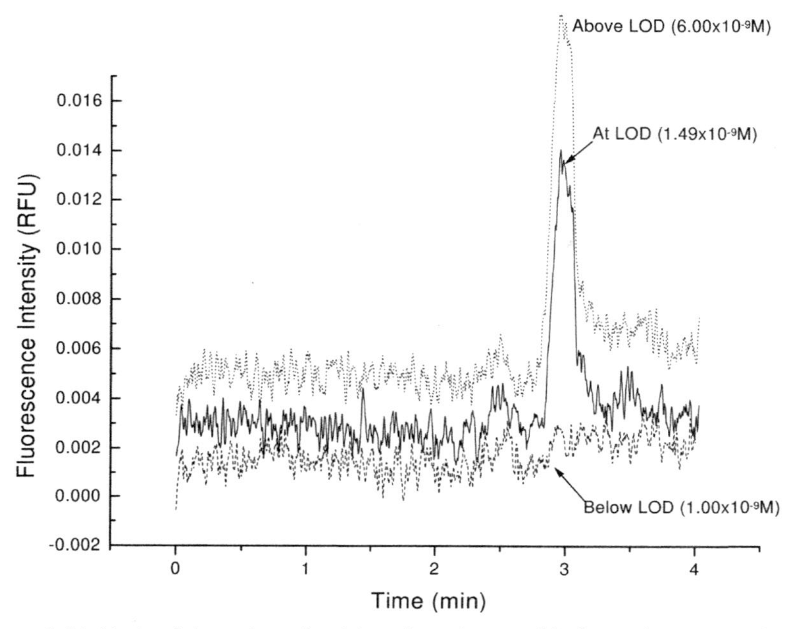

Figure 5.66 Limit of detection of a 1:1 molar mixture of iodocyanine green–human serum albumin by capillary gel electrophoresis with diode-laser-induced fluorescence detection. *With permission from E.D. Moody, P.J. Viskari, C.L. Colyer, Non-covalent labeling of human serum albumin with indocyanine green: a study by capillary electrophoresis with diode laser-induced fluorescence detection, Journal of Chromatography B Biomedical Sciences and Applications 729 (1–2) (1999) 55–64 [293].*

A SDS capillary electrophoresis technique of proteins in a sieving matrix was reported by Craig et al. [299] utilizing a two-spectral-channel detector that resolved fluorescence from the samples and standards covalently labeled by two different dyes as shown in Fig. 5.67. Others achieved picomolar detection limits with noncovalent fluorogenic labeling of proteins using Sypro Red dye [294].

Pyrilium type labeling dyes (e.g., Chromeo 503) were introduced as efficient tags for protein and peptide analysis in CE [300,301]. This dye represents a group of fluorogenic reagents, which produce highly fluorescent products for CE-LIF detection. One of the important features of these dyes is that they do not increase the charge state of the labeled protein or peptide, which is very important in the IEF analysis. They also have very low fluorescence when unconjugated, that is, requiring no purification before analysis. Szekrenyes et al. [302] reported an efficient high-throughput multicapillary SDS gel electrophoresis system with LIF

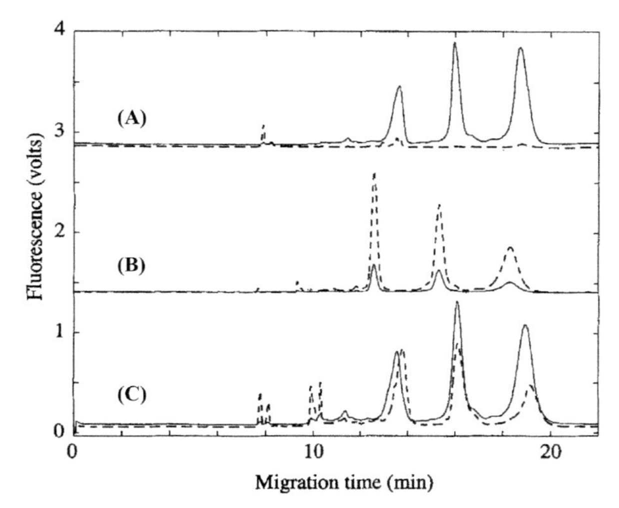

Figure 5.67 Electrophoretic separation of 3-(2-furoyl)quinoline-2-carboxaldehyde- and 6-(fluorescein-5-carboxamido)hexanoic acid succimidyl ester-labeled trypsinogen, ovalbumin and conalbumin. Excitation was at 488 nm and emission was measured at 630 (solid line) and 513 (dashed line) nm. (A) FQ-labeled proteins, (B) FX-labeled proteins, and (C) FQ-labeled proteins spiked with FX-labeled proteins. *With permission from D.B. Craig, R.M. Polakowski, E. Arriaga, J.C.Y. Wong, H. Ahmadzadeh, C. Stathakis, Sodium dodecyl sulfate-capillary electrophoresis of proteins in a sieving matrix utilizing two-spectral channel laser-induced fluorescence detection, Electrophoresis 19 (12) (1998) 2175–2178 [299].*

detection for the large-scale analysis of monoclonal antibodies. Covalent fluorophore labeling of the IgG samples with Chromeo 503 was applied. This method was capable for the rapid purity assessment and subunit characterization of IgG molecules, including the ngHC, truncated light chains (Pre-LC), and alternative splice variants (Fig. 5.68).

SDS-CGE was used with LIF detection for the purity analysis and characterization of AAV gene therapy vectors. Labeling of the viral proteins was performed with the fluorescence dye Chromeo P503 [303]. Others developed a CGE-LIF method for the determination of *S*-glutathionylated proteins (Pr-SSG) related to the pathogenesis of aging and age-related disorders, such as AD. The fluorescent labeling method was based on the specific reduction of protein-bound *S*-glutathionylation with glutaredoxin and labeling with thiol-reactive fluorescent Dylight 488 maleimide. The detection limit of the approach was in the attomolar range for Pr-SSG [304].

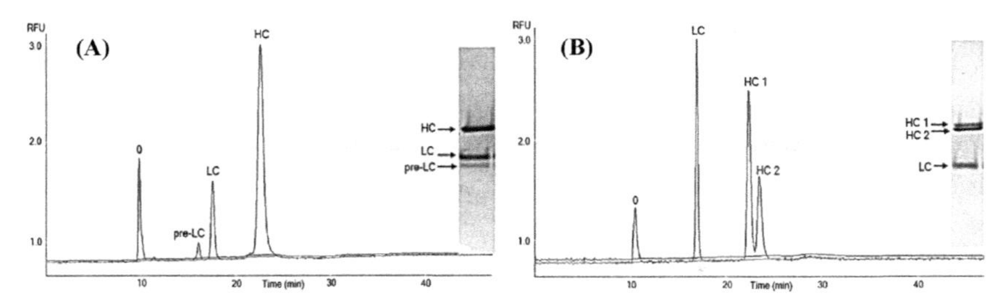

Figure 5.68 Sodium dodecyl sulfate capillary gel electrophoresis and sodium dodecyl sulfate polyacrylamide slab gel electrophoresis analysis of two model mAb samples #1 (A) and #2 (B) after fluorophore labeling. 0—remaining labeling dye, *LC*—light chain, and *HC*—(also HC1 and HC2) heavy chain. *With permission from A. Szekrenyes, U. Roth, M. Kerekgyarto, A. Szekely, I. Kurucz, K. Kowalewski, et al., High-throughput analysis of therapeutic and diagnostic monoclonal antibodies by multicapillary SDS gel electrophoresis in conjunction with covalent fluorescent labeling, Analytical and Bioanalytical Chemistry 404 (5) (2012) 1485–1494 [302].*

5.2.5 Applications in biotechnology

Most biotherapeutic products currently marketed for human or veterinary use are protein or peptide based. Therefore it is a strong demand for advanced analytical methods for their analysis in the rapidly growing biotechnology and biopharmaceutical industries [305,306]. Capillary electrophoresis-based methods are powerful and rapid techniques for protein analysis, due to high resolution, efficient separation, and high sensitivity especially when coupled to LIF or MS detection methods.

Recombinant biotechnology proteins were monitored at 214 nm in fused-silica capillary columns filled with SDS polyacrylamide gel already as early as the 1990s [219]. Fifty nanomoles of a protein was quantified and an impurity peak of MW ~1500 Da was detected besides the main protein of interest (MW 60 kDa). Characterization of monoclonal antibodies by SDS-CGE for the purity and impurity analysis was reported by Foley and coworkers for both under reduced and nonreduced conditions [307]. The method was linear, accurate, and precise in the range of 0.25–3.0 mg/mL protein concentration with the limit of quantitation (LOQ) of 0.02 mg/mL for reduced and nonreduced monoclonal antibodies. SDS-CGE method is commonly applied for mAb analysis, such as purity, quantitation of the ngHC peak, identity, and stability testing (against accelerated temperature tests, UV light exposure, and high pH condition) [232,308–311]. The high resolving power of the method readily supports the development of mAb products from early-stage clone selection to final product characterization.

Michels et al. [312] developed a sensitive capillary electrophoresis-SDS (CE-SDS) approach with LIF detection for impurity monitoring during the production of monoclonal antibodies. The therapeutic protein samples were labeled with 3-(2-furoyl)quinoline-2-carboxaldehyde. The validation process according to the guidelines of the International Committee of Harmonization indicated that the proposed method was accurate (RSD $\leq 0.8\%$) for the relative peak distribution and LOD and LOQ ($r^2 > 0.997$) in the range of 0.5–1.5 mg/mL and in the range of 10–35 ng/mL, respectively (Fig. 5.69). The sample preparation was automated with a robotic platform for SDS-CGE analysis of therapeutic proteins by Hutterer and coworkers [313]. This approach significantly

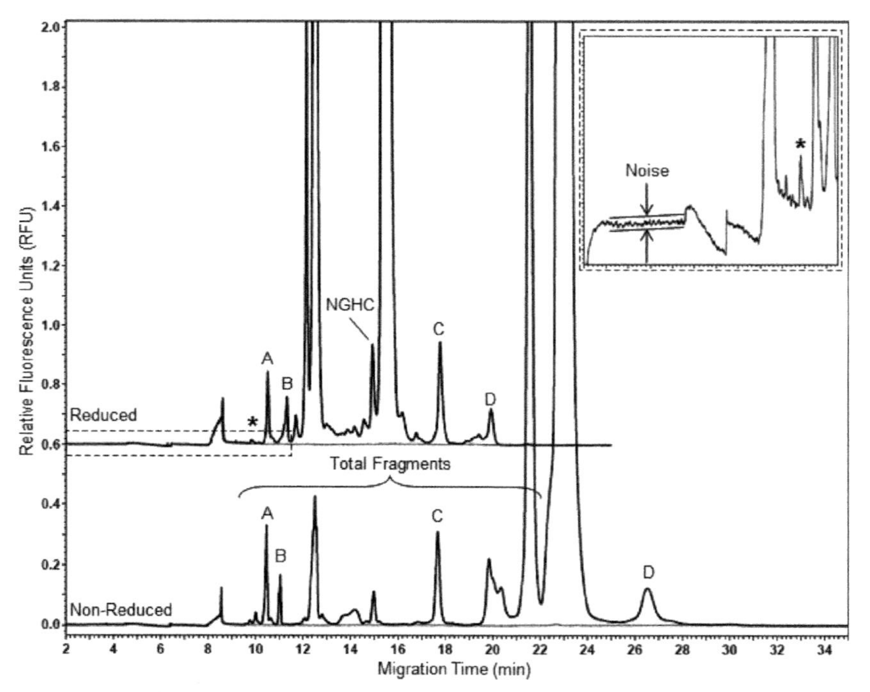

Figure 5.69 Enlarged view of the recombinant monoclonal antibody sample electropherogram artificially contaminated with 0.25% (w/w) impurities. Limit of detection (*LOD*) and limit of quantitation (*LOQ*) values were extrapolated from the signal-to-noise ratio of process impurity peaks A and B, and product impurity peaks C and D. The LOD and LOQ obtained for these impurities were 10–30 and 35–100 ng/mL, respectively. The inset shows the noise level. *With permission from D.A. Michels, M. Parker, O. Salas-Solano, Quantitative impurity analysis of monoclonal antibody size heterogeneity by CE-LIF: example of development and validation through a quality-by-design framework. Electrophoresis 33 (5) (2012) 815–826 [312].*

reduced the total analysis time. Sample recovery greater than 90% was observed, irrespectively of the composition and concentration of the sample.

Analysis and molecular mass determination of proteins of pharmaceutical interest including recombinant bovine somatotropin (molecular mass: 21,812 Da), sCD4-PE (molecular mass: 59,172), and polyclonal antithymocyte equine immune globulin (molecular mass: c.150,000) were performed by SDS-CGE [314]. The determined molecular masses of the sample proteins were comparable with the results of SDS-PAGE and with the theoretical values. The advantages of the SDS-CGE over SDS-PAGE were the rapid analysis times and improved precision of molecular mass determination.

β-Lg, α-lactalbumin (α-Lac), BSA, and IgG were separated from acid whey and whey protein concentrate by SDS-CGE [315]. Due to the complex nature of the samples, refinement of the CGE sample buffer was necessary, but the method could not separate the two β-Lg isoforms or the glycosylated form of α-Lac from the β-Lg. Others also applied SDS-CGE for detailed quantification of whey proteins in infant formulas (Fig. 5.70) [316]. SDS-CGE was also used for wheat varietal fingerprinting [317]. Addition of a small amount of organic solvent to the sieving matrix (5% MeOH + 5% glycerol) allowed to obtain excellent resolution of the high-molecular-mass glutenin subunits (80–130 kDa) that correlated with bread making quality.

SDS-CGE proved to be useful to separate mAb chimeric BR96-doxorubicin immunoconjugates [318]. Six peaks were identified corresponding

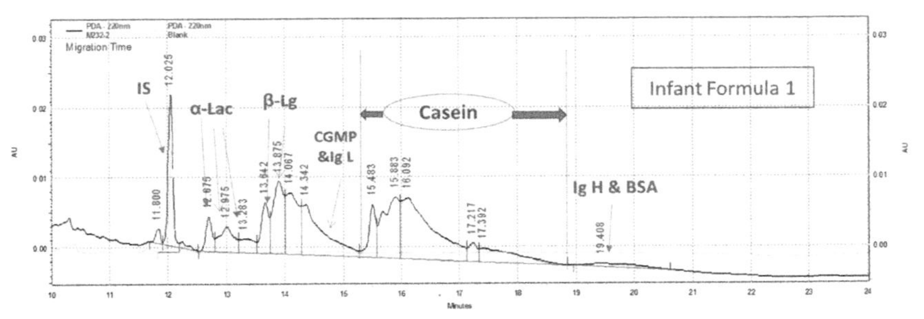

Figure 5.70 Electropherogram of infant formula with sweet whey ingredient (60% whey claim). *With permission from P. Feng, C. Fuerer, A. McMahon, Quantification of whey protein content in infant formulas by sodium dodecyl sulfate-capillary gel electrophoresis (SDS-CGE): single-laboratory validation, first action 2016.15, Journal of AOAC International 100 (2) (2017) 510–521 [316].*

to all possible conjugated species as Fig. 5.71 shows. Analysis of the trace revealed that the light, heavy, and light-heavy chain conjugates were the predominant species.

SDS-CGE was used to analyze high-density lipoproteins [255] separating the two important forms of apoA-I and apoA-II. Quantification by SDS-CGE showed good agreement with immunoassay results. Seven major erythrocyte membrane proteins were separated and identified by SDS-CGE using a commercially available replaceable gel matrix [319]. Hiraoka et al. applied SDS-CGE to evaluate the concentration ratios of albumin and α2-macroglobulin in cerebrospinal fluid and concurrent serum samples from patients with various neurological disorders [320].

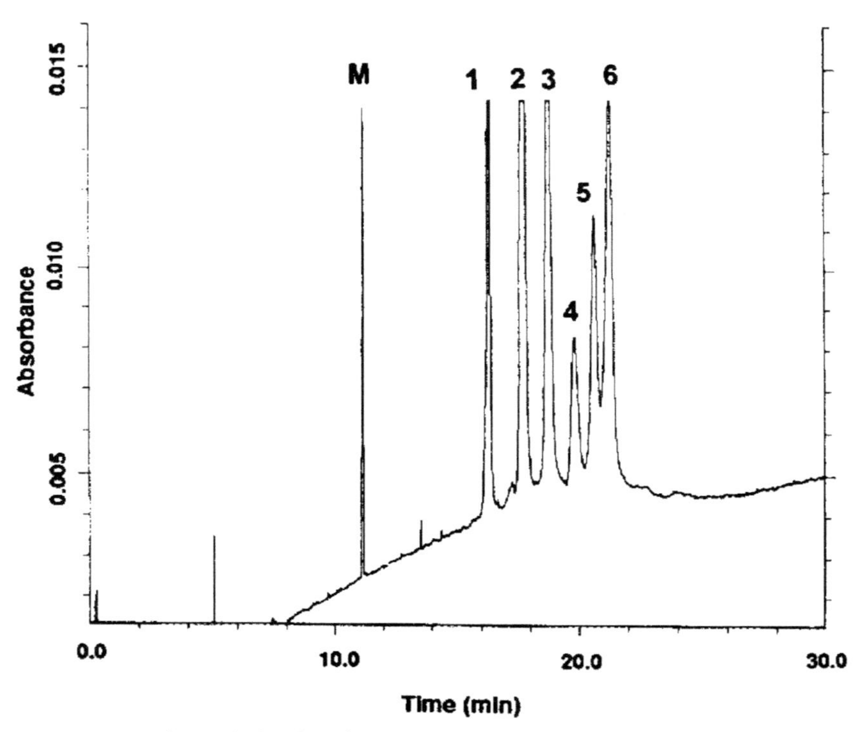

Figure 5.71 Sodium dodecyl sulfate capillary gel electrophoresis analysis of MAb BR96-DOX immunoconjugate. M: reference marker; peaks: 1 = light chain (*L*); 2 = heavy chain (*H*); 3 = 1H + 1L (HL); 4 = 2H (HH); 5 = 2H + 1L (HHL); 6 = 2H + 2L (HHLL). *With permission from J. Liu, S. Abid, M.S. Lee, Analysis of monoclonal antibody chimeric BR96-doxorubicin immunoconjugate by sodium dodecyl sulfate-capillary gel electrophoresis with ultraviolet and laser-induced fluorescence detection. Analytical Biochemistry 229 (2) (1995) 221–228 [318].*

Ten percentage of dextran solution (MW 2 M) was used to separate low-MW 2S albumin isoforms from lupins [321]. The authors found that addition of glycerol or ethylene glycol to the SDS separation buffer improved the resolution. SDS-CGE was used for the size separation of antibodies and their fragments suggesting its potential as a quality control tool [322]. Binding affinity between IgG1 and anti-IgG1 antibody can also be characterized by this technique as the affinity interaction between the two antibodies was high enough not to be affected by the denaturation process.

Diluted HPMC sieving matrix was used for tissue protein-based classification analysis of squamous-cell lung carcinomas and small-cell lung carcinomas [323]. HPMC concentration and zone length were optimized and the water-soluble proteins from human lung tissues were separated with greatly improved resolution and increased peak intensity.

A CGE method was developed for the identification and/or quantification of viral proteins (hemagglutinin fragment 1, hemagglutinin fragment 2, matrixprotein, and nucleoprotein) in influenza virus and virosome samples [324]. They optimized the separation conditions to considerably reduce analysis time as shown in Fig. 5.72. The CGE method showed better precision

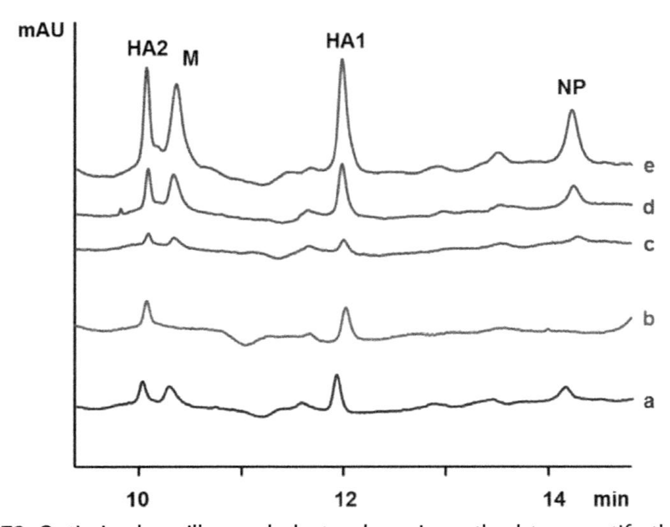

Figure 5.72 Optimized capillary gel electrophoresis method to quantify the hemagglutinin fragments (*HA*) in an inactivated influenza virus (a) and a virosome sample (b). B/Brisbane NIBSC was used as calibration standard: 7 mg/mL HA (c), 29 mg/mL HA (d), and 76 mg/mL HA (e). *With permission from E. van Tricht, L. Geurink, B. Pajic, J. Nijenhuis, H. Backus, M. Germano, et al., New capillary gel electrophoresis method for fast and accurate identification and quantification of multiple viral proteins in influenza vaccines. Talanta 144 (2015) 1030–1035 [324].*

and accuracy than RP-HPLC in the same analyte concentration range, additionally allowing the identification of different virus strains based on their specific protein profiles. The total analysis time was much shorter than in the case of radial immunodiffusion. Others developed an SDS-CGE method with head-column field-amplified sample stacking with UV detection for the analysis of AAV capsid proteins (Fig. 5.73) [230]. Capsid proteins' limit of detection of 0.2 ng/mL (3.3 pM) was achieved with convenient UV absorbance detection at 214 nm; therefore 4.3 femtomole virus particles were sufficient for the purity analysis. The former method provided three orders of magnitude enhancement in sensitivity compared to conventional SDS-CGE. Mellado et al. [325] applied SDS-CGE for the analysis of triple 2/6/7- and double 2/6-layered rotavirus-like particles (RLPs), because these species represent promising vaccine candidates against rotavirus infection. The electropherograms of RLPs contains peaks that could be assigned to the viral proteins (VP2, VP6, and VP7). The VP7 glycoprotein containing samples were also analyzed after deglycosylation, which process decreased its MW by 4–7 kDa. The presented approach allowed consistent and reproducible determination of VP6/VP7 ratio differences between the samples in addition to the quantification of proteins.

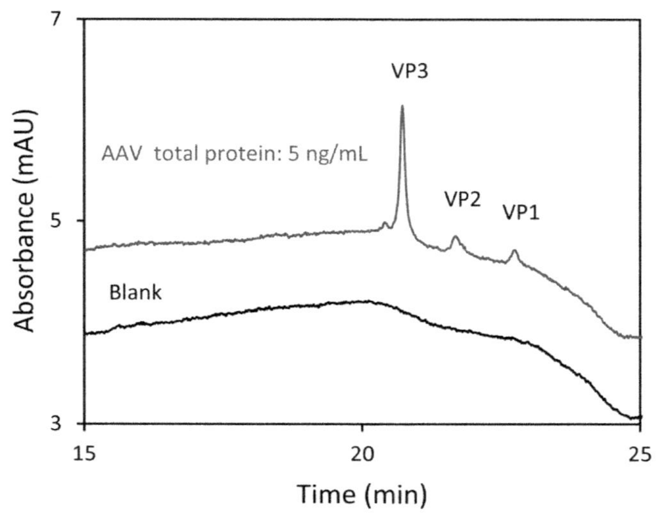

Figure 5.73 Electropherograms of a blank (PBS buffer) and an AAV8 capsid sample (5 ng/mL). *With permission from C.X. Zhang, M.M. Meagher, Sample stacking provides three orders of magnitude sensitivity enhancement in SDS capillary gel electrophoresis of adeno-associated virus capsid proteins. Analytical Chemistry 89 (6) (2017) 3285–3292 [230].*

A unique approach was developed utilizing a segmental-filling of an SDS plug and polyethylene oxide as anticonvective medium into the separation capillary to analyze microheterogeneities and protein isoforms. The separation was detected by laser-induced native fluorescence by a pulsed Nd:YAG laser at 266 nm [326]. The authors reported that the length and the concentration of the SDS plug played a significant role in determining resolution and sensitivity. Gomis et al. [327] reported on size-based separation of cider proteins by capillary SDS gel electrophoresis using LPA sieving matrix and UV detection. The cider proteins were baseline separated and their relative molecular masses were successfully determined. The SDS-CGE results showed good agreement with those obtained by SDS-PAGE. Gutierrez and coworkers used CGE to test purified cider proteins employing an LPA sieving matrix [328] after preparing the samples with ethanol precipitation, dialysis, ultrafiltration, and gel filtration. Among the tested sample preparation methods ultrafiltration was found to be the best for CGE analysis.

Olive leaves and pulps were classified according to their cultivar using protein profiles obtained by capillary SDS gel electrophoresis (CGE) [329]. Proteins were extracted using an enzyme-mediated method and different profiles were observed with 11 common peaks by SDS-CGE in all cultivars. The normalized protein peak areas from the CGE electropherograms were applied as predictors to create linear discriminant analysis models. Based on this approach, all olive leaf and pulp samples can be correctly classified for the 12 different Spanish regions, which demonstrated that protein profiles were characteristic for each cultivar. High sensitivity single cell protein analysis was attempted by CGE-LIF in the known phase of the cell cycle [330] revealing that protein content may vary from cell to cell and the major factor on protein expression is the cell cycle. Cellular proteins were mixed with denaturing SDS solution, on-column labeled and separated using 8% pullulan as sieving medium (Fig. 5.74).

Busnel et al. [331] reported on cIEF separations of hydrophobic proteins in anticonvective media versus glycerol solution-filled capillary and demonstrated the latter one as a good alternative medium. Another multifunctional separation medium is hydroxyethylcellulose-graft-PDMA copolymer that was used in protein and dsDNA separations [332]. This sieving material also provided excellent noncovalent coating, even resisting adsorption of basic proteins.

Capillary electrophoresis was introduced for microscale Western blot analysis by Anderson et al. [333] to separate and analyze SDS—protein

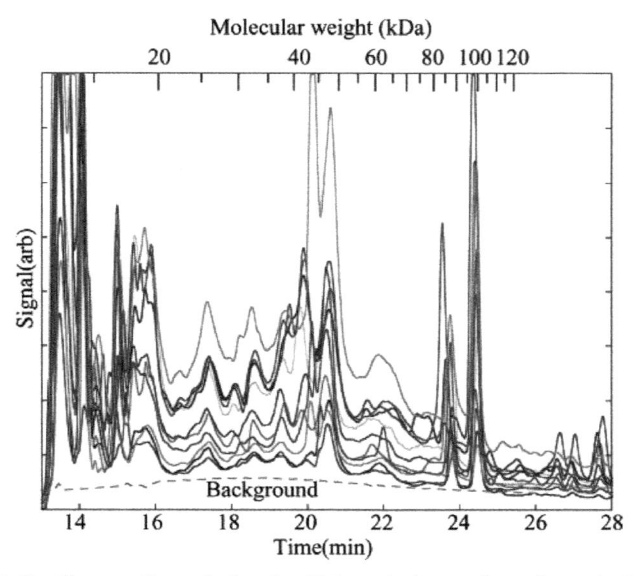

Figure 5.74 Capillary sodium dodecyl sulfate gel electrophoresis analysis of cellular proteins of 11 different HT29 human cancer cells. The background signal, generated by analysis of the cellular supernatant, is the dashed curve. *With permission from S. Hu, L. Zhang, S. Krylov, N.J. Dovichi, Cell cycle-dependent protein fingerprint from a single cancer cell: image cytometry coupled with single-cell capillary sieving electrophoresis, Analytical Chemistry 75 (14) (2003) 3495–3501 [330].*

complexes. The system utilized a gel-filled capillary that was interfaced with a blotting membrane that captured proteins during their electromigration, unnecessitating the need for electroblotting. The separation column was grounded via a sheath capillary, and a 2D translation stage moved the blotting membrane (Fig. 5.75). Adequate adsorption of the proteins to the blotting membrane was facilitated by a methanol–buffer mixture. The method substantially reduced analysis time to c.1 hour and the mass detection limit of the approach was 50 pg for lysozyme. Kodani et al. developed an anti-hepatitis C virus (HCV) assay that was performed on an automated capillary electrophoresis-based western blot platform using a fourth-generation HCV recombinant fusion protein [334]. The specificity and sensitivity of assay were 100% and 95% based on the testing of 70 human serum or plasma samples, respectively.

SDS-CGE was coupled to matrix-assisted laser desorption ionization time-of-flight mass spectrometry (MALDI-TOF-MS) using a poly-(tetrafluoroethylene) (PTFE) membrane by Lu and coworkers

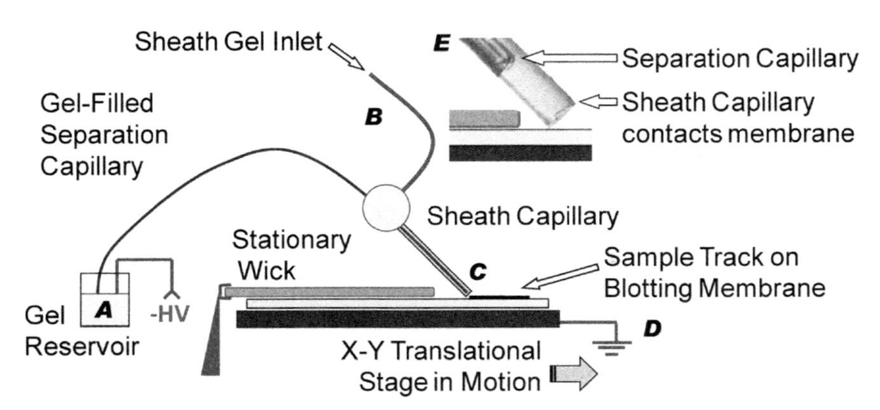

Figure 5.75 Schematics of the sodium dodecyl sulfate capillary gel electrophoresis western blot instrument. *With permission from G.J. Anderson, C.M. Cipolla, and R.T. Kennedy, Western blotting using capillary electrophoresis, Analytical Chemistry 83 (4) (2011) 1350–1355 [333].*

[335]. During the analysis, the SDS-CGE separated proteins were captured by the membrane and the SDS was easily removed prior to MALDI-TOF-MS. In brief, to prepare the proteins collected on the membrane for the MS analysis, the membrane collector part was carefully taken out from the system. The SDS was washed off by placing the PTFE membrane in 10% methanol for 60 minutes at 40°C. After SDS removal, the collected proteins were directly analyzed on the membrane by MALDI-TOF-MS. The authors also demonstrated the ability of immunoblotting and staining of the collected proteins. A capillary electrophoresis–MS method was developed for the analysis SDS-complexed mAb samples [336] based on the in-capillary removal of SDS from the sample by coinjecting cationic surfactants, such as CTAB or benzalkonium chloride in methanol as organic solvent. With this strategy, MS signal recoveries were greater than that of 94% and the method could be applied up to 10% (v/v) SDS concentration. Beckman et al. applied a longer alkyl chain detergent than SDS for CGE to analyze new modality therapeutic proteins [337]. The approach used a sodium hexadecyl sulfate containing gel matrix, which resulted in increased peak resolution via better peak shape (decreased fronting) and separation power compared to commercial SDS-CGE for the analyte molecules (Fig. 5.76). Their work highlighted the possibility of the use of detergents other than SDS to increase the resolution and separation power of CGE-based separation methods in the presence of surfactants.

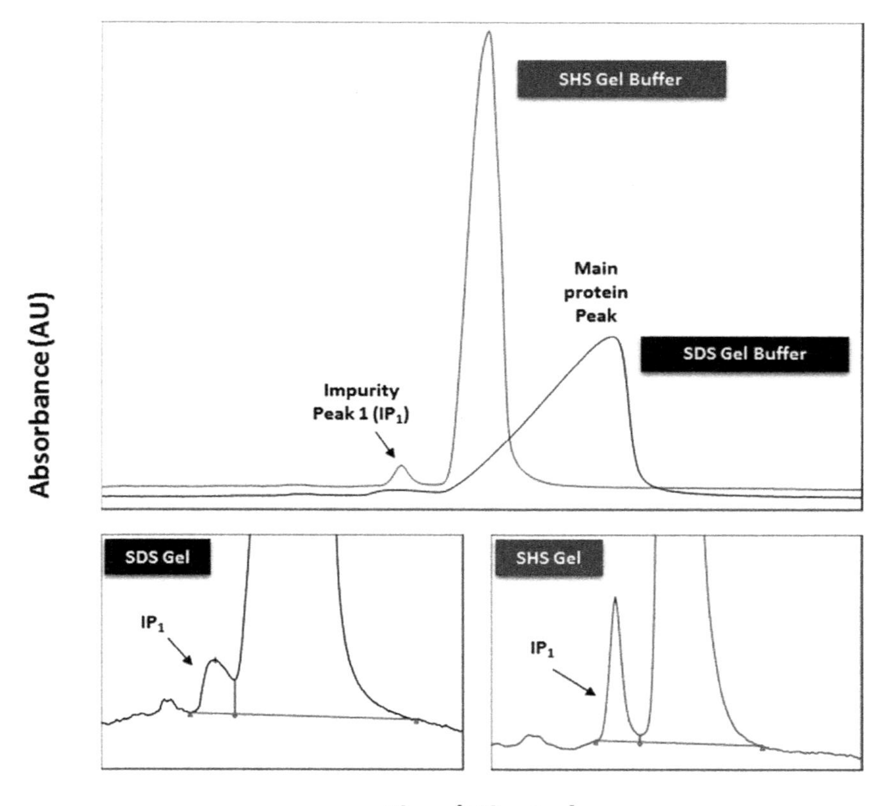

Figure 5.76 Capillary gel electrophoresis of recombinant therapeutic protein-1 in the presence of various chain length surfactants comparing results with or without 0.2% sodium hexadecyl sulfate added to the gel buffer solution. *With permission from J. Beckman, Y. Song, Y. Gu, S. Voronov, N. Chennamsetty, S. Krystek, et al., Purity determination by capillary electrophoresis sodium hexadecyl sulfate (CE-SHS): a novel application for therapeutic protein characterization, Analytical Chemistry 90 (4) (2018) 2542–2547 [337].*

5.3 Capillary gel electrophoresis of carbohydrates

The most important posttranslational modifications of proteins are phosphorylation, glycosylation, and lipidation, which all play important roles in the modulation of biological functions. Phosphorylation and GlcNAcylation play key roles in most signal transduction pathways. *N*- and *O*-glycosylation are the major cause of protein microheterogeneity and involved in cell recognition and immunoresponse among others. The oligosaccharide structures are attached to an asparagine residue containing the consensus sequence of Asp-X-Ser/Ther (X cannot be Pro) in case of *N*-glycosylation, and mostly to serine or threonine residues in case of *O*-glycosylation.

5.3.1 Analytical glycobiology

Glycans are involved in practically every area of biology ranging from embryology and developmental biology, to evolutionary development and physiology [338], which necessitates a set of powerful analytical methodologies to reveal functionally important structural features [339–341]. Analytical glycomics represents a tool for identification of specific carbohydrate patterns to discover novel glycan biomarkers, which represents one of the major interests in the biomedical fields. Glycosylation changes on the other hand may influence the clinical efficiency of therapeutic glycoproteins, for example, monoclonal antibodies, antibody drug conjugates or fusion proteins; therefore there is a great recent interest for their comprehensive characterization in the biopharmaceutical industry as well.

5.3.2 Release of *N*- and *O*-linked oligosaccharides from glycoproteins

N- and *O*-linked oligosaccharide release methods are based on either enzymatic or chemical procedures. The applied method depends on factors, such as the type of glycosylation present and the nature and amount of the sample [342]. Chemical release of *N*- and *O*-glycans from glycoproteins can be accomplished by hydrazinolysis and alkaline β-elimination especially in the case of *O*-glycans [343]. The enzymatic methods are more convenient than chemical methods and able to release glycans from glycoproteins under mild conditions, particularly in the case of *N*-glycans, where several endoglycosidases and glycoamidases are available for specific cleavage [344]. Unfortunately, for the time being, the number of endoglycanases to release *O*-linked glycans is very limited [345]. Therefore chemical methods are preferred for the quantitative and reproducible analysis of *O*-linked oligosaccharides that sometime result in peeling, making their structural determination ambiguous [346,347].

5.3.3 Fluorophore labeling of carbohydrates

Carbohydrates do not possess any chromophore of fluorophore groups and mostly not charged and therefore need derivatization with a charged fluorophore or a chromophore tag to ensure their electromigration and detectability by UV or fluorescent methods [348]. Reductive amination of the aldehyde group at the reducing end of oligosaccharides is one of the most frequently used methods for carbohydrate tagging. Common labeling reagents for capillary electrophoresis are 8-aminopyrene-1,3,6-trisulfonic acid (APTS), 8-aminonaphtalene-1,3,6-trisulfonic acid (ANTS), 2-aminoacridone [349] and CBQCA [350].

Evangelista et al. labeled mono- and oligosaccharides with APTS by reductive animation. The derivatized sugars were separated by capillary electrophoresis (CE) with LIF detection using a 488 nm Ar-ion laser [351]. Chiesa and Horváth [352] investigated the separation and electrophoretic migration behavior of maltooligosaccharides, derivatized via their reducing end with ANTS. Bonn and coworkers compared capillary electrophoresis and micellar electrokinetic chromatography of 4-aminobenzonitrile carbohydrate derivatives [353]. Honda et al. [354] tagged various aldoses by 3-methyl-l-phenyl-2-pyrazolin-5-one and analyzed them as borate complexes using UV detection. The same approach also resulted in the good separation of homologous oligoglycans with different interglycosidic linkages that is otherwise a highly challenging task. The use of CBQCA as a precolumn labeling agent for amino sugar analysis in various biological mixtures enabled low-attomole detection limits [355].

5.3.4 Capillary gel electrophoresis separation of oligosaccharides

CGE with laser-induced fluorescent detection was reported as a powerful alternative of PAGE for profiling complex carbohydrates [356]. Fig. 5.77 compares the logarithmic relative mobility as a function of the logarithmic molecular mass of oligosaccharides in cross-linked polyacrylamide slab gel and CGE separations. While the slope of the CGE plot was almost linear (upper trace) in case of PAGE, it exponentially decreased (lower trace) suggesting different migration mechanism and possible interaction with the separation matrix.

Complex carbohydrates, released from glycoproteins, were readily profiled by CGE with LIF-based detection utilizing the APTS label on the sugar molecules [253]. Fig. 5.78 compares the APTS-labeled maltooligosaccharide

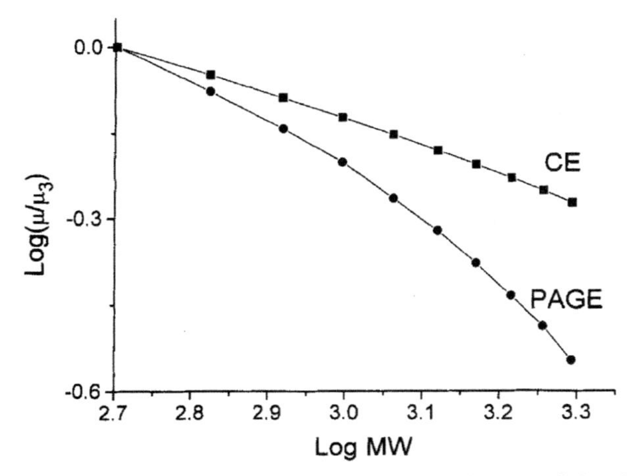

Figure 5.77 Double logarithmic plots of the relative mobilities of the different polymerization degree maltooligomers as a function of their molecular weights in capillary gel electrophoresis (■) and polyacrylamide slab gel electrophoresis (•). *With permission from A. Guttman, C. Starr, Capillary and slab gel electrophoresis profiling of oligosaccharides. Electrophoresis 16 (6) (1995) 993–997 [356].*

ladder separation (upper trace) with the released *N*-glycans from bovine fetuin (middle trace) and bovine ribonuclease B (lower trace). As one can observe, baseline separation of the various sialylated fetuin glycans and the positional isomers of the high-mannose structures (Man-7 and -8) were attained.

Novotny and coworkers introduced a sensitive, laser-assisted detection-based CE method to analyze complex oligosaccharide mixtures and developed a powerful separation method [357]. High concentration cross-linked polyacrylamide gel was used (18%T/3%C) for the analysis of autoclave hydrolyzed polygalacturonic acid. The same group also applied entangled polymer solutions to separate polysaccharides [358]. For neutral polysaccharides, complexation with borate provided the required charge, while a fluorescent tag was also needed for detection with good sensitivity. Variable electric fields, pulsed along the separation capillary at a 180-degree angle appeared to have a profound influence on molecular shape rearrangements of polysaccharides with respect to the separation. The authors demonstrated highly efficient separations of polydextrans in the range of 8,000–2,000,000 Da.

The effects of operational variables during CGE separation of ANTS-labeled oligosaccharides were investigated on the migration

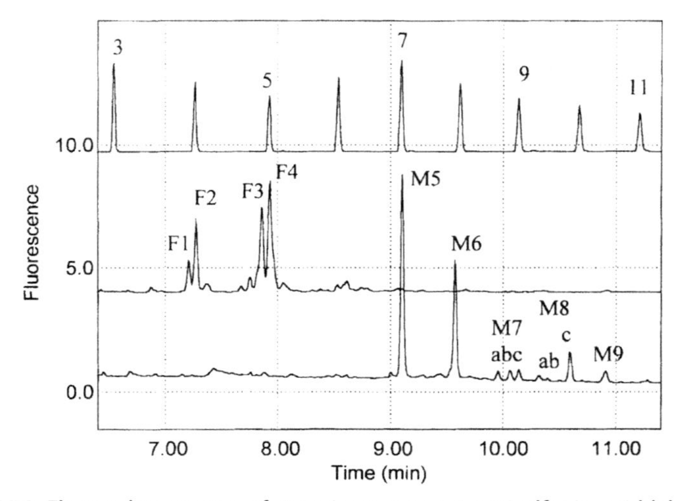

Figure 5.78 Electropherograms of 8-aminopyrene-1,3,6-trisulfonic acid-labeled glycans released from bovine fetuin (middle trace) and bovine ribonuclease B (lower trace) compared to the maltooligosaccharide ladder standard (upper trace). Numbers in the upper trace correspond to the degree of polymerization of the glucose oligomers. Peaks: F1 = tetrasialo-triantennary-2xα2,6; F2 = tetrasialo-triantennary-2xα2,3; F3 = trisialo-triantennary-2xα2,6; F4 = trisialo-triantennary-2xα2,3; M5–M9: Mannose 5–Mannose 9. Conditions: capillary: $\ell = 40$ cm (effective) neutrally coated capillary with 50 μm i.d.; buffer: 25 mM acetate, pH 4.75; detection: laser-induced fluorescence: excitation: 488 nm, emission: 520 nm; applied field strength: 500 V/cm; temperature: 20°C. *With permission from Guttman, A.,* High-resolution carbohydrate profiling by capillary gel electrophoresis. *Nature (London) 380 (6573) (1996) 461–462 [253].*

properties in respect to field strength, separation temperature, solute molecular mass and polymer matrix concentration [359]. Albeit, the concentration of the PEO sieving matrix was above the entanglement threshold, the parallel lines in the Ferguson plot suggested that the separation of ANTS-labeled oligosaccharides were not subject to molecular sieving. The same group investigated the migration shifts for linear (maltooligosaccharides) and branched (sialylated, neutral and core fucosylated biantennary IgG glycans) carbohydrates in the temperature range of 20°C – 50°C with or without polymer additives in the background electrolytes [360]. They found that with increasing temperature the glucose unit (GU) values of all IgG glycans shifted in all types of background electrolytes. The associated activation energy changes could explain this phenomenon. Szigeti et al. introduced temperature gradient CGE for enhanced separation of branched glycans released from therapeutic proteins [361]. A mixture of afucosylated, fucosylated, and high-mannose-type *N*-linked oligosaccharides showed resolution maximums at different temperatures; therefore a temperature gradient was applied to fine tune the separation efficiency. All glycans in the sample mixture were baseline separated with the optimized temperature gradient (Fig. 5.79).

CGE was used for the analysis of APTS-labeled *N*-linked oligosaccharides, enzymatically released from bovine pancreatic RNase B [362]. The released and labeled high-mannose structures were identified by spiking with corresponding commercially available individual oligosaccharide standards. Baseline separation of the three positional isomers of the mannose-7 and mannose-8 oligosaccharides was attained as shown in Fig. 5.80. Comparison of the electrophoretic mobilities of the high-mannose-type branched carbohydrates to the linear glucose oligomers were shown using different gel concentrations in the running buffer. Interestingly, increasing gel concentration resulted in an increase in the relative mobility values of the high-mannose-type carbohydrate molecules compared to the linear glucose

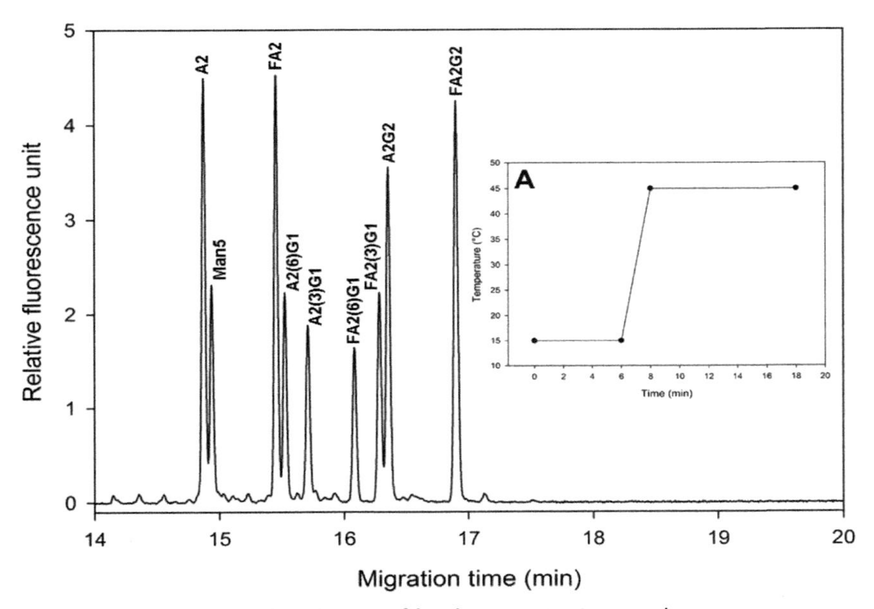

Figure 5.79 Separation of *N*-glycans of biotherapeutics interest by temperature gradient capillary gel electrophoresis. Inset A: temperature gradient profile. *With permission from M. Szigeti, A. Guttman, High-resolution glycan analysis by temperature gradient capillary electrophoresis, Analytical Chemistry 89 (4) (2017) 2201–2204 [361].*

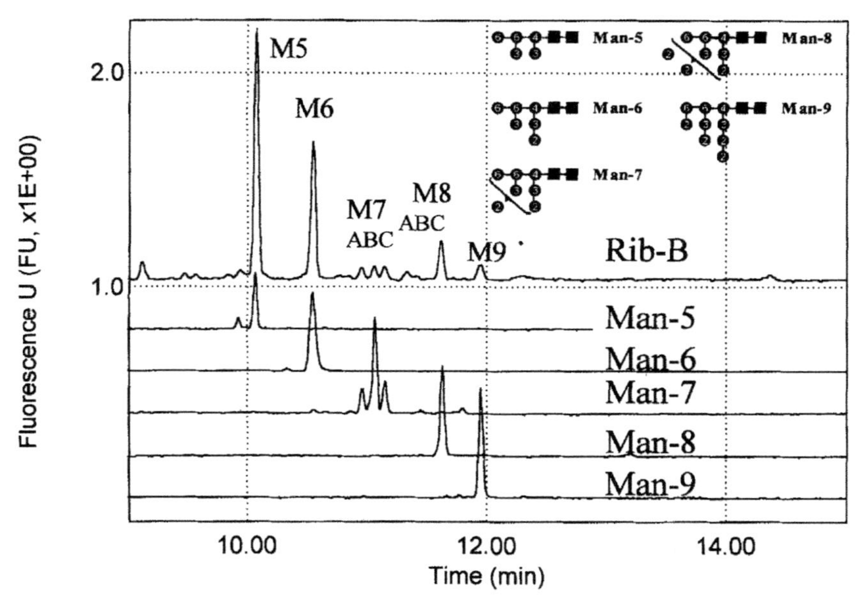

Figure 5.80 Capillary gel electrophoresis separation of 8-aminopyrene-1,3,6-trisulfonic acid-labeled high-mannose-type oligosaccharides released from bovine ribonuclease B (upper trace) and the individual standard structures (lower traces). Inset: Structural representation of the high-mannose-type *N*-linked oligosaccharides: ■: GlcNAc β1–4; •4: Man β1–4; •6: Man α 1–6; •3: Man α 1–3; •2: Man α 1–2. *With permission from A. Guttman, T. Pritchett, Capillary gel electrophoresis separation of high-mannose type oligosaccharides derivatized by 1-aminopyrene-3,6,8-trisulfonic acid, Electrophoresis 16 (10) (1995) 1906–1911 [362].*

oligomers. This increase in relative migration time was not due to sieving but seemed to be related to the hydrodynamic volume of the branched high-mannose glycans as well as to the viscosity of the separation medium.

As a continuation of this work, coinjection of fluorescently labeled RNase B glycan pool and maltooligosaccharide ladder standard provided the opportunity of high precision determination of the corresponding GU values of the individual high-mannose structures with various linkage and positional differences as well as the contribution of individual mannose residues to the resulting unique migration characteristics [363]. Table 5.5 shows the structures of all the high-mannose oligosaccharides found in the ribonuclease B glycan pool along with their molecular weights (M_r), mass-to-charge ratios (M/C), migration times (MT), as well as their measured ($GU_{measured}$) and calculated glucose unit ($GU_{calculated}$) values.

Table 5.5 The structure of the Man-5 glycan of ribonuclease B, exhibiting the possible linkage sites of additional Man(α1,2) residues forming the Man-6, Man-7, Man-8, and Man-9 structures.

```
D1—Man (α1,6)
            \
             Man (α1,6)
            /          \
D2—Man (α1,3)           Man (β1,4)-GlcNAc(β1,4)-GlcNAc-APTS
                       /
     D3—D—Man (α1,3)
```

Name	Peak	D	D1	D2	D3	M_r	M/C	MT	GU_{meas}	GU_{calc}
Man-5	1	−	−	−	−	1692	564	9.65	6.35	6.35
Man-6	2	+	−	−	−	1854	618	10.12	7.17	7.17
Man-7/D2	3	+	−	+	−	2016	672	10.51	7.95	7.95
Man-7/D3	4	+	−	−	+	2016	672	10.61	8.15	8.15
Man-7/D1	5	+	+	−	−	2016	672	10.70	8.32	8.32
Man-8/D1, 2	6	+	+	+	−	2178	726	10.88	8.67	9.10
Man-8/D2, 3	7	+	−	+	+	2178	726	10.95	8.81	8.93
Man-8/D1, 3	8	+	+	−	+	2178	726	11.15	9.19	9.30
Man-9	9	+	+	+	+	2340	780	11.47	9.78	10.08

Notes: D, D1, D2, and D3 = Man (α1, 2) and $GU_D = 0.82$; $GU_{D1} = 1.15$; $GU_{D2} = 0.78$; $GU_{D3} = 0.98$. (+) represents the presence and (−) represents the absence of the Man (α1, 2) residue at the actual position. *MT*, migration time (minute); M_r, molecular weight; *M/C*, mass-to-charge ratio; GU_{meas}, measured glucose unit value; GU_{calc}, calculated glucose unit value.
Source: With permission from A. Guttman, S. Herrick, Effect of the quantity and linkage position of mannose(a1,2) residues in capillary gel electrophoresis of high-mannose-type oligosaccharides, Analytical Biochemistry 235 (2) (1996) 236–239 [363].

High-molecular-weight poly(neuraminic acid) was separated in a neutral sieving polymer containing hundreds of monosaccharide residues [364]. *N*-acetylneuraminic acid and hyaluronic acid were analyzed by CGE by the same group [364,365], using polyethylene glycol and pullulan matrices. An interesting observation was that small oligomers of both polysaccharides migrated in reverse order of their molecular masses, that is, larger than pentamer for *N*-acetylneuraminic acid polymers and decamer for hyaluronic acid. The authors considered that these unusual migration patterns were caused by three-dimensional structures and associated volume differences of the smaller oligomers.

A highly viscous polyacrylamide matrix was used by Novotny and coworkers for high-resolution separation of hyaluronate [366], a highly polydisperse negatively charged polysaccharide. CGE assay was used to characterize the *N*-glycosylation patterns of viral glycoproteins of the influenza A virus by Schwarzer et al. [367] to measure batch-to-batch reproducibility of vaccine production. To ensure consistent quality of the products, glycosylation profiles should be tightly controlled, as glycosylation affects bioactivity and antigenicity. The method included purification of the virus from the cell culture, glycoprotein isolation, glycan release, clean up and profiling by CGE-LIF. A capillary-based DNA sequencing instrument was used for the analysis.

Guttman et al. investigated the glycosylation of the HIV envelop glycoprotein (Env) subunit gp120—an important highly glycosylated protein in vaccine development—with CGE-LIF and MALDI-MS [368]. Information about the gp120 can help to understand how to induce immune response against the HIV virus. Fig. 5.81 shows a comparative *N*-glycosylation profiling study of gp120 samples by CE-LIF from a primary isolate (CM244) and the reengineered construct of this same isolate (A244), which was used as protein boost for an RV144 vaccine trial along with mother and infant samples. The glycosylation study of the gp120 constructs showed that the addition of specific tags, such as the herpes simplex virus glycoprotein D tag, significantly influenced the overall glycosylation patterns. Thus considering the immunogenicity of various Env immunogens may change and should be considered during vaccine development.

Szigeti et al. developed a rapid *N*-glycan release method using immobilized PNGaseF microcolumns for effective removal of *N*-glycans within 10 minutes from all major types of: (1) neutral (IgG), (2) highly sialylated (fetuin), and (3) high-mannose (ribonuclease B) carbohydrate containing glycoprotein standards. The released and APTS-labeled glycans were separated by CGE-LIF utilizing 1% PEO (MW: 900 kDa) containing separation matrix [369]. The same group also developed a fully automated glycan analysis platform with excellent yield and good repeatability, where the entire magnetic bead–based sample preparation process needed less than 4 hours in a 96-well plate format. CGE separation was also optimized for rapid analysis (<3 minutes) to accommodate high-throughput processing, required by the biopharmaceutical industry [370].

A high-throughput automated glycosylation profiling platform was developed using a multi-CGE-based DNA analyzer platform with LIF detection for monitoring of Fc glycosylation of therapeutic antibodies [371]. The enzymatically released glycans were APTS labeled in a 96-well plate format and 48 samples were simultaneously separated by CGE. The developed method showed excellent accuracy and repeatability by comparing the glycan analysis results with two UPLC-based methods. Ruhaak et al. characterized the

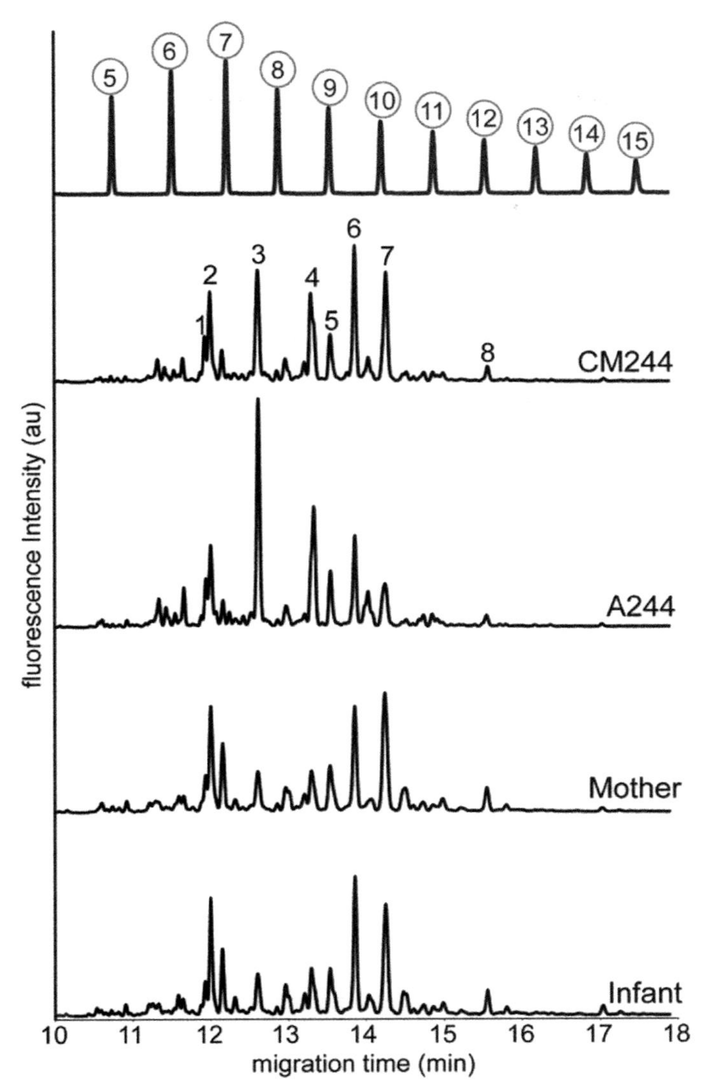

Figure 5.81 Comparative *N*-glycosylation profiling of the HIV envelop glycoprotein (Env) subunit gp120 by capillary electrophoresis-laser-induced fluorescence. The upper trace shows the maltooligosaccharide ladder. Electropherograms are aligned and shown for CM244, A244, mother, and infant. *With permission from M. Guttman, C. Varadi, K.K. Lee, A. Guttman, Comparative glycoprofiling of HIV gp120 immunogens by capillary electrophoresis and MALDI mass spectrometry, Electrophoresis 36 (11—12) (2015) 1305—1313 [368].*

N-glycosylation profiles of alpha1-antitrypsin (AAT)- and immunoglobulin A (IgA)-enriched fractions from human plasma by high-throughput multi-CGE with LIF detection [372]. They found that the *N*-glycosylation profile of the AAT-enriched fractions differed with age and gender; furthermore, some of these glycans were associated with cardiovascular and metabolic diseases; thus the glycans of AAT and IgA could be good biomarker candidates. The glycosylation pattern of IgA-enriched fractions showed differences between males and females.

Váradi et al. studied the changes in serum IgG glycoforms in patients with Crohn's disease (CD) and rheumatoid arthritis before and 2 weeks after the antitumor necrosis factor alpha (anti-TNF-α) treatment. Glycans were released from the isolated serum IgG, APTS labeled and analyzed with CGE-LIF. Significant alterations were detected between responders and nonresponders in both disease groups [373]. Fig. 5.82 shows the disease-specific changes detected in CD patients for response to anti-TNF-α therapy. Meszaros et al. studied the *N*-glycosylation profile of pooled human serum samples (90 patients each) from lung cancer (LC), chronic obstructive pulmonary disease (COPD) and comorbidity (LC with COPD) patients and compared to healthy individuals (control) by CGE with high sensitivity LIF detection. Sixty-one *N*-glycan structures were identified in the pooled control human serum sample and compared to pooled lung cancer, COPD, and comorbidity of COPD with lung cancer patient samples, but only quantitative differences of the identified glycans were observed in the patient groups [374].

5.3.5 Selected other applications

CGE was applied to detect molar ratios and degree of polymerization of oligosaccharides in food and beverage products and to reveal changes in the extent of the oligosaccharide distribution [375]. Szilagyi et al. [376] analyzed the oligosaccharide composition of different wort samples during the brewing process with various yeast types by CGE-LIF. They found that the fermentation with different yeast types produced unique oligosaccharide signatures; thus the oligosaccharide fermentation process can be precisely monitored by capillary electrophoresis. Albrecht et al. [377] applied CGE with LIF detection for the characterization of konjac glucomannan (KGM) oligosaccharide mixtures and also studied their structural changes during in vitro fermentation. The KGM oligosaccharide was digested by applying *endo*-β-(1,4)-mannosidase and *endo*-β-(1,4)-glucanase enzymes. They found that the *endo*-β-(1,4)-glucanase digest exhibited a

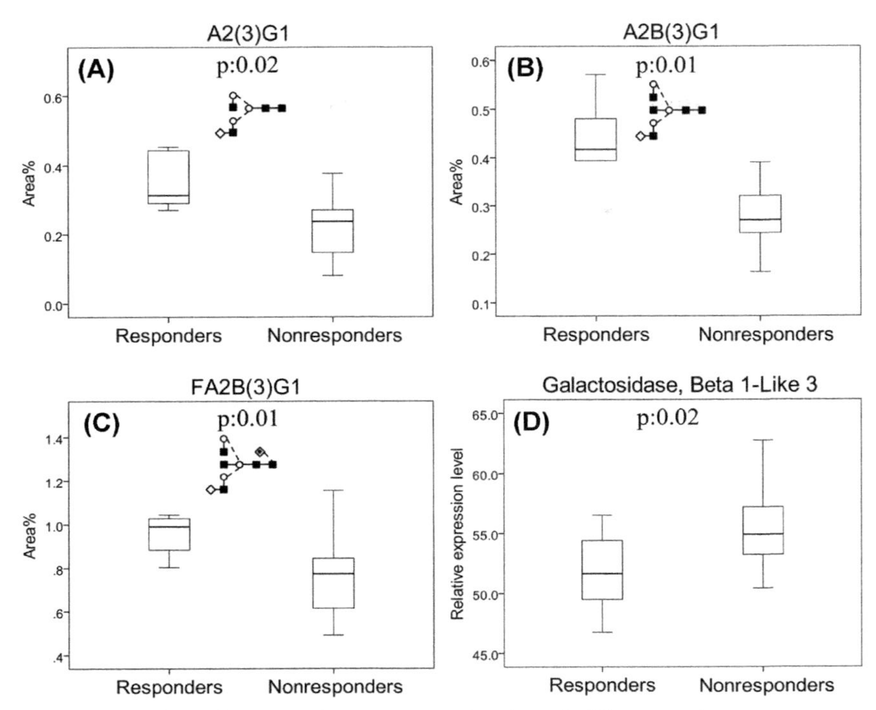

Figure 5.82 Glycosylation and gene expression-based differences between responders and nonresponders in rheumatoid arthritis for anti-TNF-α treatment. (A) A2(3)G1 (biantennary monogalactosylated), (B) A2B(3)G1 (biantennary monogalactosylated with bisecting GlcNAc), (C) FA2B(3)G1 (biantennary monogalactosylated with core-fucose and bisecting GlcNAc) and (D) Galactosidase, Beta 1-Like 3. *With permission from C. Varadi, Z. Hollo, S. Poliska, L. Nagy, Z. Szekanecz, A. Vancsa, et al., Combination of IgG N-glycomics and corresponding transcriptomics data to identify anti-TNF-alpha treatment responders in inflammatory diseases, Electrophoresis 36 (11–12) (2015) 1330–1335 [373].*

large degradability of the DP2, DP3, DP5, and DP6 components during the fermentation in human gut, whereas the *endo*-β-(1,4)-mannanase digest was degraded only slightly. Zandberg and coworkers used CGE to study bovine milk oligosaccharides (BMOs) extracted from the milk of cows fed with different diets [378]. During the lactation period of the cows, the authors identified 10 unusually anionic BMOs, which were probably phosphorylated and sulfated species.

An ultrafast high-resolution multicapillary gel electrophoresis method was developed for the separation of human milk oligosaccharides (HMO) [379]. The authors tested two different gel compositions: (1) a PEO based acetate buffered separation matrix and (2) a borate cross-linked dextran-based gel. The analysis time for a 96-well sample plate was less than 80 minutes with the non-borate buffered separation matrix and less than 1 hour with the dextran-borate gel. The same group applied CGE for monitoring bacteria-mediated production of 3-fucosyllactose, lacto-*N*-tetraose, and lacto-*N*-neotetraose as shown in Fig. 5.83 [380]. Here again, the application of borate-based gels was required to baseline separate the 2′- and 3-fucosylated lactose positional isomers. In all of the cases, the analysis of HMOs was performed within a couple of minutes with high resolution and excellent reproducibility.

Capillary array electrophoresis with anticonvective medium was implemented in a modified DNA sequencing device for large-scale analysis of APTS-labeled mono- and oligosaccharides, produced by enzymatic digestion of cellohexaose (model substrate) and lignocellulosic biomass [381–383]. Callewaert et al. [384] used APTS labeling of oligosaccharides isolated from human serum. APTS in conjunction with capillary electrophoresis-based analysis was used to characterize lipophosphoglycan polymorphisms in *Leishmania tropica* [385]. *N*-glycosylation pattern of erythropoietin produced by CHO cells in batch processes was monitored by capillary electrophoresis after APTS labeling [386].

5.4 Capillary affinity gel electrophoresis

Traditional affinity slab gel electrophoresis is utilizing the interaction between the ligands within a gel matrix and, in most instances,

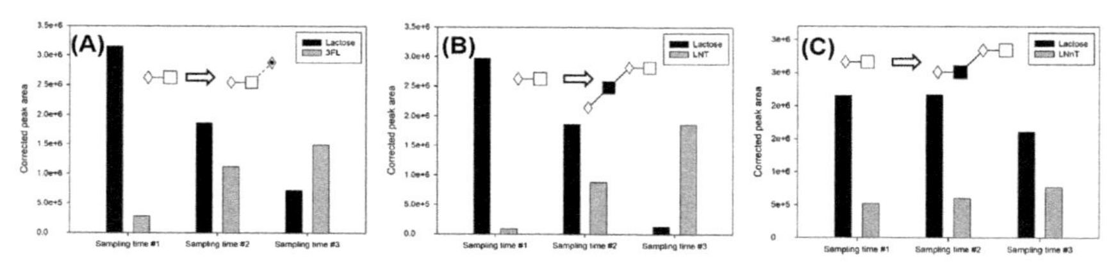

Figure 5.83 In-process analysis of bacteria-mediated production of 3-fucosyllactose, lacto-*N*-tetraose, and lacto-*N*-neotetraose (A–C, respectively) tracked by capillary gel electrophoresis. *With permission from M. Szigeti, A. Meszaros-Matwiejuk, D. Molnar-Gabor, A. Guttman, Rapid capillary gel electrophoresis analysis of human milk oligosaccharides for food additive manufacturing in-process control, Analytical and Bioanalytical Chemistry 413 (6) (2021) 1595–1603 [380].*

biomolecules of interest during the electrophoresis process. The interaction results in a ligand-mediated mobility shift. The method in polyacrylamide or agarose slab gel format was extensively used to characterize ligand binding of such biologically important polymers as DNA, proteins, or complex carbohydrates [387]. Capillary affinity gel electrophoresis is also based on electromigration differences between the free solute and the solute–affinity ligand complex. The migration shift is usually proportional to the binding affinity that allows kinetic measurement [388]. It is important to note that the high electric field that is applied in CGE separations should not alter the stringency of the complexation. Utilization of affinity interactions in CGE offers a powerful complexation-based tool to increase selectivity and also to evaluate the thermodynamic and kinetic aspects of binding. Automation of capillary affinity gel electrophoresis and calculation of the kinetic constants made the approach quite useful in biochemical and biotechnology applications [389]. Affinity interactions can also be useful in the development of specific assays for selected compounds of interest.

Affinity-type ligand molecules may be either soluble or immobilized in the gel buffer system [390]. With the use of soluble ligands (Fig. 5.84A), analyte complexes having a broad range of physical or chemical properties can be formed. If the ligand is very small compared to the analyte molecules, the mobility change of the complex will be relatively small or negligible, provided that the ligand has zero or little charge. A substantial effect on the electrophoretic mobility is expected if the ligand and the complex are comparable in size, or if the ligand is small but highly charged. In this latter case, the interaction may result in a strong decrease or increase in the migration time of the complex depending on the charge state of the ligand. An example of such selectivity manipulation by soluble affinity ligand interactions was evaluated by the addition of intercalating agents to the separation gel buffer system in dsDNA analysis [390,391]. In this instance, the ligand, EB, intercalated between the two strands of the DNA double helix during the separation process. As EB is positively charged at the separation pH, the complexation reduced the migration times of all the DNA fragments and therefore increased the separation time window, also resulting in enhanced resolution.

Alternatively, the ligand can be immobilized by physical entrapping (Fig. 5.84B) or by chemical binding (Fig. 5.84C) to the sieving polymer. The entrapping method was demonstrated for the separation of chiral compounds by incorporating cyclodextrins into a small-pore

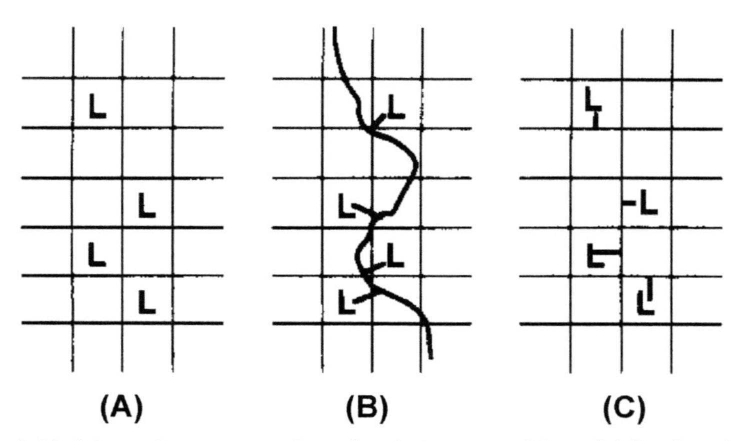

Figure 5.84 Schematic representation of techniques used for solubilized and immobilized affinity ligands in polyacrylamide gels. (A) Solubilized ligand method: the ligand can move freely in the gel buffer system. (B) Macroligand method: the solution of acrylamide contains the macroligand that becomes entrapped within the gel matrix after polymerization. (C) Chemically bound ligand: direct copolymerization of the gel matrix with the copolymerizable derivative of the ligand. *With permission from A. Guttman, N. Cooke, Capillary gel affinity electrophoresis of DNA fragments, Analytical Chemistry 63 (18) (1991) 2038–2042 [390].*

polyacrylamide gel-filled capillary column [392,393]. A comprehensive study aimed to understand the effect of ion-pairing and intercalating reagents, which interacted with the sieving polymer and modified its physical properties [394], revealing that the addition of different ion-pairing reagents resulted in different degrees of peak retentions. Intercalating agents, such as EB, on the other hand, reduced peak widths and affected the retention of all the fragments.

5.4.1 Chiral affinity additives and chiral selectors

Gel-filled capillary columns can be utilized with complexing agents to achieve unique selectivities. As described above, in capillary affinity gel electrophoresis, one can either chemically attach complexing agents to the gel matrix or noncovalently incorporate the affinity ligand into the separation gel matrix. This latter approach was used by Guttman et al. [392] to entrap the neutral inclusion compound β-cyclodextrin by polymerization in a cross-linked polyacrylamide gel in narrow bore capillaries. Cyclodextrins are nonionic cyclic oligosaccharides consisting of glucose subunits with the shape of a toroid or hollow truncated cone. The cavity is relatively hydrophobic, whereas the external surface is hydrophilic. The torus of the larger circumference holds chiral secondary hydroxyl groups. By incorporating cyclodextrins into a small-pore cross-linked polyacrylamide gel matrix, the complexing agent is immobilized by entrapment. The chiral selectivity arose from the chiral glucose moiety at the entrance of the cavity. According to the authors, this column could be considered analogous to an HPLC column, in which the complexing agent is fixed to the solid support. Fig. 5.85 shows the separation of D,L-dansylated amino acids using a β-cyclodextrin gel capillary with the theoretical plate number over 100,000 using an effective separation length of 15 cm at an electric field of 1000 V/cm. Please note that similar to that of in open tubular operation, there is a plethora of chemistries, which can be utilized in affinity gel column operations.

A homogenous polyacrylamide-allyl-β-cyclodextrin copolymer gel was also attempted to separate drug enantiomers [393]. In that instance, the large pore size defined by the composition of the gel ensured the necessary electroosmotic flow for proper capillary electrochromatography operation.

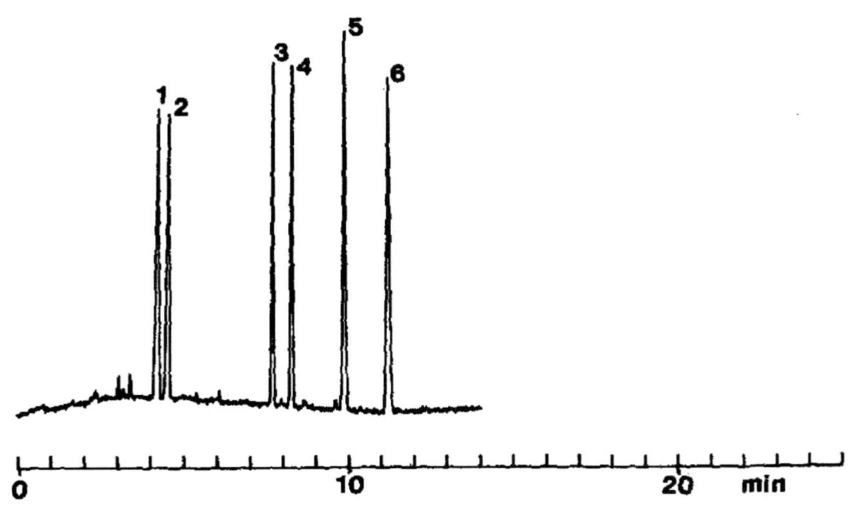

Figure 5.85 High-efficiency separation of D,L-dansylated amino acids. Peaks: 1 = Dns-L-Glu, 2 = Dns-D-Glu, 3 = Dns-L-Ser, 4 = Dns-D-Ser, 5 = Dns-l, -Leu, 6 = Dns-D-Leu buffer. Conditions: 0.1 M Tris, 0.25 M boric acid (pH 8.3), 7 M urea, gel: 5%T/3.3%C, capillary 150 mm × 0.075 mm i.d., E = 1000 V/cm. *With permission from A. Guttman, A. Paulus, A.S. Cohen, N. Grinberg, B.L. Karger, Use of complexing agents for selective separation in high-performance capillary electrophoresis. Chiral resolution via cyclodextrins incorporated within polyacrylamide gel columns, Journal of Chromatography 448 (1) (1988) 41–53 [392].*

Glutaraldehyde cross-linked BSA affinity material was utilized within a fused-silica capillary tube to separate tryptophan enantiomers [395]. Baseline resolution with high plate numbers was obtained. The approach of the authors represented a significant improvement in efficiency compared to HPLC separation of the same sample with immobilized BSA. Cruzado and Vigh [396] copolymerized allyl carbamoylated β-cyclodextrin derivatives with acrylamide and used the resulting gel as chiral selector for enantiomeric separations in capillary electrophoresis. Both cross-linked and noncross-linked gels were produced by adjusting the concentrations of the monomer (acrylamide) and crosslinker (bisacrylamide) along with the allyl cyclodextrin derivative. The chiral selectivity of the noncross-linked cyclodextrin containing gels for the separation of dansylated amino acid enantiomers depended on the cyclodextrin concentration. Diastereomeric (+)-diacetyl-L-tartaric anhydride derivatives of D- and L-tryptophan were separated with PVP as polymeric additive, considering the fact that PVP can undergo hydrophobic and dipolar interactions with the sample molecules influencing their effective mobility [397]. The effect of PVP concentration on the separation is depicted in Fig. 5.86.

Capillary affinity gel electrophoresis of enantiomers is an important separation challenge as described by Chankvetadze [398] regarding the theory and applications. In an interesting approach, covalently linked BSA to a high-molecular-mass dextran (MW 2,000,000) was used for

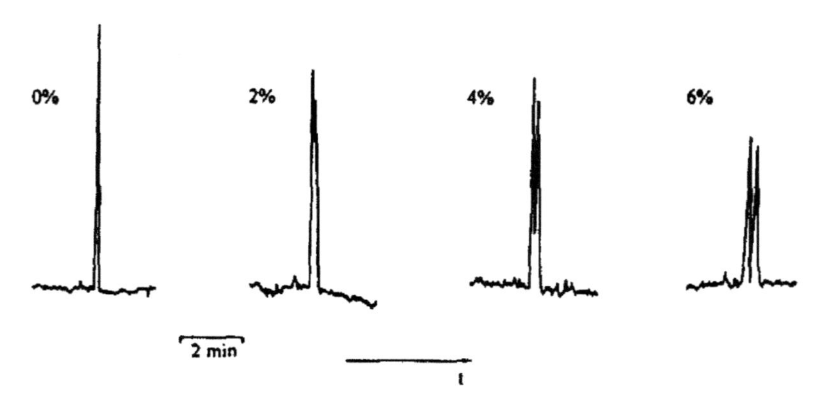

Figure 5.86 Influence of the polyvinylpirrolidone concentration in the background electrolyte on the separation of the diastereoisomeric D- and L-tryptophan diacetyl-L-tartaric acid monoamides. *With permission from W. Schuetzner, S. Fanali, A. Rizzi, E. Kenndler, Separation of diastereomeric derivatives of enantiomers by capillary zone electrophoresis with a polymer network: use of polyvinylpyrrolidone as buffer additive, Journal of Chromatography 639 (2) (1993) 375–378 [397].*

chiral analysis [399]. Baseline separation of leucovorin enantiomers was obtained proving that ligand immobilization can be generally applied to a wide variety of ligand–substrate systems. Separation of other enantiomers (ibuprofen and some amino acids) was also attained using dextran and BSA as buffer additives in capillary columns. Chiral separation of dansylated amino acids in dextran (MW 2,000,000) polymer network containing β-cyclodextrin was reported by Hartwick and coworkers [400]. Temperature effects on the chiral separation were studied and optimal efficiency and resolution for amino acid racemates were obtained at 25°C, which decreased with increasing temperature. Cyclofructans are cyclic oligosaccharides made of β-2,1-linked fructofuranose units were also utilized as chiral selectors in capillary electrophoresis [401].

CGE-based chiral separations were attempted by using a soluble β-cyclodextrin polymer. The effects of polymer concentration and buffer pH were investigated on the effective mobility and resolution [402,403]. Albeit, the β-cyclodextrin polymer exhibited better stereoselectivity than that of the monomers, the increase in pH resulted in a decrease in chiral selectivity. Nilsson and coworker proposed a new model for separation matrix formulation based on copolymerization of proteins as chiral selectors to create a gel, which had the potential to resolve different types of enantiomers in capillary affinity gel electrophoresis [404]. In their particular study, cellulase and BSA were cross-linked with glutaraldehyde to provide an affinity sieving medium for the separation of chiral β-adrenergic blocker enantiomers, such as pindolol. The enantiomers were baseline separated as depicted in Fig. 5.87. BSA was used as stabilizer for the cellulase.

Similar to cyclodextrins, chiral selection in CE is possible by the use of linear polysaccharide buffer additives as noncyclic polysaccharides also feature optical activity that support resolution of enantiomers. Nishi [405] applied dextrans, dextrins, and chondroitin sulfate for the separation of ionic and nonionic enantiomers. The latter approach can be considered as a form of affinity electrokinetic chromatography capable of separating both ionic and neutral enantiomers. Lambda-carrageenan, a linear high-molecular-weight sulfated polysaccharide, was introduced by Beck and Neau [406] as a chiral selector in capillary electrophoresis for the separation of enantiomers of weakly basic pharmaceutical compounds. Besides the separation of racemic compounds, diastereomeric pairs were also resolved.

Molecularly imprinted polymer nanoparticles were utilized for chiral separation of propanolol enantiomers as demonstrated in Fig. 5.88 [407]. The fact that the resulting peaks did not show any distortion (tailing) suggested

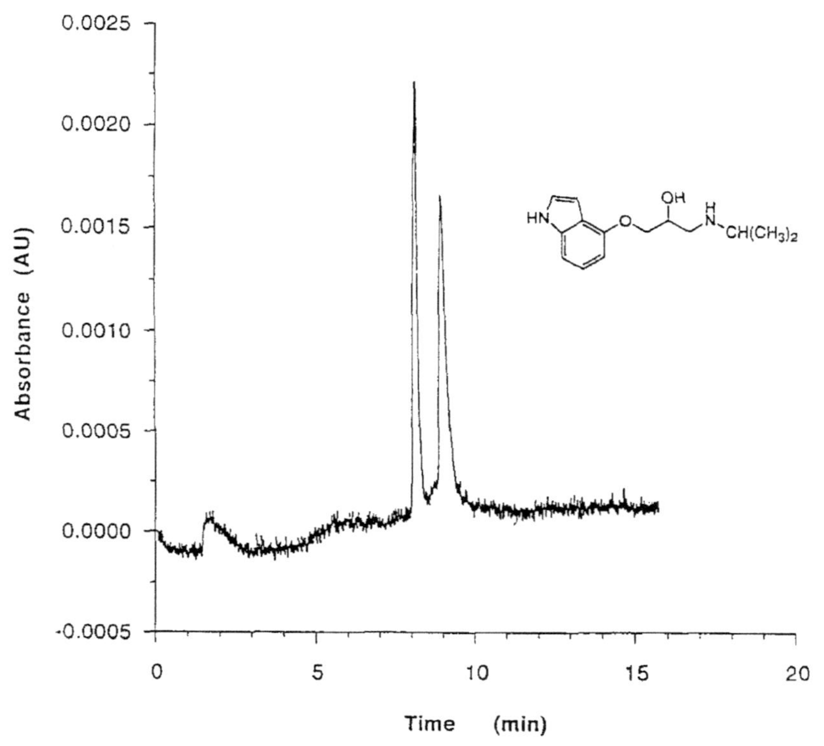

Figure 5.87 Capillary affinity gel electrophoresis separation of racemic pindolol (0.1 mM) with cellulase/bovine serum albumin gel. *With permission from H. Ljungberg, S. Nilsson, Protein-based capillary affinity gel electrophoresis for chiral separation of b-adrenergic blockers, Journal of Liquid Chromatography 18 (18–19) (1995) 3685–3698 [404].*

homogenous binding of the solute molecules to the nanoparticles. Others [408] used glycogen, an electrically neutral branched polysaccharide as enantioselective agent for CGE separation of 18 chiral compounds.

5.4.2 Capillary affinity gel electrophoresis of DNA

Incorporation of affinity ligands within the gel or polymer solution inside the capillary provides an efficient mean for selectivity manipulation. The usefulness of this approach was demonstrated by obtaining high-resolution separation of DNA restriction fragments using EB, a soluble affinity (intercalating) agent in the gel buffer system [390]. The authors also developed a migration model [Eq. (5.5)] that was used for selectivity optimization for such parameters as ligand concentration and applied electric field in terms

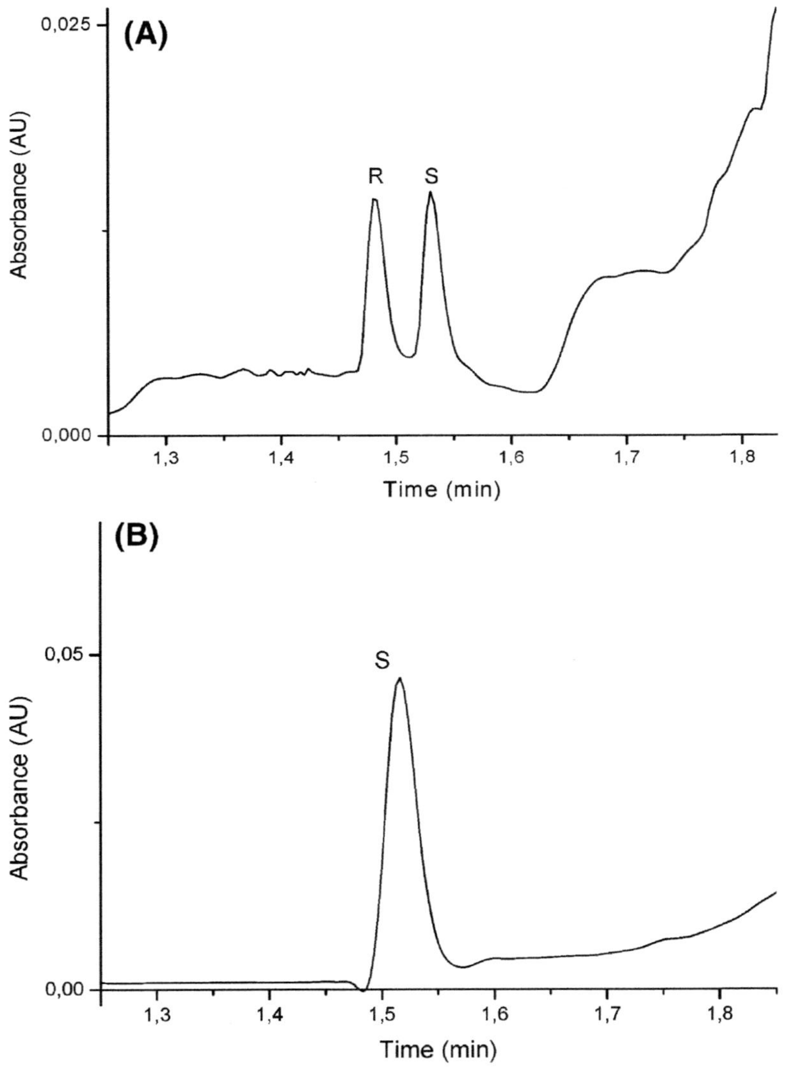

Figure 5.88 Capillary electrochromatographic analysis of (A) rac-propranolol and (B) (*S*)-propranolol (60 μM). Conditions: nanoparticle hydrodynamic injection (10 mg/mL) 0.5 psi for 10 s, sample injection 3 s at 6 kV, electrolyte 10 mM phosphate/20% (v/v) acetonitrile at pH 7, separation voltage at 16 kV, and temperature of the capillary 30°C. *With permission from F. Priego-Capote, L. Ye, S. Shakil, S.A. Shamsi, S. Nilsson, Monoclonal behavior of molecularly imprinted polymer nanoparticles in capillary electrochromatography, Analytical Chemistry 80 (8) (2008) 2881–2887 [407].*

of their influence on retention and selectivity on different-sized DNA molecules.

$$v = \mu_{\mathrm{p}} E \frac{1}{1 + K[L^{+}]^{m}} \tag{5.5}$$

where ν is the velocity of the polyion complex, μ_p is the electrophoretic mobility of the polyion, E is the applied electric field, K is the formation constant of the complex, $[L^{+}]$ is the affinity ligand concentration, and m is the number of ligands in the polyion complex. Fig. 5.89A and B compares the separations of the pBR322 DNA-Msp I digest restriction fragment mixture in the absence and in the presence of the affinity agent EB.

As one can observe, significant separation improvement was obtained with the addition of EB, especially in the 140–200-bp fragment range (peaks 8–14). Due to the decreased mobility of the DNA–EB complex, the applied electric field was doubled in the presence of EB to achieve comparable migration times. Please note that even the two identical length 147- and 160-bp fragments were separated, probably because of the shape differences due to their sequence dependent secondary structure and concomitant intercalation-mediated changes.

Capillary affinity gel electrophoresis was employed for oligodeoxynucleotide sequence recognition using poly(9-vinyladenine)-polyacrylamide hydrogel matrix [409–411]. Deoxynucleotide hexamers consisting of five thymidylic acids and one adenylic acid were retained in the order TpTpTpTpTpA > TpTpTpTpApT > TpTpTpApTpT, which suggested that the sequential thymidylic acids interacted with the affinity agent poly(9-vinyladenine) [409,410]. The same group investigated the detection of mismatch positions on DNA–polyvinyladenine hybrids using capillary affinity gel electrophoresis by applying sequence-specific recognition of oligodeoxynucleotides using the change in the hybridization process of heteroduplexes between nucleic acid analogs immobilized in the gel and oligodeoxynucleotides with a mismatch position [412]. In another study, the effect of urea concentration was examined on the base-specific separation of oligodeoxynucleotides in capillary affinity gel electrophoresis using poly(9-vinyladenine) affinity ligand [413]. Temperature programming was also attempted to increase the separation power of capillary affinity gel electrophoresis, especially for sensitive base-specific separation of oligodeoxynucleotides [414]. When the capillary temperature was changed during the separation the migration time and resolution varied accordingly as inducing alterations in the dissociation process of specific hydrogen bonding between the two

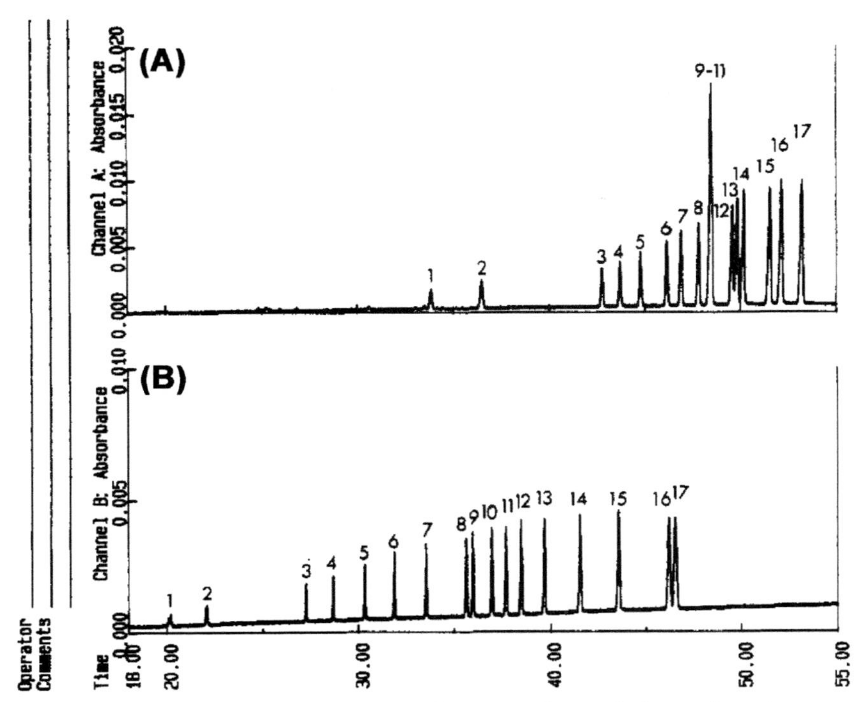

Figure 5.89 Effect of the ethidium bromide on capillary gel electrophoresis separation of pBR322 DNA-Msp I digest restriction fragment mixture: (A) No ethidium bromide added and (B) 1 μg/mL ethidium bromide added to the gel buffer system. Peaks (base pairs): 1 = 26; 2 = 34; 3 = 67; 4 = 76; 5 90; 6 = 110; 7 = 123; 8 = 147; 9 = 147; 10 = 160; 11 = 160; 12 = 180 13 = 190 14 = 201; 15 = 217; 16 = 238; and 17 = 242. Conditions: E = 100 V/cm (A) and 200 V/cm (B); effective separation length = 20 cm; running buffer, 100 mM TBE, pH 8.3; injection 3 s, 30 mW. *With permission from A. Guttman, N. Cooke, Capillary gel affinity electrophoresis of DNA fragments, Analytical Chemistry 63 (18) (1991) 2038–2042 [390].*

molecules. Temperature-programmed capillary affinity gel electrophoresis resulted in selective and sensitive base recognition, while separation efficiencies were still kept high (10^6 plates/m) as shown in Fig. 5.90 [414].

The difference between bis- and monointercalating dyes was investigated by Kim and Morris [415] on the separation of duplex DNA in HPMC solution. The authors found that while the use of bis intercalating dyes resulted in extra peaks probably due to the formation of intramolecular dye complexes, monomeric forms allowed full separation of all fragments without observing any extra peaks. Addition of glycerol to HPMC sieving matrix with borate containing background electrolyte improved

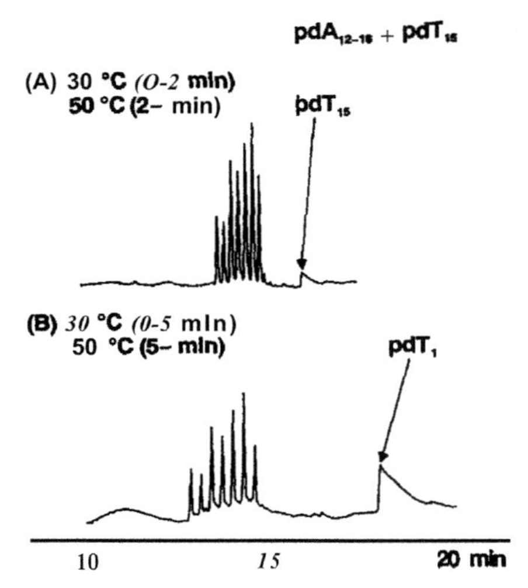

Figure 5.90 Effect of temperature programming on the capillary affinity gel electrophoresis separation of a mixture of pdA_{12-18} and pdT_{15}, (A) 30°C for 2 min, (B) 30°C for 5 min, and then step to 50°C. Gel: 8% T/5% C and 0.1% polyvinylalcohol (MW 10–30 kDa). *With permission from Y. Baba, M. Tsuhako, T. Sawa, M. Akashi, Temperature-programmed capillary affinity gel electrophoresis for the sensitive base-specific separation of oligodeoxynucleotides. Journal of Chromatography 632 (1–2) (1993) 137–142 [414].*

dsDNA separation as depicted in Fig. 5.91 [416], probably due to the borate complexation causing in this way a pore size distribution change.

Poly(9-vinyladenine) affinity additive was applied to evaluate the effect of urea concentration on the separation of oligo(thymidylic acids) in respect to migration behavior and peak efficiency [413]. While both the migration time and peak efficiency of oligothymidylic acids, which interacted with the affinity moiety, were altered, the migration time of oligodeoxyadenylic, which did not interact with the affinity moiety, simply increased with higher urea concentration, while the theoretical plate number was not affected. Tsukada et al. developed an affinity capillary electrophoresis approach using an allele-specific PNA probe modified with polyethylene glycol [417]. This method allowed to separate single-base-mutated single-stranded DNA with a stable secondary structure within 20 minutes.

5.4.3 Other capillary affinity gel electrophoresis applications

In situ prepared methacrylate-based imprinted dispersion polymers used as agglomerates (c.10 μm) of micrometer-sized globular particles possessed antibody mimetic molecular recognition [418]. The imprinted polymer

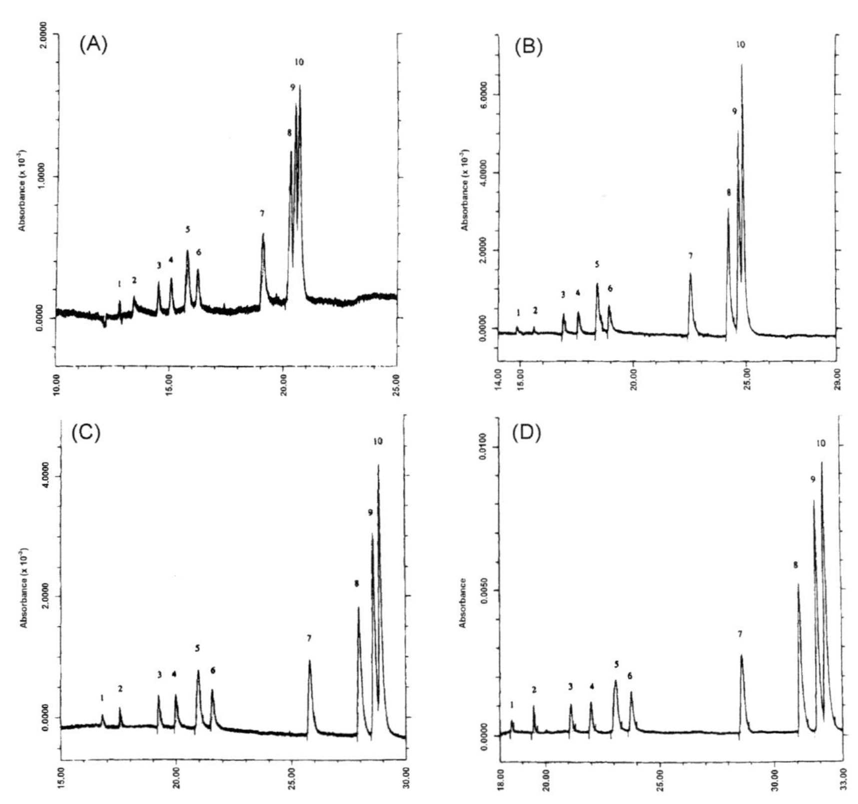

Figure 5.91 Separation of HaeIII digested ΦX DNA marker using capillary gel electrophoresis with the presence of (A) 0%, (B) 2.5%, (C) 5%, and (D) 7.5% glycerol (v/v). The DNA fragment size for the numbered peaks: 1–72; 2–118; 3–194; 4–234; 5–271/281; 6–310; 7–603; 8–872; 9–1078; and 10–1353 bp. Conditions: buffer, 90 mM Tris-borate, 2 mM EDTA (pH 8.5), 0.5% (w/v) hydroxypropylmethyl cellulose. *With permission from J. Cheng, K.R. Mitchelson, Glycerol-enhanced separation of DNA fragments in entangled solution capillary electrophoresis, Analytical Chemistry 66 (23) (1994) 4210–4214 [416].*

particles, selective for specific solute structures, were prepared in situ in the capillary with the affinity retention predictably varied by changing the pH of the background electrolyte as shown in Fig. 5.92. In addition, the electrolytes were pumped hydrodynamically through the capillaries, allowing rapid phase changes and offering microchromatography options.

CGE with immobilized glycan specific antibodies was applied to separate structurally similar carbohydrates [419]. The results indicated that weak biospecific interactions can be utilized in a capillary affinity gel electrophoresis format to obtain very selective separation of two forms of

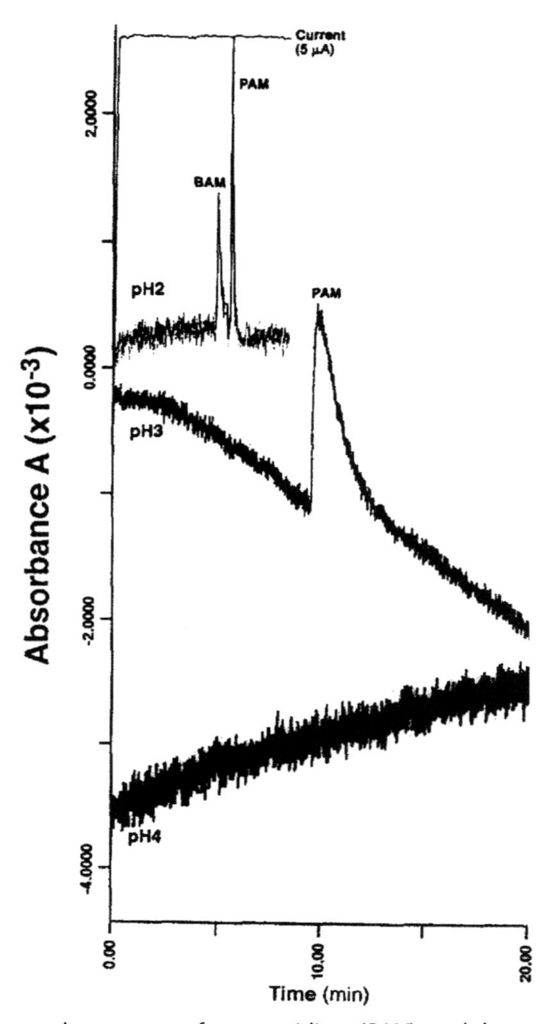

Figure 5.92 Electropherograms of pentamidine (*PAM*) and benzamidine injected onto a capillary filled with polymer network consisting of 8% noncross-linked polyacryloylaminoethoxyethanol. The column was prepared with PAM as template employing a pH 2 electrolyte. The superimposed electropherograms of PAM injected on to the same polymer network filled capillary under the same conditions are shown, but employing pH 3 and 4 electrolytes, respectively. *With permission from K. Nilsson, J. Lindell, O. Norrloew, B. Sellergren, Imprinted polymers as antibody mimetics and new affinity gels for selective separations in capillary electrophoresis, Journal of Chromatography A 680 (1) (1994) 57–61 [418].*

fluorophore labeled maltose. Electrophoretically mediated microanalysis was demonstrated by Regnier and coworkers by injecting alkaline phosphatase or/and β-galactosidase and mixed with the appropriate reagents in the capillary by electrophoretic mixing. The enzyme activity was then assessed by electromigrating the products, which formed during the enzymatic reaction, through the detection point. Field programing was used to drive the components into the capillary and then allow them to react with no field applied and electrophoretically separate the products after the reaction was complete (Fig. 5.93). Performing this assay allowed to obtain detection limits as low as 7.6 pM (52 zmol) [209].

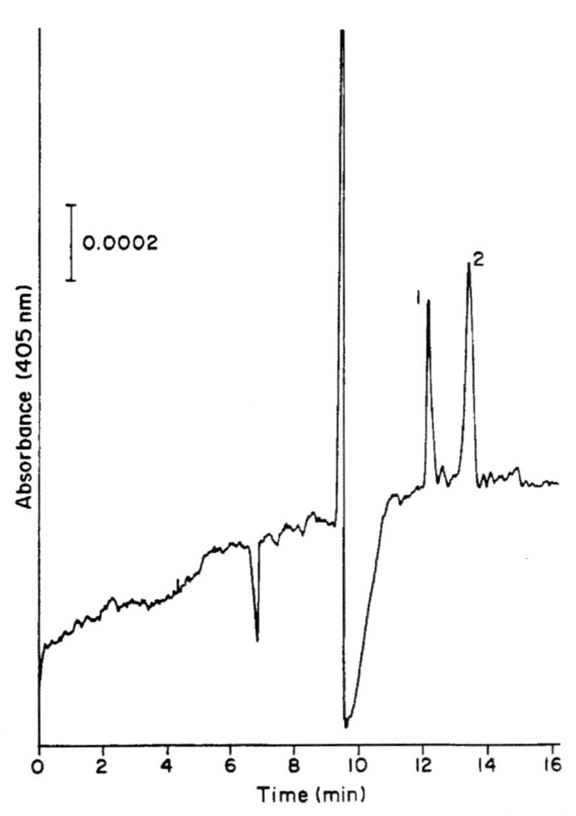

Figure 5.93 Electropherogram of dual alkaline phosphatase (*ALP*) and β-galactosidase enzymes assay in an acrylamide gel-filled capillary. ALP concentration, 0.008 mg/mL; β-galactosidase concentration, 2.3×10^{-5} mg/mL. Incubation time, 10 min; injection, 3 kV, 30 s. The two substrate concentrations were 1 mg/mL. The first peak to pass the window in a zero potential assay was from β-galactosidase and the second from ALP. *With permission from D. Wu, F.E. Regnier, Native protein separations and enzyme microassays by capillary zone and gel electrophoresis, Analytical Chemistry 65 (15) (1993) 2029–2035 [209].*

An interesting postcapillary affinity detection of protein variants utilized transfers after the separation the sample components and a fluorescein-labeled analyte-binding protein added into the reaction capillary [420]. The affinity complexes were formed with the protein variants in the reaction capillary and monitored with LIF detection. The method was capable to baseline separate IgG Fc variants with excellent resolution. The same group also studied the influence on resolution and detection limits of the analyte transfer and affinity complex formation. Capillary SDS gel electrophoresis was used to track antibody functionalization with sulfo-long chain-succinimidyl 6-[3-(2-pyridyldithio)propionamido] hexanoate (sulfo-LC-SPDP), to be used as ligands for target-specific ultrasound contrast reagents [421]. The authors showed that antibody disulfide bonds were prone to reduction as a function of the reducing agent and, using DTT with acidic pH, the intra- and interdisulfide bridges were better preserved on the modified antibodies.

References

[1] A. Smith, R.J. Nelson, Capillary electrophoresis of DNA, Current Protocols in Nucleic Acid Chemistry 13 (1) (2003) 10.9.1–10.9.16
[2] P.G. Righetti, C. Gelfi, Capillary electrophoresis of DNA for molecular diagnostics: an update, Journal of Capillary Electrophoresis and Microchip Technology 6 (3–4) (1999) 119–124.
[3] B.C. Durney, C.L. Crihfield, L.A. Holland, Capillary electrophoresis applied to DNA: determining and harnessing sequence and structure to advance bioanalyses (2009–2014), Analytical and Bioanalytical Chemistry 407 (23) (2015) 6923–6938.
[4] R.S. Dubrow, Analysis of synthetic oligonucleotide purity by capillary gel electrophoresis, American Laboratory (Shelton, CT, United States) 23 (5) (1991) 64. 66–67.
[5] H. Vu, C. McCollum, K. Jacobson, P. Theisen, R. Vinayak, E. Spiess, et al., Fast oligonucleotide deprotection. Phosphoramidite chemistry for DNA synthesis, Tetrahedron Letters 31 (50) (1990) 7269–7272.
[6] C.A. Chang, D. Ahle, M.S. Urdea, D.N. Heiger, B.L. Karger, Characterization of synthetic branched DNA (BDNA) using polyacrylamide gel-filled capillary electrophoresis, Nucleic Acids Symposium Series 24 (1991) 223.
[7] J.A. Luckey, L.M. Smith, Automated methods in DNA sequence analysis, Laboratory Robotics and Automation 3 (4–5) (1991) 175–180.
[8] L.M. Smith, High-speed DNA sequencing by capillary gel electrophoresis, Nature 349 (6312) (1991) 812–813.
[9] A. Guttman, A. Paulus, A.S. Cohen, B.L. Karger, H. Rodriguez, W.S. Hancock, High performance, capillary gel electrophoresis: high resolution and micropreparative applications, Electrophoresis'88, VCH, Weinheim, Germany, Copenhagen, 1988.
[10] W.J. Warren, G. Vella, Analysis of synthetic oligodeoxyribonucleotides by capillary gel electrophoresis and anion-exchange HPLC, Biotechniques 14 (4) (1993) 598–606.
[11] C. Gelfi, M. Perego, S. Morelli, A. Nicolin, P.G. Righetti, Analysis of antisense oligonucleotides by capillary electrophoresis, gel-slab electrophoresis, and HPLC: a comparison, Antisense & Nucleic Acid Drug Development 6 (1) (1996) 47–53.

[12] V.T. Ravikumar, D.C. Capaldi, W.F. Lima, E. Lesnik, B. Turney, D.L. Cole, Antisense phosphorothioate oligodeoxyribonucleotide targeted against ICAM-1: synthetic and biological characterization of a process-related impurity formed during oligonucleotide synthesis, Bioorganic & Medicinal Chemistry 11 (21) (2003) 4673–4679.

[13] A. Guttman, A.S. Cohen, D.N. Heiger, B.L. Karger, Analytical and micropreparative ultrahigh resolution of oligonucleotides by polyacrylamide gel high-performance capillary electrophoresis, Analytical Chemistry 62 (2) (1990) 137–141.

[14] A.S. Cohen, D.R. Najarian, A. Paulus, A. Guttman, J.A. Smith, B.L. Karger, Rapid separation and purification of oligonucleotides by high-performance capillary gel electrophoresis, Proceedings of the National Academy of Sciences of the United States of America 85 (24) (1988) 9660–9663.

[15] B.L. Karger, A. Cohen, High-performance microcapillary gel electrophoresis and method for preparing the gel-containing column, in United States 1989, Northeastern University, USA, 1989, 15 pp.

[16] A. Guttman, Capillary Column Containing a Dynamically Cross-Linked Composition and Method of Use, USA, Beckman Coulter Inc., 1993.

[17] J. Macek, U.R. Tjaden, J. Van der Greef, Resolution and concentration detection limit in capillary gel electrophoresis, Journal of Chromatography 545 (1) (1991) 177–182.

[18] A. Paulus, J.I. Ohms, Analysis of oligonucleotides by capillary gel electrophoresis, Journal of Chromatography 507 (1990) 113–123.

[19] D. Demorest, R. Dubrow, Factors influencing the resolution and quantitation of oligonucleotides separated by capillary electrophoresis on a gel-filled capillary, Journal of Chromatography 559 (1–2) (1991) 43–56.

[20] C. Gelfi, M. Perego, P.G. Righetti, Capillary electrophoresis of oligonucleotides in sieving liquid polymers in isoelectric buffers, Electrophoresis 17 (9) (1996) 1470–1475.

[21] J.A. Luckey, L.M. Smith, A model for the mobility of single-stranded DNA in capillary gel electrophoresis, Electrophoresis 14 (5–6) (1993) 492–501.

[22] D. Rodbard, A. Chrambach, Estimation of molecular radius, free mobility, and valence using polyacrylamide gel electrophoresis, Analytical Biochemistry 40 (1) (1971) 95–134.

[23] S. Guillouzic, L.C. McCormick, G.W. Slater, Electrophoresis in the presence of gradients: I. Viscosity gradients, Electrophoresis 23 (12) (2002) 1822–1832.

[24] A. Guttman, R.J. Nelson, N. Cooke, Prediction of migration behavior of oligonucleotides in capillary gel electrophoresis, Journal of Chromatography 593 (1–2) (1992) 297–303.

[25] T. Satow, T. Akiyama, A. Machida, Y. Utagawa, H. Kobayashi, Simultaneous determination of the migration coefficient of each base in heterogeneous oligo-DNA by gel filled capillary electrophoresis, Journal of Chromatography 652 (1) (1993) 23–30.

[26] Y. Cordier, O. Roch, P. Cordier, R. Bischoff, Capillary gel electrophoresis of oligonucleotides: prediction of migration times using base-specific migration coefficients, Journal of Chromatography A 680 (2) (1994) 479–489.

[27] K.D. Konrad, S.L. Pentoney Jr., Contribution of secondary structure to DNA mobility in capillary gels, Electrophoresis 14 (5–6) (1993) 502–508.

[28] J. Cheng, T. Kasuga, N.D. Watson, K.R. Mitchelson, Enhanced single-stranded DNA conformation polymorphism analysis by entangled solution capillary electrophoresis, Journal of Capillary Electrophoresis 2 (1) (1995) 24–29.

[29] T. Kasuga, J. Cheng, K.R. Mitchelson, Metastable single-strand DNA conformational polymorphism analysis results in enhanced polymorphism detection, PCR Methods and Applications 4 (4) (1995) 227–233.

[30] H.L. Cheng, S.S. Chiou, Y.M. Liao, Y.L. Chen, S.M. Wu, Genotyping of single nucleotide polymorphism in gamma-glutamyl hydrolase gene by capillary electrophoresis, Electrophoresis 32 (15) (2011) 2021–2027.
[31] I.V. Kourkine, C.N. Hestekin, A.E. Barron, Technical challenges in applying capillary electrophoresis-single strand conformation polymorphism for routine genetic analysis, Electrophoresis 23 (10) (2002) 1375–1385.
[32] P.S. Andersen, C. Jespersgaard, J. Vuust, M. Christiansen, L.A. Larsen, Capillary electrophoresis-based single strand DNA conformation analysis in high-throughput mutation screening, Human Mutation 21 (5) (2003) 455–465.
[33] A.W.H.M. Kuypers, P.M.W. Willems, M.J. van der Schans, P.C.M. Linssen, H.M. C. Wessels, C.H.M.M. de Bruijn, et al., Detection of point mutations in DNA using capillary electrophoresis in a polymer network, Journal of Chromatography Biomedical Applications 621 (2) (1993) 149–156.
[34] H.S. Hwang, G.W. Shin, H.J. Park, C.Y. Ryu, G.Y. Jung, Effect of temperature gradients on single-strand conformation polymorphism analysis in a capillary electrophoresis system using Pluronic polymer matrix, Analytica Chimica Acta 793 (2013) 114–118.
[35] E.S. Lander, L.M. Linton, B. Birren, C. Nusbaum, M.C. Zody, J. Baldwin, et al., Initial sequencing and analysis of the human genome, Nature 409 (6822) (2001) 860–921.
[36] J.C. Venter, et al., The sequence of the human genome, Science (New York, N.Y.) 291 (2001) 1304–1351.
[37] A. Guttman, Capillary electrophoresis using replaceable gels, European Patent Applications, Beckman Instruments, Inc., USA, 1992, p. 8.
[38] N.J. Dovichi, Capillary gel electrophoresis for DNA sequencing: separation and detection, Handbook of Capillary Electrophoresis, Edited by James P. Landers, CRC Press, 1994, pp. 369–387.
[39] A.S. Cohen, D.R. Najarian, B.L. Karger, Separation and analysis of DNA sequence reaction products by capillary gel electrophoresis, Journal of Chromatography 516 (1) (1990) 49–60.
[40] M.C. Ruiz-Martinez, J. Berka, A. Belenkii, F. Foret, A.W. Miller, B.L. Karger, DNA sequencing by capillary electrophoresis with replaceable linear polyacrylamide and laser-induced fluorescence detection, Analytical Chemistry 65 (20) (1993) 2851–2858.
[41] B.L. Karger, A. Guttman, DNA sequencing by CE, Electrophoresis 30 (Suppl. 1) (2009) S196–S202.
[42] O. Salas-Solano, E. Carrilho, L. Kotler, A.W. Miller, W. Goetzinger, Z. Sosic, et al., Routine DNA sequencing of 1000 bases in less than one hour by capillary electrophoresis with replaceable linear polyacrylamide solutions, Analytical Chemistry 70 (1998) 3996–4003.
[43] P. Lindberg, J. Roeraade, DNA sequencing at elevated temperature by capillary electrophoresis, Methods in Molecular Biology 163 (2001) 289–308.
[44] K.O. Voss, H.P. Roos, N.J. Dovichi, The effect of temperature oscillations on DNA sequencing by capillary electrophoresis, Analytical Chemistry 73 (6) (2001) 1345–1349.
[45] H. Cottet, P. Gareil, On the use of the activation energy concept to investigate analyte and network deformations in entangled polymer solution capillary electrophoresis of synthetic polyelectrolytes, Electrophoresis 22 (4) (2001) 684–691.
[46] M.J. Rocheleau, R.J. Grey, D.Y. Chen, H.R. Harke, N.J. Dovichi, Formamide modified polyacrylamide gels for DNA sequencing by capillary gel electrophoresis, Electrophoresis 13 (8) (1992) 484–486.
[47] V. Barbier, J.L. Viovy, Advanced polymers for DNA separation, Current Opinion in Biotechnology 14 (1) (2003) 51–57.

[48] H. He, B.A. Buchholz, L. Kotler, A.W. Miller, A.E. Barron, B.L. Karger, DNA sequencing with hydrophilic and hydrophobic polymers at elevated column temperatures, Electrophoresis 23 (10) (2002) 1421–1428.
[49] T. Nishikawa, H. Kambara, Separation of long DNA fragments by capillary gel electrophoresis with laser-induced fluorescence detection, Electrophoresis 15 (2) (1994) 215–220.
[50] M.C. Ruiz-Martinez, O. Salas-Solano, E. Carrilho, L. Kotler, B.L. Karger, A Sample purification method for rugged and high-performance DNA sequencing by capillary electrophoresis using replaceable polymer solutions: A. Development of the cleanup protocol, Analytical Chemistry 70 (8) (1998) 1516–1527.
[51] Y. Xiong, S.R. Park, H. Swerdlow, Base stacking: pH-mediated on-column sample concentration for capillary DNA sequencing, Analytical Chemistry 70 (17) (1998) 3605–3611.
[52] H. Swerdlow, R. Gesteland, Capillary gel electrophoresis for rapid, high resolution DNA sequencing, Nucleic Acids Research 18 (6) (1990) 1415–1419.
[53] H. Tan, E.S. Yeung, Automation and integration of multiplexed on-line sample preparation with capillary electrophoresis for high-throughput DNA sequencing, Analytical Chemistry 70 (19) (1998) 4044–4053.
[54] H. Drossman, J.A. Luckey, A.J. Kostichka, J. D'Cunha, L.M. Smith, High-speed separations of DNA sequencing reactions by capillary electrophoresis, Analytical Chemistry 62 (9) (1990) 900–903.
[55] D.Y. Chen, H.P. Swerdlow, H.R. Harke, J.Z. Zhang, N.J. Dovichi, Single-color laser-induced fluorescence detection and capillary gel electrophoresis for DNA sequencing, in: Proceedings of SPIE-The International Society for Optical Engineering, 1991.
[56] D.Y. Chen, H.P. Swerdlow, H.R. Harke, J.Z. Zhang, N.J. Dovichi, Low-cost, high-sensitivity laser-induced fluorescence detection for DNA sequencing by capillary gel electrophoresis, Journal of Chromatography 559 (1–2) (1991) 237–246.
[57] J.W. Chen, A.S. Cohen, B.L. Karger, Identification of DNA molecules by precolumn hybridization using capillary electrophoresis, Journal of Chromatography 559 (1–2) (1991) 295–305.
[58] H. Swerdlow, S. Wu, H. Harke, N.J. Dovichi, Capillary gel electrophoresis for DNA sequencing: laser-induced fluorescence detection with the sheath flow cuvette, Journal of Chromatography 516 (1) (1990) 61–67.
[59] J.Z. Zhang, D.Y. Chen, H. Harke, N.J. Dovichi, DNA sequencing by capillary gel electrophoresis and laser-induced fluorescence detection, Analytical Sciences 7 (Suppl., Pt. 2) (1991) 1485–1489.
[60] J.Z. Zhang, D.Y. Chen, S. Wu, H.R. Harke, N.J. Dovichi, High-sensitivity laser-induced fluorescence detection for capillary electrophoresis, Clinical Chemistry 37 (9) (1991) 1492–1496.
[61] D. Chen, H.R. Harke, N.J. Dovichi, Two-label peak-height encoded DNA sequencing by capillary gel electrophoresis: three examples, Nucleic Acids Research 20 (18) (1992) 4873–4880.
[62] S.L. Pentoney Jr., K.D. Konrad, W. Kaye, A single-fluor approach to DNA sequence determination using high performance capillary electrophoresis, Electrophoresis 13 (8) (1992) 467–474.
[63] S. Tabor, C.C. Richardson, DNA sequence analysis with a modified bacteriophage T7 DNA polymerase. Effect of pyrophosphorolysis and metal ions, The Journal of Biological Chemistry 265 (14) (1990) 8322–8328.
[64] S. Bay, H. Starke, J. Elliott, N. Dovichi, Accuracy of two-color peak height encoded DNA sequencing by capillary gel electrophoresis and laser-induced fluorescence, Proceedings of SPIE—The International Society for Optical Engineering 1891 (1993) 8–11.

[65] H. Lu, E. Arriaga, D.Y. Chen, N.J. Dovichi, High-speed and high-accuracy DNA sequencing by capillary gel electrophoresis in a simple, low cost instrument. Two-color peak-height encoded sequencing at 40 degrees C, Journal of Chromatography A 680 (2) (1994) 497–501.
[66] X.C. Huang, M.A. Quesada, R.A. Mathies, DNA sequencing using capillary array electrophoresis, Analytical Chemistry 64 (18) (1992) 2149–2154.
[67] X.C. Huang, M.A. Quesada, R.A. Mathies, Capillary array electrophoresis: an approach to high-speed, high-throughput DNA sequencing, Proceedings of SPIE—The International Society for Optical Engineering 1891 (1993) 12–20.
[68] J. Zhang, Y. Fang, J.Y. Hou, H.J. Ren, R. Jiang, P. Roos, et al., Use of non-cross-linked polyacrylamide for four-color DNA sequencing by capillary electrophoresis separation of fragments up to 640 bases in length in two hours, Analytical Chemistry 67 (24) (1995) 4589–4593.
[69] R. Tomisaki, Y. Baba, M. Tsuhako, S. Takahashi, K. Murakami, T. Anazawa, et al., High-speed DNA-sequencer using capillary gel electrophoresis with a laser-induced four-color fluorescent DNA detector, Analytical Sciences 10 (5) (1994) 817–820.
[70] B.K. Nunnally, H. He, L.C. Li, S.A. Tucker, L.B. McGown, Characterization of visible dyes for four-decay fluorescence detection in DNA sequencing, Analytical Chemistry 69 (13) (1997) 2392–2397.
[71] H. Swerdlow, J.Z. Zhang, D.Y. Chen, H.R. Harke, R. Grey, S. Wu, et al., Three DNA sequencing methods using capillary gel electrophoresis and laser-induced fluorescence, Analytical Chemistry 63 (24) (1991) 2835–2841.
[72] C. Heller, G.W. Slater, P. Mayer, N. Dovichi, D. Pinto, J.L. Viovy, et al., Free-solution electrophoresis of DNA, Journal of Chromatography A 806 (1) (1998) 113–121.
[73] J.C. Albrecht, J.S. Lin, A.E. Barron, A 265-base DNA sequencing read by capillary electrophoresis with no separation matrix, Analytical Chemistry 83 (2) (2011) 509–515.
[74] W.N. Vreeland, C. Desruisseaux, A.E. Karger, G. Drouin, G.W. Slater, A.E. Barron, Molar mass profiling of synthetic polymers by free-solution capillary electrophoresis of DNA-polymer conjugates, Analytical Chemistry 73 (8) (2001) 1795–1803.
[75] K. Lazaruk, P.S. Walsh, F. Oaks, D. Gilbert, B.B. Rosenblum, S. Menchen, et al., Genotyping of forensic short tandem repeat (STR) systems based on sizing precision in a capillary electrophoresis instrument, Electrophoresis 19 (1) (1998) 86–93.
[76] Z. Elek, Z. Kovacs, G. Keszler, M. Szabo, E. Csanky, J. Luo, et al., High throughput multiplex SNP-analysis in chronic obstructive pulmonary disease and lung cancer, Current Molecular Medicine 20 (3) (2020) 185–193.
[77] J. Ren, A. Ulvik, H. Refsum, P.M. Ueland, Chemical mismatch cleavage combined with capillary electrophoresis: detection of mutations exon 8 of the cystathionine beta-synthase gene, Clinical Chemistry 44 (10) (1998) 2108–2114.
[78] J. Bjorheim, S. Lystad, A. Lindblom, U. Kressner, S. Westring, S. Wahlberg, et al., Mutation analyses of KRAS exon 1 comparing three different techniques: temporal temperature gradient electrophoresis, constant denaturant capillary electrophoresis and allele specific polymerase chain reaction, Mutation Research 403 (1–2) (1998) 103–112.
[79] T.J. Kasper, M. Melera, P. Gozel, R.G. Brownlee, Separation and detection of DNA by capillary electrophoresis, Journal of Chromatography 458 (1988) 303–312.
[80] Y. Wang, J.M. Wallin, J. Ju, G.F. Sensabaugh, R.A. Mathies, High-resolution capillary array electrophoretic sizing of multiplexed short tandem repeat loci using energy-transfer fluorescent primers, Electrophoresis 17 (1996) 1485–1490.
[81] M. Oto, A. Koguchi, Y. Yuasa, Analysis of a polyadenine tract of the transforming growth factor-b type II receptor gene in colorectal cancers by non-gel-sieving capillary electrophoresis, Clinical Chemistry 43 (5) (1997) 759–763.

[82] N. Stellwagen, C. Gelfi, P.G. Righetti, The use of gel and capillary electrophoresis to investigate some of the fundamental physical properties of DNA, Electrophoresis 23 (2) (2002) 167–175.
[83] B.A. Siles, K.A. O'Neil, M.A. Fox, D.E. Anderson, A.F. Kuntz, S.C. Ranganath, et al., Genetic fingerprinting of grape plant (Vitis vinifera) using random amplified polymorphic DNA (RAPD) analysis and dynamic size-sieving capillary electrophoresis, Journal of Agricultural and Food Chemistry 48 (2000) 5903–5912.
[84] A.R. Tobler, S. Short, M.R. Andersen, T.M. Paner, J.C. Briggs, S.M. Lambert, et al., The SNPlex genotyping system: a flexible and scalable platform for SNP genotyping, Journal of Biomolecular Techniques 16 (4) (2005) 398–406.
[85] S. Ozawa, K. Sugano, T. Sonehara, S. Fukuzono, A. Ichikawa, N. Fukayama, et al., High resolution for single-strand conformation polymorphism analysis by capillary electrophoresis, Analytical Chemistry 76 (20) (2004) 6122–6129.
[86] K. Srinivasan, J.E. Girard, P. Williams, R.K. Roby, V.W. Weedn, S.C. Morris, et al., Electrophoretic separations of polymerase chain reaction-amplified DNA fragments in DNA typing using a capillary electrophoresis laser-induced fluorescence system, Journal of Chromatography A 652 (1) (1993) 83–91.
[87] M. Guttman, P. Fules, A. Guttman, Analysis of site-directed mutagenesis constructs by capillary electrophoresis using linear polymer sieving matrices, Journal of Chromatography A 1014 (1–2) (2003) 21–27.
[88] K. Khrapko, J.S. Hanekamp, W.G. Thilly, A. Belenkii, F. Foret, B.L. Karger, Constant denaturant capillary electrophoresis (CDCE): a high resolution approach to mutational analysis, Nucleic Acids Research 22 (3) (1994) 364–369.
[89] X.C. Li-Sucholeiki, K. Khrapko, P.C. Andre, L.A. Marcelino, B.L. Karger, W.G. Thilly, Applications of constant denaturant capillary electrophoresis/high-fidelity polymerase chain reaction to human genetic analysis, Electrophoresis 20 (1999) 1224–1232.
[90] M.Z. Xue, O. Bonny, S. Morgenthaler, M. Bochud, V. Mooser, W.G. Thilly, et al., Use of constant denaturant capillary electrophoresis of pooled blood samples to identify single-nucleotide polymorphisms in the genes (Scnn1a and Scnn1b) encoding the alpha and beta subunits of the epithelial sodium channel, Clinical Chemistry 48 (2002) 718–728.
[91] P.O. Ekstrom, R. Wasserkort, M. Minarik, F. Foret, W.G. Thilly, Two-point fluorescence detection and automated fraction collection applied to constant denaturant capillary electrophoresis, Biotechniques 29 (3) (2000) 582. 584, 586–589.
[92] M.K. Ramlee, T.D. Yan, A.M.S. Cheung, C.T.H. Chuah, S. Li, High-throughput genotyping of CRISPR/Cas9-mediated mutants using fluorescent PCR-capillary gel electrophoresis, Scientific Reports 5 (2015) 15587.
[93] H. Arakawa, K. Uetanaka, M. Maeda, A. Tsuji, Y. Matsubara, K. Narisawa, Analysis of polymerase chain reaction-product by capillary electrophoresis with laser-induced fluorescence detection and its application to the diagnosis of medium-chain acyl-coenzyme A dehydrogenase deficiency, Journal of Chromatography A 680 (2) (1994) 517–523.
[94] Y. Baba, R. Tomisaki, C. Sumita, M. Tsuhako, High-resolution separation of PCR product and gene diagnosis by capillary gel electrophoresis, Biomedical Chromatography 8 (6) (1994) 291–293.
[95] B.K. Clark, C.L. Nickles, K.C. Morton, J. Kovac, M.J. Sepaniak, Rapid separation of DNA restriction digests using size selective capillary electrophoresis with application to DNA fingerprinting, Journal of Microcolumn Separations 6 (5) (1994) 503–513.
[96] K.J. Ulfelder, H.E. Schwartz, J.M. Hall, F.J. Sunzeri, Restriction fragment length polymorphism analysis of ERBB2 oncogene by capillary electrophoresis, Analytical Biochemistry 200 (2) (1992) 260–267.

[97] Y. Liu, W.G. Kuhr, Separation of double- and single-stranded DNA restriction fragments: capillary electrophoresis with polymer solutions under alkaline conditions, Analytical Chemistry 71 (9) (1999) 1668–1673.
[98] C. Heller, Separation of double-stranded and single-stranded DNA in polymer solutions. Part 1. Mobility and separation mechanism, Electrophoresis 20 (10) (1999) 1962–1977.
[99] C. Heller, Separation of double-stranded and single-stranded DNA in polymer solutions. Part 2. Separation, peak width, and resolution, Electrophoresis 20 (10) (1999) 1978–1986.
[100] H. Cottet, P. Gareil, J.L. Viovy, The effect of blob size and network dynamics on the size-based separation of polystyrenesulfonates by capillary electrophoresis in the presence of entangled polymer solutions, Electrophoresis 19 (12) (1998) 2151–2162.
[101] S.A. Nevins, B.A. Siles, Z.E. Nackerdien, Analysis of gamma radiation-induced damage to plasmid DNA using dynamic size-sieving capillary electrophoresis, Journal of Chromatography B 741 (2000) 243–255.
[102] M.-M. Hsieh, W.-L. Tseng, H.-T. Chang, On-column preconcentration and separation of DNA fragments using polymer solutions in the presence of electroosmotic flow, Electrophoresis 21 (14) (2000) 2904–2910.
[103] H.E. Schwartz, K. Ulfelder, F.J. Sunzeri, M.P. Busch, R.G. Brownlee, Analysis of DNA restriction fragments and polymerase chain reaction products towards detection of the AIDS (HIV-1) virus in blood, Journal of Chromatography 559 (1–2) (1991) 267–283.
[104] M. Strege, A. Lagu, Separation of DNA restriction fragments by capillary electrophoresis using coated fused silica capillaries, Analytical Chemistry 63 (13) (1991) 1233–1236.
[105] K.S. Cook, J. Luo, A. Guttman, L. Thompson, Vaccine plasmid topology monitoring by capillary gel electrophoresis, Current Molecular Medicine 20 (10) (2020) 798–805.
[106] M.H. Kleemiss, M. Gilges, G. Schomburg, Capillary electrophoresis of DNA restriction fragments with solutions of entangled polymers, Electrophoresis 14 (5–6) (1993) 515–522.
[107] E.F. Rossomando, L. White, K.J. Ulfelder, Capillary electrophoresis: separation and quantitation of reverse transcriptase polymerase chain reaction products from polio virus, Journal of Chromatography, B: Biomedical Sciences and Applications 656 (1) (1994) 159–168.
[108] J. Cheng, T. Kasuga, K.R. Mitchelson, E.R.T. Lightly, N.D. Watson, W.J. Martin, et al., Polymerase chain reaction heteroduplex polymorphism analysis by entangled solution capillary electrophoresis, Journal of Chromatography A 677 (1) (1994) 169–177.
[109] M.A. Marino, L.A. Turni, S.A. Del Rio, P.E. Williams, Molecular size determinations of DNA restriction fragments and polymerase chain reaction products using capillary gel electrophoresis, Journal of Chromatography A 676 (1) (1994) 185–189.
[110] M. Nesi, P.G. Righetti, M.C. Patrosso, A. Ferlini, M. Chiari, Capillary electrophoresis of polymerase chain reaction-amplified products in polymer networks: the case of Kennedy's disease, Electrophoresis 15 (5) (1994) 644–646.
[111] M.D. Evans, J.T. Wolfe, D. Perrett, J. Lunec, K.E. Herbert, Analysis of internucleosomal DNA fragmentation in apoptotic thymocytes by dynamic sieving capillary electrophoresis, Journal of Chromatography A 700 (1–2) (1995) 151–162.
[112] C. Gelfi, P.G. Righetti, F. Leoncini, V. Brunelli, P. Carrera, M. Ferrari, CAG triplet analysis in families with androgen insensitivity syndrome by capillary electrophoresis in polymer networks, Journal of Chromatography A 706 (1–2) (1995) 463–468.

[113] Y. Wang, J. Ju, B.A. Carpenter, J.M. Atherton, G.F. Sensabaugh, R.A. Mathies, Rapid sizing of short tandem repeat alleles using capillary array electrophoresis and energy-transfer fluorescent primers, Analytical Chemistry 67 (7) (1995) 1197–1203.
[114] A.E. Barron, W.M. Sunada, H.W. Blanch, Capillary electrophoresis of DNA in uncrosslinked polymer solutions: evidence for a new mechanism of DNA separation, Biotechnology and Bioengineering 52 (2) (1996) 259–270.
[115] A.P. Buenz, A.E. Barron, J.M. Prausnitz, H.W. Blanch, Capillary electrophoretic separation of DNA restriction fragments in mixtures of low- and high-molecular-weight hydroxyethylcellulose, Industrial & Engineering Chemistry Research 35 (9) (1996) 2900–2908.
[116] W. Lu, D.S. Han, J. Yuan, J.M. Andrieu, Multi-target PCR analysis by capillary electrophoresis and laser-induced fluorescence, Nature 368 (6468) (1994) 269–271.
[117] B.R. McCord, J.M. Jung, E.A. Holleran, High resolution capillary electrophoresis of forensic DNA using a non-gel sieving buffer, Journal of Liquid Chromatography 16 (9–10) (1993) 1963–1981.
[118] F. Martin, D. Vairelles, B. Henrion, Automated ribosomal DNA fingerprinting by capillary electrophoresis of PCR products, Analytical Biochemistry 214 (1) (1993) 182–189.
[119] D. Figeys, E. Arriaga, A. Renborg, N.J. Dovichi, Use of the fluorescent intercalating dyes Popo-3, Yoyo-3 and Yoyo-1 for ultrasensitive detection of double-stranded DNA separated by capillary electrophoresis with hydroxypropylmethyl cellulose and non-cross-linked polyacrylamide, Journal of Chromatography A 669 (1–2) (1994) 205–216.
[120] T. Anazawa, H. Matsunaga, E.S. Yeung, Electrophoretic quantitation of nucleic acids without amplification by single-molecule imaging, Analytical Chemistry 74 (19) (2002) 5033–5038.
[121] N.Y. Zhang, E.S. Yeung, Simultaneous separation and genetic typing of four short tandem repeat loci by capillary electrophoresis, Journal of Chromatography A 768 (1) (1997) 135–141.
[122] V. Garcia-Canas, M. Mondello, A. Cifuentes, Combining ligation reaction and capillary gel electrophoresis to obtain reliable long DNA probes, Journal of Separation Science 34 (9) (2011) 1011–1019.
[123] B.C. Durney, B.A. Bachert, H.S. Sloane, S. Lukomski, J.P. Landers, L.A. Holland, Reversible phospholipid nanogels for deoxyribonucleic acid fragment size determinations up to 1500 base pairs and integrated sample stacking, Analytica Chimica Acta 880 (2015) 136–144.
[124] T. Guszczynski, H. Pulyaeva, D. Tietz, M.M. Garner, A. Chrambach, Capillary zone electrophoresis of large DNA, Electrophoresis 14 (5–6) (1993) 523–530.
[125] K. Hebenbrock, K. Schuegerl, R. Freitag, Analysis of plasmid-DNA and cell protein of recombinant Escherichia coli using capillary gel electrophoresis, Electrophoresis 14 (8) (1993) 753–758.
[126] Y. Kim, M.D. Morris, Rapid pulsed field capillary electrophoretic separation of megabase nucleic acids, Analytical Chemistry 67 (5) (1995) 784–786.
[127] Y. Kim, M.D. Morris, Pulsed field capillary electrophoresis of multikilobase length nucleic acids in dilute methyl cellulose solutions, Analytical Chemistry 66 (19) (1994) 3081–3085.
[128] J. Sudor, M.V. Novotny, Separation of large DNA fragments by capillary electrophoresis under pulsed-field conditions, Analytical Chemistry 66 (15) (1994) 2446–2450.
[129] N. Chen, L. Wu, A. Palm, T. Srichaiyo, S. Hjerten, High-performance field inversion capillary electrophoresis of 0.1–23 kbp DNA fragments with low-gelling, replaceable agarose gels, Electrophoresis 17 (9) (1996) 1443–1450.

[130] D.N. Heiger, A.S. Cohen, B.L. Karger, Separation of DNA restriction fragments by high performance capillary electrophoresis with low and zero crosslinked polyacrylamide using continuous and pulsed electric fields, Journal of Chromatography 516 (1) (1990) 33–48.
[131] T. Demana, M. Lanan, M.D. Morris, Improved separation of nucleic acids with analyte velocity modulation capillary electrophoresis, Analytical Chemistry 63 (23) (1991) 2795–2797.
[132] Z.Q. Li, C.C. Liu, X.M. Dou, Y. Ni, J.X. Wang, Y. Yamaguchi, Electromigration behavior of nucleic acids in capillary electrophoresis under pulsed-field conditions, Journal of Chromatography A 1331 (2014) 100–107.
[133] M.J. Navin, T.L. Rapp, M.D. Morris, Variable frequency modulation in DNA separations using field inversion capillary gel electrophoresis, Analytical Chemistry 66 (7) (1994) 1179–1182.
[134] F. Gao, C. Tie, X.-X. Zhang, Z. Niu, X. He, Y. Ma, Star-shaped polymers for DNA sequencing by capillary electrophoresis, Journal of Chromatography A 1218 (20) (2011) 3037–3041.
[135] Z. Li, X. Dou, Y. Ni, K. Sumitomo, Y. Yamaguchi, The influence of polymer concentration, applied voltage, modulation depth and pulse frequency on DNA separation by pulsed field CE, Journal of Separation Science 33 (17–18) (2010) 2811–2817.
[136] A. Guttman, N. Cooke, Denaturing capillary gel electrophoresis, American Biotechnology Laboratory 9 (4) (1991) 10.
[137] T.I. Todorov, O. De Carmejane, N.G. Walter, M.D. Morris, Capillary electrophoresis of RNA in dilute and semidilute polymer solutions, Electrophoresis 22 (12) (2001) 2442–2447.
[138] J. Khandurina, H.S. Chang, B. Wanders, A. Guttman, Automated high throughput RNA analysis by capillary electrophoresis, Biotechniques 32 (2002) 1226–1228.
[139] E. Katsivela, M.G. Hoefle, Separation of transfer RNA and 5S ribosomal RNA using capillary electrophoresis, Journal of Chromatography A 700 (1–2) (1995) 125–136.
[140] N.D. Borson, M.A. Strausbauch, P.J. Wettstein, R.P. Oda, S.L. Johnston, J.P. Landers, Direct quantitation of RNA transcripts by competitive single-tube RT-PCR and capillary electrophoresis, Biotechniques 25 (1) (1998) 130–137.
[141] G. Stanta, S. Bonin, RNA quantitative analysis from fixed and paraffin-embedded tissues: membrane hybridization and capillary electrophoresis, Biotechniques 24 (2) (1998) 271–276.
[142] G. Stanta, S. Bonin, R. Perin, RNA extraction from formalin-fixed and paraffin-embedded tissues, Methods in Molecular Biology 86 (1998) 23–26.
[143] P. Amini, J. Ettlin, L. Opitz, E. Clementi, A. Malbon, E. Markkanen, An optimised protocol for isolation of RNA from small sections of laser-capture microdissected FFPE tissue amenable for next-generation sequencing, BMC Molecular Biology 18 (1) (2017) 22.
[144] J.L. Zabzdyr, S.J. Lillard, Measurement of single-cell gene expression using capillary electrophoresis, Analytical Chemistry 73 (23) (2001) 5771–5775.
[145] F. Han, S.J. Lillard, In-situ sampling and separation of RNA from individual mammalian cells, Analytical Chemistry 72 (17) (2000) 4073–4079.
[146] J.G. Goldsmith, E.C. Ntuen, E.C. Goldsmith, Direct quantification of gene expression using capillary electrophoresis with laser-induced fluorescence, Analytical Biochemistry 360 (1) (2007) 23–29.
[147] P.L. Chang, Y.S. Chang, J.H. Chen, S.J. Chen, H.C. Chen, Analysis of BART7 microRNA from Epstein-Barr virus-infected nasopharyngeal carcinoma cells by capillary electrophoresis, Analytical Chemistry 80 (22) (2008) 8554–8560.

[148] N. Li, C. Jablonowski, H. Jin, W. Zhong, Stand-alone rolling circle amplification combined with capillary electrophoresis for specific detection of small RNA, Analytical Chemistry 81 (12) (2009) 4906–4913.
[149] R.-M. Jiang, Y.-S. Chang, S.-J. Chen, J.-H. Chen, H.-C. Chen, P.-L. Chang, Multiplexed microRNA detection by capillary electrophoresis with laser-induced fluorescence, Journal of Chromatography A 1218 (18) (2011) 2604–2610.
[150] T.J. Gray, F. Kong, P. Jelfs, V. Sintchenko, S.C. Chen, Improved identification of rapidly growing mycobacteria by a 16S-23S internal transcribed spacer region PCR and capillary gel electrophoresis, PLoS One 9 (7) (2014) e102290.
[151] C. Liu, Y. Yamaguchi, X. Zhu, Z. Li, Y. Ni, X. Dou, Analysis of small interfering RNA by capillary electrophoresis in hydroxyethylcellulose solutions, Electrophoresis 36 (14) (2015) 1651–1657.
[152] M. Barciszewska, A. Sucha, S. Balabanska, M.K. Chmielewski, Gel electrophoresis in a polyvinylalcohol coated fused silica capillary for purity assessment of modified and secondary-structured oligo- and polyribonucleotides, Scientific Reports 6 (2016) 19437.
[153] L. De Scheerder, A. Sparen, G.A. Nilsson, P.O. Norrby, E. Ornskov, Designing flexible low-viscous sieving media for capillary electrophoresis analysis of ribonucleic acids, Journal of Chromatography A 1562 (2018) 108–114.
[154] J. Gomez, S. Melon, J.A. Boga, M.E. Alvarez-Arguelles, S. Rojo-Alba, A. Leal-Negredo, et al., Capillary electrophoresis of PCR fragments with 5-labelled primers for testing the SARS-Cov-2, Journal of Virological Methods 284 (2020) 113937.
[155] T. Lu, L.J. Klein, S. Ha, R.R. Rustandi, High-Resolution capillary electrophoresis separation of large RNA under non-aqueous conditions, Journal of Chromatography A 1618 (2020) 460875.
[156] L. DeDionisio, S.M. Gryaznov, Analysis of a ribonuclease H digestion of N3'->P5' phosphoramidate-RNA duplexes by capillary gel electrophoresis, Journal of Chromatography B: Biomedical Applications 669 (1) (1995) 125–131.
[157] A.S. Cohen, M. Vilenchik, J.L. Dudley, M.W. Gemborys, A.J. Bourque, High-performance liquid chromatography and capillary gel electrophoresis as applied to antisense DNA, Journal of Chromatography 638 (2) (1993) 293–301.
[158] T.Q. Dinh, C.N. Sridhar, N. Dattagupta, Base composition analysis of phosphorothioate oligomers by capillary gel electrophoresis, Journal of Chromatography A 744 (1–2) (1996) 341–346.
[159] A.J. Bourque, A.S. Cohen, Quantitative analysis of phosphorothioate oligonucleotides in biological fluids using direct injection fast anion-exchange chromatography and capillary gel electrophoresis, Journal of Chromatography, B: Biomedical Applications 662 (2) (1994) 343–349.
[160] L. DeDionisio, Capillary gel electrophoresis and the analysis of DNA phosphorothioates, Journal of Chromatography 652 (1) (1993) 101–108.
[161] D.J. Rose, Characterization of antisense binding properties of peptide nucleic acids by capillary gel electrophoresis, Analytical Chemistry 65 (24) (1993) 3545–3549.
[162] N. Dean, R. McKay, L. Miraglia, R. Howard, S. Cooper, J. Giddings, et al., Inhibition of growth of human tumor cell lines in nude mice by an antisense oligonucleotide inhibitor of protein kinase C-a expression, Cancer Research 56 (15) (1996) 3499–3507.
[163] G.S. Srivatsa, M. Batt, J. Schuette, R.H. Carlson, J. Fitchett, C. Lee, et al., Quantitative capillary gel electrophoresis assay of phosphorothioate oligonucleotides in pharmaceutical formulations, Journal of Chromatography A 680 (2) (1994) 469–477.
[164] M. Vilenchik, A. Belenky, A.S. Cohen, Monitoring and analysis of antisense DNA by high-performance capillary gel electrophoresis, Journal of Chromatography A 663 (1) (1994) 105–113.

[165] A. Belenky, D.L. Smisek, A.S. Cohen, Sequencing of antisense DNA analogs by capillary gel electrophoresis with laser-induced fluorescence detection, Journal of Chromatography A 700 (1–2) (1995) 137–149.
[166] G.J.M. Bruin, K.O. Boernsen, D. Huesken, E. Gassmann, H.M. Widmer, A. Paulus, Stability measurements of antisense oligonucleotides by capillary gel electrophoresis, Journal of Chromatography A 709 (1) (1995) 181–195.
[167] B.A. Siles, K.A. O'Neil, D.L. Tung, L. Bazar, G.B. Collier, C.I. Lovelace, The use of dynamic size-sieving capillary electrophoresis and mismatch repair enzymes for mutant DNA analysis, Journal of Capillary Electrophoresis 5 (1998) 51–58.
[168] J. Qin, Y. Fung, B. Lin, DNA diagnosis by capillary electrophoresis and microfabricated electrophoretic devices, Expert Review of Molecular Diagnostics 3 (3) (2003) 387–394.
[169] D.-S. Lian, X.-Y. Chen, H.-S. Zeng, Y.-Y. Wang, Capillary electrophoresis based on nucleic acid analysis for diagnosing inherited diseases, Clinical Chemistry and Laboratory Medicine (CCLM) 59 (2) (2021) 249–266.
[170] A.R. Isenberg, B.R. McCord, B.W. Koons, B. Budowle, R.O. Allen, DNA typing of a polymerase chain reaction amplified D1S80/amelogenin multiplex using capillary electrophoresis and a mixed entangled polymer matrix, Electrophoresis 17 (9) (1996) 1505–1511.
[171] R. Thompson, S. Zoppis, B. McCord, An overview of DNA typing methods for human identification: past, present, and future, Methods in Molecular Biology 830 (2012) 3–16.
[172] E. Giardina, A. Spinella, G. Novelli, Past, present and future of forensic DNA typing, Nanomedicine (Lond) 6 (2) (2011) 257–270.
[173] I. Barme, G.J.M. Bruin, A. Paulus, M. Ehrat, Preconcentration and separation of antisense oligonucleotides by on-column isotachophoresis and capillary electrophoresis in polymer-filled capillaries, Electrophoresis 19 (8–9) (1998) 1445–1451.
[174] K. Khan, A. Van Schepdael, T. Saison-Behmoaras, A. Van Aerschot, J. Hoogmartens, Analysis of antisense oligonucleotides by on-capillary isotachophoresis and capillary polymer sieving electrophoresis, Electrophoresis 19 (12) (1998) 2163–2168.
[175] S.-H. Chen, J.M. Gallo, Use of capillary electrophoresis methods to characterize the pharmacokinetics of antisense drugs, Electrophoresis 19 (16–17) (1998) 2861–2869.
[176] H. Oana, R.W. Hammond, J.J. Schwinefus, S.-C. Wang, M. Doi, M.D. Morris, High-speed separation of linear and supercoiled DNA by capillary electrophoresis. Buffer, entangling polymer, and electric field effects, Analytical Chemistry 70 (3) (1998) 574–579.
[177] G. Raucci, C.A. Maggi, D. Parente, Capillary electrophoresis of supercoiled DNA molecules: parameters governing the resolution of topoisomers and their separation from open forms, Analytical Chemistry 72 (4) (2000) 821–826.
[178] L.A. Mitchenall, R.E. Hipkin, M.M. Piperakis, N.P. Burton, A. Maxwell, A rapid high-resolution method for resolving DNA topoisomers, BMC Research Notes 11 (1) (2018) 37.
[179] J. Ren, A. Ulvik, P.M. Ueland, H. Refsum, Analysis of single-strand conformation polymorphism by capillary electrophoresis with laser-induced fluorescence detection using short-chain polyacrylamide as sieving medium, Analytical Biochemistry 245 (1) (1997) 79–84.
[180] Z. Haidong, Single-strand conformational polymorphism analysis, in: J.P. Fennell, A.H. Baker (Eds.), Hypertension: Methods and Protocols, Humana Press, Totowa, NJ, 2005, pp. 149–158.
[181] Z. Kalvatchev, P. Draganov, Single-strand conformation polymorphism (SSCP) analysis: A rapid and sensitive method for detection of genetic diversity among virus population, Biotechnology & Biotechnological Equipment 19 (3) (2005) 9–14.

[182] J. Bashkin, M. Marsh, D. Barker, R. Johnston, DNA sequencing by capillary electrophoresis with a hydroxyethylcellulose sieving buffer, Applied and Theoretical Electrophoresis 6 (1996) 23–28.
[183] M.A. Quesada, Replaceable polymers in DNA sequencing by capillary electrophoresis, Current Opinion in Biotechnology 8 (1) (1997) 82–93.
[184] R.L. Goldfeder, D.P. Wall, M.J. Khoury, J.P.A. Ioannidis, E.A. Ashley, Human genome sequencing at the population scale: a primer on high-throughput DNA sequencing and analysis, American Journal of Epidemiology 186 (8) (2017) 1000–1009.
[185] D.C. Koboldt, K.M. Steinberg, D.E. Larson, R.K. Wilson, E.R. Mardis, The next-generation sequencing revolution and its impact on genomics, Cell 155 (1) (2013) 27–38.
[186] J.C. Rendon, F. Cortes-Mancera, A. Duque-Jaramillo, M.C. Ospina, M.C. Navas, Analysis of hepatitis B virus genotypes by restriction fragment length polymorphism, Biomedica: Revista del Instituto Nacional de Salud 36 (0) (2015) 79–88.
[187] B.B. Asiimwe, M.L. Joloba, S. Ghebremichael, T. Koivula, D.P. Kateete, F.A. Katabazi, et al., DNA restriction fragment length polymorphism analysis of Mycobacterium tuberculosis isolates from HIV-seropositive and HIV-seronegative patients in Kampala, Uganda, BMC Infectious Diseases 9 (2009) 12.
[188] P.G. Righetti, C. Gelfi, Capillary electrophoresis of DNA for molecular diagnostics, Electrophoresis 18 (10) (1997) 1709–1714.
[189] K. Kleparnik, P. Bocek, DNA diagnostics by capillary electrophoresis, Chemical Reviews 107 (11) (2007) 5279–5317.
[190] C. Sofocleous, A. Kolialexi, A. Mavrou, Molecular diagnosis of fragile X syndrome, Expert Review of Molecular Diagnostics 9 (1) (2009) 23–30.
[191] S.M. Roper, O.L. Tatum, Forensic aspects of DNA-based human identity testing, Journal of Forensic Nursing 4 (4) (2008) 150–156.
[192] C.C. Connon, A.K. LeFebvre, R.C. Benjamin, Development of a normalized extraction to further aid in fast, high-throughput processing of forensic DNA reference samples, Forensic Science International—Genetics 25 (2016) 112–124.
[193] W. Goodwin, S. Alimat, Analysis of four PCR/SNaPshot multiplex assays analyzing 52 SNPforID markers, Electrophoresis 38 (7) (2017) 1007–1015.
[194] C.N. Dong, Y.D. Yang, S.J. Li, Y.R. Yang, X.J. Zhang, X.D. Fang, et al., Whole genome nucleosome sequencing identifies novel types of forensic markers in degraded DNA samples, Scientific Reports 6 (2016) 26101.
[195] C. Bory, C. Chantin, D. Bozon, Capillary electrophoretic analysis of DA restriction fragments and PCR products for polymorphism and mutation studies in cystic fibrosis and Gaucher's disease, Journal of Pharmaceutical and Biomedical Analysis 13 (4/5) (1995) 511–514.
[196] Y. Baba, R. Tomisaki, C. Sumita, I. Morimoto, S. Sugita, M. Tsuhako, et al., Rapid typing of variable number of tandem repeat locus in the human apolipoprotein B gene for DNA diagnosis of heart disease by polymerase chain reaction and capillary electrophoresis, Electrophoresis 16 (8) (1995) 1437–1440.
[197] N. Woo, S.K. Kim, Y. Sun, S.H. Kang, Enhanced capillary electrophoretic screening of Alzheimer based on direct apolipoprotein E genotyping and one-step multiplex PCR, Journal of Chromatography, B 1072 (2018) 290–299.
[198] Z.Q. Li, R.L. Yang, Q. Wang, D.W. Zhang, S.L. Zhuang, Y. Yamaguchi, Electrophoresis of periodontal pathogens in poly(ethyleneoxide) solutions with uncoated capillary, Analytical Biochemistry 471 (2015) 70–72.
[199] T.Y. Pan, C.C. Wang, C.J. Shih, H.F. Wu, S.S. Chiou, S.M. Wu, Genotyping of intron 22 inversion of factor VIII gene for diagnosis of hemophilia A by inverse-shifting polymerase chain reaction and capillary electrophoresis, Analytical and Bioanalytical Chemistry 406 (22) (2014) 5447–5454.

[200] N. Woo, S.K. Kim, S.H. Kang, Voltage-programming-based capillary gel electrophoresis for the fast detection of angiotensin-converting enzyme insertion/deletion polymorphism with high sensitivity, Journal of Separation Science 39 (16) (2016) 3230–3238.

[201] B.C. Courtney, K.C. Williams, Q.A. Bing, J.J. Schlager, Capillary gel electrophoresis as a method to determine ligation efficiency, Analytical Biochemistry 228 (2) (1995) 281–286.

[202] K. Kleparnik, Z. Mala, P. Bocek, Fast separation of DNA sequencing fragments in highly alkaline solutions of linear polyacrylamide using electrophoresis in bare silica capillaries, Electrophoresis 22 (4) (2001) 783–788.

[203] J.L. Zabzdyr, S.J. Lillard, A qualitative look at multiplex gene expression of single cells using capillary electrophoresis, Electrophoresis 26 (1) (2005) 137–145.

[204] E. Esteban-Cardenosa, M. Duran, M. Infante, E. Velasco, C. Miner, High-throughput mutation detection method to scan BRCA1 and BRCA2 based on heteroduplex analysis by capillary array electrophoresis, Clinical Chemistry 50 (2) (2004) 313–320.

[205] P.O. Ekstrom, T. Bjorge, A. Dorum, A.S. Longva, K.M. Heintz, D.J. Warren, et al., Determination of hereditary mutations in the BRCA1 gene using archived serum samples and capillary electrophoresis, Analytical Chemistry 76 (15) (2004) 4406–4409.

[206] E. Velasco, M. Infante, M. Duran, L. Perez-Cabornero, D.J. Sanz, E. Esteban-Cardenosa, et al., Heteroduplex analysis by capillary array electrophoresis for rapid mutation detection in large multiexon genes, Nature Protocols 2 (1) (2007) 237–246.

[207] C.C. Wang, J.G. Chang, J. Ferrance, H.Y. Chen, C.Y. You, Y.F. Chang, et al., Quantification of SMN1 and SMN2 genes by capillary electrophoresis for diagnosis of spinal muscular atrophy, Electrophoresis 29 (13) (2008) 2904–2911.

[208] L. Guo, H. Yang, B. Qiu, X. Xiao, L. Xue, D. Kim, et al., Capillary electrophoresis with electrochemiluminescent detection for highly sensitive assay of genetically modified organisms, Analytical Chemistry 81 (23) (2009) 9578–9584.

[209] D. Wu, F.E. Regnier, Native protein separations and enzyme microassays by capillary zone and gel electrophoresis, Analytical Chemistry 65 (15) (1993) 2029–2035.

[210] H. Lindner, M. Wurm, A. Dirschlmayer, B. Sarg, W. Helliger, Application of high-performance capillary electrophoresis to the analysis of H1 histones, Electrophoresis 14 (5–6) (1993) 480–485.

[211] V. Dolnik, M.V. Novotny, Separation of amino acid homopolymers by capillary gel electrophoresis, Analytical Chemistry 65 (5) (1993) 563–567.

[212] N. Adamson, P.F. Riley, E.C. Reynolds, The analysis of multiple phosphoseryl-containing casein peptides using capillary zone electrophoresis, Journal of Chromatography 646 (2) (1993) 391–396.

[213] E. Rohde, A.J. Tomlinson, D.H. Johnson, S. Naylor, Comparison of protein mixtures in aqueous humor by membrane preconcentration—capillary electrophoresis - mass spectrometry, Electrophoresis 19 (13) (1998) 2361–2370.

[214] C.J. Herring, J. Qin, An on-line preconcentrator and the evaluation of electrospray interfaces for the capillary electrophoresis/mass spectrometry of peptides, Rapid Communications in Mass Spectrometry: RCM 13 (1) (1999) 1–7.

[215] F. Foret, E. Szoko, B.L. Karger, Trace analysis of proteins by capillary zone electrophoresis with on-column transient isotachophoretic preconcentration, Electrophoresis 14 (5–6) (1993) 417–428.

[216] B.L. Karger, A.S. Cohen, A. Guttman, High-performance capillary electrophoresis in the biological sciences, Journal of Chromatography 492 (1989) 585–614.

[217] A. Guttman, Capillary sodium dodecyl sulfate-gel electrophoresis of proteins, Electrophoresis 17 (8) (1996) 1333–1341.

[218] A.S. Cohen, B.L. Karger, High-performance sodium dodecyl sulfate polyacrylamide gel capillary electrophoresis of peptides and proteins, Journal of Chromatography 397 (1987) 409–417.
[219] K. Tsuji, *High-performance* capillary electrophoresis of proteins. Sodium dodecyl sulfate-polyacrylamide gel-filled capillary column for the determination of recombinant biotechnology-derived proteins, Journal of Chromatography 550 (1–2) (1991) 823–830.
[220] A. Guttman, C. Filep, B.L. Karger, Fundamentals of capillary electrophoretic migration and separation of SDS proteins in borate cross-linked dextran gels, Analytical Chemistry 93 (26) (2021) 9267–9276.
[221] A. Guttman, P. Shieh, J. Lindahl, N. Cooke, Capillary sodium dodecyl sulfate gel electrophoresis of proteins. II. On the Ferguson method in polyethylene oxide gels, Journal of Chromatography A 676 (1) (1994) 227–231.
[222] A. Guttman, J. Nolan, Comparison of the separation of proteins by sodium dodecyl sulfate-slab gel electrophoresis and capillary sodium dodecyl sulfate-gel electrophoresis, Analytical Biochemistry 221 (2) (1994) 285–289.
[223] K. Benedek, S. Thiede, High-performance capillary electrophoresis of proteins using sodium dodecyl sulfate-poly(ethylene oxide), Journal of Chromatography A 676 (1) (1994) 209–217.
[224] K. Benedek, A. Guttman, Ultra-fast high-performance capillary sodium dodecyl sulfate gel electrophoresis of proteins, Journal of Chromatography A 680 (2) (1994) 375–381.
[225] A. Guttman, P. Shieh, B.L. Karger, Capillary SDS gel electrophoresis of proteins, in: B.D. Hames (Ed.), Gel Electrophoresis of Proteins: A Practical Approach, Oxford University Press, Oxford, 1998, pp. 105–126.
[226] W. Li, B. Yang, D. Zhou, J. Xu, W. Li, W.C. Suen, Identification and characterization of monoclonal antibody fragments cleaved at the complementarity determining region using orthogonal analytical methods, Journal of Chromatography, B, Analytical Technologies in the Biomedical and Life Sciences 1048 (2017) 121–129.
[227] G. Chen, S. Ha, R.R. Rustandi, Characterization of glycoprotein biopharmaceutical products by Caliper LC90 CE-SDS gel technology, Methods in Molecular Biology 988 (2013) 199–209.
[228] M. Han, D. Phan, N. Nightlinger, L. Taylor, S. Jankhah, B. Woodruff, et al., Optimization of CE-SDS method for antibody separation based on multi-users experimental practices, Chromatographia 64 (5) (2006) 1–8.
[229] G. Hunt, W. Nashabeh, Capillary electrophoresis sodium dodecyl sulfate nongel sieving analysis of a therapeutic recombinant monoclonal antibody: a biotechnology perspective, Analytical Chemistry 71 (13) (1999) 2390–2397.
[230] C.X. Zhang, M.M. Meagher, Sample stacking provides three orders of magnitude sensitivity enhancement in SDS capillary gel electrophoresis of adeno-associated virus capsid proteins, Analytical Chemistry 89 (6) (2017) 3285–3292.
[231] T. Li, M. Santos, A. Guttman, Ultrahigh sensitivity analysis of adeno-associated virus (AAV) capsid proteins by sodium dodecyl sulphate capillary gel electrophoresis, Chromatography Today 13 (4) (2020) 10–12.
[232] C.E. Sanger-van de Griend, CE-SDS method development, validation, and best practice—an overview, Electrophoresis 40 (18–19) (2019) 2361–2374.
[233] I. Kustos, V. Toth, B. Kocsis, I. Kerepesi, L. Emody, F. Kilar, Capillary electrophoretic analysis of wild type and mutant Proteus penneri outer membrane proteins, Electrophoresis 21 (2000) 3020. 3007.
[234] I. Kustos, B. Kocsis, I. Kerepesi, F. Kilar, Protein profile characterization of bacterial lysates by capillary electrophoresis, Electrophoresis 19 (1998) 2317–2323.

[235] T. Manabe, H. Oota, J. Mukai, Size separation of sodium dodecyl sulfate complexes of human plasma proteins by capillary electrophoresis employing linear polyacrylamide as a sieving polymer, Electrophoresis 19 (13) (1998) 2308–2316.
[236] K. Ganzler, K.S. Greve, A.S. Cohen, B.L. Karger, A. Guttman, N.C. Cooke, High-performance capillary electrophoresis of SDS-protein complexes using UV-transparent polymer networks, Analytical Chemistry 64 (22) (1992) 2665–2671.
[237] A. Guttman, J. Horvath, N. Cooke, Influence of temperature on the sieving effect of different polymer matrixes in capillary SDS gel electrophoresis of proteins, Analytical Chemistry 65 (3) (1993) 199–203.
[238] A. Guttman, P. Shieh, D. Hoang, J. Horvath, N. Cooke, Effect of operational variables on the separation of proteins by capillary sodium dodecyl sulfate-gel electrophoresis, Electrophoresis 15 (2) (1994) 221–224.
[239] A. Guttman, C. Filep, Capillary sodium dodecyl sulfate gel electrophoresis of proteins: introducing the three dimensional *Ferguson* method, Analytica Chimica Acta 1183 (2021) 338958
[240] K. Tsuji, Factors affecting the performance of sodium dodecyl sulfate gel-filled capillary electrophoresis, Journal of Chromatography A 661 (1–2) (1994) 257–264.
[241] K. Tsuji, Evaluation of sodium dodecyl sulfate non-acrylamide, polymer gel-filled capillary electrophoresis for molecular size separation of recombinant bovine somatotropin, Journal of Chromatography 652 (1) (1993) 139–147.
[242] P.C.H. Shieh, D. Hoang, A. Guttman, N. Cooke, Capillary sodium dodecyl sulfate gel electrophoresis of proteins. I. Reproducibility and stability, Journal of Chromatography A 676 (1) (1994) 219–226.
[243] A. Widhalm, C. Schwer, D. Blaas, E. Kenndler, Capillary zone electrophoresis with a linear, non-cross-linked polyacrylamide-gel—separation of proteins according to molecular mass, Journal of Chromatography 549 (1–2) (1991) 446–451.
[244] D. Wu, F.E. Regnier, Sodium dodecyl sulfate-capillary gel electrophoresis of proteins using non-cross-linked polyacrylamide, Journal of Chromatography 608 (1–2) (1992) 349–356.
[245] Z. Deyl, I. Miksik, Separation of collagen type I chain polymers by electrophoresis in non-cross-linked polyacrylamide-filled capillaries, Journal of Chromatography A 698 (1–2) (1995) 369–373.
[246] A. Guttman, J.A. Nolan, N. Cooke, Capillary sodium dodecyl sulfate gel electrophoresis of proteins, Journal of Chromatography 632 (1–2) (1993) 171–175.
[247] W.E. Werner, D.M. Demorest, J. Stevens, J.E. Wiktorowicz, Size-dependent separation of proteins denatured in SDS by capillary electrophoresis using a replaceable sieving matrix, Analytical Biochemistry 212 (1) (1993) 253–258.
[248] M. Zhu, V. Levi, T. Wehr, Size-based separations of native proteins and SDS-protein complexes using non-gel sieving capillary electrophoresis, American Biotechnology Laboratory 11 (1) (1993) 26. 28.
[249] M.R. Karim, J.C. Janson, T. Takagi, Size-dependent separation of proteins in the presence of sodium dodecyl-sulfate and dextran in capillary electrophoresis—effect of molecular-weight of dextran, Electrophoresis 15 (12) (1994) 1531–1534.
[250] T. Takagi, M.R. Karim, A new mode of size-dependent separation of proteins by capillary electrophoresis in presence of sodium dodecyl-sulfate and concentrated oligomeric dextran, Electrophoresis 16 (8) (1995) 1463–1467.
[251] E.N. Fung, E.S. Yeung, High-speed DNA sequencing by using mixed poly(ethylene oxide) solutions in uncoated capillary columns, Analytical Chemistry 67 (13) (1995) 1913–1919.
[252] A. Guttman, On the separation mechanism of capillary sodium dodecyl sulfate-gel electrophoresis of proteins, Electrophoresis 16 (4) (1995) 611–616.

[253] A. Guttman, High-resolution carbohydrate profiling by capillary gel electrophoresis, Nature (London) 380 (6573) (1996) 461–462.
[254] S.D. Proctor, J.C. Mamo, Separation and quantification of apolipoprotein B-48 and other apolipoproteins by dynamic sieving capillary electrophoresis, Journal of Lipid Research 38 (1997) 410–414.
[255] J. Stocks, M.N. Nanjee, N.E. Miller, Analysis of high density lipoprotein apolipoproteins by capillary zone and capillary SDS gel electrophoresis, Journal of Lipid Research 39 (1) (1998) 218–227.
[256] M. Chiari, M. Nesi, P.G. Righetti, Capillary zone electrophoresis of DNA fragments in a novel polymer network: poly(N-acryloylaminoethoxyethanol), Electrophoresis 15 (5) (1994) 616–622.
[257] E. Simo-Alfonso, M. Conti, C. Gelfi, P.G. Righetti, Sodium dodecyl sulfate capillary electrophoresis of proteins in entangled solutions of poly(vinyl alcohol), Journal of Chromatography A 689 (1) (1995) 85–96.
[258] R. Lausch, T. Scheper, O.W. Reif, J. Schloesser, J. Fleischer, R. Freitag, Rapid capillary gel electrophoresis of proteins, Journal of Chromatography 654 (1) (1993) 190–195.
[259] S. Yao, D.S. Anex, W.B. Caldwell, D.W. Arnold, K.B. Smith, P.G. Schultz, SDS capillary gel electrophoresis of proteins in microfabricated channels, Proceedings of the National Academy of Sciences of the United States of America 96 (10) (1999) 5372–5377.
[260] G. Righetti Pier, R. Sebastiano, A. Citterio, Capillary electrophoresis and isoelectric focusing in peptide and protein analysis, Proteomics 13 (2) (2013) 325–340.
[261] M.R. Pergande, S.M. Cologna, Isoelectric point separations of peptides and proteins, Proteomes 5 (1) (2017) 4.
[262] S. Mack, D. Arnold, G. Bogdan, L. Bousse, L. Danan, V. Dolnik, et al., A novel microchip-based imaged CIEF-MS system for comprehensive characterization and identification of biopharmaceutical charge variants, Electrophoresis 40 (23–24) (2019) 3084–3091.
[263] J.R. Mazzeo, I.S. Krull, Capillary isoelectric-focusing of proteins in uncoated fused-silica capillaries using polymeric additives, Analytical Chemistry 63 (24) (1991) 2852–2857.
[264] B. Bjellqvist, K. Ek, P.G. Righetti, E. Gianazza, A. Gorg, R. Westermeier, et al., Isoelectric-focusing in immobilized ph gradients—principle, methodology and some applications, Journal of Biochemical and Biophysical Methods 6 (4) (1982) 317–339.
[265] S.M. Chen, J.E. Wiktorowicz, Isoelectric focusing by free solution capillary electrophoresis, Analytical Biochemistry 206 (1) (1992) 84–90.
[266] J. Johansson, D.T. Witte, M. Larsson, S. Nilsson, Real-time fluorescence imaging of isotachophoretic preconcentration for capillary electrophoresis, Analytical Chemistry 68 (17) (1996) 2766–2770.
[267] X.Z. Wu, J. Wu, J. Pawliszyn, Fluorescence imaging detection for capillary isoelectric focusing, Electrophoresis 16 (8) (1995) 1474–1478.
[268] B.K. Clark, M.J. Sepaniak, Evaluation of a spatial scanning laser fluorometric detector for capillary electrophoresis: application to studies of band migration and dispersion, Journal of Microcolumn Separations 7 (6) (1995) 593–601.
[269] A.S. Zarabadi, T. Huang, J.G. Mielke, Capillary isoelectric focusing with whole column imaging detection (iCIEF): a new approach to the characterization and quantification of salivary alpha-amylase, Journal of Chromatography, B, Analytical Technologies in the Biomedical and Life Sciences 1053 (2017) 65–71.
[270] T. Pritchett, Isoelectric focusing of proteins by capillary electrophoresis, Current Innovations in Molecular Biology 1 (1995) 127–145.

[271] U. Grossbach, Microelectrofocusing of proteins in capillary gels, Biochemical and Biophysical Research Communications 49 (3) (1972) 667–672.
[272] T.L. Huang, P.C.H. Shieh, N. Cooke, Isoelectric-focusing of proteins in capillary electrophoresis with pressure-driven mobilization, Chromatographia 39 (9–10) (1994) 543–548.
[273] M. Horka, F. Ruzicka, V. Hola, K. Slais, Separation of similar yeast strains by IEF techniques, Electrophoresis 30 (12) (2009) 2134–2141.
[274] T. Li, T. Gao, H. Chen, P. Pekker, A. Menyhart, A. Guttman, Rapid determination of full and empty adeno-associated virus capsid ratio by capillary isoelectric focusing, Current Molecular Medicine 20 (10) (2020) 814–820.
[275] T. Li, T. Gao, H. Chen, Z. Demianova, F. Wang, M. Malik, et al., Determination of full, partial and empty capsid ratios for adeno-associated virus (AAV) analysis. Sciex Technical Note, Drug Discovery and Development, 2020, p. 4.
[276] B.P. Salmanowicz, M. Langner, S. Franaszek, Charge-based characterisation of high-molecular-weight glutenin subunits from common wheat by capillary isoelectric focusing, Talanta 129 (2014) 9–14.
[277] J. Lin, Q. Tan, S. Wang, A high-resolution capillary isoelectric focusing method for the determination of therapeutic recombinant monoclonal antibody, Journal of Separation Science 34 (14) (2011) 1696–1702.
[278] P. Dou, Z. Liu, J. He, J.J. Xu, H.Y. Chen, Rapid and high-resolution glycoform profiling of recombinant human erythropoietin by capillary isoelectric focusing with whole column imaging detection, Journal of Chromatography A 1190 (1–2) (2008) 372–376.
[279] O. Salas-Solano, B. Kennel, S.S. Park, K. Roby, Z. Sosic, B. Boumajny, et al., Robustness of iCIEF methodology for the analysis of monoclonal antibodies: an interlaboratory study, Journal of Separation Science 35 (22) (2012) 3124–3129.
[280] G. Zhu, L. Sun, R. Wojcik, D. Kernaghan, J.B. McGivney IV, N.J. Dovichi, et al., A rapid cIEF–ESI–MS/MS method for host cell protein analysis of a recombinant human monoclonal antibody, Talanta 98 (2012) 253–256.
[281] C.P. Toseland, Fluorescent labeling and modification of proteins, Journal of Chemical Biology 6 (3) (2013) 85–95.
[282] K.M. Dean, A.E. Palmer, Advances in fluorescence labeling strategies for dynamic cellular imaging, Nature Chemical Biology 10 (7) (2014) 512–523.
[283] A.M. Garcia-Campana, M. Taverna, H. Fabre, LIF detection of peptides and proteins in CE, Electrophoresis 28 (1–2) (2007) 208–232.
[284] M. Lacroix, V. Poinsot, C. Fournier, F. Couderc, Laser-induced fluorescence detection schemes for the analysis of proteins and peptides using capillary electrophoresis, Electrophoresis 26 (13) (2005) 2608–2621.
[285] C. Colyer, Noncovalent labeling of proteins in capillary electrophoresis with laser-induced fluorescence detection, Cell Biochemistry and Biophysics 33 (3) (2000) 323–337.
[286] M. Sano, I. Nishino, K. Ueno, H. Kamimori, Assay of collagenase activity for native triple-helical collagen using capillary gel electrophoresis with laser-induced fluorescence detection, Journal of Chromatography, B: Analytical Technologies in the Biomedical and Life Sciences 809 (2) (2004) 251–256.
[287] K. Hu, H. Ahmadzadeh, S.N. Krylov, Asymmetry between sister cells in a cancer cell line revealed by chemical cytometry, Analytical Chemistry 76 (13) (2004) 3864–3866.
[288] E.H. Turner, K. Lauterbach, H.R. Pugsley, V.R. Palmer, N.J. Dovichi, Detection of green fluorescent protein in a single bacterium by capillary electrophoresis with laser-induced fluorescence, Analytical Chemistry 79 (2) (2007) 778–781.

[289] E.L. Gump, C.A. Monnig, Pre-column derivatization of proteins to enhance detection sensitivity for sodium dodecyl sulfate non-gel sieving capillary electrophoresis, Journal of Chromatography A 715 (1) (1995) 167–177.
[290] E.A. Arriaga, Y.N. Zhang, N.J. Dovichi, Use of 3-(P-carboxybenzoyl)quinoline-2-carboxaldehyde to label amino-acids for high-sensitivity fluorescence detection in capillary electrophoresis, Analytica Chimica Acta 299 (3) (1995) 319–326.
[291] S.D. Gilman, J.J. Pietron, A.G. Ewing, Postcolumn derivatization in narrow-bore capillaries for the analysis of amino-acids and proteins by capillary electrophoresis with fluorescence detection, Journal of Microcolumn Separations 6 (4) (1994) 373–384.
[292] I. Benito, M.L. Marina, J.M. Saz, J.C. Diez-Masa, Detection of bovine whey proteins by on-column derivatization capillary electrophoresis with laser-induced fluorescence monitoring, Journal of Chromatography A 841 (1) (1999) 105–114.
[293] E.D. Moody, P.J. Viskari, C.L. Colyer, Non-covalent labeling of human serum albumin with indocyanine green: a study by capillary electrophoresis with diode laser-induced fluorescence detection, Journal of Chromatography, B, Biomedical Sciences and Applications 729 (1–2) (1999) 55–64.
[294] M.D. Harvey, D. Bandilla, P.R. Banks, Subnanomolar detection limit for sodium dodecyl sulfate-capillary gel electrophoresis using a fluorogenic, noncovalent dye, Electrophoresis 19 (12) (1998) 2169–2174.
[295] S. Hu, Z. Zhang, L.M. Cook, E.J. Carpenter, N.J. Dovichi, Separation of proteins by sodium dodecylsulfate capillary electrophoresis in hydroxypropylcellulose sieving matrix with laser-induced fluorescence detection, Journal of Chromatography A 894 (2000) 291–296.
[296] S. Hu, L. Zhang, L.M. Cook, N.J. Dovichi, Capillary sodium dodecyl sulfate-DALT electrophoresis of proteins in a single human cancer cell, Electrophoresis 22 (2001) 3677–3682.
[297] J. Feng, E.A. Arriaga, Quantification of carbonylated proteins in rat skeletal muscle mitochondria using capillary sieving electrophoresis with laser-induced fluorescence detection, Electrophoresis 29 (2) (2008) 475–482.
[298] S. Wu, J.J. Lu, S. Wang, K.L. Peck, G. Li, S. Liu, Staining method for protein analysis by capillary gel electrophoresis, Analytical Chemistry 79 (20) (2007) 7727–7733.
[299] D.B. Craig, R.M. Polakowski, E. Arriaga, J.C.Y. Wong, H. Ahmadzadeh, C. Stathakis, Sodium dodecyl sulfate-capillary electrophoresis of proteins in a sieving matrix utilizing two-spectral channel laser-induced fluorescence detection, Electrophoresis 19 (12) (1998) 2175–2178.
[300] K.E. Swearingen, J.A. Dickerson, E.H. Turner, L.M. Ramsay, R. Wojcik, N.J. Dovichi, Reaction of fluorogenic reagents with proteins II: capillary electrophoresis and laser-induced fluorescence properties of proteins labeled with Chromeo P465, Journal of Chromatography A 1194 (2) (2008) 249–252.
[301] E.H. Turner, J.A. Dickerson, L.M. Ramsay, K.E. Swearingen, R. Wojcik, N.J. Dovichi, Reaction of fluorogenic reagents with proteins III. Spectroscopic and electrophoretic behavior of proteins labeled with Chromeo P503, Journal of Chromatography A 1194 (2) (2008) 253–256.
[302] A. Szekrenyes, U. Roth, M. Kerekgyarto, A. Szekely, I. Kurucz, K. Kowalewski, et al., High-throughput analysis of therapeutic and diagnostic monoclonal antibodies by multicapillary SDS gel electrophoresis in conjunction with covalent fluorescent labeling, Analytical and Bioanalytical Chemistry 404 (5) (2012) 1485–1494.
[303] Z. Zhang, J. Park, H. Barrett, S. Dooley, C. Davies, M.F. Verhagen, Capillary electrophoresis-sodium dodecyl sulfate with laser-induced fluorescence detection as a highly sensitive and quality control-friendly method for monitoring adeno-associated virus capsid protein purity, Human Gene Therapy 32(11–12) (2021) 628–637.

[304] C. Zhang, C. Rodriguez, M.L. Circu, T.Y. Aw, J. Feng, S-Glutathionyl quantification in the attomole range using glutaredoxin-3-catalyzed cysteine derivatization and capillary gel electrophoresis with laser-induced fluorescence detection, Analytical and Bioanalytical Chemistry 401 (7) (2011) 2165–2175.
[305] E. Tamizi, A. Jouyban, The potential of the capillary electrophoresis techniques for quality control of biopharmaceuticals—a review, Electrophoresis 36 (6) (2015) 831–858.
[306] G. Walsh, Biopharmaceutical benchmarks 2018, Nature Biotechnology 36 (12) (2018) 1136–1145.
[307] J. Zhang, S. Burman, S. Gunturi, J.P. Foley, Method development and validation of capillary sodium dodecyl sulfate gel electrophoresis for the characterization of a monoclonal antibody, Journal of Pharmaceutical and Biomedical Analysis 53 (5) (2010) 1236–1243.
[308] R.R. Rustandi, M.W. Washabaugh, Y. Wang, Applications of CE SDS gel in development of biopharmaceutical antibody-based products, Electrophoresis 29 (17) (2008) 3612–3620.
[309] A. Szekely, A. Szekrenyes, M. Kerekgyarto, A. Balogh, J. Kadas, J. Lazar, et al., Multicapillary SDS-gel electrophoresis for the analysis of fluorescently labeled mAb preparations: a high throughput quality control process for the production of QuantiPlasma and PlasmaScan mAb libraries, Electrophoresis 35 (15) (2014) 2155–2162.
[310] S. Hapuarachchi, S. Fodor, I. Apostol, G. Huang, Use of capillary electrophoresis-sodium dodecyl sulfate to monitor disulfide scrambled forms of an Fc fusion protein during purification process, Analytical Biochemistry 414 (2) (2011) 187–195.
[311] O.O. Dada, R. Rao, N. Jones, N. Jaya, O. Salas-Solano, Comparison of SEC and CE-SDS methods for monitoring hinge fragmentation in IgG1 monoclonal antibodies, Journal of Pharmaceutical and Biomedical Analysis 145 (2017) 91–97.
[312] D.A. Michels, M. Parker, O. Salas-Solano, Quantitative impurity analysis of monoclonal antibody size heterogeneity by CE-LIF: example of development and validation through a quality-by-design framework, Electrophoresis 33 (5) (2012) 815–826.
[313] M.E. Le, A. Vizel, K.M. Hutterer, Automated sample preparation for CE-SDS, Electrophoresis 34 (9–10) (2013) 1369–1374.
[314] K. Tsuji, Sodium dodecyl sulfate polyacrylamide gel- and replaceable polymer-filled capillary electrophoresis for molecular mass determination of proteins of pharmaceutical interest, Journal of Chromatography, B: Biomedical Applications 662 (2) (1994) 291–299.
[315] N.M. Kinghorn, C.S. Norris, G.R. Paterson, D.E. Otter, Comparison of capillary electrophoresis with traditional methods to analyze bovine whey proteins, Journal of Chromatography A 700 (1–2) (1995) 111–123.
[316] P. Feng, C. Fuerer, A. McMahon, Quantification of whey protein content in infant formulas by sodium dodecyl sulfate-capillary gel electrophoresis (SDS-CGE): single-laboratory validation, first action 2016.15, Journal of AOAC International 100 (2) (2017) 510–521.
[317] W.E. Werner, J.E. Wiktorowicz, D.D. Kasarda, Wheat varietal identification by capillary electrophoresis of gliadins and high molecular weight glutenin subunits, Cereal Chemistry 71 (5) (1994) 397–402.
[318] J. Liu, S. Abid, M.S. Lee, Analysis of monoclonal antibody chimeric BR96-doxorubicin immunoconjugate by sodium dodecyl sulfate-capillary gel electrophoresis with ultraviolet and laser-induced fluorescence detection, Analytical Biochemistry 229 (2) (1995) 221–228.
[319] C. Lin, F. Cotton, C. Boutique, D. Dhermy, F. Vertongen, B. Gulbis, Capillary gel electrophoresis: separation of major erythrocyte membrane proteins, Journal of Chromatography B: Biomedical Sciences and Applications 742 (2) (2000) 411–419.

[320] A. Hiraoka, I. Tominaga, K. Hori, Sodium dodecylsulfate capillary gel electrophoretic measurement of the concentration ratios of albumin and a2-macroglobulin in cerebrospinal fluid and serum of patients with neurological disorders, Journal of Chromatography A 895 (1−2) (2000) 339−344.
[321] B.P. Salmanowicz, Capillary electrophoresis of seed 2S albumins from Lupinus species, Journal of Chromatography A 894 (1−2) (2000) 297−310.
[322] H.G. Lee, High-performance sodium dodecyl sulfate-capillary gel electrophoresis of antibodies and antibody fragments, Journal of Immunological Methods 234 (1−2) (2000) 71−81.
[323] B. Deng, N. Ye, G. Luo, Y. Wang, Separation of tissue proteins of human lung carcinomas by partial-filling capillary electrophoresis, Journal of Nanoscience and Nanotechnology 5 (8) (2005) 1193−1198.
[324] E. van Tricht, L. Geurink, B. Pajic, J. Nijenhuis, H. Backus, M. Germano, et al., New capillary gel electrophoresis method for fast and accurate identification and quantification of multiple viral proteins in influenza vaccines, Talanta 144 (2015) 1030−1035.
[325] M.C.M. Mellado, C. Franco, A. Coelho, P.M. Alves, A.L. Simplicio, Sodium dodecyl sulfate-capillary gel electrophoresis analysis of rotavirus-like particles, Journal of Chromatography A 1192 (1) (2008) 166−172.
[326] W.L. Tseng, Y.W. Lin, H.T. Chang, Improved separation of microheterogeneities and isoforms of proteins by capillary electrophoresis using segmental filling with SDS and PEO in the background electrolyte, Analytical Chemistry 74 (18) (2002) 4828−4834.
[327] D.B. Gomis, S. Junco, Y. Exposito, M.D. Gutierrez, Size-based separations of proteins by capillary electrophoresis using linear polyacrylamide as a sieving medium: model studies and analysis of cider proteins, Electrophoresis 24 (9) (2003) 1391−1396.
[328] D. Blanco, S. Junco, Y. Exposito, D. Gutierrez, Study of various treatments to isolate low levels of cider proteins to be analyzed by capillary sieving electrophoresis, Journal of Liquid Chromatography & Related Technologies 27 (10) (2004) 1523−1539.
[329] M. Vergara-Barberan, M.J. Lerma-Garcia, J.M. Herrero-Martinez, E.F. Simo-Alfonso, Classification of olive leaves and pulps according to their cultivar by using protein profiles established by capillary gel electrophoresis, Analytical and Bioanalytical Chemistry 406 (6) (2014) 1731−1738.
[330] S. Hu, L. Zhang, S. Krylov, N.J. Dovichi, Cell cycle-dependent protein fingerprint from a single cancer cell: image cytometry coupled with single-cell capillary sieving electrophoresis, Analytical Chemistry 75 (14) (2003) 3495−3501.
[331] J.M. Busnel, A. Varenne, S. Descroix, G. Peltre, Y. Gohon, P. Gareil, Evaluation of capillary isoelectric focusing in glycerol-water media with a view to hydrophobic protein applications, Electrophoresis 26 (17) (2005) 3369−3379.
[332] S. Peng, R. Shi, R. Yang, D. Zhou, Y. Wang, Hydroxyethylcellulose-graft-poly (N,N-dimethylacrylamide) copolymer as a multifunctional separation medium for CE, Electrophoresis 29 (21) (2008) 4351−4354.
[333] G.J. Anderson, C.M. Cipolla, R.T. Kennedy, Western blotting using capillary electrophoresis, Analytical Chemistry 83 (4) (2011) 1350−1355.
[334] M. Kodani, M. Martin, V.L. de Castro, J. Drobeniuc, S. Kamili, An automated immunoblot method for detection of IgG antibodies to hepatitis C virus: a potential supplemental antibody confirmatory assay, Journal of Clinical Microbiology 57 (3) (2019) e01567-18.
[335] J.J. Lu, Z. Zhu, W. Wang, S. Liu, Coupling sodium dodecyl sulfate-capillary polyacrylamide gel electrophoresis with matrix-assisted laser desorption ionization time-of-flight mass spectrometry via a poly(tetrafluoroethylene) membrane, Analytical Chemistry 83 (5) (2011) 1784−1790.

[336] L. Sanchez-Hernandez, C. Montealegre, S. Kiessig, B. Moritz, C. Neususs, In-capillary approach to eliminate SDS interferences in antibody analysis by capillary electrophoresis coupled to mass spectrometry, Electrophoresis 38 (7) (2017) 1044–1052.
[337] J. Beckman, Y. Song, Y. Gu, S. Voronov, N. Chennamsetty, S. Krystek, et al., Purity determination by capillary electrophoresis sodium hexadecyl sulfate (CE-SHS): a novel application for therapeutic protein characterization, Analytical Chemistry 90 (4) (2018) 2542–2547.
[338] A. Varki, R.D. Cummings, J.D. Esko, P. Stanley, G.W. Hart, M. Aebi, et al., Essentials of Glycobiology, 3rd ed., Cold Spring Harbor Laboratory Press, Cold Spring Harbor (NY), 2015.
[339] M.V. Novotny, W.R. Alley Jr., B.F. Mann, Analytical glycobiology at high sensitivity: current approaches and directions, Glycoconjugate Journal 30 (2) (2013) 89–117.
[340] G. Zauner, A.M. Deelder, M. Wuhrer, Recent advances in hydrophilic interaction liquid chromatography (HILIC) for structural glycomics, Electrophoresis 32 (24) (2011) 3456–3466.
[341] Y. Mimura, T. Katoh, R. Saldova, R. O'Flaherty, T. Izumi, Y. Mimura-Kimura, et al., Glycosylation engineering of therapeutic IgG antibodies: challenges for the safety, functionality and efficacy, Protein Cell 9 (1) (2018) 47–62.
[342] T. Merry, S. Astrautsova, Chemical and enzymatic release of glycans from glycoproteins, Methods in Molecular Biology 213 (2003) 27–40.
[343] Y. Mechref, M.V. Novotny, Structural investigations of glycoconjugates at high sensitivity, Chemical Reviews 102 (2) (2002) 321–369.
[344] R.A. O'Neill, Enzymatic release of oligosaccharides from glycoproteins for chromatographic and electrophoretic analysis, Journal of Chromatography A 720 (1–2) (1996) 201–215.
[345] S. Yamamoto, M. Kinoshita, S. Suzuki, Current landscape of protein glycosylation analysis and recent progress toward a novel paradigm of glycoscience research, Journal of Pharmaceutical and Biomedical Analysis 130 (2016) 273–300.
[346] T.S. Mattu, L. Royle, J. Langridge, M.R. Wormald, P.E. Van den Steen, J. Van Damme, et al., O-glycan analysis of natural human neutrophil gelatinase B using a combination of normal phase-HPLC and online tandem mass spectrometry: implications for the domain organization of the enzyme, Biochemistry 39 (51) (2000) 15695–15704.
[347] W.B. Struwe, R. Gough, M.E. Gallagher, D.T. Kenny, S.D. Carrington, N.G. Karlsson, et al., Identification of O-glycan structures from chicken intestinal mucins provides insight into campylobactor jejuni pathogenicity, Molecular & Cellular Proteomics: MCP 14 (6) (2015) 1464–1477.
[348] M. Breadmore, E. Hilder, A. Kazarian, Fluorophores and chromophores for the separation of carbohydrates by capillary electrophoresis, in: N. Volpi (Ed.), Capillary Electrophoresis of Carbohydrates: From Monosaccharides to Complex Polysaccharides, Humana Press, Totowa, NJ, 2011, pp. 23–51.
[349] V. Mantovani, F. Galeotti, F. Maccari, N. Volpi, Recent advances in capillary electrophoresis separation of monosaccharides, oligosaccharides, and polysaccharides, Electrophoresis 39 (1) (2018) 179–189.
[350] N.V. Shilova, N.V. Bovin, Fluorescent labels for analysis of mono- and oligosaccharides, Russian Journal of Bioorganic Chemistry 29 (4) (2003) 309–324.
[351] R.A. Evangelista, M.-S. Liu, F.-T.A. Chen, Characterization of 9-aminopyrene-1,4,6-trisulfonate derivatized sugars by capillary electrophoresis with laser-induced fluorescence detection, Analytical Chemistry 67 (13) (1995) 2239–2245.
[352] C. Chiesa, C. Horvath, Capillary zone electrophoresis of malto-oligosaccharides derivatized with 8-aminonaphthalene-1,3,6-trisulfonic acid, Journal of Chromatography 645 (2) (1993) 337–352.

[353] H. Schwaiger, P.J. Oefner, C. Huber, E. Grill, G.K. Bonn, Capillary zone electrophoresis and micellar electrokinetic chromatography of 4-aminobenzonitrile carbohydrate derivatives, Electrophoresis 15 (1994) 941–952.
[354] S. Honda, S. Suzuki, A. Nose, K. Yamamoto, K. Kakehi, Capillary zone electrophoresis of reducing monosaccharides and oligosaccharides as the borate complexes of their 3-methyl-1-phenyl-2-pyrazolin-5-one derivatives, Carbohydrate Research 215 (1) (1991) 193–198.
[355] J.P. Liu, O. Shirota, M. Novotny, Capillary electrophoresis of amino-sugars with laser-induced fluorescence detection, Analytical Chemistry 63 (5) (1991) 413–417.
[356] A. Guttman, C. Starr, Capillary and slab gel electrophoresis profiling of oligosaccharides, Electrophoresis 16 (6) (1995) 993–997.
[357] J. Liu, O. Shirota, M.V. Novotny, Sensitive, laser-assisted determination of complex oligosaccharide mixtures separated by capillary gel electrophoresis at high resolution, Analytical Chemistry 64 (8) (1992) 973–975.
[358] J. Sudor, M. Novotny, Electromigration behavior of polysaccharides in capillary electrophoresis under pulsed-field conditions, Proceedings of the National Academy of Sciences of the United States of America 90 (20) (1993) 9451–9455.
[359] A. Guttman, N. Cooke, C.M. Starr, Capillary electrophoresis separation of oligosaccharides: I. Effect of operational variables, Electrophoresis 15 (12) (1994) 1518–1522.
[360] A. Guttman, M. Kerekgyarto, G. Jarvas, Effect of separation temperature on structure specific glycan migration in capillary electrophoresis, Analytical Chemistry 87 (23) (2015) 11630–11634.
[361] M. Szigeti, A. Guttman, High-resolution glycan analysis by temperature gradient capillary electrophoresis, Analytical Chemistry 89 (4) (2017) 2201–2204.
[362] A. Guttman, T. Pritchett, Capillary gel electrophoresis separation of high-mannose type oligosaccharides derivatized by 1-aminopyrene-3,6,8-trisulfonic acid, Electrophoresis 16 (10) (1995) 1906–1911.
[363] A. Guttman, S. Herrick, Effect of the quantity and linkage position of mannose (a1,2) residues in capillary gel electrophoresis of high-mannose-type oligosaccharides, Analytical Biochemistry 235 (2) (1996) 236–239.
[364] K. Kakehi, M. Kinoshita, S. Hayase, Y. Oda, Capillary electrophoresis of N-acetylneuraminic acid polymers and hyaluronic acid: correlation between migration order reversal and biological functions, Analytical Chemistry 71 (8) (1999) 1592–1596.
[365] S. Hayase, Y. Oda, S. Honda, K. Kakehi, High-performance capillary electrophoresis of hyaluronic acid: determination of its amount and molecular mass, Journal of Chromatography A 768 (2) (1997) 295–305.
[366] M. Hong, J. Sudor, M. Stefansson, M.V. Novotny, High-resolution studies of hyaluronic acid mixtures through capillary gel electrophoresis, Analytical Chemistry 70 (3) (1998) 568–573.
[367] J. Schwarzer, E. Rapp, U. Reichl, N-glycan analysis by CGE-LIF: profiling influenza A virus hemagglutinin N-glycosylation during vaccine production, Electrophoresis 29 (20) (2008) 4203–4214.
[368] M. Guttman, C. Varadi, K.K. Lee, A. Guttman, Comparative glycoprofiling of HIV gp120 immunogens by capillary electrophoresis and MALDI mass spectrometry, Electrophoresis 36 (11–12) (2015) 1305–1313.
[369] M. Szigeti, J. Bondar, D. Gjerde, Z. Keresztessy, A. Szekrenyes, A. Guttman, Rapid N-glycan release from glycoproteins using immobilized PNGase F microcolumns, Journal of Chromatography, B, Analytical Technologies in the Biomedical and Life Sciences 1032 (2016) 139–143.

[370] M. Szigeti, C. Lew, K. Roby, A. Guttman, Fully automated sample preparation for ultrafast N-glycosylation analysis of antibody therapeutics, Journal of Laboratory Automation 21 (2) (2016) 281–286.
[371] D. Reusch, M. Haberger, T. Kailich, A.K. Heidenreich, M. Kampe, P. Bulau, et al., High-throughput glycosylation analysis of therapeutic immunoglobulin G by capillary gel electrophoresis using a DNA analyzer, mAbs 6 (1) (2014) 185–196.
[372] L.R. Ruhaak, C.A. Koeleman, H.W. Uh, J.C. Stam, D. van Heemst, A.B. Maier, et al., Targeted biomarker discovery by high throughput glycosylation profiling of human plasma alpha1-antitrypsin and immunoglobulin A, PLoS One 8 (9) (2013) e73082.
[373] C. Varadi, Z. Hollo, S. Poliska, L. Nagy, Z. Szekanecz, A. Vancsa, et al., Combination of IgG N-glycomics and corresponding transcriptomics data to identify anti-TNF-alpha treatment responders in inflammatory diseases, Electrophoresis 36 (11–12) (2015) 1330–1335.
[374] B. Meszaros, G. Jarvas, A. Farkas, M. Szigeti, Z. Kovacs, R. Kun, et al., Comparative analysis of the human serum N-glycome in lung cancer, COPD and their comorbidity using capillary electrophoresis, Journal of Chromatography, B, Analytical Technologies in the Biomedical and Life Sciences 1137 (2020) 121913.
[375] A. Guttman, S. Brunet, N. Cooke, Capillary electrophoresis fingerprinting of carbohydrates in the biopharmaceutical and food and beverage industries, LC–GC 14 (9) (1996) 788. 790, 791–792.
[376] T.G. Szilagyi, B.H. Vecseri, Z. Kiss, L. Hajba, A. Guttman, Analysis of the oligosaccharide composition in wort samples by capillary electrophoresis with laser induced fluorescence detection, Food Chemistry 256 (2018) 129–132.
[377] S. Albrecht, G.C. van Muiswinkel, H.A. Schols, A.G. Voragen, H. Gruppen, Introducing capillary electrophoresis with laser-induced fluorescence detection (CE-LIF) for the characterization of konjac glucomannan oligosaccharides and their in vitro fermentation behavior, Journal of Agricultural and Food Chemistry 57 (9) (2009) 3867–3876.
[378] S.D. Vicaretti, N.A. Mohtarudin, A.M. Garner, W.F. Zandberg, Capillary electrophoresis analysis of bovine milk oligosaccharides permits an assessment of the influence of diet and the discovery of nine abundant sulfated analogues, Journal of Agricultural and Food Chemistry 66 (32) (2018) 8574–8583.
[379] D. Sarkozy, B. Borza, A. Domokos, E. Varadi, M. Szigeti, A. Meszaros-Matwiejuk, et al., Ultrafast high-resolution analysis of human milk oligosaccharides by multicapillary gel electrophoresis, Food Chemistry 341 (Pt 2) (2021) 128200.
[380] M. Szigeti, A. Meszaros-Matwiejuk, D. Molnar-Gabor, A. Guttman, Rapid capillary gel electrophoresis analysis of human milk oligosaccharides for food additive manufacturing in-process control, Analytical and Bioanalytical Chemistry 413 (6) (2021) 1595–1603.
[381] J. Khandurina, D.L. Blum, J.T. Stege, A. Guttman, Automated carbohydrate profiling by capillary electrophoresis: a bioindustrial approach, Electrophoresis 25 (14) (2004) 2326–2331.
[382] J. Khandurina, N.A. Olson, A.A. Anderson, K.A. Gray, A. Guttman, Large-scale carbohydrate analysis by capillary array electrophoresis: part 1. Separation and scale-up, Electrophoresis 25 (18–19) (2004) 3117–3121.
[383] J. Khandurina, A.A. Anderson, N.A. Olson, J.T. Stege, A. Guttman, Large-scale carbohydrate analysis by capillary array electrophoresis: part 2. Data normalization and quantification, Electrophoresis 25 (18–19) (2004) 3122–3127.
[384] N. Callewaert, R. Contreras, L. Mitnik-Gankin, L. Carey, P. Matsudaira, D. Ehrlich, Total serum protein N-glycome profiling on a capillary electrophoresis-microfluidics platform, Electrophoresis 25 (18–19) (2004) 3128–3131.

[385] R.P.P. Soares, T. Barron, K. McCoy-Simandle, M. Svobodova, A. Warburg, S.J. Turco, Leishmania tropica: intraspecific polymorphisms in lipophosphoglycan correlate with transmission by different Phlebotomus species, Experimental Parasitology 107 (1–2) (2004) 105–114.
[386] F. Le Floch, B. Tessier, S. Chenuet, J.M. Guillaume, P. Cans, A. Marc, et al., HPCE monitoring of the N-glycosylation pattern and sialylation of murine erythropoietin produced by CHO cells in batch processes, Biotechnology Progress 20 (3) (2004) 864–871.
[387] M. Dunn, Affinity electrophoresis, Laboratory Practice 33 (3) (1984) 13–14. 17–18.
[388] I.J. Colton, J.D. Carbeck, J. Rao, G.M. Whitesides, Affinity capillary electrophoresis: a physical-organic tool for studying interactions in biomolecular recognition, Electrophoresis 19 (3) (1998) 367–382.
[389] K. Takeo, Advances in affinity electrophoresis, Journal of Chromatography A 698 (1–2) (1995) 89–105.
[390] A. Guttman, N. Cooke, Capillary gel affinity electrophoresis of DNA fragments, Analytical Chemistry 63 (18) (1991) 2038–2042.
[391] S. Nathakarnkitkool, P.J. Oefner, G. Bartsch, M.A. Chin, G.K. Bonn, High-resolution capillary electrophoretic analysis of DNA in free solution, Electrophoresis 13 (1–2) (1992) 18–31.
[392] A. Guttman, A. Paulus, A.S. Cohen, N. Grinberg, B.L. Karger, Use of complexing agents for selective separation in high-performance capillary electrophoresis. Chiral resolution via cyclodextrins incorporated within polyacrylamide gel columns, Journal of Chromatography 448 (1) (1988) 41–53.
[393] A. Vegvari, A. Foldesi, C. Hetenyi, O. Kocnegarova, M.G. Schmid, V. Kudirkaite, et al., A new easy-to-prepare homogeneous continuous electrochromatographic bed for enantiomer recognition, Electrophoresis 21 (15) (2000) 3116–3125.
[394] R.P. Singhal, J. Xian, Separation of DNA restriction fragments by polymer-solution capillary zone electrophoresis: influence of polymer concentration and ion-pairing reagents, Journal of Chromatography 652 (1) (1993) 47–56.
[395] S. Birnbaum, S. Nilsson, Protein-based capillary affinity gel electrophoresis for the separation of optical isomers, Analytical Chemistry 64 (22) (1992) 2872–2874.
[396] I.D. Cruzado, G. Vigh, Chiral separations by capillary electrophoresis using cyclodextrin-containing gels, Journal of Chromatography 608 (1–2) (1992) 421–425.
[397] W. Schuetzner, S. Fanali, A. Rizzi, E. Kenndler, Separation of diastereomeric derivatives of enantiomers by capillary zone electrophoresis with a polymer network: use of polyvinylpyrrolidone as buffer additive, Journal of Chromatography 639 (2) (1993) 375–378.
[398] B. Chankvetadze, Contemporary theory of enantioseparations in capillary electrophoresis, Journal of Chromatography A 1567 (2018) 2–25.
[399] P. Sun, G.E. Barker, R.A. Hartwick, N. Grinberg, R. Kaliszan, Chiral separations using an immobilized protein-dextran polymer network in affinity capillary electrophoresis, Journal of Chromatography 652 (1) (1993) 247–252.
[400] P. Sun, G.E. Barker, G.J. Mariano, R.A. Hartwick, Enhanced chiral separation of dansylated amino acids with cyclodextrin-dextran polymer network by capillary electrophoresis, Electrophoresis 15 (6) (1994) 793–798.
[401] G. Hellinghausen, D.W. Armstrong, Cyclofructans as chiral selectors: an overview, Methods in Molecular Biology 1985 (2019) 183–200.
[402] B.A. Ingelse, F.M. Everaerts, J. Sevcik, Z. Stransky, S. Fanali, A further study on the chiral separation power of a soluble neutral b-cyclodextrin polymer, Journal of High Resolution Chromatography 18 (6) (1995) 348–352.

[403] B.A. Ingelse, F.M. Everaerts, C. Desiderio, S. Fanali, Enantiomeric separation by capillary electrophoresis using a soluble neutral b-cyclodextrin polymer, Journal of Chromatography A 709 (1) (1995) 89–98.
[404] H. Ljungberg, S. Nilsson, Protein-based capillary affinity gel electrophoresis for chiral separation of b-adrenergic blockers, Journal of Liquid Chromatography 18 (18–19) (1995) 3685–3698.
[405] H. Nishi, Enantiomer separation of basic drugs by capillary electrophoresis using ionic and neutral polysaccharides as chiral selectors, Journal of Chromatography A 735 (1–2) (1996) 345–351.
[406] G.M. Beck, S.H. Neau, Lambda-carrageenan: a novel chiral selector for capillary electrophoresis, Chirality 8 (7) (1996) 503–510.
[407] F. Priego-Capote, L. Ye, S. Shakil, S.A. Shamsi, S. Nilsson, Monoclonal behavior of molecularly imprinted polymer nanoparticles in capillary electrochromatography, Analytical Chemistry 80 (8) (2008) 2881–2887.
[408] J. Chen, Y. Du, F. Zhu, B. Chen, Glycogen: a novel branched polysaccharide chiral selector in CE, Electrophoresis 31 (6) (2010) 1044–1050.
[409] M. Akashi, T. Sawa, Y. Baba, M. Tsuhako, Specific separation of oligodeoxynucleotides by capillary affinity gel electrophoresis (CAGE) using poly(9-vinyladenine)-polyacrylamide conjugated gel, Journal of High Resolution Chromatography 15 (9) (1992) 625–626.
[410] Y. Baba, M. Tsuhako, T. Sawa, M. Akashi, E. Yashima, Specific base recognition of oligodeoxynucleotides by capillary affinity gel electrophoresis using polyacrylamide-poly(9-vinyladenine) conjugated gel, Analytical Chemistry 64 (17) (1992) 1920–1925.
[411] Y. Baba, H. Inoue, M. Tsuhako, T. Sawa, A. Kishida, M. Akashi, Evaluation of the selective binding ability of oligodeoxynucleotides to poly(9-vinyladenine) using capillary affinity gel electrophoresis, Analytical Sciences 10 (6) (1994) 967–969.
[412] Y. Baba, R. Tomisaki, M. Tsuhako, T. Sawa, Y. Inami, A. Kishida, et al., Detection of mismatch positions on the DNA-polyvinyladenine hybrids using capillary affinity gel electrophoresis, Nucleic Acids Symposium Series 29 (1993) 81–82.
[413] Y. Baba, M. Tsuhako, T. Sawa, M. Akashi, Effect of urea concentration on the base-specific separation of oligodeoxynucleotides in capillary affinity gel electrophoresis, Journal of Chromatography 652 (1) (1993) 93–99.
[414] Y. Baba, M. Tsuhako, T. Sawa, M. Akashi, Temperature-programmed capillary affinity gel electrophoresis for the sensitive base-specific separation of oligodeoxynucleotides, Journal of Chromatography 632 (1–2) (1993) 137–142.
[415] Y. Kim, M.D. Morris, Separation of nucleic acids by capillary electrophoresis in cellulose solutions with mono- and bisintercalating dyes, Analytical Chemistry 66 (7) (1994) 1168–1174.
[416] J. Cheng, K.R. Mitchelson, Glycerol-enhanced separation of DNA fragments in entangled solution capillary electrophoresis, Analytical Chemistry 66 (23) (1994) 4210–4214.
[417] H. Tsukada, L.M. Kundu, Y. Matsuoka, N. Kanayama, T. Takarada, M. Maeda, Quantitative single-nucleotide polymorphism analysis in secondary-structured DNA by affinity capillary electrophoresis using a polyethylene glycol-peptide nucleic acid block copolymer, Analytical Biochemistry 433 (2) (2013) 150–152.
[418] K. Nilsson, J. Lindell, O. Norrloew, B. Sellergren, Imprinted polymers as antibody mimetics and new affinity gels for selective separations in capillary electrophoresis, Journal of Chromatography A 680 (1) (1994) 57–61.
[419] H. Ljungberg, S. Ohlson, S. Nilsson, Exploitation of a monoclonal antibody for weak affinity-based separation in capillary gel electrophoresis, Electrophoresis 19 (3) (1998) 461–464.

[420] J.K. Abler, K.R. Reddy, C.S. Lee, Post-capillary affinity detection of protein micro-heterogeneity in capillary zone electrophoresis, Journal of Chromatography A 759 (1–2) (1997) 139–147.

[421] S. Cherkaoui, T. Bettinger, M. Hauwel, S. Navetat, E. Allemann, M. Schneider, Tracking of antibody reduction fragments by capillary gel electrophoresis during the coupling to microparticles surface, Journal of Pharmaceutical and Biomedical Analysis 53 (2) (2010) 172–178.

CHAPTER SIX

Related microseparation techniques

6.1 Ultrathin-layer gel electrophoresis

Ultrathin-layer gel electrophoresis is a combination of slab gel electrophoresis and capillary gel electrophoresis [1], providing a multilane separation platform (a plurality of virtual channels) with excellent heat dissipation characteristics allowing the application of high voltages necessary to obtain a rapid and efficient analysis of biopolymers. Detection of the separated bands is usually accomplished in real time by continuous laser induced fluorescence (LIF) scanning of the separation lanes. While gel and buffer compositions used in ultrathin-layer gel electrophoresis are practically the same as in conventional procedures, the efficiency is increased due to its reduced cross-sectional area, thus offers the ability to apply high separation voltages. In general, the thickness of ultrathin-layer separation platforms ranges from 0.050 to 0.25 mm. Ultrathin slab gels with a gel thickness of <0.1 mm allowed simultaneous development of several electropherograms as different lanes in the gel with efficient heat dissipation. Sample introduction to ultrathin slab gels can be accomplished in various ways. One approach permitted continuous analysis from a flowing stream discrete sampling for sequential separations [2]. Another method used membrane-mediated sample loading for DNA samples [3].

6.1.1 DNA analysis

Vertical thin-layer gels were first introduced in high-throughput automated DNA sequencing [4]. Later, horizontal, ultrathin-layer polyacrylamide gel electrophoresis was implemented by Van den Berg [5] in a multizonal format for large-scale analysis of polymerase chain reaction (PCR) products in order to reveal short sequence repeat polymorphisms. Detection of dsDNA fragments in the range of 200–3000 bp was accomplished by silver staining and up to 400 PCR samples were analyzed in 2 hours. Later, Ewing's group introduced a novel approach for DNA

Capillary Gel Electrophoresis.
DOI: https://doi.org/10.1016/B978-0-444-52234-4.00009-X

fragment analysis, which combined the parallel processing capabilities of slab gels with the advantages of sample introduction employing a capillary column [2]. Gels were fabricated by using 57-μm spacers between two quartz plates, and the single capillary was used to introduce dsDNA fragments into the gel-filled ultrathin-layer separation platform. The capillary-mediated sample introduction approach allowed multiple samples to be rapidly deposited along the edge of the ultrathin slab gels for consequent separation (Fig. 6.1). The method was applied for large-scale analysis of PCR amplified short tandem repeats (STRs) using nondenaturing polyacrylamide gels and ethidium bromide labeling with LIF detection of the

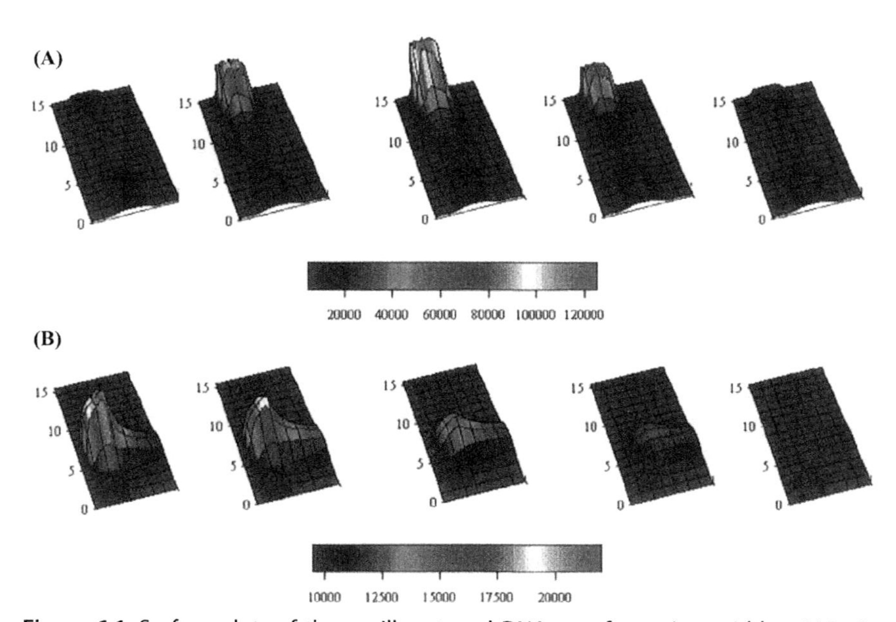

Figure 6.1 Surface plots of the capillary-to-gel DNA transfer region: width = 115 pixels; depth = 160 pixels. The capillary is at the lower/front portion of each image and the slab gel entrance at the top/back. $E_{gel} = 125$ V/cm; $E_{cap} = 80$ V/cm. The images were taken at ~25-second intervals, showing the transfer of fluorescent DNA fragments from the capillary to the gel, depending on the structural integrity of the capillary tip. (A) Capillary-to-gel alignment was correct, so the DNA migrates from the capillary directly into the gel entrance. In the last image, the DNA successfully entered the gel and was no longer visible in the viewing region. (B) Capillary tip was cracked, so the DNA exited the capillary tip and dissipated out into solution instead of migrating into the slab gel. *With permission from P.B. Hietpas, K.M. Bullard, D.A. Gutman, A.G. Ewing, Ultrathin slab gel separations of DNA using a single capillary sample introduction system, Analytical Chemistry 69 (13) (1997) 2292–2298 [2].*

amplified fragments. This technique demonstrated promising results by increasing the throughput of STR analysis [6].

Trost and Guttman reported on the development and implementation of an automated, horizontal ultrathin-layer agarose gel electrophoresis platform, equipped with an integrated scanning laser-induced fluorescence detection, for large-scale DNA fragment analysis [7,8]. The automated ultrathin-layer agarose gel electrophoresis setup consisted of a high-voltage power supply, an ultrathin-layer separation cassette with built-in buffer reservoirs, and a fiber-optic bundle-based scanning detection system. A lens set, connected to the illumination/detection block via fiber-optic bundle, scanned across the ultrathin-layer separation gel by means of a high-speed translation stage (Fig. 6.2). A laser excitation source (532 nm frequency-doubled Nd-YAG laser) and an avalanche photodiode were connected to the central excitation fiber and the surrounding collecting

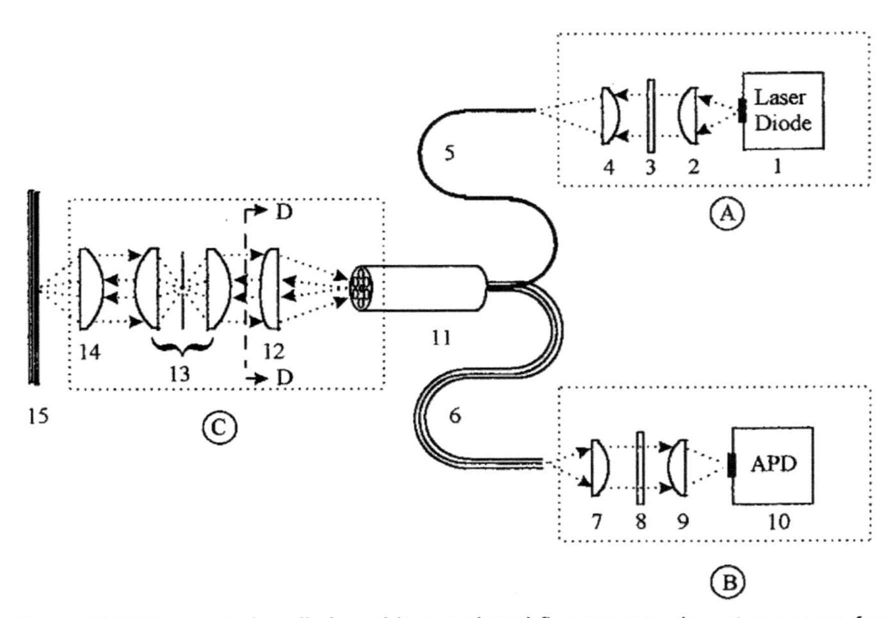

Figure 6.2 Fiber-optic bundle-based laser-induced fluorescence detection system for ultrathin-layer agarose gel electrophoresis. (A) excitation source module, (B) dection module, (C) scanning module. (1) 640 nm red diode laser, (2) aspheric lens, (3) narrow-band interference filter, (4) lens, (5, 6, and 11) fiber bundle, (7 and 9) aspheric lenses, (8) wide-band interference filter, (10) avalanche photodiode, (12) spherical lens, (13) spatial filter, (14) aspheric lens, (15) glass separation platform. *With permission from P. Trost, A. Guttman, Fiber bundle based scanning detection system for automated DNA sequencing, Analytical Chemistry 70 (18) (1998) 3930–3935 [8].*

fibers of the fiber-optic bundle, respectively [8]. Ultrathin gels were also used by Luckey et al. for rapid DNA sequence analysis [9].

In addition to the use of conventional preseparation tagging, ultrathin-layer gel electrophoresis systems readily accommodated fluorophore labeling during the separation process referred to as "in migratio" labeling by, for example, complexation of dsDNA fragments with high-sensitivity staining dyes. Sample introduction onto the ultrathin separation platform was easily accomplished by membrane-mediated loading technology, which also enabled robotic spotting of multiple samples (Fig. 6.3) [3].

The samples were injected manually or automatically (liquid handling robots) onto the surface of the loading membrane tabs, outside of the separation/detection platform. The sample spotted membrane was then placed into the injection (cathode) slot of the ultrathin-layer separation platform, in intimate contact with the straight gel edge. Under the influence of the applied electric field, the sample components migrated into the gel. This automated sample injection method can be easily applied to most high-throughput thin-layer slab gel electrophoresis-based DNA analysis applications (e.g., automated DNA sequencing, genotyping, and PCR product analysis) [10].

Simultaneous separation of various dsDNA fragment mixtures is shown in Fig. 6.4 exhibiting a broad size range from 20 up to 23,130 bp, using a single agarose gel composition [7]. The sample components were visualized by "in migratio" ethidium bromide staining (50 nM). The first five lanes depict the dilution series of the 100 bp DNA ladder representing 25, 10, 5, 2.5, and 1.25 ng total DNA (Lanes 1–5), corresponding to 863, 300, 170, 80, and 40 pg per band, respectively. Based on these results, the limit of detection of the automated, high-performance ultrathin-layer agarose gel electrophoresis system using laser-induced fluorescence/avalanche photodiode detection was 40 pg DNA per band. Lane 6 shows the separation of the φX174 DNA Hae-III restriction digest mixture ranging from 72 to 1353 bp. Lane 7 depicts a rapid (<17 minutes)

Figure 6.3 Loading membrane for ultrathin-layer gel electrophoresis. *With permission from S.M. Cassel, A. Guttman, Membrane mediated sample loading for automated DNA sequencing, Electrophoresis 19 (1998) 1341–1346 [3].*

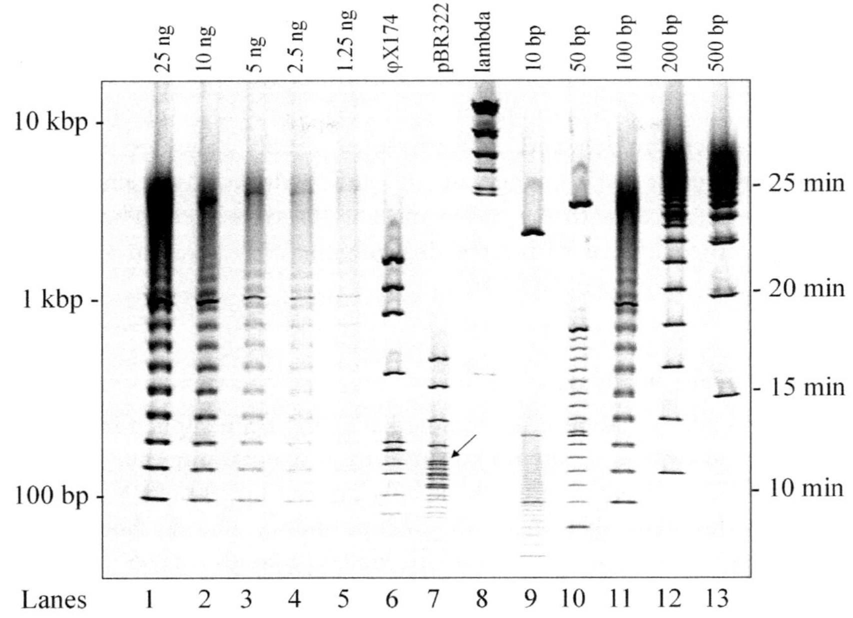

Figure 6.4 Separation of DNA sizing ladders and restriction digest fragment mixtures by automated ultrathin-layer agarose gel electrophoresis. Lanes 1–5: dilution series of a 100 bp ladder (25, 10, 5, 2.5, and 1.25 ng total DNA injected); Lane 6: φX174 DNA Hae-III restriction digest mixture; Lane 7: pBR322 DNA Msp-I restriction digest mixture; Lane 8: lambda DNA Hind-III restriction digest mixture; Lanes 9–13: 10, 50, 100, 200, and 500 bp DNA sizing ladders. Conditions: Effective separation length: 6 cm; separation matrix: 2% agarose gel in 0.5 × TBE buffer containing 25 nM ethidium bromide; running buffer: 0.5 × TBE; applied voltage: 750 V; temperature: 25°C; injection: membrane mediated, 0.5-μL sample per tab. *With permission from A. Guttman, Automated DNA fragment analysis by high performance ultrathin-layer agarose gel electrophoresis, LC-GC 17 (11) (1999) 1020, 1022, 1024, 1026 [7].*

and high-resolution separation of the pBR322 DNA Msp-I restriction digest mixture.

The high separation efficiency of the system enabled to obtain baseline resolution of the four base pair difference of the 242 and 238 bp fragments (arrow in Lane 7). Lane 8 exhibits the separation of the lambda DNA Hind-III restriction digest fragments on the same gel composition ranging up to 23,130 bp in size. Lanes 9–13 depict the separations of the 10, 50, 100, 200, and 500 bp DNA sizing ladders, respectively. The high resolving power of the system was also demonstrated by the nice separation of the ten base pair ladder (Lane 9) ranging from 20 to 320 bp. Similar, high-resolution separations were observed for other ladders.

Electrophoresis of fluorescently labeled polystyrene carboxylate particles of 46.5 nm radius was carried out by the Chrambach group using a horizontal gel electrophoresis apparatus with polymer solutions and intermittent scanning of the migration path [11]. Dextran, polyvinylpyrrolidone, polyacrylamide, and polyethylene glycol polymers with the molecular weight (MW) range of 10^6 and hydroxyethyl cellulose and polyethylene glycol with the MW range of 10^5 were used. The decrease in dispersion coefficient with increasing polymer concentration was apparent with the polymers of MW 10^6.

6.1.2 Protein analysis

Electrophoresis and isoelectric focusing have long been applied for protein mapping and to study protein extraction in horizontal ultrathin-layer format [12]. The 0.12–0.36 mm thick polyacrylamide gel layer was deposited onto tiny glass plates (e.g., microscope slides). The method enabled the analysis of 1 ng tissue culture specimens. Ultrathin-layer polyacrylamide isoelectric focusing gel was employed in two-dimensional (2D) analysis of plant and fungal proteins.

Automated ultrathin-layer sodium dodecyl sulfate (SDS)–gel electrophoresis was applied for large-scale analysis of SDS–protein complexes and for MW estimation of the separated species [13]. Fig. 6.5 compares the separations of two differently labeled protein standard mixtures for MW estimation of phosphorylase B, using ultrathin-layer SDS agarose/linear polyacrylamide gel electrophoresis. Trace A depicts the separation of a noncovalently (Sypro Red, SR) stained five-protein test mixture containing α-lactalbumin (ALA), carbonic anhydrase (CAH), ovalbumin (OVA), bovine serum albumin (BSA), and β-galactosidase (BGA). Trace B corresponds to the analysis of noncovalently SR-labeled phosphorylase B. Trace C shows the separation of the covalently fluorescein isothiocyanate (FITC) labeled five-protein test mixture of trypsin inhibitor (TRI), CAH, alcohol dehydrogenase (ADH), BSA, and BGA. Noncovalent labeling of the sample and standard proteins took place immediately prior to loading. Detection of the migrating SDS–protein complexes was accomplished in real time during the electrophoresis separation process. The actual separation distance was only 3.5 cm, still offering high resolution [14].

Fig. 6.6 depicts the standard curves constructed for MW estimation of phosphorylase B by plotting the logarithmic MWs of the SR and

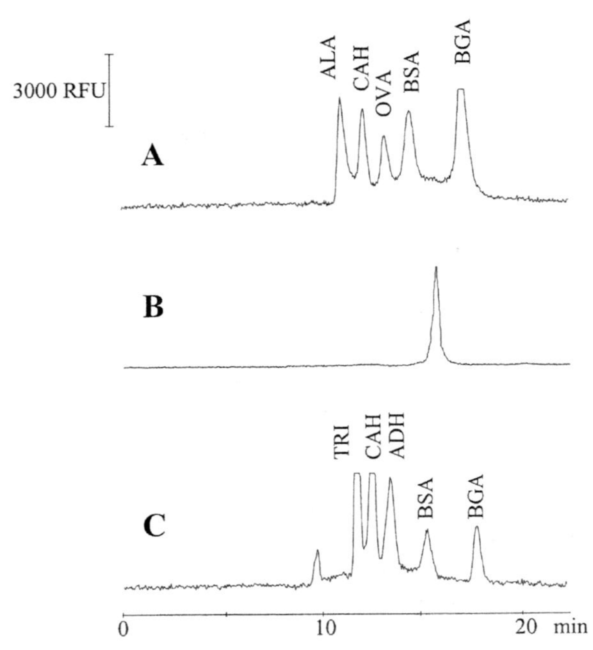

Figure 6.5 Molecular weight analysis of phosphorylase B (Trace B) and the separation of noncovalently (Sypro Red) labeled protein markers (Trace A: ALA, α-lactalbumin; BGA, β-galactosidase; BSA, bovine serum albumin; CAH, carbonic anhydrase; OVA, ovalbumin) and the covalently (FITC) labeled protein markers (Trace C: ADH, alcohol dehydrogenase; BGA, β-galactosidase; BSA, bovine serum albumin; CAH, carbonic anhydrase; TRI, trypsin inhibitor). Separation conditions: Gel: 1% agarose/2% linear polyacrylamide (LPA: MW 700,000–1,000,000) in 50 mM Tris, 50 mM TAPS, 0.05% SDS (pH 8.4); separation buffer: 50 mM Tris, 50 mM TAPS, 0.05% SDS (pH 8.4); separation voltage: 420 V, current: 5 mA; gel thickness: 190 μm; $\ell = 3.5$ cm; temperature: 25°C; Sample loading: 0.2 μL into 2.5 × 4 × 0.19 mm injection wells. Sample buffer contained 0.05% SDS and 1 × Sypro Red. *With permission from A. Guttman, Z. Ronai, Z. Csapo, A. Gerstner, M. Sasvari-Szekely, Rapid analysis of covalently and non-covalently fluorophore-labeled proteins using ultra-thin-layer sodium dodecyl sulfate gel electrophoresis, Journal of Chromatography A 894 (1–2) (2000) 329–335 [14].*

FITC-labeled test proteins against their electrophoretic mobilities (derived from the data of Fig. 6.5, Traces A and C). For the SR-labeled protein standards, a second-order polynomial function provided the best fitting of the standard curve ($r^2 = 0.9999$). This finding was in contrast to the linear relationship between the logarithmic molecular mass and electrophoretic mobility of the FITC-labeled protein standards ($r^2 = 0.9995$). The curvature of the SR plot was probably caused by the noncovalent attachment of the negatively charged staining dye that increased the charge density

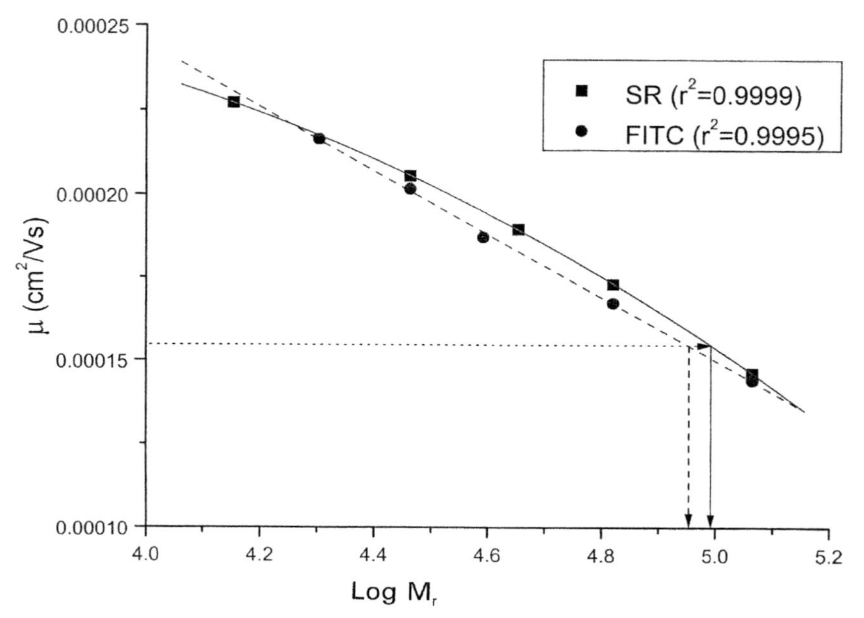

Figure 6.6 Standard curves of electrophoretic mobility versus logarithmic molecular mass (derived from the data of Fig. 6.5 Traces A and C) for molecular mass estimation using covalently (- - -) and noncovalently (—) labeled protein standards. *With permission from A. Guttman, Z. Ronai, Z. Csapo, A. Gerstner, M. Sasvari-Szekely, Rapid analysis of covalently and non-covalently fluorophore-labeled proteins using ultra-thin-layer sodium dodecyl sulfate gel electrophoresis, Journal of Chromatography A 894 (1–2) (2000) 329–335 [14].*

and concomitantly the overall electrophoretic velocity of the complex. Based on the calibration plots obtained, the MW of the phosphorylase B peak was estimated as 97.250 (0.15% error) and 89 750 (7.8% error) for the SR- and FITC-labeled samples, respectively, compared to the literature value of MW 97,400.

6.2 Multidimensional approaches

Multidimensional separation systems substantially enhance the peak capacity, thus increase the number of individual peaks that fit into a separation window. The full potential of multidimensional separation systems, however, realized only when the individual separation means are purely orthogonal. In an interesting approach to couple capillary electrophoresis (CE) with ultrathin gel electrophoresis [15], the CE capillary was translated across the entire length of the entrance of the ultrathin gel plate. Although CE and gel electrophoresis are not perfectly orthogonal, adequate differences existed in the separation mechanisms for successful 2D separation of tryptic peptides.

Sanderink and coworkers [16] used microscale 2D electrophoresis with 1.3 mm × 35 mm isoelectric focusing (IEF) gel-filled capillaries and 38 mm × 35 mm × 1 mm polyacrylamide gradient gels to separate alkaline phosphatase isoforms from human sera. They revealed that liver and bone isoforms in normal sera showed overlapping isoelectric points but different MWs. Liver disease patients showed additional isoform groups. Alkaline phosphatase isoforms were separated after lectin precipitation from serum as shown in Fig. 6.7.

A novel 2D CE-based separation system was reported for protein analysis by Liu et al. [17] utilizing a dialysis hollow-fiber membrane as the interface for online coupling of capillary isoelectric focusing (cIEF)

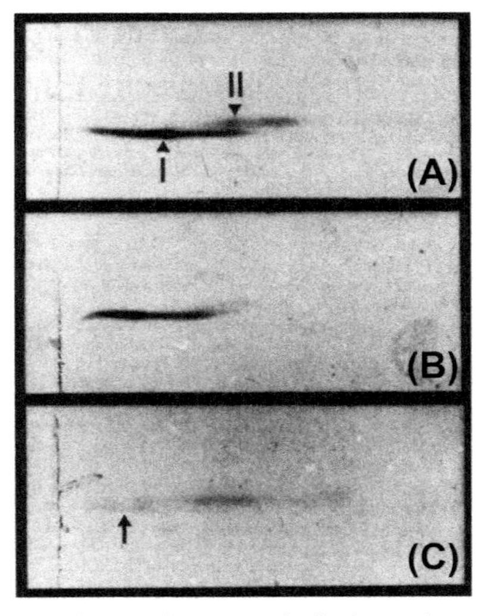

Figure 6.7 Microscale 2D electrophoresis of alkaline phosphatase isoforms after lectin precipitation. (A) From serum of a patient with extra hepatic cholestasis; (B) supernatant after wheat germ agglutinin (WGA) precipitation; (C) AP isoforms solubilized from the precipitate with *N*-acetylglucosamine. *With permission from G.J. Sanderink, Y. Artur, F. Paille, M.M. Galteau, G. Siest, Micro-scale two-dimensional electrophoresis of alkaline phosphatase from serum, Clinical Chemistry 34 (4) (1988) 730–735 [16].*

and CGE. The cIEF focused proteins were transferred to the hollow-fiber interface to complex with SDS then separated by size in the CGE dimension in a replaceable dextran sieving matrix solution. In their study, rat lung cancer proteins were studied by UV absorbance detection. Another multidimensional CE separation approach was published by Hu et al. [18] to analyze proteins from cellular samples. The first dimension employed an SDS-pullulan polymer buffer system for size separation connected through a simple gap junction to MECC/MEKC as a second dimension to study single-cell expression from native MC3T3-E1 osteoprogenitor cells. The MC3T3-E1 cells were transfected with the human transcription regulator TWIST and MCF-7 breast cancer cells before and after treatment with an apoptosis-inducing treatment.

Coupling capillary gel electrophoresis and other CE methods such as MECC/MEKC with the use of LIF detection provided a high-resolution multidimensional CE system [19]. The analysis only required 60 minutes with better than 1% migration time reproducibility in both dimensions for the 50 most intense features. In another attempt, 2D capillary electrophoresis was used to separate components based on their size and hydrophobicity [20]. The first dimension size separated the solute molecules and the fractions were electrokinetically transferred into a second capillary, where the sample components were further resolved based on their hydrophobicity. A stacking effect occurred at the interface when methyl-beta-cyclodextrin was added to the transfer buffer from the first to the second dimension.

Marlow et al. [21] reported on the use of 0.2 mm semirigid backing (polyester) supported IEF gels as the first dimension, which quickly dried on the backing after the focusing step. Then, narrow strips were cut from the dried gel and applied to the second dimension. Others [22] used cellophane foil to support ultrathin-layer isoelectric focusing gels. As the polyacrylamide gel was firmly attached to the cellophane foil, it provided good protection from mechanical damage and enabled easy handling. Since cellophane is permeable to ions, the use of this combination alleviated the difficulties of the removal of thin gels from the support. Using a similar approach, proteins were also separated under nondenaturing conditions and transferred onto a nitrocellulose membrane followed by enzyme assay—based detection.

An automated two dimensional separation/detection system was proposed later, capable of rapid 2D analysis of proteins by ultrathin-layer gel

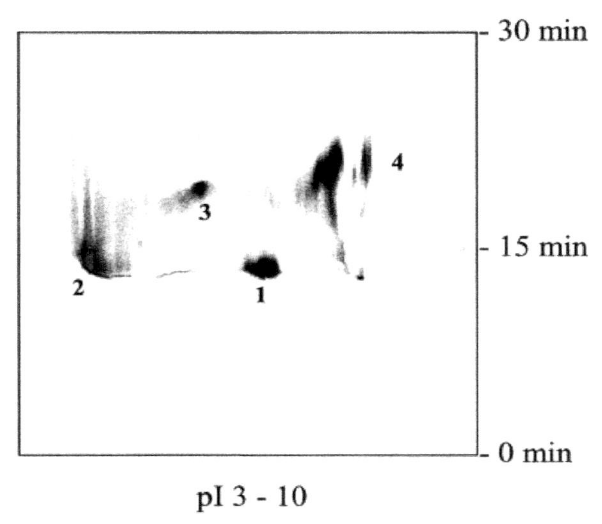

Figure 6.8 Two-dimensional separation of a protein test mixture by ultrathin-layer gel electrophoresis. Spots: (1) equine myoglobin, pI = 7.0; Mr = 17,500; (2) soybean trypsin inhibitor, pI = 4.5; Mr = 21,500; (3) bovine carbonic anhydrase, pI = 5.9, 6.0; Mr = 31,000; (4) rabbit muscle glyceraldehyde 3-phosphate dehydrogenase (GAPDH), pI = 8.3, 8.5; Mr = 36,000. Conditions: first dimension: pH 3–10 on an IPG strip (500 V for 15 minutes, 1000 V for 15 minutes, 1500 V for 15 minutes, and 3000 V for 60 minutes). The IPG strip was then equilibrated by 10 μL 50 mM Tris/TAPS, 0.05% SDS, pH 8.4 buffer containing 40 μL 50× SR for 2 minutes. Second dimension: 9% linear polyacrylamide gel in 50 mM Tris-TAPS, 0.05% SDS, pH 8.4 (100 V/cm). Detection: LIF 532 nm excitation/605 nm emission. Effective separation distance: 1.5 cm. *With permission from A. Guttman, Z. Csapo, D. Robbins, Rapid two-dimensional analysis of proteins by ultra-thin layer gel electrophoresis, Proteomics 2 (4) (2002) 469–474 [23].*

electrophoresis with real-time imaging of the separated components [23]. The setup utilized fiber optics–based laser-induced fluorescence scanning technology [8]. The samples were noncovalently labeled by "in migratio" fluorescent SR staining. The method featured a simple workflow with high speed and good detection sensitivity, as shown in Fig. 6.8 [23].

References

[1] A. Guttman, Z. Ronai, Ultrathin-layer gel electrophoresis of biopolymers, Electrophoresis 21 (18) (2000) 3952–3964.

[2] P.B. Hietpas, K.M. Bullard, D.A. Gutman, A.G. Ewing, Ultrathin slab gel separations of DNA using a single capillary sample introduction system, Analytical Chemistry 69 (13) (1997) 2292–2298.

[3] S.M. Cassel, A. Guttman, Membrane mediated sample loading for automated DNA sequencing, Electrophoresis 19 (1998) 1341–1346.

[4] L.M. Smith, J.Z. Sanders, R.J. Kaiser, P. Hughes, C. Dodd, C.R. Connell, C. Heiner, S.B.H. Kent, L.E. Hood, Fluorescence detection in automated DNA sequence analysis, Nature (London, United Kingdom) 321 (6071) (1986) 674–679.
[5] B.M. Van den Berg, Horizontal ultrathin-layer multi-zonal electrophoresis of DNA. An efficient tool for large-scale polymerase chain reaction (PCR) fragment analysis, Electrophoresis 18 (15) (1997) 2861–2864.
[6] K.M. Bullard, P. Beyer Hietpas, A.G. Ewing, Capillary sample introduction of polymerase chain reaction (PCR) products separated in ultrathin slab gels, Electrophoresis 19 (1) (1998) 71–75.
[7] A. Guttman, Automated DNA fragment analysis by high performance ultrathin-layer agarose gel electrophoresis, LC-GC 17 (11) (1999) 1020. 1022, 1024, 1026.
[8] P. Trost, A. Guttman, Fiber bundle based scanning detection system for automated DNA sequencing, Analytical Chemistry 70 (18) (1998) 3930–3935.
[9] J.A. Luckey, H. Drossman, A.J. Kostichka, D.A. Mead, J. D'Cunha, T.B. Norris, L. M. Smith, High speed DNA sequencing by capillary electrophoresis, Nucleic Acids Research 18 (15) (1990) 4417–4421.
[10] A. Guttman, Sample stacking during membrane-mediated loading in automated DNA sequencing, Analytical Chemistry 71 (16) (1999) 3598–3602.
[11] H.-T. Chang, A. Chrambach, Feasibility of electrophoresis of a subcellular-sized particle in polymer solutions, using automated horizontal gel apparatus, Applied and Theoretical Electrophoresis 5 (2) (1995) 73–77.
[12] M. Inczedy-Marcsek, E. Lindner, R. Haessler, G. Zwack-Megele, F. Roisen, G. Yorke, Extraction of proteins within ultrathin-layer polyacrylamide electrophoresis (SDS-PAGE) and isoelectric focusing (PAGIF) of cryostat sections and tissue culture specimens, Acta Histochemica, Supplementband 36 (1988) 377–394.
[13] Z. Csapo, A. Gerstner, M. Sasvari-Szekely, A. Guttman, Automated ultra-thin-layer SDS gel electrophoresis of proteins using noncovalent fluorescent labeling, Analytical Chemistry 72 (11) (2000) 2519–2525.
[14] A. Guttman, Z. Ronai, Z. Csapo, A. Gerstner, M. Sasvari-Szekely, Rapid analysis of covalently and non-covalently fluorophore-labeled proteins using ultra-thin-layer sodium dodecylsulfate gel electrophoresis, Journal of Chromatography A 894 (1–2) (2000) 329–335.
[15] Y.-M. Liu, J.V. Sweedler, Two-dimensional separations: capillary electrophoresis coupled to channel gel electrophoresis, Analytical Chemistry 68 (22) (1996) 3928–3933.
[16] G.J. Sanderink, Y. Artur, F. Paille, M.M. Galteau, G. Siest, Micro-scale two-dimensional electrophoresis of alkaline phosphatase from serum, Clinical Chemistry 34 (4) (1988) 730–735.
[17] H. Liu, C. Yang, Q. Yang, W. Zhang, Y. Zhang, On-line combination of capillary isoelectric focusing and capillary non-gel sieving electrophoresis using a hollow-fiber membrane interface: a novel two-dimensional separation system for proteins, Journal of Chromatography, B: Analytical Technologies in the Biomedical and Life Sciences 817 (1) (2005) 119–126.
[18] S. Hu, D.A. Michels, M.A. Fazal, C. Ratisoontorn, M.L. Cunningham, N.J. Dovichi, Capillary sieving electrophoresis/micellar electrokinetic capillary chromatography for two-dimensional protein fingerprinting of single mammalian cells, Analytical Chemistry 76 (14) (2004) 4044–4049.
[19] J.R. Kraly, M.R. Jones, D.G. Gomez, J.A. Dickerson, M.M. Harwood, M. Eggertson, T.G. Paulson, C.A. Sanchez, R. Odze, Z. Feng, B.J. Reid, N.J. Dovichi, Reproducible two-dimensional capillary electrophoresis analysis of Barrett's esophagus tissues, Analytical Chemistry 78 (17) (2006) 5977–5986.

[20] D. Gonzalez-Gomez, D. Cohen, J.A. Dickerson, X. Chen, F. Canada-Canada, N.J. Dovichi, Improvement in protein separation in Barretts esophagus samples using two-dimensional capillary electrophoresis analysis in presence of cyclodextrins as buffer additives, Talanta 78 (1) (2009) 193–198.
[21] G.C. Marlow, D.E. Wurst, D.C. Loschke, The use of ultrathin-layer polyacrylamide gel isoelectric focusing in two-dimensional analysis of plant and fungal proteins, Electrophoresis 9 (11) (1988) 693–704.
[22] A.V. Yakhyayev, I.M. Voronkova, V.A. Sukhanov, A simplified method of protein electroblotting after isoelectric focusing on an ultrathin polyacrylamide gel layer fixed on a cellophane support, Electrophoresis 12 (9) (1991) 680–682.
[23] A. Guttman, Z. Csapo, D. Robbins, Rapid two-dimensional analysis of proteins by ultra-thin layer gel electrophoresis, Proteomics 2 (4) (2002) 469–474.

CHAPTER SEVEN

Appendix

Manufacturers' directory

Agilent Technologies, Inc.
5301 Stevens Creek Blvd, Santa Clara, CA 95051, United States
Phone: +1-800-227-9770

BiOptic, Inc.
(23141) 5F, No. 6, Ln. 130, Minquan Rd., Hsin-Tien District, New Taipei City, Taiwan (R.O.C.)
Phone: +886-2-2218-8726, Fax: +886-2-2218-8727

Bio-Rad Laboratories Inc. Life Science
1000 Alfred Nobel Drive, Hercules, CA 94547, United States
Phone: +1-510-741-1000/ + 1-800-424-6723, Fax: +1-510-741-5800/ + 1-800-879-2289

Intabio
47370 Fremont Blvd, Fremont, CA 94538, United States

JASCO Inc.
28600 Mary's Court, Easton, MD 21601, United States
Phone: +1-410-822-1220/ + 1-800-6548483

Molex LLC (Capillary)
2222 Wellington Court, Lisle, IL 60532, United States
Phone: +1-866-733-6659, Fax: +1-630-813-9770

ProteinSimple (brand by Bio-Techne)
3001 Orchard Parkway, San Jose, CA 95134, United States
Phone: +1-888-607-9692/ + 1-408510-5500, Fax: +1-408-510-5599

SCIEX (Legal Entity: AB Sciex LLC)
500 Old Connecticut Path, Framingham, MA 01701, United States
Phone: +1-877-740-2129

Sigma-Aldrich Corp.
PO Box 14508, St. Louis, MO 63178, United States
Phone: +1-800-325-3010, Fax: +1-800-3255052

Capillary Gel Electrophoresis.
DOI: https://doi.org/10.1016/B978-0-444-52234-4.00006-4

Spectra-Physics Headquarters
1565 Barber Lane, Milpitas, CA 95035, United States
Phone: +1 408-980-4300, Fax: +1 800-775-5273
Spellman Corporate (High Voltage Power Supply)
475 Wireless Blvd, Hauppauge, NY 11788, United States
Phone: +1 631-630-3000, Fax: +1 631-435-1620
Thermo Fisher Scientific
168 3rd Ave., Waltham, MA, United States
Phone: +1 800-955-6288

Further reading

S. Kanchi, S. Sagrado, M.I. Sabela, K. Bisetty (Eds.), Capillary Electrophoresis: Trends and Developments in Pharmaceutical Research, first ed., Jenny Stanford Publishing, 2017.
T. Wehr, M. Zhu, R. Rodriguez-Diaz (Eds.), Capillary Electrophoresis of Proteins, first ed., CRC Press, 1999.
Y. Michotte, A. Van Eeckhaut (Eds.), Chiral Separations by Capillary Electrophoresis, firsted., CRC Press, 2009.
R.H.H. Neubert, H.-H. Ruttinger (Eds.), Affinity Capillary Electrophoresis in Pharmaceutics and Biopharmaceutics, first ed., CRC Press, 2003.
D. Keren, Protein Electrophoresis in Clinical Diagnosis, first ed., CRC Press, 2003.
R.H. Liu, J.G. Shewale (Eds.), Forensic DNA Analysis: Current Practices and Emerging Technologies, first ed., CRC Press, 2013.
P. Schmitt-Kopplin (Ed.), Capillary Electrophoresis, Methods and Protocols, first ed., Humana Press, 2008.
P. Thibault, S. Honda (Eds.), Capillary Electrophoresis of Carbohydrates, Methods in Molecular Biology, vol. 213, Humana Press, 2003.
J.R. Petersen, A.A. Mohammad (Eds.), Clinical and Forensic Applications of Capillary Electrophoresis, Series: Pathology and Laboratory Medicine, Humana Press, 2001.
N. Volpi, F. Maccari (Eds.), Capillary Electrophoresis of Biomolecules, Methods and Protocols, first ed., Humana Press, 2013.
C.F. Poole (Ed.), Capillary Electromigration Separation Methods, Elsevier, 2018.
R. Weinberger (Ed.), Practical Capillary Electrophoresis, second ed., Academic Press, 2000.
K.R. Mitchelson, J. Cheng (Eds.), Capillary Electrophoresis of Nucleic Acids, first ed., Humana Press, 2001.
M.A. Strege, A.L. Lagu (Eds.), Capillary Electrophoresis of Proteins and Peptides, first ed., Humana Press, 2004.
S. Makovets (Ed.), DNA Electrophoresis, Methods and Protocols, first ed., Humana Press, 2013.
C.D. García, K.Y. Chumbimuni-Torres, E. Carrilho (Eds.), Capillary Electrophoresis and Microchip Capillary Electrophoresis: Principles, Applications, and Limitations, John Wiley & Sons, Inc, 2013.
G. Hanrahan, F.A. Gomez (Eds.), Chemometric Methods in Capillary Electrophoresis, John Wiley & Sons, Inc, 2009.
J.P. Landers (Ed.), Handbook of Capillary and Microchip Electrophoresis and Associated Microtechniques, third ed., CRC Press, 2007.
Z. El Rassi (Ed.), Carbohydrate Analysis by Modern Chromatography and Electrophoresis, Journal of Chromatographic Library, vol. 66, Elsevier, 2002.

A.T. Andrews (Ed.), Electrophoresis: Theory, Techniques, and Biochemical and Clinical Applications, second ed., Clarendon Press, 1986.

S.F.Y. Li, Capillary Electrophoresis: Principles, Practice and Application, Journal of Chromatographic Library, vol. 52, Elsevier, 1992.

D. Coleman (Ed.), Directory of Capillary Electrophoresis, Trac Supplement, No. 1, Elsevier, 1994.

B.L. Karger, L.R. Snyder, C. Horvath, An Introduction to Separation Science, John Wiley & Sons, Inc, 1973.

R. Westermeier, Electrophoresis in Practice: A Guide to Methods and Applications of DNA and Protein Separations, fourth ed., John Wiley & Sons, Inc, 2004.

K.D. Altria (Ed.), Capillary Electrophoresis Guidebook, Principles, Operation and Applications, Methods in Molecular Biology, vol. 52, Humana Press., 1995.

A. Manz, H. Becker (Eds.), Microsystem Technology in Chemistry and Life Sciences, Springer., 1998.

C. Heller, Analysis of Nucleic Acids by Capillary Electrophoresis, CHROMATOGRAPHIA CE-Series, vol. 1, Springer., 1997.

K.D. Altria, Analysis of Pharmaceuticals by Capillary Electrophoresis, CHROMATOGRAPHIA CE-Series, vol. 2, Springer., 1998.

A. Paulus, A. Klockow-Beck, Analysis of Carbohydrates by Capillary Electrophoresis, CHROMATOGRAPHIA CE-Series, vol. 3, Springer., 1999.

W. Kok, Capillary Electrophoresis: Instrumentation and Operation, CHROMATOGRAPHIA CE-Series, vol. 4, Springer., 2000.

A.B. Chen, W. Nashabeh, T. Wehr (Eds.), CE in Biotechnology: Practical Applications for Protein and Peptide Analyses, CHROMATOGRAPHIA CE-Series, vol. 5, Springer, 2001.

S. Wren, The Separation of Enantiomers by Capillary Electrophoresis, CHROMATOGRAPHIA CE-Series, vol. 6, Springer, 2001.

P. Camilleri (Ed.), Capillary Electrophoresis: Theory and Practice, second ed., CRC Press, 1997.

A. Chrambach, M.J. Dunn, B.J. Radola (Eds.), Advances in Electrophoresis, vol. 6, John Wiley & Sons, Inc, 1993.

R.A. Mosher, D.A. Saville, W. Thormann, The Dynamics of Electrophoresis, John Wiley & Sons, Inc, 1992.

A.S. Rathore, A. Guttman, Electrokinetic Phenomena: Principles and Applications in Analytical Chemistry and Microchip Technology, Marcel Dekker, Inc, 2004.

M.J. Heller, A. Guttman, Integrated Microfabricated Biodevices: Advanced Technologies for Genomics, Drug Discovery, Bioanalysis, and Clinical Diagnostics, Marcel Dekker, Inc, 2001.

N.A. Guzman (Ed.), Capillary Electrophoresis Technology, Chromatographic Science Series, vol. 64, Marcel Dekker, Inc, 1993.

Z. Deyl, F. Svec (Eds.), Capillary Electrochromatography, Journal of Chromatographic Library, vol. 62, Elsevier, 2001.

A. Van Schepdael (Ed.), Microchip Capillary Electrophoresis Protocols, Methods in Molecular Biology, vol. 1274, Humana Press, 2015.

P.G. Righetti (Ed.), Capillary Electrophoresis in Analytical Biotechnology, CRC Press, 1995.

G. Lunn, Capillary Electrophoresis Methods for Pharmaceutical Analysis, John Wiley & Sons, Inc, 2000.

P. Bocek, M. Deml, P. Gebauer, V. Dolnik, Analyticla Isotachophoresis, Electrophoresis Library, John Wiley & Sons, Inc, 1987.

P.G. Righetti, Isoelectric Focusing: Theory, Methodology and Application, Laboratory Techniques in Biochemistry and Molecular Biology, vol. 11, Elsevier, 1983.

A. Chrambach, The Practice of Quantitative Gel Electrophoresis, Advanced Methods in the Biological Sciences, John Wiley & Sons, Inc, 1985.

D.R. Baker, Capillary Electrophoresis, Techniques in Analytical Chemistry, John Wiley & Sons, Inc, 1995.

R. Kuhn, S. Hoffstetter-Kuhn, Capillary Electrophoresis: Principles and Practice, Springer, 1993.

M.G. Khaledi (Ed.), High-Performance Capillary Electrophoresis: Theory, Techniques, and Applications, Chemical Analyis, vol. 146, John Wiley & Sons, Inc, 1998.

A. Van Eeckhaut, Y. Michotte (Eds.), Chiral Separations by Capillary Electrophoresis, CRC Press, 2010.

T. Wehr, R. Rodriguez-Diaz, M. Zhu (Eds.), Capillary Electrophoresis of Proteins Chromatographic Science Series, vol. 80, CRC Press, 1998.

C.D. García, K.Y. Chumbimuni-Torres, E. Carrilho (Eds.), Capillary Electrophoresis and Microchip Capillary Electrophoresis: Principles, Applications, and Limitations, John Wiley & Sons, Inc, 2013.

C. Henry (Ed.), Microchip Capillary Electrophoresis: Methods and Protocols, Methods in Molecular Biology, vol. 339, Humana Press, 2006.

S. Kanchi, S. Sagrado, M.I. Sabela, K. Bisetty (Eds.), Capillary Electrophoresis: Trends and Developments in Pharmaceutical Research, Jenny Stanford Publishing, 2017.

S. Ahuja, M. Jimidar (Eds.), Capillary Electrophoresis Methods for Pharmaceutical Analysis, Separation Science and Technology, vol. 9, Academic Press, 2008.

R. Ramautar (Ed.), Capillary Electrophoresis—Mass Spectrometry for Metabolomics, New Developments in Mass Spectrometry, vol. 6, RSC Publishing, 2018.

G. de Jong, Capillary Electrophoresis—Mass Spectrometry (CE-MS): Principles and Applications, John Wiley & Sons, Inc, 2016.

T.M. Phillips (Ed.), Clinical Applications of Capillary Electrophoresis: Methods and Protocols, Methods in Molecular Biology, vol. 1972, Humana Press, 2019.

B. Chankvetadze, Capillary Electrophoresis in Chiral Analysis, John Wiley & Sons, Inc, 1997.

P. Koscielniak, M. Trojanowicz, Flow and Capillary Electrophoretic Analysis, Analytical Chemistry and Microchemistry Series, Nova Science Publishers, 2018.

N. Volpi, F. Maccari (Eds.), Capillary Electrophoresis of Biomolecules, Methods in Molecular Biology, vol. 984, Humana Press, 2013.

H. Parvez, R.-G. Caudy (Eds.), Capillary Electrophoresis in Biotechnology and Environmental Analysis, CRC Press, 1997.

C.F. Poole (Ed.), Capillary Electromigration Separation Methods, Series: Handbooks in Separation Science, Elsevier, 2018.

R.A. Frazier, J.M. Ames, H.E. Nursten, Capillary Electrophoresis for Food Analysis: Method Development, RSC Publishing, 2000.

Abbreviations

AAEE	*N*-acryloylaminoethoxyethanol
AAP	*N*-acryloylaminopropanol
AAV	adeno-associated virus
ACM	analyte concentrator microreactor
AFMC	analyte focusing by micelle collapse
AMAC	2-aminoacridone
ANTS	8-aminonaphtalene-1,3,6-trisulfonic acid
APS	ammonium persulfate
APTS	8-aminopyrene-1,3,6-trisulfonic acid
α	separation selectivity
BDNA	branched DNA
BGE	background electrolyte
BIS	*N,N'*-methylene-bisacrylamide
bp	base pair
CAE	capillary array electrophoresis
CAGE	capillary affinity gel electrophoresis
CBQCA	3-(4-carboxybenoyl)-2-quinolinecarboxalaldehyde
CCD	charged coupled device
CDCE	constant denaturant capillary electrophoresis
cDNA	complementary DNA
C^4D	capacitively coupled contactless conductivity detection
CE	capillary electrophoresis
CE-ECL	capillary electrophoresis coupled with electrochemiluminescent detection
CE-MS	capillary electrophoresis coupled with mass spectrometry
CGE	capillary gel electrophoresis
CGEKC	capillary gel electrokinetic chromatography
CGIEF	capillary gel isoelectric focusing
CGITP	capillary gel isotachophoresis
cIEF	capillary isoelectric focusing
CTAB	cetyltrimethylammonium bromide
CZE	capillary zone electrophoresis
D	diffusion coefficient
Da	Dalton
ddTTP	2′,3′-dideoxythymidine-5′-triphosphate
DGGE	denaturing gradient gel electrophoresis
dsDNA	double-stranded DNA
E	electric field strength
E_a	activation energy of viscous flow
EB	ethidium bromide
EMMA	electrophoretically mediated microanalysis

EMD electromigration dispersion
EOF electroosmotic flow
ESI electrospray ionization
EST expressed sequence tag
ε relative permittivity
f friction coefficient
FASI field-amplified sample injection
FASS field-amplified sample stacking
F_e electrical force
FESI field-enhanced sample injection
FESS field-enhanced sample stacking
F_f frictional force
FFPE formalin fixed paraffin embedded
FICGE field inversion capillary gel electrophoresis
FTICR Fourier transform ion cyclotron resonance
Φ^* entanglement threshold
GC gas chromatography
GFP green fluorescent protein
GMP guanosine-5′-monophosphate
GU glucose unit
HC heavy chain
HEC hydroxyethyl cellulose
HMW-GS high-molecular-weight glutenin subunits
HPLC high-performance liquid chromatography
HPMC hydroxypropylmethyl cellulose
IACE immunoaffinity capillary electrophoresis
iCIEF imaged capillary isoelectric focusing
i.d. internal diameter
IEF isoelectric focusing
IgG immunoglobulin G
IS-PCR inverse-shifting polymerase chain reaction
ITME in-tube microextraction
ITP isotachophoresis
ITS internal transcribed spacer
K_R retardation coefficient
LC liquid chromatography
LDA linear discriminant analysis
LED light-emitting diode
LESA liquid extraction surface analysis
LIF laser-induced fluorescence
LPA linear polyacrylamide
LPME liquid-phase microextraction
LOD limit of detection
LOQ limit of quantitation
LVSS large volume sample stacking
mAb monoclonal antibody
MALDI matrix-assisted laser desorption/ionization

MCAD	medium-chain acyl-coenzyme A dehydrogenase
MECC/MEKC	micellar electrokinetic capillary chromatography
mRNA	messenger RNA
miRNA	microRNA
MS	mass spectrometry
MSS	micelle to solvent stacking
MW	molecular weight
μ	electrophoretic mobility
μ_0	free-solution mobility
N	number of theoretical plates
n	chain length expressed in nucleotide units
NACE	nonaqueous capillary electrophoresis
ngHC	nonglycosylated heavy chain
η	viscosity
PAA	polyacrylamide
PAD	pulsed amperometric detection
PAGE	polyacrylamide gel electrophoresis
PCR	polymerase chain reaction
PDMA	poly(*N*, *N*-dimethylacrylamide)
PEG	polyethylene glycol
PEO	polyethylene oxide
pI	isoelectric point
PMG	poly(methylglutamate)
PNA	peptide nucleic acid
PNIPAM-g-PEO	poly(*N*-isopropylacrylamide)-γ-polyethylene oxide
PPS	polystyrenesulfonate
PSC	polystyrene carboxylate
PTFE	poly-(tetrafluoroethylene)
PVA	polyvinyl alcohol
PVP	polyvinylpyrrolidone
Q	net charge
QI	injection amount
QC-RT-PCR	quantitative competitive reverse transcription PCR
QD	quantum dot
Q_j	Joule heat
RCA	rolling circle amplification
RFLP	restriction fragment length polymorphism
RGM	rapidly growing mycobacteria
rhEPO	recombinant human erythropoietin
RL	resolution length
RP-HPLC	reversed phase high-performance liquid chromatography
R_s	resolution
RSD	relative standard deviation
RT-PCR	reverse transcriptase polymerase chain reaction
SDME	single-drop microextraction
SDS	sodium dodecyl sulfate
SDS-CGE	sodium dodecyl sulfate capillary gel electrophoresis

SDS-PAGE	sodium dodecyl sulfate polyacrylamide slab gel electrophoresis
SHS	sodium hexadecyl sulfate
SIBA	sequential injection before analysis
siRNA	small interfering RNA
SNP	single-nucleotide polymorphism
SPR	surface plasmon resonance
ssDNA	single-stranded DNA
SSCP	single-stranded conformation polymorphism
SSR	short sequence repeat
STR	short tandem repeat
SWDMI	separation window dependent multiple injection
σ^2	variance
σ^2_{inj}	injection variance
σ^2_{det}	detection variance
σ^2_{diff}	diffusion variance
TEMED	*N,N,N′,N′*-tetramethylethylenediamine
TGGE	temperature gradient gel electrophoresis
tITP	transient isotachophoresis
TOF-MS	time-of-flight mass spectrometry
UPLC	ultra-performance liquid chromatography
UV	ultraviolet
v	migration velocity
WCID	whole column imaging detection
ξ	average mesh size

Index

Note: Page numbers followed by "*f*" and "*t*" refer to figures and tables, respectively.

S

T

U

V

W

Z

9780444522344